Crystals, X-rays and Proteins

Comprehensive Protein Crystallography

Dennis Sherwood
MA (Hons), MPhil, PhD

Jon Cooper
BA (Hons), PhD

OXFORD
UNIVERSITY PRESS

OXFORD

UNIVERSITY PRESS

Great Clarendon Street, Oxford OX2 6DP

Oxford University Press is a department of the University of Oxford.
It furthers the University's objective of excellence in research, scholarship,
and education by publishing worldwide in

Oxford New York

Auckland Cape Town Dar es Salaam Hong Kong Karachi
Kuala Lumpur Madrid Melbourne Mexico City Nairobi
New Delhi Shanghai Taipei Toronto

With offices in

Argentina Austria Brazil Chile Czech Republic France Greece
Guatemala Hungary Italy Japan Poland Portugal Singapore
South Korea Switzerland Thailand Turkey Ukraine Vietnam

Oxford is a registered trade mark of Oxford University Press
in the UK and in certain other countries

Published in the United States
by Oxford University Press Inc., New York

© D. Sherwood and J. Cooper 2011

British Library Cataloguing in Publication Data

Data available

Library of Congress Cataloging in Publication Data

Sherwood, Dennis.
Crystals, X-rays, and proteins :
comprehensive protein crystallography / Dennis Sherwood, Jon Cooper.
p. ; cm.
Update to: Crystals, X-rays, and proteins / Dennis Sherwood. 1976.
Includes bibliographical references and index.
ISBN 978-0-19-955904-6 (hbk. : alk. paper)
1. Proteins–Structure. 2. X-ray crystallography. I. Cooper, Jon (Jonathan B.)
II. Sherwood, Dennis. Crystals, X-rays, and proteins. III. Title.
[DNLM: 1. Proteins—chemistry. 2. Crystallography, X-Ray—methods.
3. Protein Conformation. QU 55]
QP551.S535 2011
612.8—dc22   2010032106

Typeset by SPI Publisher Services, Pondicherry, India
Printed in Great Britain
on acid-free paper by
CPI Antony Rowe, Chippenham, Wiltshire

ISBN 978–0–19–955904–6 (Hbk)

1 3 5 7 9 10 8 6 4 2

Preface

This book presents a complete account of the theory of the diffraction of X-rays by crystals, and whilst this book is aimed mainly at structural biologists, it will be of interest to any student of the properties of crystals, or of the molecules which form crystals. Study of some aspect of the crystalline state affects virtually all branches of science. Physicists, engineers, metallurgists, geologists and mineralogists are primarily interested in packing in the solid state, and the interpretation of physical behaviour in terms of crystal structure; chemists and life scientists, on the other hand, are usually more concerned with the structure of the molecules within a crystal.

The impact which X-ray diffraction has made upon the molecular life sciences is indeed most significant, for it is the most powerful known technique which enables the realisation of one of the primary goals of molecular biology—the interpretation of biological function in terms of the structures of those molecules taking part. The extent to which the results of X-ray crystallography have revolutionised molecular biology is covered in numerous excellent biochemistry texts, which many readers will be familiar with. However, it is hoped that the information contained within this book will enable the student to read and critically assess the specialised literature on methods for structure solution and on individual structure determinations. Almost all of the chapters are followed by a bibliography or references section to allow the reader to pursue further detailed aspects of each topic covered, and a general bibliography is available at the end of the book.

The book is aimed primarily at life scientists and chemists who need a complete and thorough explanation of the theory of X-ray diffraction. Since this readership usually lacks the mathematical and physical background of the physical scientist, all the relevant mathematics, crystallography and wave theory are developed from first principles. The book assumes no knowledge beyond elementary integral calculus, and may be used as a text at any post-sixth-form/high-school level. Although the book is directed at life scientists and chemists, it is hoped that it may be found useful by physical scientists who desire a refresher course in fundamental principles, or perhaps wish to have a book which contains the entire theory in detail. In places, some acquaintance with basic protein biochemistry is assumed—terms such as 'amino acid' are used without definition. Whilst this book is unashamedly 'proteinist', the methods described for structure solution apply as well to any type of macromolecule, such as nucleic acids, as they do to proteins: accordingly, we ask the reader to take every occurrence of the word 'protein' in its broadest context!

The sixteen chapters fall into three parts.

Part I: Fundamentals: Chapters 1–6

The theory of X-ray diffraction presupposes a certain knowledge of crystallography and the physics of waves, with the attendant mathematics. Part I presents this necessary background material. Chapter 1 is a brief introduction to the science of crystals, and explains why X-ray diffraction is the most useful technique for detailed structural studies. Elementary vector analysis and complex algebra form the content of Chapter 2, and vector techniques are applied directly in Chapter 3, which discusses the geometry of crystals. Some of the proofs need not be studied extensively, but the concepts of the description of the lattice and of lattice symmetry are of prime importance. Complex number theory is invoked in Chapter 4, where aspects of wave theory are explained.

Two significant chapters then follow. Fundamental to the theory of X-ray diffraction is Fourier theory, which is the subject of Chapter 5. Assuming no prior knowledge, the concepts of Fourier transforms and convolutions are explained and illustrated. This knowledge is essential for an understanding of Part II. Part I closes with a qualitative discussion of diffraction. Diffraction is by no means an obvious phenomenon, and it is hoped that a careful reading of the many illustrative paragraphs of Chapter 6 will give the student an intuitive feeling for what is happening during diffraction, and convince him or her of the significance of the Fourier transform.

In that this book is written for students from all backgrounds, it is quite possible that various aspects of the material covered in Part I will be familiar to some students. Indeed, the authors would be surprised if any reader found it necessary to study all six chapters, and equally surprised if any student found none of them of any value. Each chapter is quite self-contained, and any may be skipped without loss of continuity. Each chapter concludes with a summary of the significant points which gives a useful overview of the chapter so that the reader can be reminded of crucial parts of the argument. The student, however, is alerted to the importance of Chapters 5 and 6, and is advised to study them carefully.

Part I ends with a brief review of the most important concepts discussed in the first six chapters.

Part II: Diffraction theory: Chapters 7–9

Chapters 7, 8 and 9 form the principal section of the theory covered in the book. As an introduction to the way structural information is contained within diffraction patterns, Chapter 7 investigates the properties of some simple one-dimensional examples. This chapter demonstrates clearly the nature of diffraction patterns, and is intended to guide the student so that he or she knows what to look for in the more complicated three-dimensional case. Chapter 8, probably the most difficult in the book, contains the theory of diffraction by a three-dimensional lattice. Here are to be found the reciprocal-lattice concept, the Ewald sphere and Bragg's law. Although the geometry of the reciprocal lattice may be somewhat tedious, it is essential for an understanding of the practical methods of data collection that the relationships between the reciprocal lattice, the Ewald sphere and Bragg's law be grasped.

The effect of the contents of the unit cell on the diffraction pattern is discussed in Chapter 9. Those who have struggled through Chapter 8 will be relieved to find that Chapter 9 is somewhat easier—for it is by applying the theory explained in this chapter that molecular structures are revealed.

A second review contains the overall argument.

Part III: Structure solution: Chapters 10–16

In Chapter 10, we outline the experimental methods for expressing, purifying and crystallising proteins, as well as the procedures for preparing crystals for X-ray data collection. In Chapter 11, we describe the generation of X-rays and the methods by which diffraction data are collected and processed.

In Chapter 12, we describe the concept of the Patterson function, which is pivotal in many methods of structure analysis. Chapter 13 covers one such method, known as molecular replacement, in which we essentially borrow a known X-ray structure in order to solve an unknown one. Of course, the known and unknown structures must be similar for this method to work.

When no structures similar to the one we are trying to solve are available, then we must resort to experimental approaches for solving the unknown structure. The relevant theory and practice are covered in Chapter 14, and the following chapter on refinement (Chapter 15) describes methods for optimizing the agreement between the structure that we derive and the experimental data. Finally, in Chapter 16, we look at a number of complementary diffraction methods which yield important and extraordinarily detailed information on how proteins perform their biological functions.

A final review follows Chapter 16.

Dennis Sherwood

In a review entitled 'Desert island crystallography', published in *Nature* (**260**, 463 (1976)), the eminent crystallographer U. W. Arndt wrote, 'A highly intelligent school leaver about to take up an appointment as an X-ray crystallographer in a biochemical laboratory and wrecked on a desert island on his way, would find the present volume invaluable in equipping him for his new post by the time of his rescue'. When I first read those words, I was immensely pleased, and I thank the late Dr Arndt for them, for they precisely encapsulate why I wrote the original volume—to provide a complete, comprehensive and comprehensible explanation of the theory and practice of this most important technique. Thirty-plus years on, the theory is pretty well the same, but the practice has (inevitably, and wonderfully) changed out of all recognition. So, it was a great surprise, and indeed pleasure, when Professor Cooper suggested this revised edition. I therefore thank him for his initiative, his energy and his expertise—it has been a pleasure to work with him to revise the 'old' text, as well as to introduce much new material to bring the book fully up to date. And there's one other matter I'd like to bring up to date too: the original edition was 'to Anny', whom, at the time, I had just met and married—so now it's 'To Anny, Torben and Torsten', adding the names of two young men who weren't around then!

Jon Cooper

I would like to thank Dennis Sherwood for allowing me to update the original *Crystals, X-rays and Proteins*, which was published in 1976. My contribution to the current book is mainly in the form of the last half-dozen or so chapters, where the experimental methods of protein crystallography are described. I would like to sincerely thank Dr Ian Tickle (Astex Therapeutics, Cambridge) for his very helpful comments on my drafts of Chapters 11, 13, 14 and 15 and his untiring assistance in their preparation. I would also like to thank Dr Peter Erskine (UCL) and Dr Mark Montgomery (UCL) for commenting on Chapter 10, and Professor Leighton Coates (Oak Ridge National Laboratory, USA) for reviewing Chapter 16. I am indebted to Professor Tom Blundell, FRS, Professor Steve Wood, Professor Peter Shoolingin-Jordan and Professor Mark Pepys, FRS, for giving me the opportunity to work in their laboratories over the past 25 years. I must also thank Dr Peter Erskine for patient help in the final stages of completing the revised manuscript.

Contents

List of symbols

Characters in **bold type** are vectors.
Characters in *italic type* are scalar quantities (or complex numbers).

A	absorption correction factor, A face-centred cell, area
a, a	crystallographic unit cell vector, acceleration
B	temperature factor, B face-centred cell
b, b	crystallographic unit cell vector
C	C face-centred unit cell
c, c	crystallographic unit cell vector, convolution function
d	resolution (interplanar distance), derivative
E, E	electric field energy, normalized structure factor
e	base of natural logarithms (2.71828), electronic charge, primary extinction correction factor
F, F	force
F	structure factor (complex number)
f	scattering factor
f_e	electronic scattering factor
f_j	atomic scattering factor
g	glide plane
H, H	magnetic field
h	reciprocal-lattice index, reflection index, plane index, Planck's constant
I	intensity
i	$\sqrt{-1}$, inversion centre
i	Cartesian unit vector
J	joule
j	Cartesian unit vector
K	general scale factor
k	reciprocal-lattice index, reflection index, plane index, general scale factor
k, k	wave vector
k	Cartesian unit vector
L	Lorentz factor

l	reciprocal-lattice index, reflection index, plane index
m	mirror plane, mass
n	arbitrary integer
n_j	unitary atomic scattering factor
n_t	n -fold screw axis
P	Patterson function, reflecting power
p	polarisation factor, real-lattice index
q	charge, real-lattice index
r	real-lattice index
$\mathbf{r}, r$	general position vector
$\mathbf{S}, S$	reciprocal-lattice vector, scattering vector
s	sign
T	Fourier transform operator, temperature
t	time
U	unitary structure factor
$\mathbf{u}, u$	displacement vector, arbitrary vector
V	volume of unit cell
$\mathbf{v}, v$	arbitrary vector, velocity
$\mathbf{w}, w$	arbitrary vector
X	absolute coordinate within unit cell
x	general Cartesian coordinate, fractional coordinate within unit cell
Y	absolute coordinate within unit cell
y	general Cartesian coordinate, fractional coordinate within unit cell
Z	absolute coordinate within unit cell
z	general Cartesian coordinate, fractional coordinate within unit cell

Greek letters

α	crystallographic unit cell angle, Eulerian angle
β	crystallographic unit cell angle, Eulerian angle
γ	crystallographic unit cell angle, Eulerian angle
Δ	finite difference
δ	delta function
δ	arbitrary phase angle
Θ	Debye temperature
θ	Bragg angle

2θ scattering angle

λ wavelength

μ linear absorption coefficient

ν frequency

π 3.14159

ρ electron density function

$\sum$ summation

τ periodic time

ϕ phase angle, spherical polar angle

χ spherical polar angle

ψ spherical polar angle, wave amplitude

ω angular velocity, wave frequency $(2\pi/\tau)$

Subscripts

$c, calc$ calculated

d diffracted

e electron

hkl reflection indices, plane indices

i, in incident, general index

j general index

o, obs observed

rel relative

$scat$ scattered

T corrected for temperature and other disorder effects

t time

x component on x axis

y component on y axis

z component on z axis

Miscellaneous

$*$ convolution, complex conjugate, reciprocal-lattice vector

$|\ldots|$ amplitude or magnitude

$\langle\ldots\rangle$ mean or average

$\cdot$ scalar or dot product

$\wedge$ vector or cross product

$\geq$ greater than

$\leq$ less than

$\approx$ approximately equals

∂ partial derivative

Overlines

$\rightarrow$ vector

—   negative index, mean or average

Part I
Fundamentals

1
The crystalline state and its study

1.1 States of matter

In order to describe the materials which surround us, we classify matter into the three broad categories of solid, liquid and gas. These three states are useful because they associate classes of matter which, although very diverse in chemical composition, exhibit similar overall physical behaviour. A first observation of the three states of matter results in the following generalisations:

Gases. Gases fill the entire volume available to them, and readily change their volume in response to changes in pressure. They have a relatively low density, and flow freely.

Liquids. Liquids flow, but at a particular temperature they occupy a fixed volume. They assume the shape of the vessel in which they are contained, and are only slightly compressible. The density of a liquid is usually intermediate between that of a gas and a solid.

Solids. Solids have both fixed size and fixed shape, have a relatively high density, and are virtually incompressible. When subject to small shearing forces, solids in general deform, but return to their original size and shape when the force is relaxed.

These observations are commonplace, and the reader is undoubtedly aware of them. But one should not disregard these results: familiar they may be, trivial they are not. If we wish to investigate the structure of matter, then the descriptions above contain a surprising amount of information. Firstly, consider the response of each state of matter to pressure. Since solids are virtually incompressible, we infer that the molecules (or atoms or ions as appropriate) within a solid are as close together as they can be, and that there is a minimum of void space between the molecules. It is only by the use of extreme pressures that the molecules can be pressed any closer together. Gases, on the other hand, readily change their volume with pressure, implying that between the molecules in a gas there is considerable 'free' space. Liquids are rather more compressible than solids, but markedly less so than gases. This may be interpreted by suggesting that in a liquid there is only a small degree of 'free' space.

Secondly, let us examine the effect on each state of matter of small shearing stresses. The fact that a solid deforms reversibly implies that the solid is able to exert forces which counter the applied stress, so that on relaxation of the applied forces, the solid returns to its original size and shape. Gases, however, show very little resistance to shear stresses and flow very easily. Liquids, once again, show intermediate behaviour. They flow as a response to shear stresses, but the rate of flow depends on the liquid considered, the resistance to shear stress being determined by the liquid's viscosity.

These observations may be interpreted by envisaging molecular models which behave in the appropriate manner. Thus a gas may be considered as an assembly of molecules which are relatively far from one another, free and independent, so that they may flow past one another very easily. Solids, on the other hand, exhibit properties which are consistent with a model in which the molecules are close together, and tightly bound to one another with relatively strong intermolecular interactions. Liquids have properties somewhere between those of a gas and a solid, and so the molecules are fairly close together, with somewhat weaker interactions than those exhibited by a solid. With these models, it is simple to rationalise the fact that gases fill the container they occupy, whereas solids have a size and shape independent of any container. Under the influence of thermal energy, gas molecules are not constrained to one another, and so will expand as much as they are able; in solids, the thermal energy is insufficient to overcome the intermolecular interactions, and so the size and shape of the solid remain more or less constant.

As scientists, we seek to investigate the natural universe as fully as possible. An obvious problem which presents itself is 'What are the molecular structures of gases, liquids and solids?' This question is asking for information on the way in which the molecules are associated with one another in each of the three states of matter. So far, all we know is the fairly vague impression that the molecules in the solid state are somewhat closer together and more tightly bound to each other than in the gaseous state. But can we deduce anything rather more specific?

1.2   Anisotropy

In this section we will consider the results of some experiments concerned with the measurement of various physical parameters of solids.

If we measure the electrical conductivity of graphite, a most surprising result is obtained. When the graphite is placed in one direction, the electrical conductivity has a particular value, but on rotating the graphite, the electrical conductivity assumes a different value. In other words, the electrical conductivity of graphite is not a constant in the general sense, but depends on the direction in which the conductivity is measured. It is the conductivity *in a particular direction* which is the true physical constant. This variation of a physical property with direction is referred to as *anisotropy*, and graphite is said to be anisotropic with respect to electrical conductivity. If a material has a physical property which has the same value in all directions, then the material is called *isotropic*. The example of the conductivity of graphite concerns one particular physical property in one material, but the phenomenon of anisotropy occurs in very many branches of the physics of nearly all solid matter. Some examples of anisotropy in general are discussed below.

Mechanical properties

It is well known that the solid mica (a silicate mineral) can be cleaved very easily into fine layers. If we try to fracture mica in a direction other than parallel to the natural layer structure, then we find that the material is comparatively strong. In this way, mica shows very pronounced anisotropy in its mechanical strength. Anisotropy with respect to cleavage is very common in minerals. If virtually any mineral is sharply struck, it will be found to cleave in a number of well-defined directions, showing that the material has weaknesses in its structure associated with particular directions. Mechanical anisotropy is also demonstrated in the response of solids to tensile stresses. When a solid is subject to a tensile stress it will in general deform differently in different directions, depending on the direction of the applied stress. Mechanical anisotropy therefore exists in the cleavage and the elastic properties of solids.

Thermal properties

On heating, all materials expand. Careful measurements on some solids, however, show that the amount of expansion varies with direction in the solid. Hence thermal expansion shows anisotropy. Anisotropy may also be seen in the thermal conductivity of some solids.

Optical properties

If a light beam is incident on the mineral calcite (a naturally occurring form of calcium carbonate), then instead of there are being one refracted beam, there are in fact two refracted beams. This phenomenon is known as *birefringence* and is exhibited by many translucent materials. Investigation of the two refracted beams shows that only one of them, the *ordinary ray*, obeys Snell's law of refraction, whilst the other beam, the *extraordinary ray*, does not. Moreover, the two beams are polarised in different directions, and it is found that the velocity of light in the material varies with the direction of propagation of the light within the mineral. This is an example of optical anisotropy in the solid state. When crystals are examined with a microscope, birefringence causes the apparent colour and brightness of a crystal to change as the polariser is rotated.

Electrical properties

The anisotropy of the electrical conductivity of graphite has already been mentioned. Another anisotropic effect may be observed on measuring the *dielectric constant* of a solid. The dielectric constant is related to the strength of an electric field within the solid, and is determined by the dipole moments of the molecules in the material. For many solids, it is found that the magnitude of the dielectric constant varies with direction.

Magnetic properties

Materials which are spontaneously magnetised at room temperature are called *ferromagnetic*, a common example of a ferromagnetic solid being iron itself. Iron may be treated in such a way as to cause it to lose or increase its spontaneous magnetisation, and if an unmagnetised piece of iron is placed in an electric coil, it may be remagnetised. Careful experiment shows that the energy required to magnetise iron (and ferromagnetic materials in general) depends on the direction in which the material is placed with respect to the coil. Hence ferromagnetic materials may be magnetised more easily in some directions than in others, showing that these materials exhibit magnetic anisotropy.

Coupled effects

Several minerals, such as quartz and tourmaline, show the *pyroelectric* and *piezoelectric* effects. In the first case, an electrical potential difference is generated across the solid when the material is heated. Piezoelectricity is the appearance of an electrical potential difference in response to mechanical pressure on the material. Both of these effects show anisotropy, in that the magnitude and direction of the potential difference vary according to the direction of the heat flow or the pressure. These effects also work in reverse: for example, passing current through a piezoelectric crystal will cause it to deform very slightly, and sound can be generated by application of an oscillating current.

We have now met anisotropy in all branches of solid state physics, and the more we investigate the behaviour of solids, the more we find that far from being exceptional, the general phenomenon of anisotropy is the rule. Of course, very few solids are anisotropic in all their physical properties but, conversely, it is rare to find solids which are entirely isotropic. In general, most solids are anisotropic with respect to some physical parameters, but isotropic with respect to others. For example, solid sodium chloride is optically isotropic but mechanically anisotropic.

The situation is quite different when we consider the physical properties of gases. There is no known gas which exhibits macroscopic anisotropy. All gases are isotropic in all their physical behaviour, even those gases derived from solids which are themselves anisotropic. Liquids, as usual, show intermediate behaviour. Most liquids are isotropic, but some anisotropic liquids do exist.

The picture emerges that solids in general exhibit a variation in their physical properties with respect to direction. In that physical phenomena as diverse as those concerning mechanical, thermal, optical, electrical and magnetic effects all exhibit anisotropy, it is unlikely that this anisotropy is a result of a chance physical event. Rather, anisotropy appears to be a property of the solid state by virtue of the fact that it is indeed solid.

1.3   The significance of order

Given that anisotropy is a fundamental characteristic of many solids, can we make any deductions which are relevant to our understanding of the structure of solids? The question we wish to answer is, 'What feature of the structure of the solid state will give rise to anisotropy?'

To answer this question we argue as follows. Any physical property of a macroscopic solid is determined in the final analysis by the properties of the molecules contained within the solid, and by the way in which the individual molecules are arranged in a multi-molecular array. Before we investigate the anisotropy of solid bodies, let us first study the properties of individual molecules—in particular, their spatial properties as might be inferred from their symmetry, and their structure.

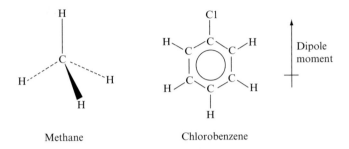

Methane   Chlorobenzene

Fig. 1.1 *Molecular structure can give rise to anisotropy.* Asymmetries in chemical structure can give rise to directional properties, as shown by the dipole moment of chlorobenzene.

Consider, for example, the two small molecules shown in Fig. 1.1, methane and chlorobenzene. Methane is a highly symmetric molecule, both spatially (the molecule can be represented as a tetrahedron) and structurally (all of the four 'arms' of the tetrahedron are identical, each being a single C–H bond). We would therefore expect the properties of methane to be the same in all directions. Chlorobenzene, however, is different in one very important respect: since the chlorine atom is more electronegative than the benzene group, chlorobenzene has a dipole moment directed along the benzene ring C–Cl bond. The direction parallel to this dipole moment therefore defines a 'special' direction in space, a direction determined by the structure of the chlorobenzene molecule, and a direction which defines anisotropy on a molecular scale. The structural complexity of individual molecules can therefore enable them to have particular directional properties, and we see that anisotropy is readily explained on a molecular scale as being fundamentally due to molecular structure.

Let us now turn to a multi-molecular aggregate of molecules as in a solid. We already know that the molecules in a solid are very close together, with strong forces between each other, but we have as yet no specific information relating to the way in which the molecules are arranged with respect to one another. Let us consider two ways of packing chlorobenzene molecules together, as shown in Fig. 1.2. Figure 1.2(a) shows a random two-dimensional arrangement of molecules in which the molecules are in close proximity to one another, but there is no particular specification of their relative positions and orientations. In Fig. 1.2(b), there is an array in which the molecules are packed in a regularly spaced manner with their dipole moments parallel. When we consider the properties of the random array, despite the fact that each individual molecule has a dipole moment, the randomness of the array will cause the effect of the dipole moment of one molecule to be cancelled by that of another molecule. If we measure the net dipole moment due to the entire array, the random nature of the structure implies that the overall dipole moment will be zero. There will be no dipole moment in any direction, and so the dielectric constant of the solid will be the same in all directions. The solid will therefore be isotropic in this respect. For the regular array of Fig. 1.2(b), however, the situation is quite different. Since all the dipole moments are parallel, they will reinforce each other, implying that the solid as a whole will have

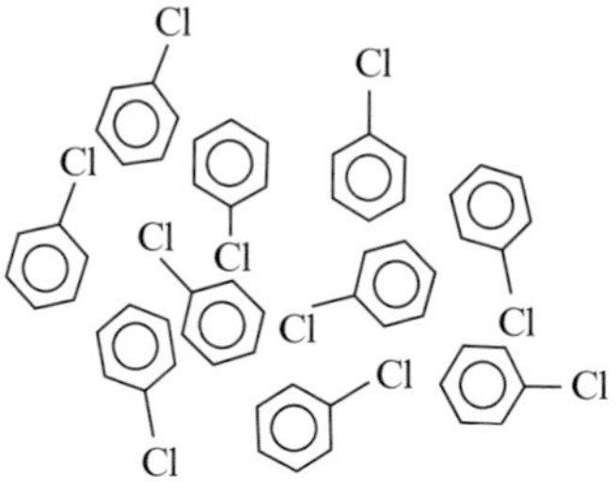

(a)

A random array,
no net dipole moment

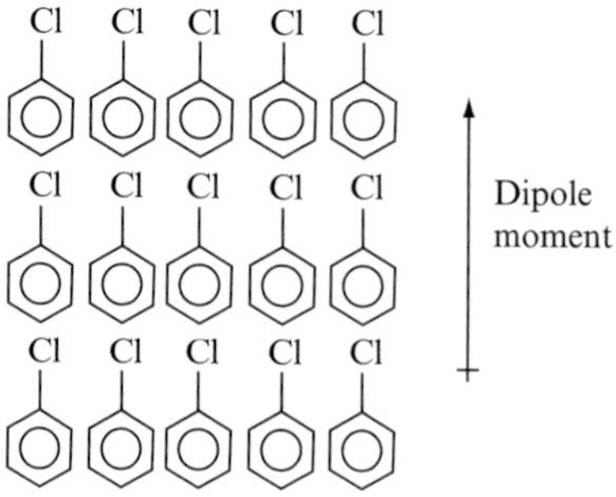

(b)

A regular array,
a net dipole moment exists

Fig. 1.2 *Which of these structures is anisotropic?* Both structures are of closely arranged chlorobenzene molecules, but the randomness of (a) implies that there will be no net dipole moment for the structure (a) as a whole. Only an ordered, regular structure such as (b) can give rise to a net dipole moment and hence macroscopic anisotropy.

a net dipole moment, and this will define a 'special' direction in space. The dielectric constant of the solid will now vary with direction, and the solid will be anisotropic.

What is the difference between the array which gives rise to anisotropic effects, and the array which is isotropic? The latter array was random, whereas the former array was just the opposite—a highly regular array in which the molecules were accurately parallel to one another. The random array is characterised by disorder, but the second, regular, array may be described as *ordered*. It is this ordering which is the clue to the significance of anisotropy. Since the physical properties of the solid state necessarily reflect the properties of very large numbers of individual molecules, it is only when these molecules are arranged in a definite, well-defined, *ordered* array that any directional properties may become apparent. If the arrangement is random, then any directional property of the component molecules will be averaged to zero on account of the random, irregular orientation and position of one molecule with respect to the next. Anisotropy, therefore, is possible *only* when the molecules are arranged with regularity and order.

This is not to say that all well-ordered arrays necessarily exhibit anisotropy. Indeed, we can easily conceive of an array of chlorobenzene molecules which is ordered, but

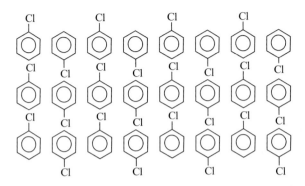

Fig. 1.3 *Order, but no anisotropy.* This array of chlorobenzene molecules is ordered, but since the dipole moments cancel in pairs, the structure will be isotropic with respect to its dielectric constant.

which will display no net dipole moment, as in Fig. 1.3. The array is definitely ordered, in that the molecules are aligned in alternate directions. But since the dipole moments of the molecules cancel in pairs, there will be no net dipole moment for the aggregate as a whole, and so the dipole moment will be zero in all directions. This implies that the solid will be isotropic with respect to its dielectric constant. It is therefore fallacious to say that all ordered arrays will be anisotropic, but it is undoubtedly true to state the converse, namely, that all anisotropic materials necessarily have an ordered structure.

1.4   Crystals

The significance of the previous discussion has been to demonstrate that the existence of anisotropy is possible if and only if the molecules in a material are ordered in some systematic manner. But it had previously been stated that anisotropy was a feature of the solid state. Consequently, we may infer that a further feature of the structure of the anisotropic solid state is that not only are the molecules in close proximity so that there are relatively strong intermolecular forces, but the molecules are arranged in an ordered manner. Since no gas is anisotropic, we may infer that the gaseous state is characterised by essentially complete disorder and randomness. This agrees with our concepts concerning the independence of molecules in the gaseous state. If the molecules are free, then there is no reason why they should be arranged with respect to one another in any particular manner, and so we expect all gases to be isotropic. This is just what is observed. Once again, liquids are intermediate in behaviour, exhibiting some degree of order and some degree of disorder.

Let us now try to consider a possible model which will distinguish between the liquid and the solid states. We believe that solids are much more ordered than liquids in view of the occurrence of anisotropy in solids. But how do we determine what is meant by 'much more ordered'? What we are asking is a question concerning a criterion by which we may assess relative degrees of order. Since order is a concept which concerns assemblies of molecules, it is reasonable to propose that any assembly which maintains its order over a greater distance is more ordered than one which is ordered over only a comparatively short distance. A significant measure of distance for a molecular system is the average intermolecular spacing, and so we may say that an ordered solid preserves the ordering of its structure over many more intermolecular spacings than does a liquid. To state exactly how many spacings are required to define an ordered solid is not easy, but if an aggregate of molecules is perfectly ordered over, say, 10^6 intermolecular spacings, then we are entering macroscopically observable domains. A liquid, on the other hand, is characterised by short-range order, in that in a liquid the molecules are quite close, with any one molecule having about the same number of nearest neighbours as any other molecule. With increasing distance, however, the presence of void spaces disrupts the ordering, and at a distance of about ten molecular spacings there is no definite manner in which the molecules are arranged and oriented. In contrast, the ordered solid state is sometimes characterised by a long-range order which extends over literally millions of molecules, so that the environment of any one molecule is identical to that of any other molecule, no matter how far apart these two molecules are. Our physical models of the structure of these three states of matter are now more complete, as shown in Fig. 1.4.

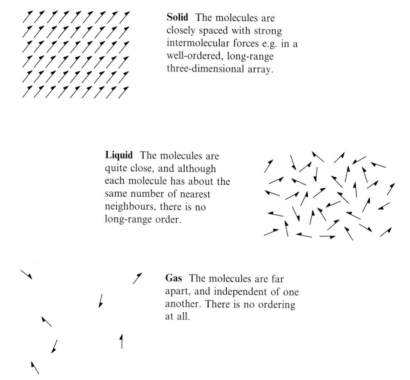

Solid The molecules are closely spaced with strong intermolecular forces e.g. in a well-ordered, long-range three-dimensional array.

Liquid The molecules are quite close, and although each molecule has about the same number of nearest neighbours, there is no long-range order.

Gas The molecules are far apart, and independent of one another. There is no ordering at all.

Fig. 1.4 *The three states of matter.* These model structures are consistent with the physical behaviour of solids, liquids and gases.

Solids which possess this long-range, three-dimensional ordering are known as *crystals*. A crystal may thus be defined as a solid which possesses long-range, three-dimensional molecular order.

A direct result of the three-dimensional ordering of molecules in a crystal is the appearance of plane faces. Indeed, perhaps the most obvious property of a 'crystal' is its macroscopic geometrical shape. The plane faces are a consequence of the regular stacking of molecules in layers, so that any face represents a plane parallel to a molecular layer. Whereas we have predicted that a crystal is likely to have plane faces, historically it was the existence of the plane faces of certain minerals which gave rise to the idea that the molecular structure of a crystal might be a regular array of 'building blocks'. The first explicit proposal of this kind came from the French mineralogist Abbe Haüy (1743–1822), who had studied the cleavage properties of calcite. He realised that when a calcite crystal was struck a sharp blow, it would cleave along a well-defined plane, leading him to surmise that should the cleavage process be continued indefinitely, one would ultimately obtain a single, indivisible unit which had a precise geometrical shape. Nowadays, we realise that the external shape of a crystal is just one more aspect of the three-dimensional regularity of the structure. It should be noted that it is quite possible for a solid to be crystalline but not to show any obvious external geometrical properties. Most metals are a case in point. Because the structure of the metal is a regular, three-dimensional array, we say that metals are crystalline, yet in general, they may assume any external shape. In fact, it is sometimes possible to see a geometrical macroscopic form for some metals, but this requires special treatment and is not the usual case. The scientific use of the word 'crystal' is therefore rather more specific than the everyday usage. In normal speech, by 'crystal' we usually mean a solid with salient plane faces such as a cut gemstone. But to the scientist, a crystal is a solid which has a well-defined, long-range, three-dimensional molecular order.

1.5 Solids which are not crystals

The above discussion has been very general, pointing out several aspects of the solid state. We might now ask the question 'Are all solids crystalline?' Specifically, we may enquire as to whether a material such as glass is a crystalline solid in the same sense that calcite is. Both calcite and glass are hard and transparent to light. But although glass may fracture, it does so in an irregular manner. In fact, the structure of glass (to be strictly accurate, we must refer to 'glasses', as there are many types of glass) comprises long macromolecules of silicon dioxide which have cooled in a random manner. Glasses, as shown in Fig. 1.5, do not have a regular, three-dimensional structure, and so they cannot be referred to as crystalline. This is further verified by the fact that glasses do not show a sharp melting point, but become progressively more fluid. When a glass forms from the molten state, the long macromolecules adopt a configuration which is not perfectly regular. Since the thermal energy available as the glass cools is not sufficient to allow the polymer to form a regular configuration, the randomness of the liquid state is 'frozen in'. Clearly, glass-like states can be formed only from long polymeric molecules which are 'too unwieldy' to respond to

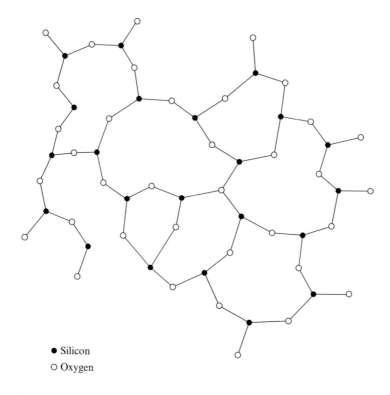

Fig. 1.5 *A schematic representation of the structure of a glass.* A silicon dioxide polymer cools to form a random molecular configuration. Although the material is 'hard', the randomness of the molecular structure implies that glasses are not crystalline. Each silicon atom will form covalent bonds to four oxygen atoms, and each oxygen will form bonds with two atoms of silicon. Only a very thin section of the glass is shown here, so, for clarity, not all of the chemical bonds are drawn.

thermal energies. Usually we refer to glasses as being *supercooled liquids*, to emphasise the randomness of the arrangement of the macromolecules. Other molecules, notably polymers, can solidify in a partially ordered state, which may exhibit anisotropy in its physical properties.

Solids in which there is no long-range order in the positions of the constituent atoms or molecules are referred to as amorphous. Amorphous solids can be made from solids that normally crystallise by rapidly cooling molten material. A high cooling rate prevents the atoms or molecules from packing in a more thermodynamically favourable crystalline state. Amorphous solids can also be produced by the use of additives which prevent the primary constituent from crystallising. A good example of this, which we will return to later on, is the addition of glycols or, more commonly, glycerol to water. On rapid cooling, the presence of the glycerol prevents the formation of ice crystals, resulting in a so-called 'vitrified' solid. This method is universally used for cooling protein crystals to sub-zero temperatures.

1.6   Crystal defects

We now know that a crystal is characterised by long-range order, and in this section we enquire as to the upper limit of 'long-range'. We have already decided that 'long-range' implies an order over about 10^6 intermolecular distances, but as yet we do not know what happens after, say, 10^{12} spacings. In fact, it is rare to find crystals which preserve perfect ordering over macroscopic distances such as may be measured with ease using ordinary laboratory equipment. The relative rarity of perfect geometrical crystals vouches for this. Once a crystal is regular over 10^6 spacings, it will exhibit the properties of a crystalline solid, but thereafter it is possible for various defects to be present as long as the perturbing effect of these defects does not have too large an effect. For instance, it is quite possible for an array of 10^6 molecules to have one vacant molecular site without disturbing the overall structure significantly. In general, defects may take many forms, but it is not the purpose of this volume to discuss crystal defects. They are mentioned solely in order not to present a misleading view of the crystalline state. If a solid is composed of many aggregates of a volume such that the order is perfect over about a million spacings, then each of these volumes is termed a *crystallite*. Metals are generally of this form, as may be shown by etching the surface of a metal and observing it under a microscope. In general, this book is concerned with the structure of perfect crystals, which are assumed to have ideal three-dimensional order throughout their entire volume. This is a fairly rare event, and probably impossible for crystals of large molecules like proteins, but once the theory for a perfect crystal is understood, then it is not difficult to investigate the effects of crystal defects.

1.7   Analysing the structure of crystals and molecules

The evidence so far presented for the regular, three-dimensional structure of crystals has been based on physical arguments which demonstrate that the macroscopic properties of crystalline solids can be interpreted only in terms of a model in which the molecules are stacked in a regular, three-dimensional array. Since the beginning of the twentieth century, several techniques have been developed which allow for direct investigation of the microstructure of solids, the more important of which are mentioned below.

Electron microscopy

An electron microscope is the analogue of a light microscope, except that electrons are used instead of light. For reasons which will become apparent after the next section, light can give information only on the structure of objects which are large relative to the wavelength of the light used (the wavelength of a wave motion is the distance between two successive wave peaks). So, using light of wavelength 600 nm (corresponding to yellow light), we cannot examine anything smaller than about 600 nm. Since the intermolecular spacing in a crystal is typically of the order of 1–10 nm, light microscopy is useless for the scrutiny of the ultrastructure of solids. To observe molecular scales, we need to use 'light' with a 'wavelength' of at most

1 nm, which is indeed now possible by exploiting the quantum mechanical properties of the electron. With the discovery of quantum mechanics, it is now known that under certain circumstances matter appears to behave as a wave. In view of this, it is possible to accelerate electrons through a potential of some 100 kV such that they act as a wave motion of wavelength of the order of 0.005 nm. Since electrons carry a negative charge, they may be deflected by magnetic fields, and so the analogue of the optical lens is the magnetic lens, as shown in Fig. 1.6. The technology of magnetic lenses allows a practical resolution of about 0.1–10 nm (1–100 Å). With an electron microscope it is possible to resolve atomic planes in crystals of small molecules, and the technique has proved to be of great use in analysing large molecules such as proteins and nucleic acids. Electron microscope pictures of crystals verify the three-dimensional stacking of molecules in planes.

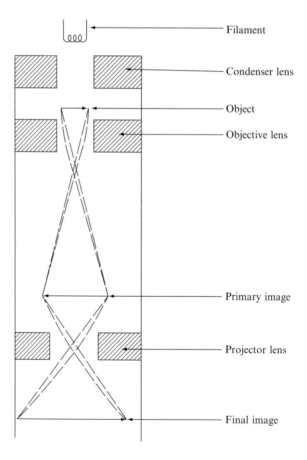

Filament

Condenser lens

Object

Objective lens

Primary image

Projector lens

Final image

Fig. 1.6 *Schematic representation of an electron microscope.* The filament is at a high negative potential and emits electrons. These are charged particles and are deflected by the magnetic lenses to form an image on a phosphorescent screen.

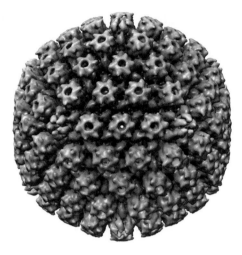

Fig. 1.7 *A cryo-electron microscope image reconstruction of the proteins making up the core of the herpes simplex virus (HSV).* The image is at a resolution of approximately 2.8 nm (28 Å). (Courtesy of Dr David Bhella, MRC Virology Unit, Glasgow.) The proteins form a highly symmetrical shell surrounding the viral DNA.

To visualise biological materials by electron microscopy (EM) requires either that they are stained using electron-dense elements such as lead, uranium or tungsten or that they are frozen very rapidly to liquid nitrogen temperatures. This causes the water in the sample to vitrify rather than form crystalline ice, which would disrupt the sample. Thus in cryo-EM, the sample is spread on a special grid and is snap-frozen by plunging it into liquid ethane. This cools the sample rapidly and preserves it close to its native state. Low electron doses are required to minimise radiation damage to the sample and, as a result, the images are noisy. However, this can be overcome for some biological systems by averaging the images of individual molecules or particles to increase the signal-to-noise ratio, so yielding detailed information about the specimen, as shown in Fig. 1.7. Occasionally proteins form two-dimensional crystals, and this allows a technique known as electron crystallography to be used. This is similar to X-ray crystallography.

Atomic force microscopy (AFM)

This method uses a probe which effectively scans the surface of the material being analysed, and gives a very high-resolution image (below 0.1 nm). The atomic force microscope consists of a very small cantilever with a sharp tip or probe that is deflected by contact forces between it and the atoms or molecules making up the surface (Fig. 1.8). The deflection of the cantilever can measured by a variety of methods, such as reflecting a laser off it onto a photodiode array or by measuring the piezoelectric effect in the cantilever as the needle scans the surface.

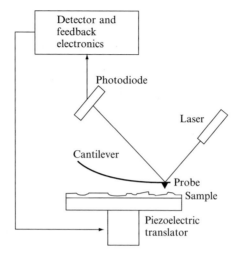

Fig. 1.8 *The atomic force microscope.* In this form of microscopy, a very small cantilever with a sharp tip or probe is used to 'image' the atoms or molecules making up the surface. The deflection of the cantilever can be measured by reflecting a laser off it onto a photodiode array, and the sample is moved by a piezoelectric tube or translator.

To avoid damaging the sample, the probe should generally not touch the surface, and so a feedback mechanism is employed which maintains a constant force between the probe and the surface. The deflection of the cantilever as the sample is moved is maintained at a constant value by constantly adjusting the height of the sample. To image the surface of the sample, piezoelectric crystals are used to move the sample in three dimensions. The force required to maintain a constant deflection of the probe as the surface of the sample is scanned gives the topography of the sample. There are also other ways to study the surface which include oscillating the probe or 'tapping' it on the surface. These methods have had many successes in studying whole cells, nucleic acids and proteins.

X-ray diffraction

The most powerful technique for investigating both the geometrical nature of the three-dimensional array of molecules within a crystal and the nature of the molecules themselves is that of X-ray diffraction. As will be discussed in detail in Chapter 6, *diffraction* is the phenomenon which occurs whenever a wave motion interacts with an obstacle. From an analysis of the nature of the diffraction, we may obtain information on the structure of the obstacle. X-rays are a form of electromagnetic radiation with a wavelength typically between 0.1 and 1.0 nm (1–10 Å). Since this is comparable to the molecular spacing in a crystal, Max von Laue (1879–1960) suggested in 1912 that a crystal may act as a diffraction grating to X-rays just as a normal diffraction grating acts on visible light. When X-rays interact with a crystal, diffraction does occur, and we may obtain information on the structure of the crystal and the molecules within

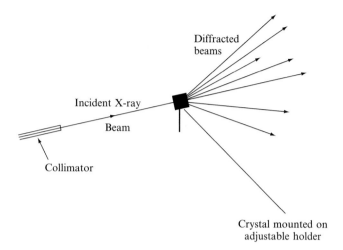

Fig. 1.9 *Schematic diagram of X-ray diffraction apparatus.* A beam of X-rays from a source is made parallel by passage through a collimator. The incident beam impinges on the crystal, which is held in a special mount which allows the crystal to be rotated with respect to the incident beam. Diffraction occurs, and the diffracted beams are recorded on photographic film or by a sensitive X-ray detector.

the crystal. X-ray diffraction was first observed by two of von Laue's students, Walter Friedrich (1883–1968) and Paul Knipping (1883–1935), and was used to determine the structure of a crystal by William Henry Bragg (1862–1942) and his son William Lawrence Bragg (1890–1971). For their contributions to physics, a Nobel Prize was awarded to von Laue in 1914, and to the two Braggs in 1915.

A typical X-ray diffraction apparatus is drawn schematically in Fig. 1.9 along with a representative diffraction image from a protein crystal in Fig. 1.10. It is clear that the information obtained by X-ray diffraction is less self-evident than the direct images formed by electron or atomic force microscopy. Hence it is necessary to understand the theory of X-ray diffraction in order to interpret the experimental data. It is with the development of this theory that the present volume will deal. Whilst there have been great advances in the field of nuclear magnetic resonance for studying molecular structures, X-ray diffraction remains probably the most powerful method and is therefore of the utmost importance to any scientist who desires to obtain knowledge of the structure of complicated molecules.

1.8 Why do we use X-rays?

As mentioned above, X-rays are a form of electromagnetic radiation. Light is itself a form of electromagnetic radiation, of wavelength between about 400 nm (violet) and 800 nm (red). Since light is a wave motion, we might again ask the question 'Why can we not use light waves to study the structure of a crystal?'

Some more insight into the answer to this question can be obtained by consideration of the following analogy, which should be familiar to anyone who has been in a small

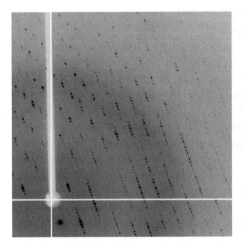

Fig. 1.10 *An X-ray diffraction pattern of a typical protein.* Although the information concerning the structure of the crystal is less direct than that which could be obtained by electron microscopy, as we shall see during the course of this book, the information from X-ray diffraction is much more detailed.

boat. A small boat, say 5 m in length, is on the sea. Waves are propagating through the water; these waves may be classified into essentially two types. The first are those waves which come in from the ocean, and are usually of long wavelength, about 20 or 30 m. The second class of waves are surface ripples such as may be produced by throwing a stone over the side of the boat. When a long-wavelength ocean wave comes to the boat, the wave merely passes underneath the boat, which rides on the wave. The short-wavelength ripple, on the other hand, is reflected by the boat, as shown in Fig. 1.11. An observer who cannot see the boat (if, for instance, there is a fog) but can detect the water waves will record that the short-wavelength surface ripples are

Fig. 1.11 *The interaction of waves with an obstacle.* The boat rides the long-wavelength ocean wave, but reflects the small-wavelength surface ripple. An observer who wishes to detect the presence of the boat can do so only by observing waves which have wavelengths smaller than or comparable to the length of the boat.

reflected but that the long-wavelength ocean waves pass on undisturbed. In that the presence of the boat does not affect the long-wavelength ocean waves appreciably, an observer who detects only these long waves will be unaware of the presence of the boat. If the observer can detect the surface ripples, however, he or she will observe the reflection and will infer the presence of the boat.

The physical significance of this example may be explained as follows. Our system consists of an obstacle (the boat), subject to two wave motions, of which one (the ocean wave) has a wavelength long in comparison with the length of the boat, and the other (the surface ripple) has a wavelength short compared with the length of the boat. The long-wavelength disturbance passes by the obstacle with essentially no alteration in its behaviour, but the short-wavelength disturbance is perturbed by the obstacle, in that reflection occurs. What is of significance is the comparison between the size of the obstacle (that is, a significant linear dimension of the obstacle) and the wavelength of the wave motion with which the obstacle interacts. Our example allows us to make the generalisation that whenever an obstacle interacts with a wave motion, then the nature of the interaction is dependent on the relative magnitudes of the wavelength of the wave motion and the dimensions of the obstacle. If the wavelength is much greater than a significant linear dimension of the obstacle, then the obstacle has only a small effect on the behaviour of the wave. If the wavelength of the wave motion is smaller than or comparable to a significant linear dimension of the obstacle, then the behaviour of the wave is affected to a large extent, and events such as reflection may occur.

This generalisation has an important application. If we wish to study an object by means of its effect upon a wave motion, then we must choose a wave form whose wavelength is smaller than or comparable to a significant linear dimension in the obstacle. Only in this event will the object cause a pronounced effect on the wave motion, and it is from this effect that we derive information about the obstacle.

Let us now apply this idea to our case of a crystal. A crystal contains two significant linear dimensions. One of these is the macroscopic length of the crystal itself, which we may take as typically 1 mm. The other significant linear dimension is the spacing between the molecules in the crystal, which we expect to be in the range of molecular dimensions, of the order of 1 nm (10 Å). We now place the crystal in the path of a light beam of wavelength 600 nm, corresponding to yellow light. The wavelength of the light is approximately 10^3 times smaller than the macroscopic crystal size, but of the order of 10^3 times greater than the intermolecular spacing. In terms of the concepts introduced above, the light wave will be affected by the macroscopic crystal size, but will not be noticeably perturbed by the intermolecular spacing. The light therefore 'sees' only the crystal as a whole, and not the molecular-scale structure. So, by looking at the effect of the crystal on a beam of yellow light, we can detect only the macroscopic features, and can gain no information on molecular structure. Light is therefore not a suitable wave motion for investigating the detailed molecular architecture of crystals.

But if we put the crystal in a beam of X-rays, the situation is changed. The wavelength of the X-ray is of the order of 0.1 nm (1 Å), quite comparable to the molecular spacing within the crystal. Hence we may predict that the molecular-scale structure will have a large effect on the X-rays. Furthermore, the order of the three-dimensional array of molecules implies that the effect of any one molecule on the X-rays is repeated in a regular manner throughout the crystal. This makes the effect of an individual molecule much amplified, so that relatively large overall effects are observable. This is indeed the case. The effect of the molecular structure and organisation of a crystal is to cause diffraction of the X-ray beam, and from the nature of the diffraction, the molecular structure of the crystal may be determined.

1.9   Why do we use diffraction?

In the previous section, we saw that X-radiation was a form of electromagnetic radiation which has a wavelength range suitable for the study of the structure of molecules within crystals. In this section, we shall answer the question 'Why do we use diffraction?'

As we have already seen, diffraction is the event which occurs when a wave motion interacts with an obstacle. But there is a form of wave–obstacle interaction which is much more obvious than diffraction, and that is the interaction which takes place in, for example, a light microscope. The question posed above then becomes 'Why do we not use an X-ray microscope?'

The answer to this question is very simple. A light microscope operates by means of a series of lenses which diverge and focus the light passing through them. The magnetic lenses in an electron microscope behave in an analogous manner. Unfortunately, at the present time, no means is known by which X-rays may be focused well enough, and so no X-ray microscope exists, because we have no X-ray lenses that would be suitable. Whilst an X-ray beam may be focused by reflecting it off a curved surface (and this is used for generating the narrow X-ray beams which crystallographers use, and also in X-ray astronomy), direct visualisation of molecular structures as would be performed in an X-ray microscope is simply beyond present technology. For this reason, we must resort to the rather less obvious phenomenon of diffraction. As will be shown in Chapter 6, the information content of a diffraction pattern is potentially the same as that in a direct image, but the information is less easy to interpret in direct structural terms. It is the purpose of this book to explain how the data derived from a diffraction pattern may be processed in order to derive fundamental information on molecular structure.

1.10   Protein crystals

Proteins are large, polymeric molecules composed of amino acids which are linked together by peptide bonds. The sequence, or order, of the amino acids is governed by the gene for the protein and dictates the three-dimensional structure of the molecule. To crystallise a protein, it must first be purified, and this can be relatively straightforward for those proteins which are very abundant in living organisms. However, the majority of proteins do not occur naturally in sufficient quantities. To overcome this problem requires the molecule to be produced in large amounts, often in bacterial cells, by a range of techniques which are broadly referred to as 'molecular biology', and we will look at these methods briefly later in the book. Once the protein has been expressed and purified, it can be persuaded to crystallise by finding suitable conditions of, for example, salt concentration, pH and various precipitants, and often much time (and protein) has to be spent finding these optimum conditions.

Protein crystals contain large amounts of water, typically 50–60% (or more) by weight, and this means that they must be kept hydrated at all times or they will dry out and lose their internal molecular order. Ways to prevent this drying-out include mounting the crystal in a very thin glass capillary which is sealed at both ends, and freezing the crystal in liquid nitrogen. Both methods require special steps to be taken to prevent the crystal from being damaged owing to the high water content. Unlike the majority of the crystals that we are familiar with, protein crystals are extremely fragile (owing again to the high water content), and they can very easily be broken by touching them, even with the tip of a needle. In spite of these special problems, the internal order of protein crystals is usually good enough for their molecular structures to be analysed by X-ray diffraction methods. One subtle advantage of the high solvent content of protein crystals, which is similar to that of human tissue, is that we are able to analyse these molecules close to their native, hydrated state, as found in living systems.

Summary

The *crystalline state* of matter is characterised by a structure in which the component atoms, molecules or ions are arranged such that

- (a) they are in close proximity to one another
- (b) there are relatively strong interactions between them, and
- (c) there is a long-range, well-defined, three-dimensional order.

Only by this model can we explain the overall physical behaviour of the crystalline state.

The principal technique which provides detailed information on both

- (a) the nature of the three-dimensional ordering of the crystal, and
- (b) the structure of the component molecules

is X-ray diffraction. X-rays are a form of electromagnetic radiation of wavelength of the order of 0.1 nm–1.0 nm (1–10 Å). Since this is comparable to the intermolecular spacing in a crystal, *diffraction* may be observed when an X-ray beam is incident on a crystal. The regularity of the three-dimensional structure acts to reinforce the diffraction caused by any one molecule, so that a detectable diffraction effect can be recorded and analysed. From the nature of the diffraction, we may derive both the crystal and the molecular structure.

Bibliography

On physical properties of the solid state

Kittel, C. *An Introduction to Solid State Physics*, 8th edn. Wiley, New York, 2004.
   Justifiably one of the most successful textbooks in print.

On electron microscopy

Bozzola, J. and Russell, L. D. *Electron Microscopy: Principles and Techniques for Biologists*. Jones and Bartlett, Sudbury, MA, 1999.

Frank, J. *Three-Dimensional Electron Microscopy of Macromolecular Assemblies: Visualization of Biological Molecules in Their Native State*. Oxford University Press, Oxford, 2006.

On atomic force microscopy

Morris, V. J., Kirby, A. R. and Gunning, A. P. *Atomic Force Microscopy for Biologists*. Imperial College Press, London, 1999.

2

Vector analysis and complex algebra

The purpose of the present chapter, and also that of Chapter 5, is to discuss the various mathematical techniques which are required in the theory of crystals and X-ray diffraction. The approach taken in the presentation of this material will be with a view to giving the reader a feeling for what the mathematics *means*, as opposed to explaining why the mathematics operates as it does. The treatment will therefore be non-rigorous and is directed towards setting out 'recipes' so that the mathematical manipulations may be carried out in the correct manner.

VECTORS

2.1   What is a vector?

One of the most significant aspects of mathematics is that its symbolism provides a pertinent shorthand for concepts which are sometimes quite complicated, and most unwieldy when expressed in a non-mathematical form. In fact, there is relatively little mathematics which cannot be expressed in verbal terms, but the great advantage of mathematical symbolism is that a single symbol may replace an entire verbal sentence. For example, the information contained in the mathematical symbol '5' may be stated verbally as 'that set of objects which may be matched with the number of fingers on one hand'. There is nothing particularly complicated about the symbol '5', but this example demonstrates that it is a most effective shorthand for a concept which can be expressed verbally only in a rather long-winded manner. Clearly, the more complex the concept we wish to describe, the more sophisticated is the appropriate mathematical symbolism and attendant formalism.

The next level of mathematical complexity after arithmetic and the handling of numerals is represented by ordinary algebra. We now use an abstract symbol such as x to signify a number which at present is not determined. We may then use the rules of algebra to perform various operations on our unknown x in order to find its numerical value. The significance of ordinary algebra, however, is that each of the symbols used stands for a *single* number, and so has magnitude only. Mathematical quantities which represent magnitudes only are called *scalars*. Scalar algebra, therefore, is a technique whereby each mathematical symbol contains only a single piece of numerical information. If we wish to convey information about more than a single number in just one symbol, then the structure of scalar algebra is inadequate, and we need a different mathematical formalism.

One technique which performs the function of handling symbols which contain information on more than one number is called *vector algebra*. For the purposes of this volume, a *vector* is a mathematical quantity which represents both a magnitude *and* a direction. In that vectors are associated with information of two sorts, any vector contains information on at least two numbers, one for the magnitude of the effect, and the others for directional properties. For a two-dimensional system, such as the geometry of a plane, the vector representation will correspond to the specification of a given magnitude in a particular direction in the plane. In a three-dimensional case, the vector will correspond to a magnitude in a specified direction in space. In fact, the formalism of vector algebra may be extended to hypothetical 'vector spaces' which have more than three dimensions. This is a problem for the mathematicians, and the present chapter will deal with only a two- and three-dimensional representation. A vector quantity will always be written in bold type, as for example $\mathbf{v}$. The magnitude associated with the vector $\mathbf{v}$ is written using the modulus notation as $|\mathbf{v}|$, or the corresponding italic letter v. Since the symbol $\mathbf{v}$ contains more information than the equivalent symbol x from scalar algebra, the rules of vector algebra are different from those of scalar algebra, and these new rules will be explained in subsequent sections of this chapter.

Clearly, vector algebra is particularly suited to the description of any system or problem which is associated with directional properties. This applies to any system which exists in three dimensions, and in the light of the discussion of anisotropy in the previous chapter, vector analysis is especially useful in describing the behaviour of crystals.

Since a vector corresponds to a directed quantity, we may represent a vector quantity graphically by an arrow. The arrow extends from an arbitrary origin, and

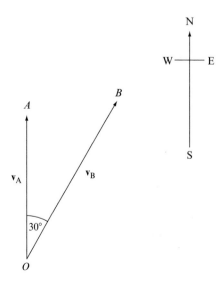

Fig. 2.1 *The representation of vectors.* The vector $\mathbf{v}_A$ represents a velocity of magnitude of 30 miles an hour directed due north, and $\mathbf{v}_B$ represents a velocity of magnitude of 40 miles an hour in a direction N 30° E.

has a length which is proportional to the magnitude of the effect to be described. The direction of the arrow as signified by the arrow head corresponds to the direction of the appropriate physical phenomenon. Note that the actual position of the vector is immaterial. What is important is that the arrow should have the correct magnitude and direction. As an example, consider the representation of the motion of two cars, A and B, the velocities $\mathbf{v}_A$ and $\mathbf{v}_B$ of which we wish to describe. Our information is that car A has a velocity of 30 miles an hour due north, and that car B has a velocity of 40 miles an hour in a direction N 30° E. These velocities may be represented by the vector diagram in Fig. 2.1, in which car B's vector is four-thirds the length of car A's vector, and there is an angle of 30° between the two vectors.

If the ends of the arrow used to represent a vector are named, as for instance in the case of the points O, A and B in Fig. 2.1, then a vector quantity may also be written in the form $\overrightarrow{OA}$. The arrow over the letters signifies that the vector is directed from O to A, and the magnitude is represented by the length of the line segment, written (in italics) as OA.

2.2   Vector addition

The sum of the vector $\mathbf{v}$ and the vector $\mathbf{w}$ is itself a vector $\mathbf{u}$, and the addition operation is written as

$$\mathbf{u} = \mathbf{v} + \mathbf{w} \tag{2.1}$$

Graphically, in two dimensions, addition may be represented as follows. From an arbitrary origin O, draw a line segment $\overrightarrow{OV}$ whose length is proportional to the magnitude of the vector $\mathbf{v}$, such that the direction of the line segment is the same as that of $\mathbf{v}$. From V, draw a second line segment $\overrightarrow{VW}$ to represent the vector $\mathbf{w}$ in both magnitude and direction. The line segment $\overrightarrow{OW}$ is then *defined* as the vector $\mathbf{u}$, which is the vector sum of $\mathbf{v}$ and $\mathbf{w}$. This sequence of operations is shown in Fig. 2.2. For a three-dimensional case, the vectors are drawn in the appropriate spatial direction according to the above 'head-to-tail' recipe. The rule of vector addition is used in the famous 'triangle of forces' familiar to those with a basic knowledge of mechanics. All physical forces have both magnitude and direction and so can be represented by vectors. When we have two forces acting on a body, the net effect of the two forces acting simultaneously is equal to the effect of a single force which corresponds to the vector sum of the original two forces. This equivalent single force is called the *resultant force*, and given two forces $\mathbf{v}$ and $\mathbf{w}$, the resultant force is the vector $\mathbf{u}$ such that

$$\mathbf{u} = \mathbf{v} + \mathbf{w}$$

More complicated vector additions

$$\mathbf{u} = \mathbf{v} + \mathbf{w} + \mathbf{s} + \dots$$

may be carried out by extending the 'head-to-tail' addition rule as shown in Fig. 2.3.

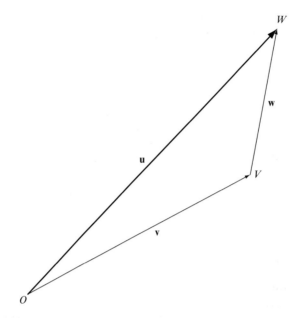

Fig. 2.2 *Vector addition.* The vectors **v** and **w** represent known vectors in both magnitude and direction. The vector **u** is defined as the sum **v** + **w**.

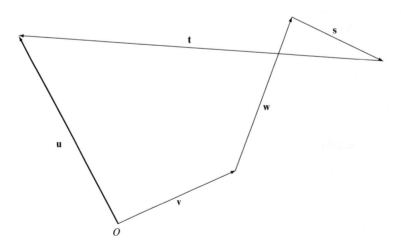

Fig. 2.3 *Extended vector addition.* Any number of vectors may be added using the 'head-to-tail' rule. In this figure, **u** is equal to the sum **v** + **w** + **s** + **t**.

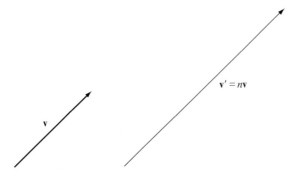

Fig. 2.4 *Multiplication by a scalar.* Both **v** and **v′** have the same direction, but the magnitude of **v′** is n times that of **v**. In this example, $n = 2$.

2.3 Multiplication by a scalar

As will be shown during the course of this chapter, there are three different binary multiplication operations defined in vector algebra. (The word 'binary' implies the manipulation of only two symbols.) This may at first sight seem strange, since up till now multiplication has meant such expressions as $2 \times 3 = 6$. However, since vector algebra is a more sophisticated technique than scalar algebra, it is not altogether surprising that the mathematical formalism which deals with the manipulation of vectors is rather more involved than that which is applicable to scalars.

The first vector multiplication operation we will discuss is the multiplication of a vector **v** by a scalar n. **v** represents both a magnitude and a direction, whereas n represents a magnitude only. The product is a new vector **v′** given by

$$\mathbf{v'} = n\mathbf{v} \tag{2.2}$$

The form of the product vector is shown in Fig. 2.4. Multiplication of a vector by a scalar leaves the direction of the vector unchanged, but alters its magnitude: **v′** has the same direction as **v**, but has a magnitude n times that of **v**.

2.4 Unit vectors

A vector which has unit magnitude is termed a *unit vector*. The unit vector **i** therefore has magnitude unity, and a direction appropriate to the system under consideration. Often, unit vectors are particularly useful in the representation of coordinate systems. For instance, in two-dimensional Cartesian coordinates we may represent the x direction by the unit vector **i**, and the y direction by the unit vector **j**. In this case, as shown in Fig. 2.5(a), we define the two unit vectors **i** and **j** to be mutually perpendicular, and directed according to the usual positions of the x and y axes.

This may easily be extended to three dimensions by using the unit vector **k** to represent the z direction, perpendicular to the plane defined by **i** and **j**. As soon as we enter three dimensions, a subtlety arises, as may be seen by comparing Figs. 2.5(b)

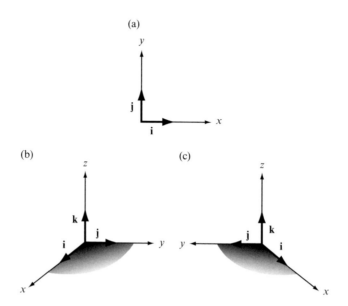

Fig. 2.5 *Vectors and Cartesian coordinates.* (a) represents a two-dimensional system. The unit base vectors **i** and **j** are along the x and y axes, respectively. (b) is a right-handed three-dimensional system. The unit base vectors **i**, **j** and **k** are along the x, y and z axes, respectively. (c) shows a left-handed three-dimensional system. The unit base vectors **i**, **j** and **k** are along the x, y and z axes, respectively. In all cases, the unit base vectors are orthogonal; (b) and (c) are mirror images. By convention, all three-dimensional Cartesian systems used in practice are right handed, as in (b).

and (c). Both diagrams show a set of three mutually perpendicular unit vectors **i**, **j** and **k**, but the two systems are mirror images of one another and are not superposable, in the same sense that a right hand cannot be placed exactly over a left hand. In that the schemes of Figs. 2.5(b) and (c) are different, the system in Fig. 2.5(b) is termed a *right-handed system*, and that in Fig. 2.5(c) a *left-handed system*. To distinguish between the two, the following sequence of operations is used. Imagine the vectors **i** and **j** to be in a plane, and rotate the vector **i** onto the vector **j** by the shortest possible rotation.

A right-handed system is defined as having the following properties:

(a) If the rotation is clockwise, the vector **k** is directed away from the observer.
(b) If the rotation is anticlockwise, the vector **k** is directed towards the observer.

Conversely, a left-handed system is such that:

(a) If the rotation is clockwise, the vector **k** is directed towards the observer.
(b) If the rotation is anticlockwise, the vector **k** is directed away from the observer.

For further clarification, refer to Fig. 2.6. By convention, Cartesian coordinates are always defined with reference to a right-handed system, as shown in Fig. 2.5(b).

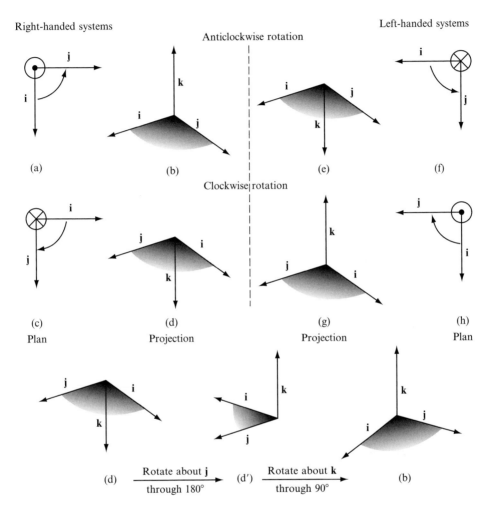

Fig. 2.6 *Right- and left-handed systems.* Depicted are examples of the operation of the rules to distinguish between right- and left-handed systems. Note that **i** is always rotated onto **j**. (a), (c), (f) and (h) show plans in which **i** and **j** are in the plane of the paper. In (a) and (f), a rotation of **i** onto **j** is anticlockwise and, according to the rules given in the text, **k** is outwards for a right-handed system and inwards for a left-handed system. Projected three-dimensional views are given in (b) and (e). (c) and (d), and (g) and (h) are likewise for a clockwise rotation. Note that (b) and (g) and also (d) and (e) are direct mirror images.

A circle containing a dot indicates an axis pointing towards the reader, and a circle containing a cross indicates an axis pointing into the page.

The sequence (d), (d′), (b) demonstrates that (b) is the same as (d). From (d), rotate about **j** through 180° to obtain (d′), and then rotate about **k** through 90° to obtain (b). Hence (d) and (b) are the same. Similarly, (e) and (g) are the same. In general, right-handed systems are the mirror images of left-handed systems.

The above use of unit vectors to define a three-dimensional Cartesian coordinate system is a special case of the use of unit vectors: in general, any three non-coplanar unit vectors may be used to define a corresponding set of three, usually not mutually perpendicular, axes. In general, we choose the representation most appropriate to our system. A set of unit vectors used to define a coordinate system is referred to as a set of *base vectors*. If the three vectors chosen are mutually perpendicular, as in the case of Cartesian coordinates, we speak of a set of *orthogonal* base vectors.

2.5   Components

Using the vector addition rule, we may represent *any* three-dimensional vector **v** by the sum of three vectors **a**, **b** and **c** as long as the 'head-to-tail' recipe is followed. Let us choose to use three vectors which are parallel to the three orthogonal Cartesian axes as defined by the unit vectors **i**, **j** and **k**. Reference to Fig. 2.7 shows that **a** lies along the x axis, **b** is parallel to the y axis, and **c** is parallel to the z axis, such that these three vectors add to give the vector **v**. In general, any vector **v** may be represented by a similar sum of three vectors, each of which is parallel to one of the unit base vectors in the Cartesian system. We now invoke the rule of the multiplication of a vector by a scalar. Since **a** is parallel to, and so has the same direction as, **i**, we may define a scalar v_x such that

$$\mathbf{a} = v_x \mathbf{i}$$

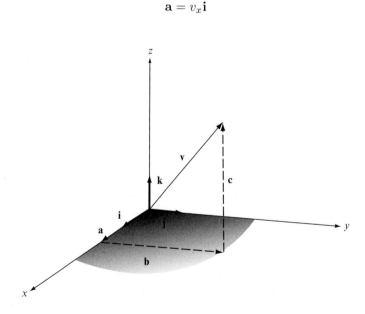

Fig. 2.7 *Components.* A vector **v** in a Cartesian system may be represented by the sum **v** $= \mathbf{a} + \mathbf{b} + \mathbf{c}$, in which **a**, **b** and **c** are parallel to **i**, **j** and **k**, respectively. Hence we may write $\mathbf{a} = v_x \mathbf{i}$, $\mathbf{b} = v_y \mathbf{j}$, $\mathbf{c} = v_z \mathbf{k}$. Therefore $\mathbf{v} = v_x \mathbf{i} + v_y \mathbf{j} + v_z \mathbf{k}$.

Similarly, we may find scalars v_y and v_z such that

$$\mathbf{b} = v_y\mathbf{j}$$

$$\mathbf{c} = v_z\mathbf{k}$$

Since we know that

$$\mathbf{v} = \mathbf{a} + \mathbf{b} + \mathbf{c}$$

then

$$\mathbf{v} = v_x\mathbf{i} + v_y\mathbf{j} + v_z\mathbf{k} \tag{2.3}$$

The vectors $v_x\mathbf{i}$, $v_y\mathbf{j}$ and $v_z\mathbf{k}$ are known as the *components* of the vector $\mathbf{v}$. If a system of base vectors $\mathbf{i}$, $\mathbf{j}$ and $\mathbf{k}$ has previously been defined, then we may use the shorthand expression in equation (2.4) to define the vector $\mathbf{v}$:

$$\mathbf{v} = (v_x, v_y, v_z) \tag{2.4}$$

The word 'components' may now refer to the scalars v_x, v_y and v_z alone. This is akin to the specification of a point in space by the use of three-dimensional coordinates. Note that the representation in equation (2.4) relies on the assumption that the appropriate set of unit base vectors $\mathbf{i}$, $\mathbf{j}$ and $\mathbf{k}$ are known. Clearly, a representation such as equation (2.4) of an ordered triplet of scalars will signify a different vector according to the system of unit vectors employed. In general, unless otherwise stated, this chapter will use a right-handed orthogonal Cartesian coordinate system in which $\mathbf{i}$ is parallel to the x axis, $\mathbf{j}$ is parallel to the y axis and $\mathbf{k}$ is parallel to the z axis.

The representation (2.4) has expressed a single vector quantity $\mathbf{v}$ in terms of three scalar quantities v_x, v_y and v_z. In this way, a three-dimensional vector contains three pieces of numerical information. This clarifies the point made earlier that vector algebra deals with the manipulation of symbols which contain more than one piece of numerical information. Similarly, a two-dimensional vector $\mathbf{v}$ given by

$$\begin{aligned} \mathbf{v} &= v_x\mathbf{i} + v_y\mathbf{j} \\ &= (v_x, v_y) \end{aligned}$$

contains information on two numbers.

We may now formulate the vector addition rule, equation (2.1), in terms of components. For two vectors $\mathbf{v}$ and $\mathbf{w}$, if

$$\mathbf{v} = (v_x, v_y, v_z)$$

and

$$\mathbf{w} = (w_x, w_y, w_z)$$

then the vector sum $\mathbf{v} + \mathbf{w}$ is, in longhand,

$$\mathbf{v} + \mathbf{w} = v_x\mathbf{i} + v_y\mathbf{j} + v_z\mathbf{k} + w_x\mathbf{i} + w_y\mathbf{j} + w_z\mathbf{k}$$

But the two vectors $v_x\mathbf{i}$ and $w_x\mathbf{i}$ are in the same direction, and may be added directly to give $(v_x + w_x)\mathbf{i}$. Similarly, we may write

$$\mathbf{v} + \mathbf{w} = (v_x + w_x)\mathbf{i} + (v_y + w_y)\mathbf{j} + (v_z + w_z)\mathbf{k}$$

$$= (v_x + w_x, v_y + w_y, v_z + w_z) \tag{2.5}$$

Hence the sum of two vectors may be represented by the sums of their components as shown by equation (2.5). The rule for multiplication of a vector by a scalar, equation (2.2), may be expressed as

$$\mathbf{v}' = n\mathbf{v} = n(v_x\mathbf{i} + v_y\mathbf{j} + v_z\mathbf{k})$$

$$= nv_x\mathbf{i} + nv_y\mathbf{j} + nv_z\mathbf{k}$$

$$= (nv_x, nv_y, nv_z) \tag{2.6}$$

2.6 Vector subtraction

The negative of a vector $\mathbf{v}$ is represented as the vector $-\mathbf{v}$, and has the same magnitude as the original vector but exactly the opposite direction, as shown in Fig. 2.8(a). In

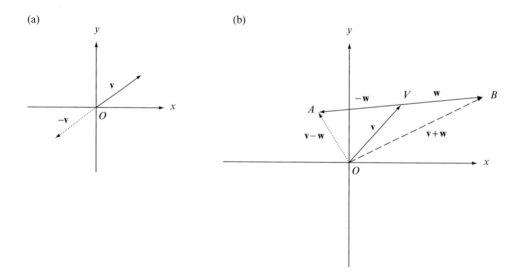

Fig. 2.8 *Negative vectors.* (a) The vector $-\mathbf{v}$ has the same magnitude as but the opposite direction to the vector $\mathbf{v}$. (b) $\overrightarrow{VA}$ represents the negative of the vector $\overrightarrow{VB}$. The subtraction operation $\mathbf{v} - \mathbf{w}$ is the same as the addition operation $\mathbf{v} + (-\mathbf{w})$. The vectors corresponding to both $\mathbf{v} + \mathbf{w}$ and $\mathbf{v} - \mathbf{w}$ are shown. A two-dimensional representation is used for clarity.

component form, if the vector **v** is given by

$$\mathbf{v} = (v_x, v_y, v_z)$$

then the vector $-\mathbf{v}$ is

$$-\mathbf{v} = (-v_x, -v_y, -v_z)$$

The operation of vector subtraction may be regarded as the addition of a negative vector. By the addition rule, equation (2.5), we may write

$$\mathbf{v} - \mathbf{w} = \mathbf{v} + (-\mathbf{w}) = (v_x - w_x, v_y - w_y, v_z - w_z)$$

The graphical representation is shown in Fig. 2.8(b), in which the vector **v** is added to the negative of the vector **w**.

2.7   Multiplication by a vector to give a scalar

A second binary vector multiplication operation is the multiplication of a vector by a second vector to give a scalar as a result. This product is called the *scalar* or *dot product*, and for two vectors **v** and **w**, it is written as $\mathbf{v} \cdot \mathbf{w}$. The definition of the scalar product is

$$\mathbf{v} \cdot \mathbf{w} = vw \cos \theta \tag{2.7}$$

in which v and w are the magnitudes of the vectors and θ is the angle between them, as shown in Fig. 2.9. Note that v, w and $\cos \theta$ are all scalars so that the product is, as required, a scalar. By definition, the product $\mathbf{w} \cdot \mathbf{v}$ is given by

$$\mathbf{w} \cdot \mathbf{v} = wv \cos \theta$$

but since scalar multiplication is commutative (this being the fact that $xy = yx$), we may write

$$wv \cos \theta = vw \cos \theta$$

$$\therefore \mathbf{w} \cdot \mathbf{v} = \mathbf{v} \cdot \mathbf{w}$$

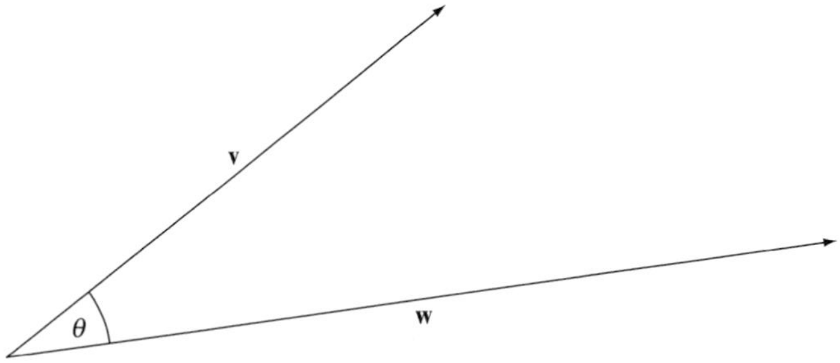

Fig. 2.9 *The scalar product* $\mathbf{v} \cdot \mathbf{w}$. By definition, the scalar product of **v** and **w** is given by $\mathbf{v} \cdot \mathbf{w} = |\mathbf{v}||\mathbf{w}| \cos \theta = vw \cos \theta$.

This shows that the operation of the scalar product is commutative. This may seem trivial, but in the next section the reader will be introduced to a multiplication operation which is not commutative, so that the order in which the operation is carried out is important.

In component form, we may write the scalar product in longhand as

$$\mathbf{v} \cdot \mathbf{w} = (v_x \mathbf{i} + v_y \mathbf{j} + v_z \mathbf{k}) \cdot (w_x \mathbf{i} + w_y \mathbf{j} + w_z \mathbf{k})$$

On multiplying out the brackets we obtain

$$\mathbf{v} \cdot \mathbf{w} = v_x w_x \mathbf{i} \cdot \mathbf{i} + v_x w_y \mathbf{i} \cdot \mathbf{j} + v_x w_z \mathbf{i} \cdot \mathbf{k}$$
$$+ v_y w_x \mathbf{j} \cdot \mathbf{i} + v_y w_y \mathbf{j} \cdot \mathbf{j} + v_y w_z \mathbf{j} \cdot \mathbf{k}$$
$$+ v_z w_x \mathbf{k} \cdot \mathbf{i} + v_z w_y \mathbf{k} \cdot \mathbf{j} + v_z w_z \mathbf{k} \cdot \mathbf{k} \tag{2.8}$$

The above expression contains scalar products of the unit base vectors such as $\mathbf{i} \cdot \mathbf{i}$ and $\mathbf{i} \cdot \mathbf{j}$. The great use of an orthogonal set of unit base vectors is that, according to equation (2.7), we may write

$$\mathbf{i} \cdot \mathbf{i} = |\mathbf{i}|^2 = 1$$
$$\mathbf{j} \cdot \mathbf{j} = |\mathbf{j}|^2 = 1$$
$$\mathbf{k} \cdot \mathbf{k} = |\mathbf{k}|^2 = 1$$

This is true because the self-scalar product involves the cosine of a zero angle, which is unity. All products of the form $\mathbf{i} \cdot \mathbf{j}$ require the cosine of the angle between two unit vectors. Since the vectors are orthogonal, this is 90°, the cosine of which is zero, so

$$\mathbf{i} \cdot \mathbf{j} = \mathbf{j} \cdot \mathbf{i} = \mathbf{i} \cdot \mathbf{k} = \mathbf{k} \cdot \mathbf{i} = \mathbf{j} \cdot \mathbf{k} = \mathbf{k} \cdot \mathbf{j} = 0$$

The nine possible scalar products of the unit base vectors $\mathbf{i}$, $\mathbf{j}$ and $\mathbf{k}$ may be presented neatly as

$$\mathbf{i} \cdot \mathbf{i} = 1 \quad \mathbf{i} \cdot \mathbf{j} = 0 \quad \mathbf{i} \cdot \mathbf{k} = 0$$
$$\mathbf{j} \cdot \mathbf{i} = 0 \quad \mathbf{j} \cdot \mathbf{j} = 1 \quad \mathbf{j} \cdot \mathbf{k} = 0$$
$$\mathbf{k} \cdot \mathbf{i} = 0 \quad \mathbf{k} \cdot \mathbf{j} = 0 \quad \mathbf{k} \cdot \mathbf{k} = 1$$

Hence, in the case of orthogonal base vectors, equation (2.8) reduces to

$$\mathbf{v} \cdot \mathbf{w} = v_x w_x + v_y w_y + v_z w_z \tag{2.9}$$

Equation (2.9) has a very handy form, and is the recipe for calculating scalar products in terms of components. It is imperative to note that equation (2.9) is applicable *if and only if* the base vectors are mutually orthogonal. When this is not the case, then the scalar product has to be evaluated by the use of equation (2.8).

A special case of equation (2.9) is the self-scalar product $\mathbf{v} \cdot \mathbf{v}$,

$$\mathbf{v} \cdot \mathbf{v} = v_x^2 + v_y^2 + v_z^2$$

and, since the angle between the vectors **v** and **v** is zero, we also know from equation (2.7) that

$$\mathbf{v} \cdot \mathbf{v} = v^2$$

implying that

$$v = \sqrt{v_x^2 + v_y^2 + v_z^2} \tag{2.10}$$

Equation (2.10) enables us to calculate the magnitude v of any vector **v** from its components (v_x, v_y, v_z).

The physical significance of the scalar product is essentially twofold:

i. Projections

In Fig. 2.10(a) are an arbitrary vector **v** and a unit vector **i**. The unit vector **i** defines a particular direction, and if we drop a perpendicular from the point V onto the direction defined by **i**, we obtain the right-angled triangular figure OVA. The length of the line segment OA is called the *projection* of the vector **v** in the direction defined by **i**. Geometrically, we see that the length of the projection is just $v \cos \theta$, where θ is the angle between **v** and **i**. Since the magnitude of **i** is unity, the scalar product $\mathbf{v} \cdot \mathbf{i}$ is

$$\mathbf{v} \cdot \mathbf{i} = v \cdot 1 \cdot \cos \theta = v \cos \theta$$

Hence the scalar product is useful in giving the projection of a vector in any direction. Also, if we require the projection along a unit base vector, then the result is just the appropriate component. The proof of this is

$$\mathbf{v} \cdot \mathbf{i} = v \cos \theta$$
$$= (v_x \mathbf{i} + v_y \mathbf{j} + v_z \mathbf{k}) \cdot \mathbf{i}$$
$$= v_x \mathbf{i} \cdot \mathbf{i} + v_y \mathbf{j} \cdot \mathbf{i} + v_z \mathbf{k} \cdot \mathbf{i}$$
$$= v_x$$
$$\therefore \mathbf{v} \cdot \mathbf{i} = v \cos \theta = v_x$$

and similarly for the other components.

ii. Work

When a force **F** moves through a distance **r**, *work* is said to be done. Since the force **F** and the distance **r** have directional properties, they can be represented by vectors; the work done, however, does not have directional properties and so is a scalar. The magnitude of the work done is defined as the scalar quantity resulting from multiplying the magnitude of the component of the force in the direction of the motion by the magnitude of the distance moved. Figure 2.10(b) shows a vector **r** representing the motion, and a vector **F** signifying the force. The component of the force **F** along

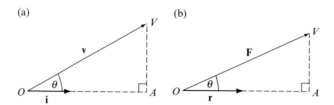

Fig. 2.10 *The use of the scalar product.* (a) The projection of **v** in the direction of **i** is the line segment OA, whose length is $v \cos \theta = \mathbf{v} \cdot \mathbf{i}$. (b) The work done by the force **F** moving through the distance **r** is equal to the magnitude of the component of the force **F** in the direction of the motion, multiplied by the magnitude of the distance moved. The component of **F** in the direction of **r** is represented by the line segment $\overrightarrow{OA}$ and has magnitude $F \cos \theta$. The work done is therefore $Fr \cos \theta = \mathbf{F} \cdot \mathbf{r}$.

the direction of motion is simply $F \cos \theta$, and so the magnitude of the work done is given by

$$\text{work} = F \cos \theta \cdot r = Fr \cos \theta = \mathbf{F} \cdot \mathbf{r}$$

2.8 Multiplication by a vector to give a vector

The third binary vector multiplication operation is the multiplication of a vector by a second vector to give a vector as a result. For two vectors **v** and **w**, this is called the *vector* or *cross product*, and is written as $\mathbf{v} \times \mathbf{w}$ or $\mathbf{v} \wedge \mathbf{w}$—this book will use the latter symbol $\wedge$. Since the product is a vector, we must specify both its magnitude and its direction. The magnitude of the product is defined as

$$|\mathbf{v} \wedge \mathbf{w}| = vw \sin \theta \tag{2.11}$$

in which θ is the angle between the vectors. The direction of the product vector is defined as follows. Rotate the vector **v** onto the vector **w** by the shortest possible route. The product vector is perpendicular to the plane defined by **v** and **w**, and is directed in a right-handed sense. So:

(a) If the rotation is clockwise, the product vector is directed away from the observer.

(b) If the rotation is anticlockwise, the product vector is directed towards the observer.

This is shown in Fig. 2.11. A convenient way to remember in which direction the product vector points is to imagine that a screw were being rotated in the same manner as defined above. Most screws are manufactured with a right-handed thread, so that the direction of motion of the screw is the same as that of the product vector.

The order of operations in forming a vector product is very important. Reference to Fig. 2.12 will show that the result of the operation $\mathbf{v} \wedge \mathbf{w}$ is a vector which is in a direction *opposite* to the vector obtained from the operation $\mathbf{w} \wedge \mathbf{v}$. This is a result of the fact that by convention, vector products are always taken in the right-handed

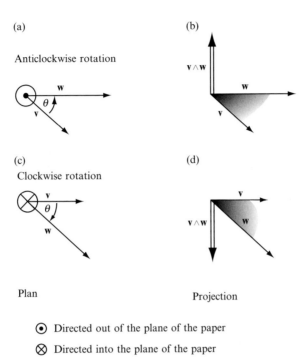

(a)

Anticlockwise rotation

(b)

(c)
Clockwise rotation

(d)

Plan

Projection

⊙ Directed out of the plane of the paper

⊗ Directed into the plane of the paper

Fig. 2.11 *The vector product* **v** ∧ **w**. The vector product **v** ∧ **w** is a vector of magnitude $vw \sin \theta$ directed perpendicular to the plane defined by **v** and **w** in the right-handed sense; (a) and (c) are plans of the plane defined by **v** and **w**. According to the rules in the text, a right-handed system is such that an anticlockwise rotation of **v** onto **w** gives a vector coming out of the plane of the paper, whereas a clockwise rotation gives a vector going into the plane of the paper. Three-dimensional representations are given in (b) and (d). See also Fig. 2.6.

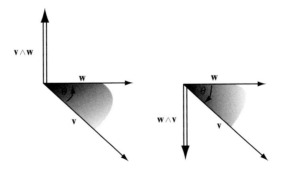

Fig. 2.12 *The vector product is non-commutative.* Both **v** ∧ **w** and **w** ∧ **v** have the same magnitude, but they are in opposite directions. Hence **v** ∧ **w** = −**w** ∧ **v**.

sense. Both vectors have the same magnitude $vw \sin \theta$, but they are in exactly opposite directions. Hence

$$\mathbf{v} \wedge \mathbf{w} = -\mathbf{w} \wedge \mathbf{v}$$

The operation of taking the vector product is therefore not commutative, and it is essential to perform the operation in the correct order.

Reference to Fig. 2.5(b) shows that the coordinate system depicted is right-handed, since the unit vectors form a right-handed set. A product such as $\mathbf{i} \wedge \mathbf{j}$ is therefore in the direction of $\mathbf{k}$. Furthermore, the magnitude of the vector product is given by

$$|\mathbf{i} \wedge \mathbf{j}| = ij \sin \theta = 1$$

This is because $\mathbf{i}$, $\mathbf{j}$ and $\mathbf{k}$ are mutually orthogonal, and so θ is 90°, the sine of which is unity. Since $\mathbf{k}$ itself is a unit vector, we have the result that

$$\mathbf{i} \wedge \mathbf{j} = \mathbf{k}$$

and, similarly,

$$\mathbf{j} \wedge \mathbf{k} = \mathbf{i}$$

$$\mathbf{k} \wedge \mathbf{i} = \mathbf{j}$$

Notice the cyclic order of the unit base vectors. Also, since the self-vector product $\mathbf{v} \wedge \mathbf{v}$ involves taking the sine of the angle zero, we have for any vector $\mathbf{v}$

$$\mathbf{v} \wedge \mathbf{v} = 0$$

Using the commutation relations for the unit base vectors, the set of nine possible vector products is

$$
\begin{array}{lll}
\mathbf{i} \wedge \mathbf{i} = 0 & \mathbf{i} \wedge \mathbf{j} = \mathbf{k} & \mathbf{i} \wedge \mathbf{k} = -\mathbf{j} \\
\mathbf{j} \wedge \mathbf{i} = -\mathbf{k} & \mathbf{j} \wedge \mathbf{j} = 0 & \mathbf{j} \wedge \mathbf{k} = \mathbf{i} \\
\mathbf{k} \wedge \mathbf{i} = \mathbf{j} & \mathbf{k} \wedge \mathbf{j} = -\mathbf{i} & \mathbf{k} \wedge \mathbf{k} = 0
\end{array}
\qquad (2.12)
$$

We may now write the component form of the vector product $\mathbf{v} \wedge \mathbf{w}$. In longhand, we have

$$\mathbf{v} \wedge \mathbf{w} = (v_x\mathbf{i} + v_y\mathbf{j} + v_z\mathbf{k}) \wedge (w_x\mathbf{i} + w_y\mathbf{j} + w_z\mathbf{k})$$

Multiplying out the brackets, being careful to keep the order of operations correct, we obtain

$$
\begin{aligned}
\mathbf{v} \wedge \mathbf{w} = {} & v_x w_x \mathbf{i} \wedge \mathbf{i} + v_x w_y \mathbf{i} \wedge \mathbf{j} + v_x w_z \mathbf{i} \wedge \mathbf{k} \\
& + v_y w_x \mathbf{j} \wedge \mathbf{i} + v_y w_y \mathbf{j} \wedge \mathbf{j} + v_y w_z \mathbf{j} \wedge \mathbf{k} \\
& + v_z w_x \mathbf{k} \wedge \mathbf{i} + v_z w_y \mathbf{k} \wedge \mathbf{j} + v_z w_z \mathbf{k} \wedge \mathbf{k}
\end{aligned}
\qquad (2.13)
$$

Using the relations of equation (2.12), equation (2.13) simplifies to

$$
\begin{aligned}
\mathbf{v} \wedge \mathbf{w} &= (v_y w_z - v_z w_y)\mathbf{i} + (v_z w_x - v_x w_z)\mathbf{j} + (v_x w_y - v_y w_x)\mathbf{k} \\
&= (v_y w_z - v_z w_y, \, v_z w_x - v_x w_z, \, v_x w_y - v_y w_x)
\end{aligned}
$$

This equation is rather unwieldy, and a neater equivalent mathematical form is obtained by using a determinant:

$$\mathbf{v} \wedge \mathbf{w} = \begin{vmatrix} \mathbf{i} & \mathbf{j} & \mathbf{k} \\ v_x & v_y & v_z \\ w_x & w_y & w_z \end{vmatrix} \tag{2.14}$$

If the reader is unfamiliar with determinants, a brief review of their properties is given in the appendix to this chapter. The determinantal form has the automatic property that

$$\mathbf{v} \wedge \mathbf{w} = -\mathbf{w} \wedge \mathbf{v}$$

for, using the determinant for $\mathbf{w} \wedge \mathbf{v}$, we see that

$$\mathbf{w} \wedge \mathbf{v} = \begin{vmatrix} \mathbf{i} & \mathbf{j} & \mathbf{k} \\ w_x & w_y & w_z \\ v_x & v_y & v_z \end{vmatrix}$$

Since two rows of the determinant are interchanged, the value of the above determinant is the negative of that in equation (2.14). Equation (2.14) is the appropriate equation to memorise in order to calculate vector products.

The predominant use of the vector product is that it contains within its formalism an innate 'three-dimensionalness', in that the result of the operation $\mathbf{v} \wedge \mathbf{w}$ is to give a vector perpendicular to the plane defined by the vectors $\mathbf{v}$ and $\mathbf{w}$. The vector product is therefore very helpful in a geometrical situation when we wish to define a plane. Consider a plane such as that defined by the two vectors $\mathbf{v}$ and $\mathbf{w}$ in Fig. 2.12. The vector $\mathbf{v} \wedge \mathbf{w}$ is perpendicular to this plane, and a little thought will show that the *only* explicit way of defining a plane is to specify that direction which is perpendicular (alternatively, 'normal') to the plane. In crystal geometry, we will often be dealing with the specification of planes within a crystal, and so the vector product will be invoked. The vector product is also used extensively in physics to describe three-dimensional vector effects such as moments, and the relationships between electric and magnetic fields.

2.9   The scalar triple product and the vector triple product

Since the product $\mathbf{v} \wedge \mathbf{w}$ is itself a vector, it may be multiplied by a third vector $\mathbf{u}$ to give either a scalar or another vector. The product $\mathbf{u} \cdot (\mathbf{v} \wedge \mathbf{w})$ is called the *scalar triple product*, and the *vector triple product* is written as $\mathbf{u} \wedge (\mathbf{v} \wedge \mathbf{w})$. The scalar triple product has a special geometrical significance, so we shall investigate its properties here; the vector triple product is less important, and will be mentioned at the end of this section.

To investigate the scalar triple product, let us take the three vectors as

$$\mathbf{u} = (u_x, u_y, u_z)$$
$$\mathbf{v} = (v_x, v_y, v_z)$$
$$\mathbf{w} = (w_x, w_y, w_z)$$

The vector product $\mathbf{v} \wedge \mathbf{w}$ may be written as

$$\mathbf{v} \wedge \mathbf{w} = (v_y w_z - v_z w_y, v_z w_x - v_x w_z, v_x w_y - v_y w_x)$$

Taking the scalar product with $\mathbf{u}$, we obtain

$$\mathbf{u} \cdot (\mathbf{v} \wedge \mathbf{w}) = u_x(v_y w_z - v_z w_y) + u_y(v_z w_x - v_x w_z) + u_z(v_x w_y - v_y w_x)$$

This equation assumes a neater form as a determinant:

$$\mathbf{u} \cdot (\mathbf{v} \wedge \mathbf{w}) = \begin{vmatrix} u_x & u_y & u_z \\ v_x & v_y & v_z \\ w_x & w_y & w_z \end{vmatrix} \tag{2.15}$$

Equation (2.15) is the normal recipe for calculating the value of a scalar triple product from the components of the individual vectors. Two aspects of equation (2.15) are worthy of notice. Firstly, the brackets in $\mathbf{u} \cdot (\mathbf{v} \wedge \mathbf{w})$ are redundant, since the alternative, namely $(\mathbf{u} \cdot \mathbf{v}) \wedge \mathbf{w}$, is meaningless. This is because the quantity $(\mathbf{u} \cdot \mathbf{v})$ is a scalar, whereas the operation of taking a vector product can connect only two vectors. Secondly, it is a property of a determinant that any cyclic permutation of either the rows or the columns leaves the value of the determinant unchanged. Hence

$$\mathbf{u} \cdot (\mathbf{v} \wedge \mathbf{w}) = \begin{vmatrix} u_x & u_y & u_z \\ v_x & v_y & v_z \\ w_x & w_y & w_z \end{vmatrix} = \begin{vmatrix} v_x & v_y & v_z \\ w_x & w_y & w_z \\ u_x & u_y & u_z \end{vmatrix} = \begin{vmatrix} w_x & w_y & w_z \\ u_x & u_y & u_z \\ v_x & v_y & v_z \end{vmatrix}$$

$$\therefore \mathbf{u} \cdot (\mathbf{v} \wedge \mathbf{w}) = \mathbf{v} \cdot \mathbf{w} \wedge \mathbf{u} = \mathbf{w} \cdot \mathbf{u} \wedge \mathbf{v}$$

The relevance of the scalar triple product may be seen from Fig. 2.13. In the figure are shown three non-coplanar vectors $\mathbf{u}$, $\mathbf{v}$ and $\mathbf{w}$, which define a parallelepiped. The vector product has a magnitude $vw \sin \theta$, which, geometrically, corresponds to the area of the parallelogram defined by $\mathbf{v}$ and $\mathbf{w}$. The direction of the vector $\mathbf{v} \wedge \mathbf{w}$ is perpendicular to the plane defined by $\mathbf{v}$ and $\mathbf{w}$, and is shown in the figure. When this vector forms a scalar product with $\mathbf{u}$, we obtain a scalar of magnitude $u|\mathbf{v} \wedge \mathbf{w}| \cos \varphi$, where φ represents the angle between $\mathbf{u}$ and the vector $\mathbf{v} \wedge \mathbf{w}$. But $u \cos \varphi$ is just the vertical height of the parallelepiped above its base. We have already seen that $|\mathbf{v} \wedge \mathbf{w}|$ is equal to the area of the base of the parallelepiped, and so the value of the expression $\mathbf{u} \cdot \mathbf{v} \wedge \mathbf{w}$ is the area of the base of the parallelepiped multiplied by its vertical height. Hence the scalar triple product of $\mathbf{u}$, $\mathbf{v}$ and $\mathbf{w}$ represents the volume of the appropriate parallelepiped. A special case arises when the three vectors $\mathbf{u}$, $\mathbf{v}$ and $\mathbf{w}$ are coplanar. In this case the vector $\mathbf{v} \wedge \mathbf{w}$ will be perpendicular to the plane of $\mathbf{v}$ and $\mathbf{w}$, and hence perpendicular to $\mathbf{u}$ as well. The scalar product of $\mathbf{u}$ with $\mathbf{v} \wedge \mathbf{w}$ therefore vanishes,

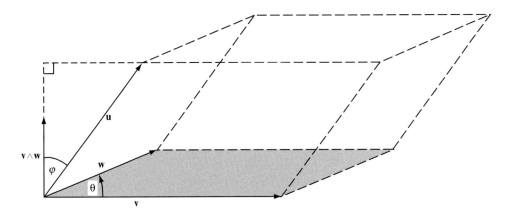

Fig. 2.13 *The scalar triple product.* The three vectors **u**, **v** and **w** define a parallelepiped. The base area is equal to $|\mathbf{v} \wedge \mathbf{w}| = |\mathbf{v}|\,|\mathbf{w}| \sin \theta$. The vertical height above the base defined by the plane containing **v** and **w** is $|\mathbf{u}| \cos \varphi$, and is parallel to the direction of $\mathbf{v} \wedge \mathbf{w}$.

Volume of parallelepiped = base area × vertical height

$$= |\mathbf{v} \wedge \mathbf{w}|\,|\mathbf{u}| \cos \varphi = \mathbf{u} \cdot \mathbf{v} \wedge \mathbf{w}$$

and the value of the scalar triple product $\mathbf{u} \cdot \mathbf{v} \wedge \mathbf{w}$ is zero. This agrees with physical reality, since there is no parallelepiped defined by three coplanar vectors. This result also follows from the determinant, equation (2.15). If all three of **u**, **v** and **w** are coplanar, then we may choose that they are all in the plane defined by the unit base vectors **i** and **j**. In this case, all three of u_z, v_z and w_z are zero and the determinant will vanish.

Just a brief note on the vector triple product $\mathbf{u} \wedge (\mathbf{v} \wedge \mathbf{w})$. The vector defined by $\mathbf{v} \wedge \mathbf{w}$ is orthogonal to the plane defined by **v** and **w**, and the vector defined by $\mathbf{u} \wedge (\mathbf{v} \wedge \mathbf{w})$ is orthogonal to both **u** and $\mathbf{v} \wedge \mathbf{w}$. This implies that $\mathbf{u} \wedge (\mathbf{v} \wedge \mathbf{w})$ lies in the plane defined by **v** and **w**—which is true even if **u**, **v** and **w** are not themselves mutually orthogonal. The vector $\mathbf{u} \wedge (\mathbf{v} \wedge \mathbf{w})$ will therefore be of the form

$$\mathbf{u} \wedge (\mathbf{v} \wedge \mathbf{w}) = \alpha \mathbf{v} + \beta \mathbf{w}$$

where α and β are scalars. A few moments of careful algebra will show that this is indeed the case, with α and β being the scalar products $\alpha = \mathbf{u} \cdot \mathbf{w}$ and $\beta = -\mathbf{u} \cdot \mathbf{v}$:

$$\mathbf{u} \wedge (\mathbf{v} \wedge \mathbf{w}) = (\mathbf{u} \cdot \mathbf{w})\,\mathbf{v} - (\mathbf{u} \cdot \mathbf{v})\,\mathbf{w}$$

We shall use this result in Chapter 8.

This completes our study of vector analysis. Many topics have been omitted, as the purpose of this section has been to familiarise the reader with those aspects of vector algebra which are necessary for the theory of the X-ray diffraction of crystals. A list of useful mathematical results will be found in the summary at the end of the chapter.

COMPLEX ALGEBRA

The first part of this chapter has discussed vector analysis, and we have seen that the vector representation is a very succinct means of handling mathematical quantities which have directional properties. The great use of vectors is therefore in the description of three-dimensional concepts such as directions in space and the associated physical phenomena. This part of the chapter will deal with another mathematical technique which is particularly useful in dealing with physical quantities associated with two numerical properties. To anticipate slightly, it will be shown in Chapter 4 that a wave motion may be described in terms of two particular parameters, called the *maximum amplitude* and the *phase*. For the moment, we may regard the maximum amplitude as a measure of the maximum height of the wave crests, and the phase as a measure of the particular part of the wave we are interested in, determining whether the wave is going 'up' or 'down', as shown in Fig. 2.14.

In that a wave has two distinctive properties, a suitable mathematical description of a wave will contain information on both the maximum amplitude and the phase of the wave motion. We therefore require a technique which is specifically suited to the handling of two pieces of numerical information in a way which is directly applicable to waves. This technique will be examined in the next few sections.

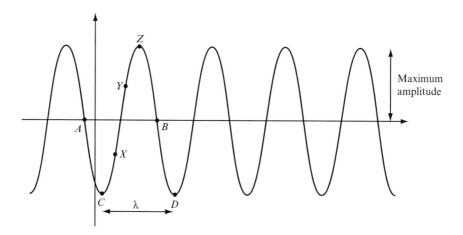

Fig. 2.14 *A wave.* At an instant in time, a cross-section through a wave motion has the appearance shown. The pattern is a periodic repetition of the section between A and B. The maximum amplitude corresponds to the maximum height of the wave crests. The phase of a point in the wave is a measure of the position along the wave relative to an arbitrary origin. Thus the points X, Y and Z have different phases with respect to A. The wavelength (λ) of the wave corresponds to the distance AB, CD or any equivalent distance across an entire cycle.

2.10   What is a complex number?

The normal process of arithmetic deals with numbers such as 5, 74, -6.59, $\sqrt{10}$ and 25.72. Numbers such as these are called *real* numbers. The significance of the word 'real' concerns the fact that any physical experiment we care to do will always result in the measurement of a number such as 5, 74, -6.59, $\sqrt{10}$ or 25.72. Since these numbers are directly measurable, they are, so to speak, 'real'. In case the fact that negative numbers can be measured in an experiment is worrying, the reader is reminded that any physical measurement, such as that of length, requires us to position an origin of coordinates at an arbitrary location in space. Once this has been done, we then define all lengths to the right of the origin to be positive, and those to the left to be negative. Hence, mathematically speaking, negative numbers are just as measurable as positive ones, and so both are real.

Suppose that we set up a physical experiment, work out the appropriate theory and find that the number x we wish to measure is a solution of the quadratic equation

$$x^2 = 9$$

This presents no problem, and we can immediately state that the value of x is either $+3$ or -3. But suppose a number z is a solution of the equation

$$z^2 = -9$$
$$= 9 \times (-1)$$
$$\therefore z = \sqrt{9} \times \sqrt{(-1)}$$
$$= \pm 3 \times \sqrt{(-1)}$$

This equation presents a problem, in that there is no known real number which, when multiplied by itself, results in any negative number in general, or -1 in particular. There is therefore no real solution to this equation, and we appear to be at an impasse. The fact that no real solution exists implies that the number z can never be measured in any physical experiment. This may deter the physicist, but to the mathematician this is but a mere detail. Equations such as $z^2 = -9$ are the delight of mathematicians, for if no solution is known to exist, then they may invent a formalism which will allow the equation to be solved. Let us define the quantity expressed by the symbol i as

$$i = \sqrt{(-1)} \tag{2.16}$$

The insoluble equation may now be written as

$$z^2 = -9$$

with the solution

$$z = \pm\, 3i$$

We have now obtained a formal solution to the equation in terms of the newly defined quantity i. Since the value of z cannot be measured in any physical experiment, we refer to the number $3i$, and also any number which is of the form i times a real number, as an *imaginary* number.

If we enquire as to the nature of the general solution to the quadratic equation

$$Az^2 + Bz + C = 0$$

then scalar algebra tells us that the answer is given by

$$z = \frac{-B \pm \sqrt{B^2 - 4AC}}{2A}$$

If $B^2 > 4AC$, then the term $B^2 - 4AC$ is positive, and the square root is a real number. The overall value of z is then real. But if $B^2 < 4AC$, we are required to take the square root of a negative number, which is imaginary. In this case the solution takes the form

$$z = \frac{-B \pm \sqrt{(4AC - B^2)(-1)}}{2A}$$

$(4AC - B^2)$ is now positive, and the solution becomes

$$z = \frac{-B}{2A} \pm \frac{i\sqrt{4AC - B^2}}{2A}$$

This has the general form

$$z = a + ib \tag{2.17}$$

in which a and b are real scalars. The number z is composed of two parts: a *real part* a, and an *imaginary part* ib. A number of the general form of equation (2.17) is called a *complex number*. It can be seen that the real numbers are a special case of complex numbers corresponding to $b = 0$ and, similarly, the imaginary numbers correspond to those complex numbers for which $a = 0$.

2.11   The Argand diagram

Equation (2.17) is the definition of a complex number. Since complex numbers include all the real numbers and all the imaginary numbers as special cases, then the set of complex numbers represents a generalisation of the concept of the real number. Note that the symbol z as defined by equation 2.17 contains numerical information on two scalars, namely a and b. In this way, the complex-number formalism is a means of handling symbols which contain two pieces of numerical information.

Complex numbers may be represented graphically on a two-dimensional diagram termed an *Argand diagram*, as shown in Fig. 2.15. An Argand diagram looks deceptively like a two-dimensional Cartesian coordinate system, but whereas Cartesian coordinates represent information concerning two sets of real numbers, the Argand diagram deals with one set of real numbers and one set of imaginary numbers. In

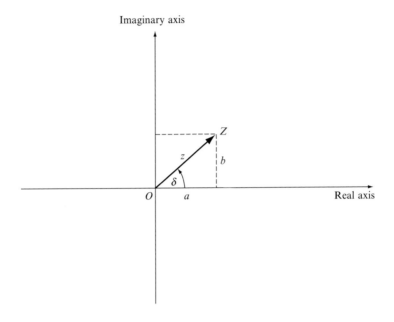

Fig. 2.15 *The Argand diagram.* Complex numbers may be represented by a line segment in the Argand diagram, the horizontal axis of which represents real numbers, and the vertical axis imaginary numbers. The point Z corresponds to the complex number $z = a + ib$, and the line segment OZ is a vector in the complex plane. The magnitude $|z|$ of z corresponds to the length of the line segment OZ. The phase angle δ is measured anticlockwise. Important geometrical relationships are

$$|z|^2 = a^2 + b^2$$
$$a = |z| \cos \delta$$
$$b = |z| \sin \delta$$
$$\tan \delta = \frac{b}{a}$$

the Argand diagram, the real numbers are plotted along the horizontal axis, and the imaginary ones along the vertical axis. Any point in the plane of the diagram corresponds to a single real number and a single imaginary number simultaneously, and so any complex number may be identified as a point in the plane of the diagram. This plane is usually referred to as the *complex plane*. The length of the line segment OZ corresponds to the *magnitude, amplitude* or *modulus* of the complex number z, and this is written as $|z|$; the angle δ between the complex number z and the real axis is referred to as the *phase, phase angle* or *argument* of the complex number z, and by convention is always measured anticlockwise; in general, this book will use the terms *magnitude* and *phase*. As is evident from Fig. 2.15, the representation of a complex number z on the Argand diagram is very similar to the representation of a two-dimensional vector $\mathbf{v} = (v_x, v_y)$ in Cartesian coordinates. In view of this, we often refer to a complex number as a *vector in the complex plane*. Also, from Fig. 2.15 we may identify the following geometrical relationships:

$$|z|^2 = a^2 + b^2$$

$$a = |z| \cos \delta$$

$$b = |z| \sin \delta$$

$$\tan \delta = \frac{b}{a}$$

A note about angles

Whilst we are most familiar with angles being measured in *degrees*, mathematicians prefer to use a different unit, namely *radians*, and some of the ensuing formulae require that radians are used. Degrees and radians can be interconverted easily by considering a circle of unit radius ($r = 1$). The circumference of such a circle, which we know subtends an angle of 360° at the centre, is therefore 2π. Mathematicians refer to 360° as being 2π radians, and therefore 180° is equal to π radians and 90° is $\pi/2$ radians. The following simple relationships apply:

$$\delta(\text{radians}) = \frac{\pi}{180} \times \delta(\text{degrees})$$

$$\delta(\text{degrees}) = \frac{180}{\pi} \times \delta(\text{radians})$$

2.12   The addition of complex numbers

Consider two complex numbers z_1 and z_2, defined as

$$z_1 = a_1 + ib_1$$

$$z_2 = a_2 + ib_2$$

The sum of these two complex numbers is written as $z_1 + z_2$ and is defined by

$$z_1 + z_2 = a_1 + ib_1 + a_2 + ib_2$$

$$= (a_1 + a_2) + i(b_1 + b_2) \tag{2.18}$$

When one is adding two complex numbers, the real and imaginary parts are handled independently. The addition of complex numbers in the Argand diagram is shown in Fig. 2.16.

The similarity in formalism between complex numbers and two-dimensional vectors is not surprising, since both techniques deal with the properties of symbols which contain two pieces of numerical information. There is a subtle difference between the two, however. The two-dimensional vector $\mathbf{v}$ given by $\mathbf{v} = (v_x, v_y)$ is a quantity which contains information on two real numbers, since both v_x and v_y may be measured directly in a physical experiment. The complex number $z = a + ib$, on the other hand, connects a real number a with an imaginary number ib. It is not possible to measure

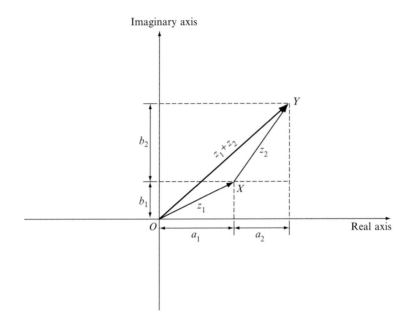

Fig. 2.16 *The addition of complex numbers.* Two complex numbers $z_1 = a_1 + ib_1$ and $z_2 = a_2 + ib_2$ are added to give a sum defined by $z_1 + z_2 = (a_1 + a_2) + i(b_1 + b_2)$. This corresponds to 'head-to-tail' addition analogous to two-dimensional vector addition.

both a and b directly in a physical experiment, and so although the formalisms of two-dimensional vectors and complex numbers are similar in appearance, each deals with a rather different concept.

2.13 Multiplication of complex numbers

The product of the two complex numbers $z_1 = a_1 + ib_1$ and $z_2 = a_2 + ib_2$ is written as $z_1 z_2$, and is defined by

$$z_1 z_2 = (a_1 + ib_1)(a_2 + ib_2) \tag{2.19}$$
$$= a_1 a_2 + ia_1 b_2 + ib_1 a_2 + i^2 b_1 b_2$$

Since $i = \sqrt{(-1)}$, then

$$i^2 = -1$$

$$\therefore z_1 z_2 = (a_1 a_2 - b_1 b_2) + i(a_1 b_2 + b_1 a_2) \tag{2.20}$$

It is left as an exercise for the reader to prove that multiplication is commutative, implying that $z_1 z_2 = z_2 z_1$.

A special case of the multiplication of two complex numbers is the self-product $zz = z^2$. So, if

$$z = a + ib$$

then, from equation (2.20) we have

$$z^2 = a^2 - b^2 + 2iab$$

2.14 The complex conjugate

If a complex number z is defined as

$$z = a + ib$$

then we may define the *complex conjugate* of z, written as z^*, as

$$z^* = a - ib \tag{2.21}$$

The distinction between a complex number and its complex conjugate is solely that the imaginary part has changed sign. Clearly, all real numbers are their own conjugates, and for a complex number z, the relationship between z and z^* is shown in the Argand diagram in Fig. 2.17: as can be seen, z^* is the mirror image of z in the real axis.

An important product is that formed by a complex number z and its complex conjugate z^*. Since

$$z = a + ib$$

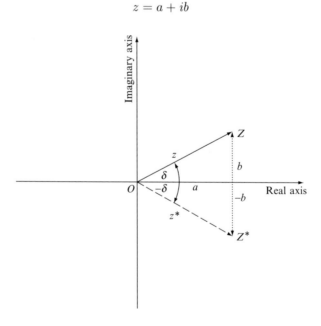

Fig. 2.17 *The complex conjugate.* The complex conjugate z^* of $z = a + ib$ is defined as $z^* = a - ib$. In the Argand diagram, this corresponds to the reflection of the line segment OZ in the real axis to give the line segment OZ^*. Note that the amplitudes of z and z^* are the same, and that the phase angle of z^* is the negative of that of z.

and

$$z^* = a - ib$$

$$zz^* = (a + ib)(a - ib)$$

$$= a^2 + b^2$$

$$= |z|^2 \qquad (2.22)$$

We see that the product zz^* is a real number, equal to $a^2 + b^2$. Reference to Fig. 2.15 will show that this is equal to the square of the magnitude $|z|$ of the complex number z. Equation (2.22) is the usual method of computing the square of the magnitude, and hence the magnitude itself, of a complex number. The importance of the product zz^* is that since it is a real number, it may be measured in a physical experiment: although z itself is not measurable, $zz^* = |z|^2$ is measurable. However, since $|z|^2 = a^2 + b^2$, knowledge of $|z|^2$ alone is not sufficient to give us the unambiguous values of a and b individually. Hence, the fact that we can measure only $|z|^2$ as opposed to z means that there is always an ambiguity in measurements concerning complex numbers. This is of major significance in the theory of X-ray diffraction, and will be discussed at greater length later in this book.

2.15   The complex exponential representation

As shown in Fig. 2.15, the real and imaginary parts of any complex number $z = a + ib$ can be expressed using the magnitude $|z|$ and the phase δ such that

$$a = |z| \cos \delta$$

$$b = |z| \sin \delta$$

$$\therefore z = |z| \, (\cos \delta + i \sin \delta) \qquad (2.23)$$

Let us examine the term $(\cos \delta + i \sin \delta)$. Assuming that δ is in radians (rather than degrees), the function $\cos \delta$ may be expressed by an infinite series as

$$\cos \delta = 1 - \frac{\delta^2}{2!} + \frac{\delta^4}{4!} - \frac{\delta^6}{6!} + \dots$$

where, for example, $6! = 6 \cdot 5 \cdot 4 \cdot 3 \cdot 2 \cdot 1 = 720$ (note that '6!' is referred to as 'six factorial'). Similarly, $\sin \delta$ is defined by the series

$$\sin \delta = \delta - \frac{\delta^3}{3!} + \frac{\delta^5}{5!} - \frac{\delta^7}{7!} + \dots$$

$$\therefore i \sin \delta = i\delta - i\frac{\delta^3}{3!} + i\frac{\delta^5}{5!} - i\frac{\delta^7}{7!} + \dots$$

$$\therefore \cos \delta + i \sin \delta = 1 + i\delta - \frac{\delta^2}{2!} - i\frac{\delta^3}{3!} + \frac{\delta^4}{4!} + i\frac{\delta^5}{5!} - \frac{\delta^6}{6!} + \dots \qquad (2.24)$$

The imaginary number i has certain special properties, given by $i = \sqrt{-1}$, $i^2 = -1$, $i^3 = -i$, $i^4 = +1$ and so on. We may use these relationships to rewrite equation (2.24) as

$$\cos \delta + i \sin \delta = 1 + i\delta + \frac{(i\delta)^2}{2!} + \frac{(i\delta)^3}{3!} + \frac{(i\delta)^4}{4!} + \frac{(i\delta)^5}{5!} + \frac{(i\delta)^6}{6!} + \dots \tag{2.25}$$

Another well-known series is that for the irrational number e^x:

$$e^x = 1 + x + \frac{x^2}{2!} + \frac{x^3}{3!} + \frac{x^4}{4!} + \frac{x^5}{5!} + \dots$$

Comparison of the expression for e^x with equation (2.25) shows that if we identify

$$x = i\delta$$

then

$$\cos \delta + i \sin \delta = e^{i\delta} \tag{2.26}$$

Using equation (2.26), we may rewrite equation (2.23) in the form

$$z = |z| \, e^{i\delta} \tag{2.27}$$

Equation (2.27) expresses a complex number in terms of its magnitude $|z|$ and its phase δ, and is referred to as the *complex exponential* form. The representations

$$z = a + ib \tag{2.17}$$

and

$$z = |z| \, e^{i\delta} \tag{2.27}$$

are entirely equivalent, despite their somewhat different appearances. The use of the complex exponential form is particularly appropriate to the description of wave motion. As was mentioned in the introductory paragraph to the discussion of complex algebra, waves are associated with two particular parameters, the maximum amplitude and the phase, and so complex exponentials of the form $|z|e^{i\delta}$ will be met throughout the rest of this book in connection with the description of waves, where $|z|$ will represent the wave's maximum amplitude and δ its phase.

The operations of addition and multiplication are defined in terms of the complex exponential representation in a manner exactly analogous to equations (2.18) and (2.19). If

$$z_1 = |z_1| \, e^{i\delta_1}$$

and

$$z_2 = |z_2| \, e^{i\delta_2}$$

then

$$z_1 + z_2 = |z_1| \, e^{i\delta_1} + |z_2| \, e^{i\delta_2} \tag{2.28}$$

and

$$z_1 z_2 = |z_1| \, e^{i\delta_1} \, |z_2| \, e^{i\delta_2}$$
$$\therefore z_1 z_2 = |z_1| \, |z_2| \, e^{i(\delta_1 + \delta_2)} \tag{2.29}$$

Note that equation (2.28) cannot be put into a more simple form. Comparison of equation (2.18) with (2.28) and of equation (2.20) with (2.29) shows that the form $z = a + ib$ is rather more compact when one is dealing with addition operations, whereas the complex exponential form $z = |z|e^{i\delta}$ is neater to use for multiplication operations.

In the complex exponential form, the complex conjugate z^* of

$$z = |z|e^{i\delta}$$

is obtained by negating the sign of i and is given by

$$z^* = |z|e^{-i\delta} \tag{2.30}$$

The sign of the phase δ has changed, but the magnitude $|z|$ is unaltered. Since phase angles are by convention measured anticlockwise, a negative phase angle corresponds to an angle measured clockwise. z^* therefore represents the mirror image of z in the real axis, as shown in Fig. 2.17. The form of equation (2.30) may also be verified as follows. From equations (2.26) and (2.27),

$$|z|e^{i(-\delta)} = |z|[\cos(-\delta) + i\sin(-\delta)]$$

But

$$\cos(-\delta) = \cos\delta$$

and

$$\sin(-\delta) = -\sin\delta$$
$$|z|e^{-i\delta} = |z|(\cos\delta - i\sin\delta) \tag{2.31}$$

Equation (2.31) has the form $a - ib$, corresponding to the complex conjugate of $a + ib$.

The product zz^* in complex exponential form becomes

$$zz^* = |z|e^{i\delta}|z|e^{-i\delta}$$
$$= |z|^2 e^{i(\delta-\delta)}$$

But

$$\delta - \delta = 0$$

and

$$e^0 = 1$$
$$\therefore zz^* = |z|^2$$

This agrees with equation (2.22).

2.16   Complex exponentials and trigonometric functions

We have already seen that the complex exponential function $e^{i\delta}$ is related to the trigonometric functions $\cos\delta$ and $\sin\delta$ by equation (2.26),

$$e^{i\delta} = \cos\delta + i\sin\delta \tag{2.26}$$

Furthermore, the complex conjugate of $e^{i\delta}$, namely $e^{-i\delta}$, may be expressed in terms of $\cos\delta$ and $\sin\delta$ if we apply equation (2.31) to the case of $|z| = 1$:

$$e^{-i\delta} = \cos\delta - i\sin\delta \tag{2.32}$$

Adding equations (2.26) and (2.32), we obtain

$$e^{i\delta} + e^{-i\delta} = 2\cos\delta$$

$$\therefore \cos\delta = \frac{e^{i\delta} + e^{-i\delta}}{2} \tag{2.33}$$

And, on subtraction, we derive an expression for $\sin\delta$,

$$e^{i\delta} - e^{-i\delta} = 2i\sin\delta$$

$$\therefore \sin\delta = \frac{e^{i\delta} - e^{-i\delta}}{2i} \tag{2.34}$$

Equations (2.33) and (2.34),

$$\cos\delta = \frac{e^{i\delta} + e^{-i\delta}}{2} \tag{2.33}$$

$$\sin\delta = \frac{e^{i\delta} - e^{-i\delta}}{2i} \tag{2.34}$$

are important and will be used elsewhere in the text. Often, we will come across expressions containing the sum or the difference of two complex exponential terms, which may be simplified by the substitution of the appropriate trigonometric function.

Summary

Vectors

A *vector* is a mathematical quantity having both magnitude and direction. If the vector has unit magnitude, it is called a *unit vector*. A set of *unit base vectors* may be used to define a coordinate system. If we choose three unit vectors **i**, **j** and **k** to define a right-handed, orthogonal, three-dimensional Cartesian coordinate system, then any vector **v** may be represented as

$$\mathbf{v} = v_x\mathbf{i} + v_y\mathbf{j} + v_z\mathbf{k} \tag{2.3}$$

or by the ordered triplet

$$\mathbf{v} = (v_x, v_y, v_z) \tag{2.4}$$

The three scalars v_x, v_y and v_z are the *components* of the vector **v** in the chosen coordinate system.

For three vectors **u**, **v** and **w** defined by

$$\mathbf{u} = (u_x, u_y, u_z)$$
$$\mathbf{v} = (v_x, v_y, v_z)$$
$$\mathbf{w} = (w_x, w_y, w_z)$$

the following vector operations are defined.

Multiplication by a scalar:

$$n\mathbf{v} = (nv_x, nv_y, nv_z) \tag{2.6}$$

Addition:

$$\mathbf{v} + \mathbf{w} = (v_x + w_x, v_y + w_y, v_z + w_z) \tag{2.5}$$

Scalar product:

$$\mathbf{v} \cdot \mathbf{w} = vw \cos \theta \tag{2.7}$$

$$\mathbf{v} \cdot \mathbf{w} = v_x w_x + v_y w_y + v_z w_z \tag{2.9}$$

This operation is commutative.

The self-scalar product $\mathbf{v} \cdot \mathbf{v}$ is the square v^2 of the magnitude v of the vector **v**, implying that

$$v = \sqrt{v_x^2 + v_y^2 + v_z^2} \tag{2.10}$$

Vector product:

$$\mathbf{v} \wedge \mathbf{w} = \begin{vmatrix} \mathbf{i} & \mathbf{j} & \mathbf{k} \\ v_x & v_y & v_z \\ w_x & w_y & w_z \end{vmatrix} \tag{2.14}$$

$$|\mathbf{v} \wedge \mathbf{w}| = vw \sin \theta \tag{2.11}$$

This operation is non-commutative:

$$\mathbf{v} \wedge \mathbf{w} = -\mathbf{w} \wedge \mathbf{v}$$

Scalar triple product:

$$\mathbf{u} \cdot (\mathbf{v} \wedge \mathbf{w}) = \begin{vmatrix} u_x & u_y & u_z \\ v_x & v_y & v_z \\ w_x & w_y & w_z \end{vmatrix} \tag{2.15}$$

The value of the scalar triple product corresponds to the volume of the parallelepiped defined by **u**, **v** and **w**.

The use of three-dimensional vectors is that they are a convenient way of handling physical quantities associated with a spatial direction. Vectors will be used to define crystal geometry, and also to signify the direction of waves in space.

Complex numbers

A *complex number* z contains both a real and an imaginary part, and may be expressed in the form of equation (2.17) or (2.27):

$$z = a + ib \tag{2.17}$$

$$z = |z|\, e^{i\delta} \tag{2.27}$$

$|z|$ is known as the *magnitude, amplitude* or *modulus* of z, and δ is known as the *phase, phase angle* or *argument* of z.

A complex number may be represented on an *Argand diagram*, the horizontal axis of which represents the real numbers, and the vertical axis the imaginary numbers.

For two complex numbers z_1 and z_2, the following operations are defined:

$$z_1 = a_1 + ib_1 \quad z_1 = |z_1|\, e^{i\delta_1}$$

$$z_2 = a_2 + ib_2 \quad z_2 = |z_2|\, e^{i\delta_2}$$

Addition:

$$z_1 + z_2 = (a_1 + a_2) + i(b_1 + b_2) = |z_1|\, e^{i\delta_1} + |z_2|\, e^{i\delta_2} \tag{2.18 and 2.28}$$

Multiplication:

$$z_1 z_2 = (a_1 a_2 - b_1 b_2) + i(a_1 b_2 + b_1 a_2) = |z_1|\,|z_2|\, e^{i(\delta_1 + \delta_2)} \tag{2.20 and 2.29}$$

The *complex conjugate* z^* of $z = a + ib = |z| e^{i\delta}$ is defined as

$$z^* = a - ib = |z| e^{-i\delta} \tag{2.21 and 2.30}$$

A useful identity is

$$zz^* = |z|^2 \tag{2.22}$$

zz^* is a real, and hence measurable, number.

Complex exponentials and trigonometric functions are related as follows:

$$e^{i\delta} = \cos\delta + i\sin\delta \tag{2.26}$$

$$e^{-i\delta} = \cos\delta - i\sin\delta \tag{2.32}$$

$$\cos\delta = \frac{e^{i\delta} + e^{-i\delta}}{2} \tag{2.33}$$

$$\sin\delta = \frac{e^{i\delta} - e^{-i\delta}}{2i} \tag{2.34}$$

Appendix

Determinants

It is left as an exercise for the reader to show that the solutions to the pair of linear scalar algebraic equations

$$a_1 x + a_2 y + a_3 = 0$$
$$b_1 x + b_2 y + b_3 = 0$$

are given by

$$x = \frac{a_2 b_3 - a_3 b_2}{a_1 b_2 - a_2 b_1}$$

$$y = \frac{a_3 b_1 - a_1 b_3}{a_1 b_2 - a_2 b_1} \tag{2.35}$$

As can be seen, expressions of the type $a_1 b_2 - a_2 b_1$ arise 'naturally'. Since this form is so common in the theory of linear algebra, the determinantal representation has been introduced. A *determinant* is a square array of symbols, called elements, enclosed by two vertical lines. A 2×2 determinant is defined as

$$\begin{vmatrix} a_1 & a_2 \\ b_1 & b_2 \end{vmatrix} = a_1 b_2 - a_2 b_1 \tag{2.36}$$

The value of the determinant is found by multiplying the diagonal elements and subtracting the values according to the rule given in equation (2.36). Imagining diagonal lines across the determinant linking the elements that are to be multiplied will help the reader to remember the operation. Using the determinant notation, the solution to the above pair of linear simultaneous equations becomes

$$\frac{x}{\begin{vmatrix} a_2 & a_3 \\ b_2 & b_3 \end{vmatrix}} = \frac{-y}{\begin{vmatrix} a_1 & a_3 \\ b_1 & b_3 \end{vmatrix}} = \frac{1}{\begin{vmatrix} a_1 & a_2 \\ b_1 & b_2 \end{vmatrix}}$$

The form of this equation is quite symmetrical. If we wish to obtain the determinant associated with the variable x, we 'cover up' the terms in x in the pair of equations, and write down the determinant of the remaining coefficients as they appear in the equations. Note that the expression containing the variable y has a minus sign associated with it.

Determinants have a number of interesting properties. Firstly, a determinant is symmetrical with respect to interchange of both the rows and the columns (also known as transposition):

$$\begin{vmatrix} a_1 & a_2 \\ b_1 & b_2 \end{vmatrix} = a_1 b_2 - a_2 b_1 = \begin{vmatrix} a_1 & b_1 \\ a_2 & b_2 \end{vmatrix}$$

Secondly, if we interchange either the rows or the columns, then the value of the determinant changes sign:

$$\begin{vmatrix} b_1 & b_2 \\ a_1 & a_2 \end{vmatrix} = b_1 a_2 - b_2 a_1 = - \begin{vmatrix} a_1 & a_2 \\ b_1 & b_2 \end{vmatrix}$$

Determinantal notation may be extended to three sets of elements. When we wish to solve a set of three linear simultaneous equations, it is found that expressions of the form

$$a_1(b_2 c_3 - b_3 c_2) - a_2(b_1 c_3 - b_3 c_1) + a_3(b_1 c_2 - b_2 c_1) \tag{2.37}$$

arise. Note that on multiplying out the brackets, the terms associated with positive signs are $a_1 b_2 c_3$, $a_2 b_3 c_1$ and $a_3 b_1 c_2$, whereas those with negative signs are $a_1 b_3 c_2$, $a_2 b_1 c_3$ and $a_3 b_2 c_1$. The significant point is that the positive terms contain the symbols a, b, c and 1, 2, 3 in the 'correct' order (allowing for 1 to come after 3), whereas the negative terms have the symbols a, b, c in the 'correct' order but 1, 2, 3 in the 'wrong' order.

Using 2×2 determinants, equation (2.37) may be written as

$$a_1 \begin{vmatrix} b_2 & b_3 \\ c_2 & c_3 \end{vmatrix} - a_2 \begin{vmatrix} b_1 & b_3 \\ c_1 & c_3 \end{vmatrix} + a_3 \begin{vmatrix} b_1 & b_2 \\ c_1 & c_2 \end{vmatrix} = \begin{vmatrix} a_1 & a_2 & a_3 \\ b_1 & b_2 & b_3 \\ c_1 & c_2 & c_3 \end{vmatrix} \tag{2.38}$$

and the definition of the 3×3 determinant is shown in equation (2.38). The expressions (2.37) and (2.38) are equivalent, and it can be seen that the determinant is a much more elegant notation than the ugly representation (2.37).

To evaluate a 3×3 determinant, we use the following procedure:

(1) Choose the element in the first row and the first column. Multiply this by the value of the 2×2 determinant formed by those elements obtained by 'covering up' the row and column containing the chosen initial element.
(2) Take the element in the first row and second column, and multiply this by the determinant formed by 'covering up' the first row and second column.
(3) Choose the element in the first row and third column, and multiply this by the determinant of the elements not in the first row and third column.
(4) Add the results of (1) and (3) and subtract the result of (2).

This procedure is an explicit explanation of the meaning of equation (2.38).

As can be seen from the theory of vector analysis, the determinant notation is useful in expressing vector products and scalar triple products. From the equations

$$\mathbf{v} \wedge \mathbf{w} = (v_y w_z - v_z w_y)\mathbf{i} + (v_z w_x - v_x w_z)\mathbf{j} + (v_x w_y - v_y w_x)\mathbf{k}$$
$$= (v_y w_z - v_z w_y)\mathbf{i} - (v_x w_z - v_z w_x)\mathbf{j} + (v_x w_y - v_y w_x)\mathbf{k}$$

and

$$\mathbf{u} \cdot (\mathbf{v} \wedge \mathbf{w}) = u_x(v_y w_z - v_z w_y) + u_y(v_z w_x - v_x w_z) + u_z(v_x w_y - v_y w_x)$$
$$= u_x(v_y w_z - v_z w_y) - u_y(v_x w_z - v_z w_x) + u_z(v_x w_y - v_y w_x)$$

we recognise the determinants

$$\mathbf{v} \wedge \mathbf{w} = \begin{vmatrix} \mathbf{i} & \mathbf{j} & \mathbf{k} \\ v_x & v_y & v_z \\ w_x & w_y & w_z \end{vmatrix} \tag{2.14}$$

$$\mathbf{u} \cdot (\mathbf{v} \wedge \mathbf{w}) = \begin{vmatrix} u_x & u_y & u_z \\ v_x & v_y & v_z \\ w_x & w_y & w_z \end{vmatrix} \tag{2.15}$$

Bibliography

Jeffrey, A. *Mathematics for Engineers and Scientists*, 6th edn. CRC Press, Boca Raton, FL, 2005. A very comprehensive reference work.

Stephenson, G. *Mathematical Methods for Science Students*. Longman, London, 1973. A textbook designed specifically to explain relevant mathematics to scientists, and very successful in its aim.

3
Crystal systematics

3.1 What is a crystal?

In Chapter 1, we learnt that the crystalline state of matter was characterised by a long-range, three-dimensional order, for only by assuming this model are we able to interpret the properties exhibited by crystals. In this chapter we shall analyse this model more closely, so that we may categorise crystals and crystal geometry in a mathematical manner.

The first task is to investigate the nature of the three-dimensional order of crystals. The existence of long-range order implies that there is some fundamental, basic structure which is repeated over and over again in space. Specifically, the question we wish to answer is: 'How is the structure of a crystal related to the fundamental repeating unit?'

The answer to this question is more easily appreciated if we examine some particular examples of the way in which ordered structures are built up. For clarity in representing patterns on a printed page, we shall for the moment deal with two-dimensional examples, and then extend the argument to three dimensions. Consider the patterns shown in Fig. 3.1. Each is formed by the regular arrangement of the 'molecule' represented by the arrow in a two-dimensional ordered array. The molecular representation was deliberately chosen to be asymmetrical so that we may concentrate on the structure and properties of the overall arrangement in space without being confused by any special properties of the molecule itself. From the patterns in Fig. 3.1 we notice that they are all different, but that they have the same underlying geometrical rectangular structure. Pattern (a) is a regular repeat of the single arrow, pattern (b) is formed from two arrows side by side, and pattern (c) from a trefoil arrangement of the arrows. Suppose we now concentrate not on the repeat unit itself, but on the way the repeat units are arranged with respect to each other. To do this, we may replace each repeat unit by a point, and if we place the point in exactly the same position relative to each repeat unit, then we may remove the repeat units, leaving a two-dimensional array of points which represents the underlying geometrical relationship of the repeat units. If we do this for each of the patterns of Fig. 3.1, we derive the same two-dimensional array of points, as shown in Fig. 3.2.

We may re-create any of the patterns of Fig. 3.1 from the array of points merely by positioning the repeat unit in exactly the same manner with respect to each point. Earlier, we had noticed that the diagrams of Fig. 3.1 had the same underlying geometrical form, and it is this that is emphasised when we consider the appropriate array of points. We call the structure that is regularly repeated in space the *motif*,

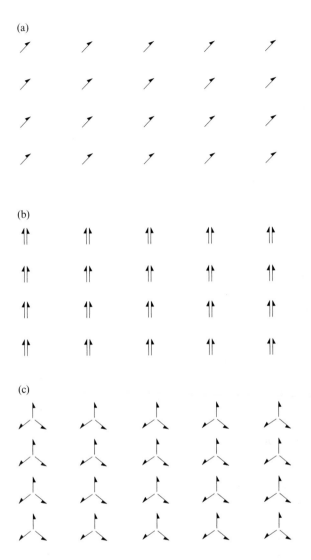

Fig. 3.1 *Three two-dimensional 'crystals'.* Each pattern is different, but they have the same underlying rectangular structure.

and the conceptual array of points which defines the geometrical relation between the motifs is called the *lattice*.

The *motif* is the structural unit which is repeated regularly in space.

The *lattice* is a conceptual array of points in space which serves to define the geometrical relationship between the motifs in a structure.

For a two-dimensional system, we speak of a *plane lattice*. There are two particular features of the motif and lattice which are important.

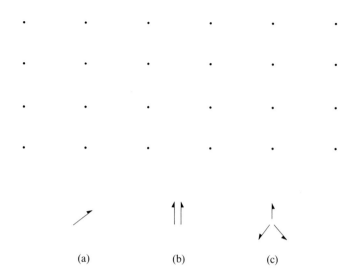

(a)   (b)   (c)

Fig. 3.2 *Lattices and motifs.* Each of the diagrams of Fig. 3.1 may be reproduced by associating the motif (a), (b) or (c) with the rectangular array of lattice points.

(a) *The position of the lattice points with respect to the motif is arbitrary, provided that we are consistent.* When we seek to represent a structure in terms of a lattice, it does not matter how we associate a lattice point with one particular motif, but once we have chosen a relationship between a lattice point and the motif, we must continue in an identical manner for all lattice points. Similarly, when we reconstruct the structure from the lattice and the motif, we must ensure that we position the motif in relation to every lattice point in exactly the same way. This is illustrated in Fig. 3.3.

(b) *The motif may be quite complex structurally.* The nature of the motif is determined by the overall pattern formed in space, and it is not necessary for the motif to be just a single molecule. For example, in Fig. 3.1(a) the motif is the single arrow, whereas in Fig. 3.1(c), the motif is the more complex trefoil form. The relevance of this to crystals is that in a crystal, the motif may be a group of several molecules, or groups of ions, or whatever is appropriate to describe the overall geometrical arrangement.

When we consider three dimensions, the situation is exactly similar, but rather more difficult to represent on a printed page. Any crystal has a well-defined repeat unit, the motif, and each motif is associated with a lattice point so that the overall lattice is a three-dimensional array, usually called a *space lattice*. In this way, we may conceptually regard a crystal to be of the form

$$\text{crystal structure} = \text{lattice} * \text{motif} \tag{3.1}$$

where the symbol $*$ may, for the moment, be taken to mean 'associated with'.

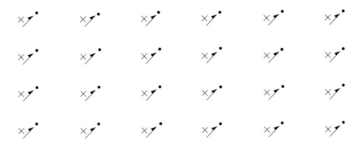

Fig. 3.3 *The position of the motif with respect to the lattice points is arbitrary as long as we are consistent.* The lattice of points is identical to the lattice of crosses, but we must not mix the two different relative positions.

Whereas the lattice is a geometrical abstraction relating to the exact geometry of the crystal form, the motif is a highly specific entity which concerns the local arrangement of molecules, and also the structure of the molecules themselves. It is the lattice, however, that determines the overall structure of the crystal, and so it is the geometry of the lattice that is used to categorise crystal structures, as will be reviewed in a later section of this chapter.

The breakdown of crystal structure into the lattice and motif as signified by the conceptual 'equation' (3.1) is important from an informational point of view. Essentially, there are two questions relevant to the study of crystals, and these are 'What is the structure of the molecules within a crystal?' and 'What is the nature of the geometrical array which defines the way in which the molecules are arranged in space?'

The first question is asking about the motif, whereas the second is enquiring after the nature of the lattice. These two questions are quite independent of one another, and knowing the answer to one can give no information as regards the answer to the other. In practice, most biologists are interested, for example, in the structure of protein molecules, and so are more directly concerned with the first question; physicists and mineralogists may be more curious as to the second. The fact that the two questions are independent means that, in any experiment designed to investigate crystal structure, it is likely that the experimental data from which we may derive the answer to one question will be different from that which will allow us to resolve the other. In other words, when we look at the results of an X-ray diffraction experiment, the information concerning the lattice will be presented in a different manner from that which regards the motif. This tells us that by looking at certain features of the diffraction pattern, we will be able to describe the lattice, and only by investigating other features of the diffraction pattern will we be able to determine the motif.

3.2 Symmetry

One of the most important properties of well-ordered, regular geometrical objects is that they are *symmetrical*. For example, consider the cube depicted in Fig. 3.4.

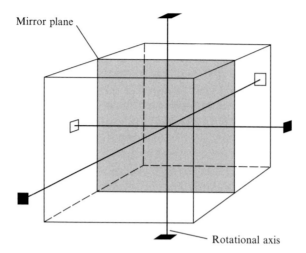

Mirror plane

Rotational axis

Fig. 3.4 *The symmetry of a cube.* A cube contains many symmetry elements, including three rotational axes and a mirror plane. Representations of other symmetry elements are excluded in order not to obscure the diagram.

If the cube is rotated through 90° about an axis perpendicular to, and centrally placed in, a pair of opposite faces, then the new position of the cube is quite indistinguishable from its former position. This is an example of rotational symmetry. Furthermore, the cube contains certain planes such that the form of the cube on one side of the plane is the mirror image of that on the other. These mirror planes are parallel to each pair of opposite faces and contain the centre of the cube, and also those planes which cut diagonally through the cube. A cube, therefore, contains many properties which confer upon the structure a high degree of symmetry. Mirror planes, and axes giving rise to rotational symmetry are just two examples of *symmetry elements*, and we shall have more to say about symmetry and symmetry elements in a later section. For the moment, the important point is that any geometrical form which is built up in a regular systematic manner is capable of containing, and usually does contain, symmetry elements of various types. Since a lattice is a well-defined geometrical form, it will have certain symmetry properties. If we refer to the two-dimensional examples shown in Fig. 3.5, we see that Fig. 3.5(a) has the symmetry of a square, that Fig. 3.5(b) has rectangular symmetry, and that Fig. 3.5(c) is representative of the symmetry shown by a hexagon (or, alternatively, an equilateral triangle). The symmetry of two- and three-dimensional lattices has been studied extensively, and we shall investigate this more thoroughly in due course.

3.3 The description of the lattice

We now know that the basic geometrical structure of a crystal may be represented by a three-dimensional array of points called the space lattice, which will probably

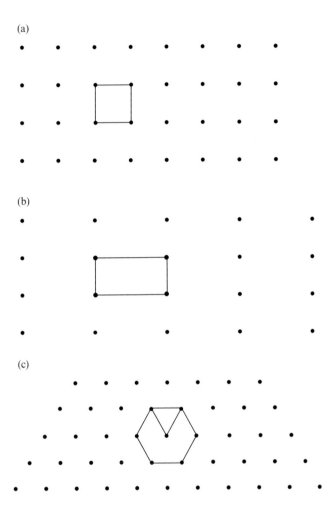

Fig. 3.5 *Lattices and symmetry.* (a) has the symmetry of a square, (b) that of a rectangle, and (c) that of a hexagon or an equilateral triangle.

have certain symmetry properties. In this section, we shall find out how to define and categorise lattices in mathematical terms.

Plane lattices

We shall first deal with the two-dimensional lattice shown in Fig. 3.6(a). If we choose an arbitrary lattice point as the origin, then we may join the origin to two different lattice points, which are chosen as the nearest lattice points to the origin in any two directions. We may define these directions by two vectors **a** and **b**, which are at an angle φ. For the moment we will call these vectors the *fundamental lattice vectors*. They will define a parallelogram, which we may call the *fundamental lattice cell*. Using

(a)

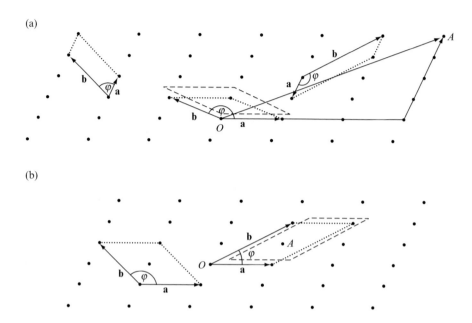

(b)

Fig. 3.6 *Lattice vectors and lattice cells.* (a) shows choices of fundamental lattice vectors which define a primitive fundamental lattice cell, as shown when a cell is displaced. Any position **r** in the lattice satisfies the equation $\mathbf{r} = p\mathbf{a} + q\mathbf{b}$, in which p and q are integers. For instance, the point A relative to the origin O may have, according to the choice of **a** and **b**, $p = 3$ and $q = 4$. (b) shows choices which give rise to a double fundamental cell. In this case, any lattice point **r** satisfies the equation $\mathbf{r} = p\mathbf{a} + q\mathbf{b}$, but p and q are now no longer integers; for instance, the point A relative to the origin O has $p = \frac{1}{2}$, $q = \frac{1}{2}$. The displaced cell now contains two lattice points.

the fundamental lattice vectors of Fig. 3.6(a), we may express any lattice point **r** in terms of **a** and **b** as

$$\mathbf{r} = p\mathbf{a} + q\mathbf{b} \tag{3.2}$$

in which p and q are integers corresponding to the number of 'steps' required to reach **r** from the origin. As shown in Fig. 3.6(a), the choice of fundamental lattice vectors **a** and **b** is quite arbitrary, and for each of the three pairs of choices shown in the figure, equation (3.2) is valid for any lattice point **r** in terms of integral values of p and q. If the fundamental lattice vectors have already been stated, we may define the point **r** by the ordered doublet (p, q). Since the choice of origin is arbitrary, p and q may take negative integral values too, which are conventionally written as $(\bar{p}, \bar{q})$, pronounced as 'bar p' and 'bar q'.

In order to define a lattice, all scientists must agree to use two conventionally specified fundamental lattice vectors so that everybody has a common description of the same lattice. The way in which we choose to define the particular pair of vectors with which to specify the lattice will be determined by the usefulness of any given representation. We will return to this matter in a paragraph or two, but for the moment, let us investigate another aspect of the fundamental lattice vectors of Fig. 3.6(a), and the appropriate fundamental lattice cells.

A means of categorising fundamental lattice cells is to determine the number of lattice points associated with a given fundamental lattice cell. To do this, we count the number of lattice points contained within the cell, using appropriate fractional values for points at corners or on edges. For the lattice in Fig. 3.6(a), we see that four different fundamental lattice cells meet at any lattice point, and so we may associate values of 1/4 with each corner lattice point. There are four corners, and we see that each fundamental lattice cell may be associated with $1/4 \times 4 = 1$ lattice point. All fundamental lattice cells associated with a single lattice point are called *primitive cells*, or *P cells*. The association with a single lattice point may be seen by displacing the fundamental lattice cell slightly so as to move the corners away from the lattice points, leaving a single lattice point entirely within the cell, as also shown in Fig. 3.6(a).

This situation may be contrasted with that shown in Fig. 3.6(b). In this case the vectors **a** and **b** define a fundamental lattice cell which, as well as having lattice points at each corner, also contains a lattice point within the area of the cell. Once again, four fundamental lattice cells meet at each corner lattice point, and so each fundamental lattice cell of the type shown in Fig. 3.6(b) is associated with $1/4 \times 4 + 1 = 2$ lattice points. In general, fundamental lattice cells associated with more than one lattice point are called *multiple cells*, and in this case we have an example of a double cell. We may also see how to represent any lattice point **r** according to the choice of fundamental lattice vectors **a** and **b** shown in Fig. 3.6(b). In this case we have

$$\mathbf{r} = p\mathbf{a} + q\mathbf{b}$$

but p and q are no longer solely integers—the point at the centre of the fundamental lattice cell, for example, is given by $p = 1/2$, $q = 1/2$. Whereas the choice of vectors giving rise to a primitive cell as in Fig. 3.6(a) allowed only integral values of p and q, when two vectors **a** and **b** are chosen such that the fundamental lattice cell is non-primitive, then we must also use certain fractional values for p and q. It is obviously simpler to choose a primitive cell as the conventional representation of a lattice, but as we shall see later, in particular circumstances the choice of a multiple cell is more informative.

Another important property of the fundamental lattice cell is that no matter whether the cell is multiple or primitive, the cell chosen must have the property that a close-packed assembly of cells fills all space with no 'gaps' between cells. Since we may always choose a primitive cell in any lattice, then only if that primitive cell fills all space will we have a regular lattice with one lattice point for every primitive-cell corner. Consider, for example, the close packing of pentagons as shown in Fig. 3.7.

An array of pentagons just cannot fit together to fill all space, and so no two-dimensional lattice based on a pentagonal primitive cell may exist. The close-fitting requirement places a very limiting restriction on the geometrical nature of two-dimensional lattices. In fact, it has been proved that only five different two-dimensional lattices may exist: these are shown in Fig. 3.8, and their properties are listed in Table 3.1.

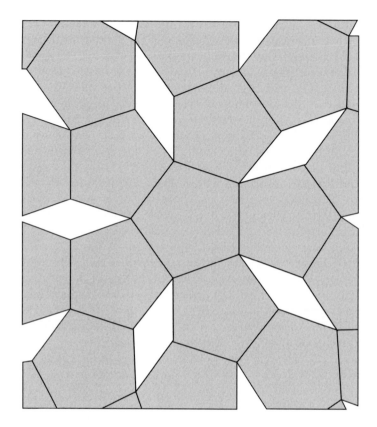

Fig. 3.7 *No plane lattice can have a pentagonal unit cell.* Since pentagons cannot fill all space, no lattice can exist with a pentagonal unit cell. If each corner of the pentagons is associated with a point, then the array of points does *not* form a regular two-dimensional lattice. (From Charles Kittel, *Introduction to Solid State Physics*, 8th edn, Wiley, New York, 2004.)

The five plane lattices of Fig. 3.8 may be divided into four *systems*, depending on the shape of the unit cell. One lattice is based on the square, one on the hexagon (or, alternatively, the equilateral triangle), one on a general oblique parallelogram and two on the rectangle.

We shall now discuss the criteria by which we determine the conventional choice of fundamental lattice vectors for any lattice. Those vectors which are conventionally used to define a lattice are called *crystallographic unit vectors*, and the cell defined by them is termed the *crystallographic unit cell*. In many respects the use of the word 'unit' is misleading, since the magnitude of the vectors is not necessarily unity, nor is the conventional cell always primitive, as we might tacitly assume from the word 'unit'. The terms 'base vector' and 'base cell' are probably more appropriate, but historically, the terminology using the word 'unit' has evolved: for clarity, this book will therefore use the term 'crystallographic unit vector' to define the fundamental lattice vector for a crystal lattice, and 'unit vector' for any vector of unit magnitude.

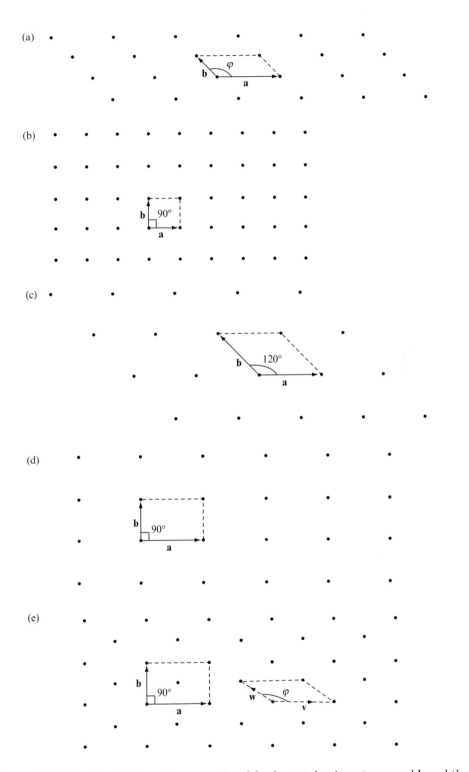

Fig. 3.8 *The five plane lattices.* The conventional fundamental unit vectors **a** and **b** and the corresponding unit cells are shown as follows: (a) oblique *P*, (b) square *P*, (c) hexagonal *P*, (d) rectangular *P*, (e) rectangular *I*. In case (e), the cell defined by the vectors **v** and **w** does not show the rectangular symmetry of the lattice; hence an incentred cell is chosen as the conventional unit cell, as shown.

Table 3.1 Plane lattices.

Lattice system	Lattice type	Conventional representation	Representative points
Oblique	P	$a \neq b$ $\varphi > 90°$	$(0, 0)$
Square	P	$a = b$ $\varphi = 90°$	$(0, 0)$
Hexagonal	P	$a = b$ $\varphi = 120°$	$(0, 0)$
Rectangular	P	$a \neq b$ $\varphi = 90°$	$(0, 0)$
	I	$a \neq b$ $\varphi = 90°$	$(0, 0), (^1/_2, ^1/_2)$

In general, the crystallographic unit vectors **a** and **b** are chosen such that they define a primitive unit cell, and of the many choices of primitive cells available, we choose that cell for which the magnitudes of the vectors **a** and **b** are most nearly equal. This last condition implies that the unit cell chosen will not have a bizarre shape. The angle φ is usually taken to be equal to or greater than 90°, and in this book, we shall call the longer vector **a**, so that $a > b$.

In certain cases, however, it is found more convenient to choose a non-primitive cell as the unit cell. This occurs when the choice of a primitive cell is less useful in describing the symmetry of the lattice as a whole. Such a case arises in the description of the rectangular lattice shown in Fig. 3.8(e). The primitive cell is that defined by the vectors **v** and **w** according to the rules given immediately above. An overview of the lattice as a whole clearly demonstrates the rectangular symmetry, whereas the choice of this particular primitive cell gives no direct information as to the presence of this special symmetry property. If we were to look at the primitive cell alone, we would probably not immediately realise that it was in fact a cell which would generate a lattice with rectangular symmetry. An alternative choice of unit cell is that defined by the crystallographic unit vectors **a** and **b**, and it is this cell which is the conventional crystallographic unit cell. The cell is a double cell with a lattice point at the centre, and it clearly represents the rectangular symmetry of the lattice. It is in order to communicate the fundamental lattice symmetry that we choose to take a non-primitive cell as the conventional crystallographic unit cell. Thus we may now state the formal definitions of the conventional crystallographic unit cell and crystallographic unit vector:

- A *crystallographic unit cell* is a parallelogram (or parallelepiped in three dimensions) defined by the plane lattice (or space lattice in three dimensions) which serves to display the symmetry of the lattice in a convenient manner.

- *Crystallographic unit vectors* are lattice vectors which define the sides of the conventionally chosen crystallographic unit cell.

If the lattice symmetry can be conveyed adequately by a primitive cell, then an appropriate primitive cell is chosen as the conventional crystallographic unit cell. But if a primitive cell is not suitable, then we choose the next simplest multiple cell. The price we pay for the choice of a multiple cell is that we must allow the use of certain fractional values for p and q in equation (3.2), but the convenience of integral values of p and q is readily sacrificed for the increased utility of a unit cell with the appropriate symmetry. A crystallographic unit cell such as that shown in Fig. 3.8(e) which contains a lattice point at the centre is called an *incentred* or *I cell*.

By definition, a primitive cell has one lattice point at each corner, which, relative to the two vectors **a** and **b**, may be denoted as $(0, 0)$, $(1, 0)$, $(1, 1)$ and $(0, 1)$. But we have already seen that a primitive cell is associated particularly with only a single lattice point, which we may refer to as the *representative point* of the unit cell. If we select the point $(0, 0)$ to be the representative point of the unit cell, then all the other lattice points may be generated by integral additions. The concept of the representative point may be extended to multiple cells, and there are as many representative points as the multiplicity demands. For example, the incentred cell of Fig. 3.8(e) has a multiplicity of two, and so there are two representative points, namely $(0, 0)$ and $(1/2, 1/2)$.

The descriptions of the five plane lattices are to be found in Table 3.1. The lattices are referred to as oblique P, square P, hexagonal P, rectangular P and rectangular I.

Space lattices

The concepts involved in the description of a two-dimensional plane lattice are readily extended to deal with a three-dimensional space lattice. The details are more complicated, but the fundamentals are the same. Firstly, we require a third crystallographic unit vector **c**, non-coplanar with **a** and **b**, which may be used to describe the third dimension. The directions specified by the crystallographic unit vectors **a**, **b** and **c** are called the *crystallographic axes*. The crystallographic unit cell is in general a parallelepiped, and the three *crystallographic angles* α, β and γ are defined as in Fig. 3.9.

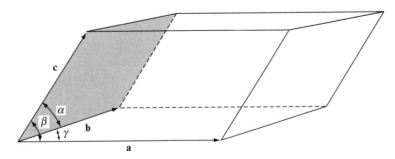

Fig. 3.9 *The general three-dimensional crystallographic unit cell.* Showing crystallographic axes **a**, **b** and **c**, and crystallographic angles α, β and γ. The volume V of the unit cell is given by $V = \mathbf{a} \cdot \mathbf{b} \wedge \mathbf{c}$.

Any lattice point $\mathbf{r}$ may be described in terms of $\mathbf{a}$, $\mathbf{b}$ and $\mathbf{c}$ as

$$\mathbf{r} = p\mathbf{a} + q\mathbf{b} + r\mathbf{c} \tag{3.3}$$

where p, q and r are integers only if $\mathbf{a}$, $\mathbf{b}$ and $\mathbf{c}$ define a primitive cell. If the conventional choice of unit cell is multiple, certain fractional values are allowed. Once the conventional crystallographic unit vectors have been stated, a point $\mathbf{r}$ may be represented by the ordered triplet (p, q, r).

In order to fulfil the space-filling requirement, there is once again a restriction on the number of possible lattices which may exist. Only 14 lattices are allowed, and these are called the *Bravais lattices*, as shown in Fig. 3.10. These fall into seven *crystal systems* depending on the overall symmetry of the unit cell. The systems are named as in Table 3.2.

As in the two-dimensional case, we choose primitive cells as the conventional crystallographic unit cells unless there is a specific advantage in choosing a multiple cell on the grounds that such a cell displays the overall symmetry of the lattice in a more meaningful way. The following conventional crystallographic unit cells are defined:

- The *primitive cell*, or *P cell*. This has a lattice point in each corner only. Since eight unit cells meet at any lattice point, any single unit cell has $1/8$ of a lattice point per corner. For a primitive cell, the number of lattice points associated with each cell is given by $1/8 \times 8 = 1$. Hence, as in two dimensions, a primitive cell is associated with a single lattice point, the representative point being $(0, 0, 0)$. If the unit cell is displaced slightly so as to move the corners away from lattice points, a single lattice point is contained within the volume of the cell.
- The *rhombohedral*, or *R cell*. This is really a special case of the primitive cell, in that only a single lattice point is associated with each unit cell. The R cell is so called because it is a parallelepiped formed by six rhombuses. The representative point is $(0, 0, 0)$.
- The *incentred*, or *I cell*. This has lattice points at each corner, plus one at the volume centre of the cell. There are $1/8 \times 8 + 1 = 2$ lattice points per cell, which may be represented as $(0, 0, 0)$ and $(1/2, 1/2, 1/2)$.
- The *face-centred*, or *F cell*. This cell has lattice points at each corner, plus one in the centre of each of its six faces. Each face-centred lattice point is shared by two unit cells, and so there are $8 \times 1/8 + 6 \times 1/2 = 4$ lattice points per unit cell. The representative points are $(0, 0, 0)$, $(1/2, 1/2, 0)$, $(1/2, 0, 1/2)$ and $(0, 1/2, 1/2)$.
- The *C-centred cell*. This cell has one lattice point at each corner and one in the centre of each face of one pair of opposite faces. This type of unit cell is possible only in the monoclinic and orthorhombic systems; the latter is such that the $\mathbf{c}$ vector is perpendicular to the plane defined by the $\mathbf{a}$ and $\mathbf{b}$ vectors. We conventionally take the lattice points to be centred in the planes perpendicular to the $\mathbf{c}$ direction (hence the terminology C-centred cell). There are $1/8 \times 8 + 2 \times 1/2 = 2$ lattice points per cell, which may be represented as $(0, 0, 0)$ and $(1/2, 1/2, 0)$.

The properties of the seven crystal classes and 14 Bravais lattices are shown in Table 3.3 and Fig. 3.10.

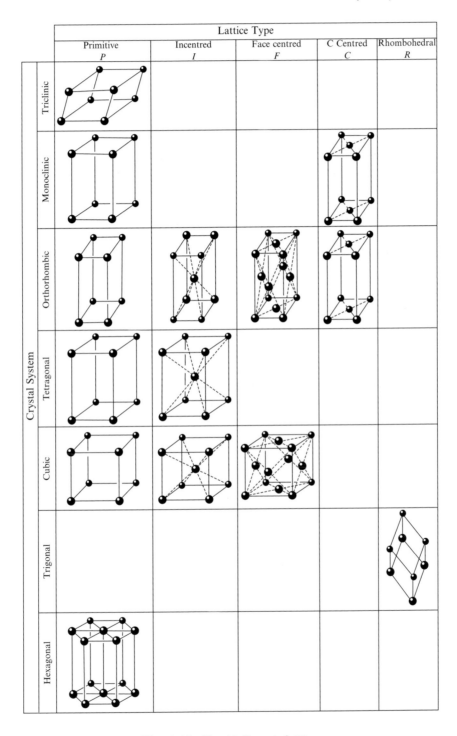

Fig. 3.10 *The 14 Bravais lattices.*

Table 3.2 The seven crystal systems.

Cubic	Tetragonal
Hexagonal	Triclinic
Monoclinic	Trigonal
Orthorhombic	

For any unit cell defined by the vectors **a**, **b** and **c**, the volume V of the unit cell is given by the scalar triple product

$$V = \mathbf{a} \cdot \mathbf{b} \wedge \mathbf{c} \tag{3.4}$$

3.4 Crystal directions

Any lattice point **r** may be represented by the equation

$$\mathbf{r} = p\mathbf{a} + q\mathbf{b} + r\mathbf{c} \tag{3.3}$$

in which p, q and r are integers if **a**, **b** and **c** define a primitive lattice, or may be fractional if the unit cell is multiple. A vector from the origin to the lattice point **r** defines a direction in space, and since this direction is related to the numbers p, q and r, we may define directions in the crystal based on their values. In Chapter 2, we saw that any pair of vectors **v** and n**v**, where n is a scalar, are parallel. In component form, this means that the vectors represented by (u, v, w) and (nu, nv, nw) are parallel, and are therefore in the same direction. Since n may be any scalar, there are clearly an infinite number of different representations of a given spatial direction depending on the value of n. When we choose to define a direction in space, we do so by quoting an ordered triplet of numbers such that they have no common factor. In this way, we avoid the redundancy introduced by the fact that the two vectors **v** and n**v** are in the same direction. As an example, consider the lattice point (6, 9, 3). A common factor is 3, implying that the vectors (6, 9, 3) and (2, 3, 1) are parallel. The direction in space corresponding to these vectors is written in square brackets, without using commas to separate the digits, as [2 3 1]. The ordered triplet is read as 'two, three, one', and not as 'two hundred and thirty-one'. If any fractions occur, then the fraction is first factorised out, leaving a set of three integers with no common factor: for example, the direction through the lattice point [$\frac{1}{2}$ $\frac{3}{2}$ $\frac{1}{2}$] is [1 3 1]. Should a negative number arise, then instead of writing for example [3 −2 −1], we write [3 $\bar{2}$ $\bar{1}$], read as 'three, bar two, bar one'.

3.5 Lattice planes

The regular geometry of the crystal lattice implies that there are many planes which may be drawn through sets of lattice points. The geometry of *lattice planes* is important in the overall description of crystal structure, and so this section presents the rules by which lattice planes are conventionally defined.

Table 3.3 The 14 Bravais lattices.

Crystal system	Lattice type	Conventional representation	Representative points
Triclinic	P	$a \neq b \neq c$ $\alpha \neq \beta \neq \gamma$ $\varphi > 90°$	$(0, 0, 0)$
Monoclinic	P	$a \neq b \neq c$ $\alpha = \gamma = 90° \neq \beta$	$(0, 0, 0)$
	C		$(0, 0, 0), (^1/_2, ^1/_2, 0)$
Orthorhombic	P	$a \neq b \neq c$ $\alpha = \beta = \gamma = 90°$	$(0, 0, 0)$
	C		$(0, 0, 0), (^1/_2, ^1/_2, 0)$
	I		$(0, 0, 0), (^1/_2, ^1/_2, ^1/_2)$
	F		$(0, 0, 0), (^1/_2, ^1/_2, 0)$ $(^1/_2, 0, ^1/_2), (0, ^1/_2, ^1/_2)$
Tetragonal	P	$a = b \neq c$ $\alpha = \beta = \gamma = 90°$	$(0, 0, 0)$
	I		$(0, 0, 0), (^1/_2, ^1/_2, ^1/_2)$
Cubic	P	$a = b = c$ $\alpha = \beta = \gamma = 90°$	$(0, 0, 0)$
	F		$(0, 0, 0), (^1/_2, ^1/_2, 0)$ $(^1/_2, 0, ^1/_2), (0, ^1/_2, ^1/_2)$
	I		$(0, 0, 0), (^1/_2, ^1/_2, ^1/_2)$
Trigonal	R	$a = b = c$ $\alpha = \beta = \gamma < 120° \neq 90°$	$(0, 0, 0)$
Hexagonal	P	$a = b \neq c$ $\alpha = \beta = 90°$ $\gamma = 120°$	$(0, 0, 0)$

As can be seen from Fig. 3.11, there are many planes which are parallel to any chosen plane. Since we have no way of defining any particular plane as being in some way 'special', when we talk of lattice planes we refer to that set of planes which are parallel to one another. A set of parallel planes is defined by three integers h, k and l, which do not share a common factor, known as *plane indices* or *Miller indices*. A triplet of Miller indices is conveniently represented as (hkl), using round brackets

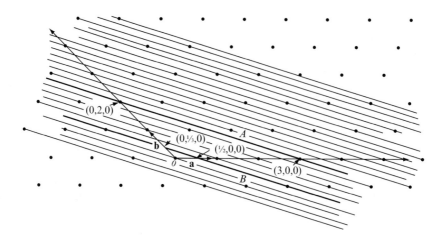

Fig. 3.11 *Lattice planes and Miller indices.* Shown, for clarity, is a section through a three-dimensional crystal, in which a set of planes is represented. The planes are perpendicular to the paper, and parallel to the **c** crystallographic axis. Note that each plane must contain lattice points, and that they are equally spaced. Some planes cut the crystallographic axes in lattice points, others do not, but in general the intercepts on the crystallographic axes are integers or simple fractions. Plane A has intercepts (3, 0, 0), (0, 2, 0) and (0, 0, ∞), and plane B, which is the closest to the origin for this set of planes, has intercepts ($1/2, 0, 0$), (0, $1/3$, 0) and (0, 0, ∞). The reciprocals of the intercepts determine the Miller indices (2 3 0) for this set of planes. For the plane closest to the origin, plane B, the reciprocals of $1/2$, $1/3$ and ∞ give the Miller indices (2 3 0) directly. For plane A, however, the reciprocals of 3, 2 and ∞ result in the fractions ($1/3\,1/2\,0$), which can be transformed into the integers (2 3 0) by multiplying by 6. It is not a coincidence that plane A is the sixth plane from the origin.

(as opposed to square brackets, which are used to represent directions), and without commas so as to distinguish the plane (hkl) from the vector (h, k, l). The significance of the integers h, k and l is that they are related to the intercepts of the planes with each of the three crystallographic axes such that the nth plane from the origin intercepts the **a** crystallographic axis at ($n/h, 0, 0$), the **b** crystallographic axis at ($0, n/k, 0$) and the **c** crystallographic axis at ($0, 0, n/l$), where n is any positive or negative integer.

As illustrated in Fig. 3.11, any set of planes with Miller indices (hkl) therefore generates a series of equally spaced intercepts along the **a** crystallographic axis at ($1/h, 0, 0$), ($2/h, 0, 0$), ($3/h, 0, 0$), ... , and similarly for the other two crystallographic axes. If the three ratios $n/h, n/k$ and n/l for any given value of n simultaneously become integers, then the corresponding intercepts ($n/h, 0, 0$), ($0, n/k, 0$) and ($0, 0, n/l$) correspond to lattice points; otherwise, a given plane will intercept the crystallographic axes between lattice points, but these intermediate intercepts can usually be expressed as simple fractions. The special case of $n = 0$ corresponds to a plane through the origin; $n = 1$ corresponds to the plane closest to the origin, with intercepts ($1/h, 0, 0$), ($0, 1/k, 0$) and ($0, 0, 1/l$); higher values of n correspond to planes progressively further from the origin.

Knowledge of the Miller indices (*hkl*) of any set of planes allows us to state that the intercepts of any plane in the set are of the form $(n/h, 0, 0)$, $(0, n/k, 0)$ and $(0, 0, n/l)$; conversely, knowledge of the intercepts of a plane $(\alpha, 0, 0)$, $(0, \beta, 0)$ and $(0, 0, \gamma)$ allows determination of the Miller indices by taking the reciprocals $(1/\alpha 1/\beta 1/\gamma)$. In general, $1/\alpha, 1/\beta$ and $1/\gamma$ are likely to be fractions, which can be converted into integral Miller indices by multiplying by an appropriate factor.

This is best illustrated by example, so consider plane A shown in bold in Fig. 3.11, which intercepts the **a** crystallographic axis at the lattice point (3, 0, 0) and the **b** crystallographic axis at the lattice point (0, 2, 0). Since the plane is parallel to the **c** crystallographic axis, the corresponding intercept is $(0, 0, \infty)$. The Miller indices of this plane are therefore $(1/3\,1/2\,0)$, which can be transformed into the simplest integral form by multiplying by 6 to give (2 3 0), indicating, as can be seen from Fig. 3.11, that plane A is the sixth from the origin. Similarly, the parallel plane B, also shown

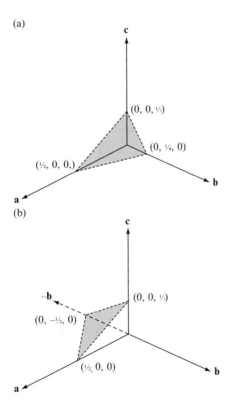

Fig. 3.12 *Negative Miller indices*. (a) shows a plane which intersects the **a** crystallographic axis at $(1/2, 0, 0)$, the **b** crystallographic axis at $(0, 1/4, 0)$ and the **c** crystallographic axis at $(0, 0, 1/3)$. This plane therefore has Miller indices (2 4 3). (b) shows a plane which intersects the **a** crystallographic axis at $(1/2, 0, 0)$, the negative direction of the **b** crystallographic axis at $(0, -1/2, 0)$ and the **c** crystallographic axis at $(0, 0, 1/3)$. This plane has Miller indices (2 $\bar{2}$ 3).

in bold, has intercepts $(1/2, 0, 0)$, $(0, 1/3, 0)$ and $(0, 0, \infty)$, from which we derive the same Miller indices (2 3 0) directly. This example illustrates a result we have already met: for any set of planes with Miller indices (hkl), the plane within that set which is closest to the origin has intercepts $(1/h, 0, 0)$, $(0, 1/k, 0)$ and $(0, 0, 1/l)$.

Note that since the crystallographic axes are right-handed by convention, any plane that intersects an axis in its negative direction is given a negative index, as shown in Fig. 3.12. And one final point about lattice planes and Miller indices: if, and only if, a crystal is cubic, then the direction $[hkl]$ is orthogonal to the set of planes with Miller indices (hkl); for all other crystal systems, there is no particular relationship between the Miller indices (hkl) and the direction $[hkl]$.

3.6 Symmetry operations and symmetry elements

As has already been stated, crystals and crystal lattices possess certain symmetry properties. In addition, the motif itself may be symmetrical, as, for instance, because of symmetry inherent in the molecules which comprise the motif. We shall now investigate the types of symmetry present, and the way in which a symmetry property acts on the crystal lattice or on an object associated with a lattice point.

When an operation is performed on a body with the result that the body assumes a new disposition in space which is totally indistinguishable from the original disposition, then the body is said to be *symmetrical*. Specifically, the body is symmetrical with respect to the operation which gives rise to that particular change in disposition. The operation in question is known as a *symmetry operation*. All symmetry operations involve a dispositional change in space, such as rotation, reflection in a plane and others which will be defined below. The geometrical locus about which the symmetry operation acts comprises a *symmetry element*. Thus for every symmetry operation, there is a corresponding symmetry element. The symmetry operation is an instruction for altering the spatial disposition of a body, and the symmetry element tells us where in the body we are to perform the appropriate symmetry operation. Consequently, any body has associated with it a set of symmetry elements about which we may perform certain symmetry operations. Since any symmetry element is a property of a body itself, we may classify all objects according to the symmetry elements which they possess.

We shall now define the various symmetry elements and operations. The effect of each operation will be described with reference to a general crystal lattice or to an object associated with a lattice point. It will be shown that the regular spacing of the lattice points in a three-dimensional array has the effect of restricting certain symmetry operations to particular cases, as, for example, demonstrated by the proper rotation operation.

Proper rotations

Symmetry operation: proper rotation
Symmetry element: proper rotation axis
Symbols: n; ❶ ▲ ◆ ⬢

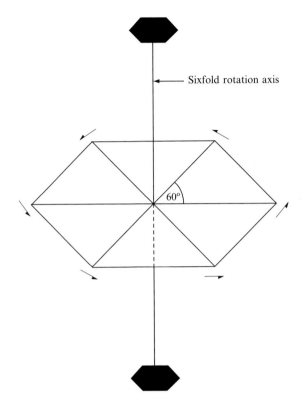

Fig. 3.13 *A sixfold proper rotation axis*; $\alpha = 360°/n = 60°$.

If an axis exists within a body such that a rotation of $\alpha°$ gives rise to a disposition indistinguishable from the original, the body is said to contain an *n-fold proper rotation axis*, where n is an integer given by $n = 360/\alpha$. A sixfold proper rotation axis is shown in Fig. 3.13. *n*-fold proper rotation axes are symbolised in writing by the numeral corresponding to n, whereas in diagrams they are represented by the appropriate shaded polygon: for example, a fourfold axis is drawn as ◆.

An important feature of the operation of proper rotation as applied to the crystal lattice is that the regular spacing of lattice points restricts the possible types of rotation. We may see this by consideration of Fig. 3.14.

A and A' represent two lattice points, which we shall assume are separated by the minimum possible lattice spacing for a particular lattice. When we deal with an infinite lattice, each point is in an exactly identical environment. This implies that if we have an n-fold proper rotation axis perpendicular to a given plane and through a particular lattice point, then since no lattice points are 'special', all lattice points in that plane must also contain n-fold proper rotation axes. A and A' are two such points in a lattice plane, and if we assume that there is an n-fold proper rotation axis through A, then there must also be one through A'. When we consider the axis through A, the point A' must obey the symmetry operation. Similarly, A must obey the symmetry operation

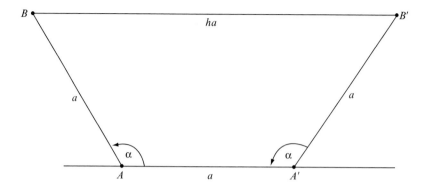

Fig. 3.14 *Rotations and lattices. A* and *A'* are two lattice points separated by the minimum lattice spacing *a*. Through *A* and *A'* pass *n*-fold rotational axes corresponding to the angle *α*. Rotating *A'* about *A* gives rise to *B*. *B'* has the position shown, since when a rotation occurs about *A'*, *B'* will fall on the point *A*. The requirement to form a regular lattice restricts the allowed values of *α*, and hence those of *n*. *n* can equal only 1, 2, 3, 4 and 6.

for the axis through *A'*. Let us suppose that the *n*-fold rotation axis corresponds to an angle of rotation *α°*. For the axis through *A*, there must be another lattice point *B* corresponding to a rotation of the point *A'* through *α°* in an anticlockwise direction. But there must also be a point *B'* which, when rotated through *α°* in an anticlockwise direction about the axis centred on *A'*, will fall onto the position of *A*. The existence of the *n*-fold symmetry axis therefore implies the presence of two more lattice points *B* and *B'* in the positions shown in Fig. 3.14.

From geometry, we see that *B* and *B'* are in a line parallel to *A* and *A'*. Hence the vector from *B* to *B'* is parallel to that from *A* to *A'*. Now, *A*, *A'*, *B* and *B'* are all lattice points, and one of the initial assumptions was that the points *A* and *A'* were separated by the minimum lattice spacing, *a*. Hence *B* and *B'* cannot be any closer than this spacing, and in fact must be separated by a distance which is an integral multiple of *a*:

$$BB' = ha$$

where *h* is an integer. Furthermore, the distances *AB* and *A'B'* must equal *a*, since they correspond to rotation operations. By geometry, the distances *AA'* and *BB'* are related by

$$AA' = BB' + 2a \cos \alpha$$

$$\therefore \ a = ha + 2a \cos \alpha$$

But if *h* is an integer, then $(1 - h)$ must also be an integer *p*:

$$2a \cos \alpha = pa$$

$$\cos \alpha = p/2$$

Hence the angle *α* must be such that its cosine is half-integral or integral.

$$\therefore \cos \alpha = 0, \pm^{1}/_{2}, \pm 1$$

and hence

$$\alpha = 0°, 60°, 90°, 120°, 180° \text{ or } 360°$$

For an n-fold axis,

$$n = 360/\alpha$$

$$\therefore \ n = 1, 2, 3, 4 \text{ or } 6$$

We see that the effect of the lattice periodicity is to restrict the possible proper rotation axes to be twofold (diad), threefold (triad), fourfold (tetrad) or sixfold (hexad), which are indicated by the symbols, ⬮ ▲ ◆ ⬢ respectively. The onefold axis rotates an object through a complete rotation and is redundant. This restriction of rotational symmetry is the reason that there are only a limited number of plane lattice types. If the reader refers back to Fig. 3.7, they will see that a pentagonal plane unit cell does not fill space, implying that no lattice may be built up using pentagons. This is equivalent to stating that no lattice may have a fivefold proper rotational symmetry axis, as proved above. It is interesting to note, in connection with Fig. 3.14, that with a sixfold (hexad) axis, points B and B' would be coincident, and this is an allowed operation.

Reflections

Symmetry operation: reflection
Symmetry element: mirror plane
Symbol: m

Any plane which is such that the disposition of an object to one side of the plane is the mirror image of that on the other is known as a *mirror plane*, as shown in Fig. 3.15. Mirror planes are symbolised by the letter m. Note that the object generated by a reflection is non-superposable on the original object. Objects which are mirror images of each other are known as *enantiomorphs* if macroscopic; microscopic mirror images are known as *enantiomers*. The reflection operation may be contrasted with the effect of a proper rotation, which generates a series of superposable images. As we shall see, one way of categorising symmetry elements depends on whether or not objects are related as enantiomorphic (or enantiomeric, as appropriate) pairs.

Inversions

Symmetry operation: inversion
Symmetry element: inversion centre, centre of symmetry
Symbol: i

The operation of *inversion*, symbolised as i, is a rotation through 180° followed by a reflection in a plane perpendicular to the rotation axis, as shown in Fig. 3.16.

If the object in Fig. 3.16 has the Cartesian coordinates (x, y, z), the operation of inversion is equivalent to transforming the object to $(-x, -y, -z)$. During this

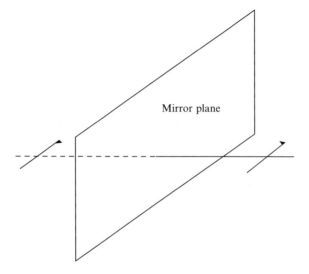

Fig. 3.15 *Reflection.* The mirror plane gives rise to an enantiomorphic pair.

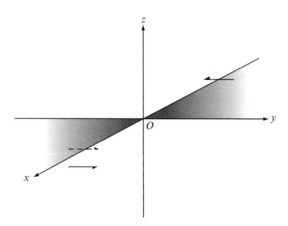

Fig. 3.16 *Inversion.* Rotate the arrow at the back through 180° about the z axis to the position shown by the dashed arrow and reflect it in the xy plane. An enantiomorphic pair is produced. The point O is the inversion centre, or centre of symmetry.

operation, only a single point, known as the *inversion centre* or *centre of symmetry*, remains unaltered. This point is the intersection of the rotation axis with the mirror plane. In view of the reflection operation, objects are formed in enantiomorphic (or enantiomeric) pairs. Structures with an inversion centre are said to be *centrosymmetric*.

Screw rotations

Symmetry operation: screw rotation
Symmetry element: screw axis
Symbols: n_t; ! ▲ ◆ ↯

The *screw rotation* is a combination of a rotation and a translation parallel to the axis of rotation. The rotation is through an angle $\alpha = 360°/n$ and, as in the case of proper rotations, the crystal lattice restricts the values of n to 2, 3, 4 or 6 ($n = 1$ is redundant). The translation, symbolised by an integer t, is parallel to the axis of rotation, and corresponds to a fraction t/n of the lattice spacing along the direction of the rotation axis. Since a total rotation of 360° generates a lattice point, as shown in Fig. 3.17, t must be an integer satisfying the inequality

$$1 < t < n$$

In writing, a *screw axis* is denoted by n_t, and in diagrams, the symbols listed in Table 3.4 are used.

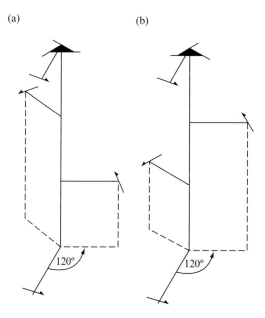

(a) (b)

Fig. 3.17 *Threefold screw axes.* (a) is the screw axis 3_1. A rotation of 120° is combined with a translation of one-third of the lattice spacing along the axis. (b) is a 3_2 screw axis. This time, a rotation of 120° is associated with a translation of two-thirds of the lattice spacing along the axis. The intermediate object is generated by a second rotation and translation. Note that after a total rotation of 360°, the object is back to the original position, but translated. This translation must be one lattice spacing, and hence intermediate translations are fractions of the value of n corresponding to the n-fold rotation axis.

Table 3.4 Symbols for screw axes.

Written screw axis symbol	Drawn screw axis symbol
2_1	
$3_1\ 3_2$	
$4_1\ 4_2\ 4_3$	
$6_1\ 6_2\ 6_3\ 6_4\ 6_5$	

Glides

> Symmetry operation: glide
> Symmetry element: glide plane
> Symbol: g

A *glide* is a combination of a reflection and a translation parallel to the plane of the mirror plane. If an object such as that shown in Fig. 3.18 is subject to a glide, it generates an enantiomorph (or enantiomer) as shown. This itself acts as an object for the glide operation, and regenerates an object superposable on, and on the same side of the mirror plane as, the original object. These two must be spaced by a lattice spacing, and so the glide operation must be such that the translation corresponds to one-half of the lattice spacing in the appropriate direction.

Improper rotations

> Symmetry operation: improper rotation
> Symmetry element: improper rotation axis
> Symbols: $\bar{n}$ (rotoinversion) and $\tilde{n}$ (rotoreflection)

An *improper rotation* is the combination of a rotation with either an inversion, or a reflection in the plane perpendicular to the rotation axis. The former is known as a *rotoinversion*, symbolised by $\bar{n}$, and the latter is called *rotoreflection*, denoted by $\tilde{n}$. Once again, the rotation is through an angle $\alpha = 360°/n$, and the values of n are restricted to (1), 2, 3, 4 and 6. In Fig. 3.19 are a threefold rotoinversion and a threefold rotoreflection.

Rotoinversions and rotoreflections are very similar, and it can be proved that any rotoinversion may be considered as a rotoreflection through an appropriate angle. This is easily demonstrated by the threefold rotoinversion of Fig. 3.19, which may be thought of as a sixfold rotoreflection; likewise, a threefold rotoreflection is equivalent to a sixfold rotoinversion. For this reason, rotoinversions and rotoreflections are discussed together. In general, when specifying a particular improper rotation, we usually refer to the rotoinversion. Note that an inversion centre corresponds to a onefold rotoinversion axis 1, and that a mirror plane is equivalent to a twofold rotoinversion axis 2. Hence there are only three unique rotoinversion operations, denoted by 3, 4 and 6.

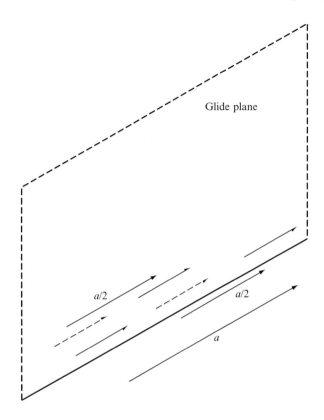

Glide plane

a/2

a/2

a

Fig. 3.18 *The glide.* A glide combines a reflection with a translation parallel to the mirror plane, known as the glide plane. Two successive glides result in an object in a position equivalent to the original object. Hence the translation for each glide must be half the associated lattice spacing.

We have now defined the six different symmetry operations and the corresponding symmetry elements. There are various ways of categorising these operations. One way is to classify them into *single operations* and *double operations* as follows.

Single operations:

  proper rotation
  reflection

Double operations:

  inversion = rotation + reflection
  screw rotation = rotation + translation
  glide = reflection + translation
  improper rotation = rotation + inversion

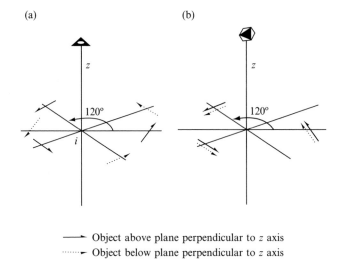

(a)   (b)

——▶ Object above plane perpendicular to *z* axis
⋯⋯▶ Object below plane perpendicular to *z* axis

Fig. 3.19 *Improper rotations.* (a) shows a threefold rotoinversion axis. A rotation by 120°
about the *z* axis is followed by inversion through the inversion centre *i*. (b) shows a threefold
rotoreflection axis. Rotation by 120° about the *z* axis is followed by reflection in the plane
perpendicular to this axis. Note that (a) is equivalent to a sixfold rotoreflection, and that (b)
is equivalent to a sixfold rotoinversion.

A second method is to associate those operations which produce solely superposable
objects, called *operations of the first kind,* and those which produce enantiomorphic
(or enantiomeric) pairs, termed *operations of the second kind.*

Operations of the first kind:

 proper rotation
 screw rotation

Operations of the second kind:

 reflection
 inversion
 glide
 improper rotation

3.7   Point groups and Laue groups

A third way of classifying symmetry groups is according to whether or not the operation involves a translation. Those which do not involve translation are proper rotation,
reflection, inversion and improper rotation. Those which do include a translation are
glide and screw rotation. The significance of this categorisation concerns the fact that
any set of symmetry operations which does not invoke translation may be thought
of as acting at a point in space and, importantly, that point in space is unchanged
by the operation. Hence any effect which is a combination of those operations which

are not associated with translations may be deduced by considering the effect of each symmetry operation acting in turn on one single point.

It has been proved that in three dimensions there are 32 different operations corresponding to combinations of proper rotations, reflections, inversions and improper rotations. These 32 arrangements are called the 32 *point groups*. This nomenclature is used to emphasise that we are concerned with only those symmetry operations which act on a point, and the word 'group' is used since the mathematical analysis of symmetry operations is encompassed by a branch of mathematics known as *group theory*. In two dimensions, it may be shown that there are 10 different point groups.

Since a crystal lattice is an array of points, it is useful to know to which point group a structure belongs, since this will give us information about the arrangement of the motif about each lattice point. Since there are 32 point groups, we speak of the 32 *crystal classes*. Every crystalline structure must belong to one of these classes. Very often, the point group manifests itself in the physical shape of a crystal, for example as shown by the relative orientations of the crystal faces. This is also true for crystals whose symmetry involves translations, since microscopic translations of a fraction of the unit cell will not be detectable in the gross appearance of the crystal, but the other symmetry operations can be manifest in the crystal shape.

The point groups are given symbols indicating their symmetry. For example, a twofold axis in a monoclinic crystal is indicated by the symbol '2'. A mirror plane orthogonal to the twofold axis is indicated by the symbol '$2/m$'. In orthorhombic crystals, the presence of twofold axes parallel to the three unit cell axes would be indicated by the symbol '222', and likewise three orthogonal mirror planes by the

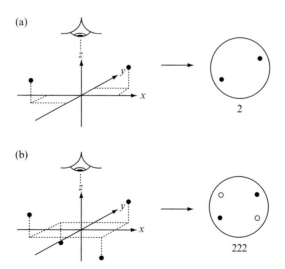

Fig. 3.20 *Point group symmetry represented using stereograms.* The reader has to imagine viewing symmetry-equivalent points, usually along the axis of highest symmetry. (a) shows the effect of a twofold axis and its associated stereogram (point group 2). (b) shows three orthogonal twofold axes (point group 222). Points below the horizontal plane in the stereogram are shown as open circles.

symbol '*mmm*'. In tetragonal crystals, the presence of a fourfold axis parallel to z is indicated by the symbol '4', and a mirror plane perpendicular to this axis would be indicated by the symbol '4/*m*'. A tetragonal crystal with twofold axes perpendicular to the tetrad axis would be indicated by the symbol '422', and the presence of mirror planes perpendicular to these axes would generate the point group '4/*mmm*'. A hexagonal crystal with only a sixfold axis would belong to the point group '6', and one with a mirror plane perpendicular to the hexad axis would belong to the point group '6/*m*'. A hexagonal crystal with twofold axes perpendicular to the hexad would belong to the point group '622', and addition of a mirror plane perpendicular to the hexad would give the point group '6/*mmm*'.

Point groups can be elegantly represented in two dimensions by drawings known as stereograms, which can be interpreted as shown in Fig. 3.20. With a stereogram, the viewer has to imagine looking down on a group of symmetry-related points along the axis of highest symmetry (usually z). Points above the horizontal plane (usually the x, y plane) are drawn as solid circles in the stereogram; points below the horizontal plane are drawn as open circles. If the horizontal plane contains a mirror, then for every point above the plane, there will be a symmetry-related one exactly below it. This is indicated in stereograms by an open circle, filled with a solid circle. A selection of stereograms for the commonly occurring point groups is given in Fig. 3.21. Stereograms were originally used to indicate the directions of the normals to crystal faces. These

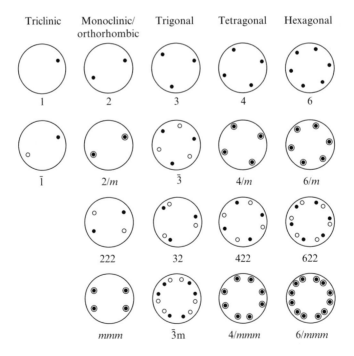

Fig. 3.21 *A selection of point groups and Laue groups commonly encountered with protein crystals and their diffraction patterns.* For a complete description of the 32 point groups and 11 Laue groups, the reader is referred to the *International Tables for Crystallography*, Volume A.

Table 3.5 The relationship between the 32 point groups and the 11 Laue groups.

Crystal class	Possible point groups, with the corresponding Laue group last
Triclinic	1, $\bar{1}$
Monoclinic	2, m, 2/m
Orthorhombic	222, $mm2$, mmm
Trigonal	3, $\bar{3}$
	32, 3m, $\bar{3}m$
Tetragonal	4, $\bar{4}$, 4/m
	422, 4mm, $\bar{4}2m$, 4/mmm
Hexagonal	6, $\bar{6}$, 6/m
	622, 6mm, $\bar{6}m2$, 6/mmm
Cubic	23, $m3$
	432, $\bar{4}3m$, $m3m$

were projected outwards onto the surface of a sphere effectively surrounding the crystal and then onto the equatorial plane towards the opposite pole. All the points on this plane are within the equatorial, or primitive, circle created where the sphere intersects the horizontal plane. Accordingly, the symmetry-related points in a stereogram are drawn enclosed within an outer circle.

Adding an inversion centre to each of the 32 point groups reduces the number of distinct groups down to 11, which are known as *Laue groups*. For reference, the relationship between the 32 point groups and the 11 Laue groups is shown in Table 3.5.

Laue groups are important because they describe the symmetry of the diffraction pattern and something known as the 'Patterson function', which is key to structure determination and will be covered in more detail in a later chapter.

3.8 Space groups

When we consider the effect of the translational symmetry of the lattice and the two operations which include translation, namely glides and screw rotations, we must enquire as to how these operations combine with each of the 32 point groups. The effect of the translations is to generate motifs in all space, and so we may refer to the types of patterns built up as *space groups*. It has been shown that there are 17 plane space groups and 230 three-dimensional space groups. Any pattern whatsoever in two dimensions must correspond to one of these 17 plane groups and, likewise, any regular pattern in three dimensions must fall into one of the 230 space groups.

The fact that the amino acids making up proteins are chiral (essentially all naturally occurring amino acids in proteins are L-enantiomers, as opposed to D-amino acids) has an important consequence. It means that proteins can crystallise only in one of the enantiomorphic space groups—namely, those space groups which do not involve mirror or inversion symmetry elements. This reduces the number of space

Table 3.6 The 65 enantiomorphic space groups in which proteins can crystallise.

Crystal system	Lattice	Minimum symmetry of unit cell	Unit cellparameters	Laue group	Space groups
Triclinic	P	None	$a \neq b \neq c$ $\alpha \neq \beta \neq \gamma$	$\bar{1}$	$P1$
Monoclinic	P	Twofold axis parallel to **b**	$a \neq b \neq c$ $\alpha = \gamma = 90°$ $\beta \neq 90°$	$2/m$	$P2, P2_1$
	C				$C2$
Orthorhombic	P	Three orthogonal twofold axes	$a \neq b \neq c$ $\alpha = \beta = \gamma = 90°$	mmm	$P222, P2_12_12_1, P222_1, P2_12_12$
	C				$C222, C222_1$
	I				$[I222, I2_12_12_1]$
	F				$F222$
Tetragonal	P	Fourfold axis parallel to **c**	$a = b \neq c$ $\alpha = \beta = \gamma = 90°$	$4/m$	$P4, (P4_1, P4_3), P4_2$
	I				$I4, I4_1$
				$4/mmm$	$P422, (P4_122, P4_322), P4_222,$ $P42_12, (P4_12_12, P4_32_12), P4_22_12$ $I422, I4_122$
Trigonal	P	Threefold axis parallel to **c**	$a = b \neq c$ $\alpha = \beta = 90°$ $\gamma = 120°$	$\bar{3}$	$P3, (P3_1, P3_2)$
				$\bar{3}m$	$[P321, P312]$ $[(P3_121, P3_221), (P3_112, P3_212)]$

Rhombohedral	R	Threefold axis parallel to diagonal	$a = b = c$ $\alpha = \beta = \gamma \neq 90°$	$\bar{3}$ $\bar{3}m$	R3 R32
Hexagonal	P	Sixfold axis parallel to **c**	$a = b \neq c$ $\alpha = \beta = 90°$ $\gamma = 120°$	$6/m$ $6/mmm$	$P6$, $(P6_1, P6_5)$, $P6_3$, $(P6_2, P6_4)$ $P622$, $(P6_122, P6_522)$, $P6_322$, $(P6_222, P6_422)$
Cubic	P I F	Threefold axes along cube diagonals	$a = b = c$ $\alpha = \beta = \gamma = 90°$	$m3$	$P23$, $P2_13$ $[I23, I2_13]$ $F23$
	P I F			$m3m$	$P432$, $(P4_132, P4_332)$, $P4_222$ $I432$, $I4_132$ $F432$, $F4_132$

Pairs of space groups in round brackets ( ) are enantiomorphic, i.e. mirror images. Space groups enclosed in square and round brackets cannot be distinguished by the diffraction pattern.
Note: the $\neq$ symbol should be interpreted as meaning 'is not constrained by symmetry to equal'.
Adapted from Cantor, C. R. and Schimmel, P. R., *Biophysical Chemistry*, Part II: *Techniques for the Study of Biological Structure and Function*. Freeman, San Francisco, 1980.

groups that proteins can crystallise in down to 65. The 65 space groups available to proteins are shown in Table 3.6. In practice, the space groups of most protein crystals can be determined without ambiguity from the diffraction data alone.

There is a nomenclature and symbolism for each of the space groups. For further information, the reader is referred to a more advanced text on geometrical crystallography, especially Volume A of the *International Tables for Crystallography*, which contains full details on all aspects of crystal symmetry. A very common space group for proteins is $P2_1$, which, by convention, signifies a 2_1 screw axis along the y axis. Using the y axis as the reference is unusual and should be remembered, since in most other space groups it is the z axis which has the highest symmetry. Another common crystal system for proteins is the orthorhombic system, and the associated primitive space groups are symbolised as $P222$, $P2_12_12$, $P222_1$ and $P2_12_12_1$. In this nomenclature, the numbers refer to the symmetry elements parallel to the crystallographic x, y and z axes, respectively: so, for example, $P2_12_12$ has 2_1 screw axes parallel to x and y and a proper twofold axis parallel to z. Tetragonal space groups are defined by a representation such as $P4_32_12$, in which the first symbol refers to the tetrad, which is always parallel to z; the second symbol refers to the x or y axis (note that both are equivalent in the tetragonal system); and the third symbol refers to the diagonal direction at 45° to x and y. Similar conventions apply for the trigonal, hexagonal and cubic systems.

Figure 3.22 shows the space group diagram for $P2_12_12_1$, which is a very commonly encountered orthorhombic space group for protein crystals. In the left-hand diagram, the symmetry-related points are shown as open circles with + or - symbols to indicate their position above or below the horizontal plane, which in this case is the (x, y) plane. The origin is the upper left-hand corner, and the x axis points downwards, with the y axis pointing towards the right. The symbols '$1/2+$' and '$1/2-$' indicate the effects of the three half-unit-cell screw translations parallel to x, y and z in this space group. The positions of the screw axes are indicated with the standard symbolism

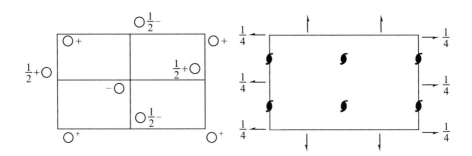

Fig. 3.22 *A space group diagram for $P2_12_12_1$. In the diagram on the left, the open circles indicate the general equivalent positions, with + and - symbols indicating whether they are above or below the plane of the page. The screw-axis shifts of half a unit cell dimension towards the viewer (z axis) are also indicated. In the diagram on the right, the positions of the 2_1 screw axes of this space group are shown. Reproduced with permission from the International Tables for X-ray Crystallography, Volume I.*

in the right-hand diagram. The '$1/4$' adjacent to the arrow symbols indicates that the screw axis occurs one quarter of a unit cell translation along the z axis towards the viewer. Note that the three screw axes do not intersect: they are parallel to the main crystallographic axes, but not coincident with them. Their effect is to generate from the point (x, y, z) three other equivalent points with fractional coordinates $(1/2 - x, -y, 1/2 + z)$, $(1/2 + x, 1/2 - y, -z)$ and $(-x, 1/2 + y, 1/2 - z)$. These are referred to as the *general equivalent positions* (GEPs), and each space group has a characteristic set of these, as given in the *International Tables*.

Summary

A *crystal structure* may be expressed by the conceptual 'equation'

$$\text{crystal structure} = \text{lattice} * \text{motif.}$$

The *motif* is that structural unit which is repeated regularly in space.

The *lattice* is a conceptual array of points in space which serves to define the geometrical relationships between the motifs in a structure.

Two-dimensional *plane lattices* and three-dimensional *space lattices* are defined in terms of *crystallographic unit cells* and *crystallographic unit vectors*, which are chosen according to the convention described below.

The conventional *crystallographic unit cell* is a parallelepiped (or parallelogram) defined by the space (or plane) lattice of a crystal which serves to display the symmetry of the lattice in a convenient manner.

The conventional *crystallographic unit vectors* are a set of three non-coplanar (or, in two dimensions, two non-collinear) lattice vectors which define the conventional crystallographic unit cell.

The directions defined by the conventional crystallographic unit vectors specify the *crystallographic axes*.

The volume V of the crystallographic unit cell defined by the crystallographic unit vectors $\mathbf{a}$, $\mathbf{b}$ and $\mathbf{c}$ is given by

$$V = \mathbf{a} \cdot \mathbf{b} \wedge \mathbf{c}$$

Any point $\mathbf{r}$ in the lattice may be defined with respect to the crystallographic unit vectors as

$$\mathbf{r} = p\mathbf{a} + q\mathbf{b} + r\mathbf{c} \tag{3.3}$$

If the crystallographic unit vectors have previously been defined, then a lattice point may be defined by the ordered triplet (p, q, r), where p, q, and r are the positive or negative values appropriate to equation (3.3).

Conventional crystallographic unit cells associated with a single lattice point are known as *primitive*; those associated with more than one are known as *multiple*. If the unit cell is primitive, then p, q and r are necessarily integers for all lattice points; for multiple cells, certain fractional values may be allowed.

The symmetry of lattices is such that only five plane lattices exist, as defined in Table 3.1. The 14 space lattices, the *Bravais lattices*, are defined in Table 3.3.

Directions in crystals are symbolised as $[pqr]$, in which p, q and r are integers which have no common factor, and correspond to a vector pointing in the appropriate direction.

Sets of parallel *lattice planes* in crystals are represented by the *Miller indices* (hkl), in which h, k and l are integers which have no common factor. The nth member of the (hkl) set intercepts the three crystallographic axes $\mathbf{a}$, $\mathbf{b}$ and $\mathbf{c}$ at $(n/h, 0, 0)$, $(0, n/k, 0)$ and $(0, 0, n/l)$, respectively.

Crystal structures as a whole may be characterised by a set of *symmetry elements* about which we may perform certain *symmetry operations*.

When an operation is performed on a body with the result that the body assumes a new disposition in space which is totally indistinguishable from the original disposition, then that operation is known as a *symmetry operation*.

Those geometrical loci which are unchanged, or invariant, with respect to a particular symmetry operation comprise a *symmetry element*.

Those symmetry operations which do not involve translation may all act at a single point. The possible combinations of these symmetry operations in three dimensions give rise to the 32 *point groups*, corresponding to the 32 *crystal classes*. When we consider the effect of the symmetry operations involving translation on each of the 32 point groups, we derive the 230 *space groups*. In two dimensions, there are 10 point groups and 17 plane groups.

Crystals and crystal structures are systematised as follows:

(a) 7 *crystal systems*, based on the geometrical structure of the conventional crystallographic unit cell.
(b) 14 *Bravais lattices*, based on the symmetry properties of the lattice.
(c) 32 *crystal classes*, based on the 32 point groups, which show that there are only 32 ways of associating a motif around a lattice point.
(d) 230 *space groups*, based on the fact that there are only 230 ways in which to generate a three-dimensional regular pattern from a motif associated with a lattice.

Of the 230 possible space groups, 65 enantiomorphic space groups are available to proteins.

Bibliography

Bishop, A. C. *An Outline of Crystal Morphology*. Hutchinson, London, 1967. A text on the external geometry of crystals.

Buerger, M. J. *Elementary Crystallography*. Wiley, New York, 1963. A very detailed discussion of crystal geometry and crystal symmetry.

Cantor, C. R. and Schimmel, P. R. *Biophysical Chemistry*, Part II: *Techniques for the Study of Biological Structure and Function*. Freeman, San Francisco, 1980.

Phillips, F. C. *An Introduction to Crystallography*. Longman, London, 1971. A very suitable introduction to the science of crystallography.

Stout, G. H. and Jensen, L. H. *X-ray Structure Determination: A Practical Guide*, 2nd edn. Wiley, New York, 1989.

International Tables for Crystallography. Volume A: *Space-Group Symmetry*. International Union of Crystallography, Chester, UK, 2005.

4
Waves and electromagnetic radiation

In this chapter, we shall leave the study of crystals and crystal structure and turn to the mathematical description of waves.

4.1 Mathematical functions

Before starting on our analysis of wave motion, it is necessary to introduce and explain an aspect of mathematical terminology. When a scientist performs an experiment, the ultimate aim is to obtain a mathematical expression which describes a phenomenon as fully as possible. The way in which we go about finding this expression is by performing experiments to determine what happens to a specific phenomenon when we change various conditions. If we find that on changing a particular condition the effect on the phenomenon of interest is consistently reproducible, then we have reason to infer a causal relationship between the condition which was changed and the phenomenon of interest. For example, a physicist may be interested in the flow of heat down an unlagged metal bar. If we vary the temperature difference between the ends of the bar, we will observe that the flow of heat will vary in a manner consistent with the variation of the temperature difference, and in a perfectly reproducible manner. We therefore infer that the flow of heat along the bar depends on the temperature difference between the ends of the bar, and possibly on other factors which we have not as yet investigated. We now know that the mathematical equation we seek must contain a term which involves an expression for the temperature difference between the ends of the bar.

In general, a specific phenomenon may be found to depend on various quantities called *variable parameters*, or simply *variables* or *parameters*. Exhaustive experiment will determine all the relevant variables which affect the given phenomenon. The scientist may then write a list of the relevant variables, and he or she will know that the equation will contain algebraic symbols representing each variable. In the heat flow example, the relevant variables are:

(a) the temperature difference between the ends of the bar, $\Delta\theta$;
(b) the cross-sectional area of the bar, A;
(c) the length of the bar, l;
(d) the thermal conductivity of the material of the bar, k.

Other variables such as the mass of the bar, the air pressure and the colour of the experimenter's hair will be found not to affect the heat flow, and need not be considered in the equation for heat flow.

Having determined which parameters are relevant to the system, the scientist then knows that the equation for heat flow will contain certain algebraic symbols representing the different variables. We may then say that the phenomenon in question is a *function* of the relevant variables. This means that the equation for the phenomenon contains information about the relevant variables. For the heat flow case, we know that the heat flow J is a function of the variables $\Delta\theta, A, l$ and k defined above. Mathematically, we express this by writing an expression such as $J(\Delta\theta, A, l, k)$. This tells us that the heat flow J depends on the variables contained in parentheses, and on those variables only. As yet, the explicit relationship between the variables is unknown, but further experiment will elucidate this. As it happens, for heat flow the correct relationship is

$$J(\Delta\theta, A, l, k) = \frac{kA\,\Delta\theta}{l}$$

The representation of mathematical functions therefore enables us to see at a glance which are those variables relevant to a given phenomenon, and hence the notation $f(x, y, z, \ldots)$ will be met often, meaning that some phenomenon represented by f depends on those variables represented by $x, y, z, \ldots$ in an as yet unknown manner.

4.2   What is a wave?

Our analysis of the behaviour of waves will start with an investigation of waves generated on the surface of water. Our system is a long, shallow tank of water, into which dips a rule. If the rule is parallel to the shorter dimension of the tank, on causing the rule to oscillate regularly in a vertical plane, a disturbance is propagated over the surface of the water. An instantaneous picture of the surface of the water will be rather like that shown in Fig. 4.1. The surface of the water has a characteristic

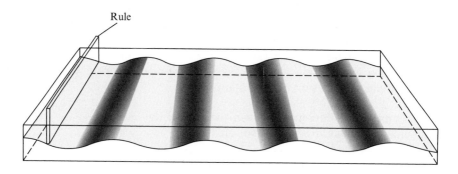

Fig. 4.1 *Water waves.* If the rule executes regular oscillations in a vertical plane, a wave will be propagated on the surface of the water in the shallow tank.

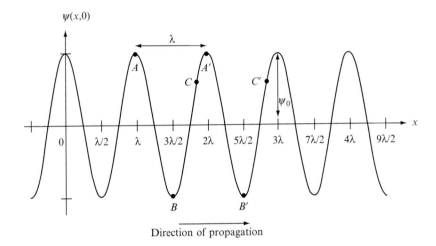

Fig. 4.2 *The graph of wave amplitude against distance at a given time.* The equilibrium level is the x axis, and the height above (or depth below) the x axis is the wave amplitude. Wave crests are at A and A'; wave troughs at B and B'. The wavelength λ corresponds to the distances AA', BB', CC' and so on. The equation of this graph is $\psi(x,0) = \psi_0 \cos(2\pi x/\lambda)$.

sinusoidal profile, represented by the cross-sectional view in Fig. 4.2. The disturbance propagated through the water is known as a *wave*.

With reference to the instantaneous cross-sectional view shown in Fig. 4.2, we may specify a number of parameters which define the wave:

The *direction of propagation* is the direction in which the waves are travelling.
The *amplitude* of the wave is the height of a point on the wave above or below the equilibrium position.
A *wave crest* or *wave peak* is a position on the wave where the wave amplitude is a maximum.
A *wave trough* is a position on the wave where the wave amplitude is a minimum.
The *wavelength* of the wave is the distance between two successive wave crests. Since the waveform is perfectly regular, this distance is also that between two wave troughs, or any two successive points at which both the amplitude and the slope (the first derivative) of the curve at that point have the same values. Note that in general, two successive points with the same amplitude are not separated by a wavelength, for they are on opposite sides of the same wave peak or trough.
A *wavefront* is a line or a surface through one wave crest, perpendicular to the direction of propagation.
A *ray* is a line drawn perpendicular to a wavefront showing the direction of propagation.

Figure 4.2 therefore represents how the amplitude of the wave varies with distance at a given time, and may be thought of as a graph in which the vertical axis represents wave amplitude, and the horizontal axis distance, at a given time.

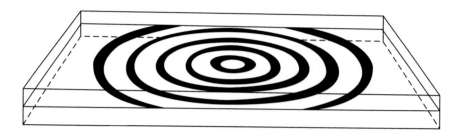

Fig. 4.3 *Circular water waves.* As formed by a point source.

For the waves shown in Fig. 4.1 and 4.2, the direction of propagation is from left to right, and the wavefronts are a series of parallel lines. This type of wave is called a *plane wave.* If, instead of using a rule as the source of the waves, we had dropped a stone into the centre of the tank, the wave pattern would have been a series of waves spreading outwards from the point of impact of the stone with the water surface. In this case the direction of propagation is radial, and the wavefronts are a set of concentric circles. This form of wave motion is called a *circular wave,* as shown in Fig. 4.3.

Let us now consider the motion of a small piece of paper which floats on the surface of the water in the tank. If plane waves are propagated through the water, experiment shows that the piece of paper will stay in the same position relative to the surface of the water, but will bob up and down in a vertical plane. The fact that the paper moves only in the vertical direction does not lack significance. We know that the direction of propagation of the waves is in the horizontal plane, yet this experiment demonstrates that the paper moves not in a direction parallel to that of propagation, but rather in a direction perpendicular, or transverse, to that of propagation. If the piece of paper were made very small, it would still behave in the same manner. If the piece of paper were so small that it had only molecular dimensions, the motion of the paper would be the same as that of the water molecules. The paper would move solely in the direction transverse to that of wave propagation, and this tells us that when a wave is propagated through water, then the individual water molecules move in a direction transverse to the direction of propagation. Any wave motion in which the local motion, which in this case is that of the water molecules, is transverse to the direction of propagation is called a *transverse wave.* Other examples of transverse waves are waves on a string, and electromagnetic waves such as light and X-rays, as we shall see later in this chapter. When the local motion in a wave is parallel to the direction of propagation, we speak of *longitudinal waves.* Sound waves are longitudinal, but since this book is primarily concerned with the behaviour of X-rays, all subsequent discussion will be directed towards the properties of transverse waves.

As the paper moves up and down on the surface of the water, we may measure the vertical height of the paper above or below the equilibrium position, and this will enable us to determine how the amplitude of a wave varies with time at a given point on the wave. If a series of these measurements is taken, we may plot a graph of wave amplitude against time, as shown in Fig. 4.4.

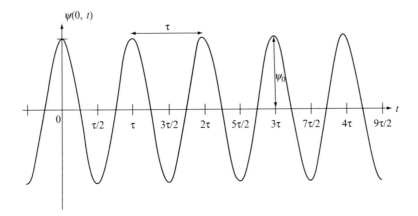

Fig. 4.4 *The graph of wave amplitude against time at a given distance.* The periodic time is the time interval between two successive amplitude peaks. The equation of this graph is $\psi(0, t) = \psi_0 \cos(2\pi t/\tau)$.

We see that this graph, like that of amplitude against distance at any given time, is also sinusoidal:

The *periodic time* τ of a wave motion is the time interval corresponding to two successive amplitude peaks of the wave at any position.

The *frequency* ν of a wave motion is the reciprocal of the periodic time τ.

For a wave of periodic time τ, the number of wave oscillations occurring per unit time is simply $1/\tau$. This is just the frequency ν, and so we have an alternative way of interpreting the frequency ν of a wave:

The *frequency* ν of a wave motion is the number of wave cycles executed per unit time.

4.3   The mathematical description of a wave

Let us reconsider the graphs shown in Fig. 4.2 and 4.4. The former shows how the amplitude of a plane wave varies with distance at a given time, whereas the latter indicates the variation of wave amplitude with time at a given distance. We see that the wave amplitude therefore varies with both time and distance, and so any mathematical description of a wave must contain information on both time and distance. In this section, we shall derive intuitively an expression which is the mathematical description of a plane wave. In the next section, we shall show how to derive the same result by more rigorous mathematics.

We shall define the direction of propagation as the x direction. Since the waves are plane, the wave amplitude is constant at any given value of x, and so the only relevant spatial parameter is therefore the distance x from an arbitrary origin. The system is therefore essentially one-dimensional. We shall later extend the argument to three dimensions.

Let us consider first the variation of the wave amplitude with distance x at a given time, which we may arbitrarily take as the time $t = 0$. With reference to Fig. 4.2, we define the following symbols:

Let the wave amplitude at a given distance x from the source at time $t = 0$ be $\psi(x, 0)$.
Let the maximum amplitude be ψ_0.
Let the wavelength be λ.

The graph in Fig. 4.2 shows that the amplitude $\psi(x, 0)$ in general varies sinusoidally with the distance x, implying that $\psi(x, 0)$ may be represented by either a sine or a cosine function. Since there is a maximum when $x = 0$, we choose the cosine, and so

$$\psi(x, 0) = \psi_0 \cos f(x, \lambda) \tag{4.1}$$

Equation (4.1) is a general expression for a sinusoid with maximum amplitude ψ_0; $f(x, \lambda)$ is some as yet unknown function of the variables x and λ. To find this function, we argue as follows. Firstly, $f(x, \lambda)$ must be dimensionless. The reason for this is that the cosine function is defined only for pure dimensionless numbers. Since both x and λ have the dimensions of length, the only general function of both x and λ which is dimensionless is

$$f(x, \lambda) = \alpha \left(\frac{x}{\lambda}\right)^n$$

in which α and n are numerical constants. Since the graph of Fig. 4.2 is perfectly regular in x, the value of n must be unity. If the reader does not believe this, look at Fig. 4.5, which shows the graph of $\cos \theta^2$. We now have

$$\psi(x, 0) = \psi_0 \cos \left(\alpha \frac{x}{\lambda}\right) \tag{4.2}$$

To determine α, we use the fact that the amplitude $\psi(x, 0)$ is a regularly periodic function, for we know that for any point x, the amplitudes at the points x and $x + \lambda$ are the same:

$$\psi(x, 0) = \psi(x + \lambda, 0)$$

But

$$\psi(x + \lambda, 0) = \psi_0 \cos \left(\alpha \frac{x + \lambda}{\lambda}\right)$$

$$= \psi_0 \cos \left(\alpha \frac{x}{\lambda} + \alpha\right)$$

$$= \psi_0 \cos \left(\alpha \frac{x}{\lambda}\right) \cos \alpha - \psi_0 \sin \left(\alpha \frac{x}{\lambda}\right) \sin \alpha$$

$$= \psi(x, 0)$$

But, since according to equation (4.2),

$$\psi(x, 0) = \psi_0 \cos \left(\alpha \frac{x}{\lambda}\right) \tag{4.2}$$

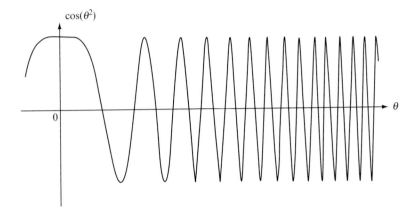

Fig. 4.5 *The graph of* $\cos\theta^2$. Compare with Fig. 4.2. The only expression of the form $\cos\theta^n$ which will give a regular graph such as Fig. 4.2 is that for which n is unity.

then

$$\cos\alpha = 1$$

and

$$\sin\alpha = 0$$

These last two equations permit the solution

$$\alpha = 0 \text{ or } 2\pi$$

Clearly, the solution $\alpha = 0$ is physically meaningless, and hence

$$\alpha = 2\pi$$

$$\therefore \psi(x,0) = \psi_0 \cos 2\pi\frac{x}{\lambda} \qquad (4.3)$$

Equation (4.3) is the mathematical expression for the graph of Fig. 4.2. It has a maximum amplitude ψ_0, repeats periodically with wavelength λ and has a maximum when $x = 0$.

We now consider the variation of the wave amplitude with time t at a given distance, which we may arbitrarily put as $x = 0$. We wish to find the equation corresponding to the graph of Fig. 4.4, and to do this let us define the following symbols:

Let the amplitude at time t and at the position $x = 0$ be $\psi(0,t)$.
Let the maximum amplitude be ψ_0. Let the periodic time be τ.

Note that the maximum amplitude ψ_0 is the same in this case as it was in the previous discussion. This is because the maximum amplitude of a wave is independent of distance and time.

The graph of Fig. 4.4 is sinusoidal in time, and since there is a maximum at $t = 0$, we may choose a cosine function

$$\psi(0, t) = \psi_0 \cos f'(t, \tau) \tag{4.4}$$

in which $f'(t, \tau)$ is another undetermined function, this time of t and τ. Once again, $f'(t, \tau)$ must be dimensionless, giving

$$f'(t, \tau) = \beta \left(\frac{t}{\tau}\right)^m$$

in which β and m are as yet undetermined numerical constants.

The argument now proceeds exactly as before. Since $\psi(0, t)$ is a perfectly regular function of t, the value of m must be unity. Also, $\psi(0, t)$ must be the same as $\psi(0, t + \tau)$,

$$\psi(0, t + \tau) = \psi_0 \cos \left(\beta \frac{t + \tau}{\tau}\right)$$

$$= \psi_0 \cos \left(\beta \frac{t}{\tau} + \beta\right) \tag{4.5}$$

But

$$\psi(0, t) = \psi_0 \cos \beta \frac{t}{\tau} \tag{4.6}$$

Equations (4.5) and (4.6) are equal, and so

$$\beta = 2\pi$$

implying that

$$\psi(0, t) = \psi_0 \cos 2\pi \frac{t}{\tau} \tag{4.7}$$

Equation (4.7) is the mathematical representation of the variation of the amplitude of a plane wave with time at a given distance.

To find the full expression for the variation of the wave amplitude $\psi(x, t)$ with both distance x and time t, we must combine equations (4.3) and (4.7) in a consistent manner.

We know that when $t = 0$,

$$\psi(x, 0) = \psi_0 \cos 2\pi \frac{x}{\lambda} \tag{4.3}$$

and when $x = 0$,

$$\psi(0, t) = \psi_0 \cos 2\pi \frac{t}{\tau} \tag{4.7}$$

Let us try

$$\psi(x, t) = \psi_0 \cos 2\pi \left(\frac{x}{\lambda} - \frac{t}{\tau}\right) \tag{4.8}$$

When $t = 0$, equation (4.8) reduces to equation (4.3). When $x = 0$, we have

$$\psi(0, t) = \psi_0 \cos 2\pi \left(-\frac{t}{\tau} \right)$$

$$= \psi_0 \cos 2\pi \frac{t}{\tau} \qquad (4.9)$$

and so we have equation (4.7). But the equation

$$\psi(x, t) = \psi_0 \cos 2\pi \left(\frac{x}{\lambda} + \frac{t}{\tau} \right) \qquad (4.10)$$

also behaves in the correct manner when $x = 0$ and when $t = 0$. How do we determine which of equations (4.8) and (4.10) is the correct description of a plane wave?

This question is concerned with the nature of the sign associated with the time variable t or, more specifically, whether the signs of the time variable t and space variable x should be the same (as in equation (4.10)) or different (as in equation (4.8)). Now, mathematically, the sign of a variable such as x or t is used to denote the direction associated with that variable. As we shall see in a few paragraphs, the expression

$$\psi(x, t) = \psi_0 \cos 2\pi \left(\frac{x}{\lambda} - \frac{t}{\tau} \right) \qquad (4.8)$$

represents a wave which, as time increases, moves in the direction of increasing x, whereas

$$\psi(x, t) = \psi_0 \cos 2\pi \left(\frac{x}{\lambda} + \frac{t}{\tau} \right) \qquad (4.10)$$

represents a wave which, as time increases, moves in the direction of decreasing x. Thus both equations (4.8) and (4.10) are correct representations of a wave, for they differ only with respect to the direction in which the wave is travelling. Since we are usually more interested in waves which move in the direction of increasing x, it is an equation of the form of equation (4.8) which is conventionally used to represent a wave.

Rewriting equation (4.8) as

$$\psi(x, t) = \psi_0 \cos \left(2\pi \frac{x}{\lambda} - 2\pi \frac{t}{\tau} \right) \qquad (4.11)$$

let us define two new variables

$$k = \frac{2\pi}{\lambda} \qquad (4.12)$$

$$\omega = \frac{2\pi}{\tau} \qquad (4.13)$$

Equation (4.8) now assumes the form

$$\psi(x, t) = \psi_0 \cos \left(kx - \omega t \right) \qquad (4.14)$$

Equation (4.14) is the conventional form for the amplitude of a plane wave. The variable k is called the *wave vector* (the reason for this will become apparent when we deal with waves in three dimensions), and the variable ω is called the *wave frequency*. Unfortunately, both ν and ω are called frequencies, although they are different. The relationships between the two frequencies ν and ω and the periodic time τ are

$$\omega = 2\pi\nu = \frac{2\pi}{\tau} \tag{4.15}$$

When using the word 'frequency', we must always specify whether we are referring to ν or ω.

Since k is defined as $2\pi/\lambda$, its units are those of reciprocal length, for instance m^{-1}. Similarly, the units of ω are those of reciprocal time, for instance s^{-1}. In accordance with the recommendations of the 11th General Conference on Weights and Measures, the unit of frequency is defined to be the *hertz* (Hz), where

$$1\text{Hz} = 1\text{s}^{-1}$$

The reader may also come across frequencies expressed in cycles per second (cps). These are the same as s^{-1}, and hence the number associated with a frequency ν is the same in units of cps, s^{-1} or Hz. The angular frequency ω is measured in radians per second and also has dimensions of s^{-1}, since radians can be regarded as being dimensionless.

It should be noted that an expression of the form

$$\psi(x,t) = \psi_0 \sin(kx - \omega t) \tag{4.16}$$

is also a valid description of a plane wave. The choice of using equation (4.14) or equation (4.16) rests on the value of the wave amplitude at $t = 0$ and $x = 0$.

Let us now consider the velocity v with which the waves are propagated. Note that the symbol v, which we are using to represent the wave velocity, is different from the somewhat similar-looking symbol ν, which represents the wave frequency. To determine the wave velocity v, we concentrate on one wave crest, and find out the time interval Δt which is taken for that particular wave crest to move a distance Δx. The ratio $\Delta x/\Delta t$ will be equal to the velocity v of the wave. Mathematically, we proceed as follows. Let us consider a wave crest which is positioned at $x = 0$ at time $t = 0$. This implies that the amplitude $\psi(0,0)$ is a maximum, so we will use equation (4.14). At a time t, the wave crest will be in a position x, and to locate this position, we require

$$\psi(x,t) = \psi_0$$

But, from equation (4.14),

$$\psi(x,t) = \psi_0 \cos(kx - \omega t) \tag{4.14}$$

Taking the principal solution, we have

$$kx - \omega t = 0$$
$$\therefore \frac{x}{t} = \frac{\omega}{k} \tag{4.17}$$

Our wave crest started off at $x = 0$, and moved to $x = x$, and hence the interval Δx is equal to x. Likewise, the time interval Δt is equal to t.

Substituting in equation (4.17),

$$\frac{\Delta x}{\Delta t} = \frac{\omega}{k} = v$$

from which we see that v is determined by the ratio of the frequency ω to the wave vector k.

The wave velocity v can also be expressed in terms of the variables ν, λ and t:

$$v = \frac{\Delta x}{\Delta t} = \frac{\omega}{k} = \frac{2\pi\nu}{2\pi/\lambda} = \lambda\nu = \frac{\lambda}{\tau} \tag{4.18}$$

This discussion allows us to verify the statement made above that a wave represented by $\psi_0 \cos(kx - \omega t)$ corresponds to a wave moving in the direction of increasing x, whereas the form $\psi_0 \cos(kx + \omega t)$ corresponds to a wave moving in the direction of decreasing x. With the waveform of equation (4.14), the variables x and t used in the derivation of equation (4.17) are positive. This implies that the increments Δx and Δt are positive, and the velocity v of equation (4.18) is also positive. Hence, the waveform of equation (4.14) is associated with a positive velocity, implying that as time increases, the wave moves in the direction of increasing x. Had we started, however, with the waveform $\psi_0 \cos(kx + \omega t)$, then a negative sign would have appeared in equation (4.17), implying that the velocity v identified in equation (4.18) would have been negative. This demonstrates that such a waveform is associated with a negative velocity, and so, as time increases, the wave moves in the direction of decreasing x.

To be strictly accurate, the wave velocity v defined by equation (4.18) is known as the *phase velocity*:

$$\text{phase velocity} = v = \frac{\omega}{k} \tag{4.18}$$

In certain circumstances, it is useful to define a second type of velocity, known as the *group velocity*:

$$\text{group velocity} = g = \frac{d\omega}{dk}$$

When the ratio $\omega/k = v$ is constant, then the phase and group velocities are equal. This will always be the case for the situations described in this book, and we shall continue to use the term 'phase velocity'. For further information of the uses of phase and group velocities, the reader is referred to the two texts on waves mentioned in the bibliography.

It should be noted, however, that the phase velocity v is constant for the propagation of waves in a single medium, but will vary from medium to medium. The general equation $v = \omega/k$ is therefore only applicable as long as ω, k and v refer to the frequency, wave vector and phase velocity all in the same medium.

Equation (4.14) may take on many different but quite equivalent forms depending on which of the variables λ, τ, v, ω, k and ν we choose to use:

$$\psi(x,t) = \psi_0 \cos (kx - \omega t) \tag{4.14}$$

$$= \psi_0 \cos 2\pi \left(\frac{x}{\lambda} - \nu t \right)$$

$$= \psi_0 \cos 2\pi \left(\frac{x}{\lambda} - \frac{t}{\tau} \right)$$

$$= \psi_0 \cos k \, (x - vt)$$

4.4   The wave equation

The derivation of equation (4.14),

$$\psi(x,t) = \psi_0 \cos (kx - \omega t) \tag{4.14}$$

for the amplitude of a plane wave as described in the previous section was based on an intuitive feeling for the form of a plane wave. That equation (4.14) is indeed correct will now be justified.

Waves are a natural phenomenon, important in many branches of physics. We have discussed water waves and light waves, but there are many other examples, such as waves on a string and sound waves. What is the feature that phenomena as diverse as light waves and sound waves have in common? To the physicist, the answer to this question concerns an important aspect of our analytical technique. When a physicist investigates a natural phenomenon, his or her task is to derive the relationship between the appropriate parameters as fully as possible. To do this, the physicist often investigates how a *change* Δx in one chosen variable x causes a *change* Δy in another variable y, whilst holding—as far as possible—all other variables constant. Much experimental data is therefore naturally expressed in terms of measurements such as Δx and Δy, and their ratio $\Delta y/\Delta x$. Equations containing terms such as $\Delta y/\Delta x$ are known as *difference equations*, and those that contain the corresponding infinitesimals dy/dx or $\partial y/\partial x$ are known as *differential equations*. One of the major ways in which a physicist works is to try to find a differential equation to describe the system of interest. Once the differential equation is obtained, the physicist will attempt to solve it to find the explicit expression which gives the direct relationship between the relevant variables.

Now, the solution of differential equations in general has an important feature. If we take the elementary differential equation

$$\frac{dy}{dx} = x$$

we know that the solution is given by

$$y = \text{\textonehalf} x^2 + c$$

where c is an arbitrary constant. An infinity of values of c satisfy the differential equation, and until we know which particular value of the variable y corresponds to a specific value of the variable x, then the constant c is undetermined. The evaluation of constants of integration and related quantities is known as the *boundary value problem*, for *boundary values* are defined as specified values of variables which are known. For instance, if we know that a car starts from rest at the origin of a coordinate system at zero time, our boundary values for the analysis of the motion of the car are '$v = 0$ when $x = 0$ and $t = 0$', where v is the car's velocity, x is distance and t is time. The relevance of this discussion is that when we wish to solve a differential equation, the solution obtained is never unique, but will contain various constants or functions of integration. The exact details of the solution will depend upon the boundary conditions appropriate to the problem under consideration. The more complex the differential equation, the more complex will be the solution, and the more knowledge we require of the system to determine all the boundary conditions. In general, if we know the differential equation which describes a given phenomenon, then any function whatsoever that satisfies that equation is an allowed description of the system. Only when we know the boundary conditions in detail can we select that particular expression which is the correct one for the system in which we are interested. Any differential equation therefore has a *general solution*, which is formed from all those expressions which satisfy the differential equation. When the boundary conditions are known, we may then particularise the general solution to give the 'right' answer. As an example, consider the general solution of the differential equation

$$\frac{dy}{dx} = x$$

which is

$$y = \frac{1}{2} x^2 + c$$

If, for a chosen system, when $x = 0$, $y = 1$, we may state that the particular solution for this system is

$$y = \frac{1}{2} x^2 + 1$$

The significance of this examination of the properties of differential equations is that a theoretical analysis of the propagation of waves in one dimension gives rise to a second-order partial differential equation, which may be written as

$$\frac{\partial^2 \psi(x,t)}{\partial x^2} = \frac{1}{v^2} \frac{\partial^2 \psi(x,t)}{\partial t^2} \tag{4.19}$$

in which $\psi(x,t)$ is the wave amplitude, x is the spatial direction, t is the time, and v is the phase velocity of the wave. The universality of this equation is that *all* wave phenomena in one dimension may be expressed by exactly equation (4.19). In the case of water waves, $\psi(x,t)$ is the amplitude of a water wave, but in the case of electromagnetic waves, $\psi(x,t)$ represents an electric or magnetic field. Because many different phenomena all obey the same mathematical equation, they are all classed as wave phenomena. Equation (4.19) is known as the one-dimensional wave equation.

Note the use of partial derivatives. We know that the expression for the wave amplitude must contain information on both space and time, and so the expression for the wave amplitude is a function of at least two variables x and t. We therefore must use the formalism of partial differentials. A form of the two-dimensional equation is

$$\frac{\partial^2 \psi(x, y, t)}{\partial x^2} + \frac{\partial^2 \psi(x, y, t)}{\partial y^2} = \frac{1}{v^2} \frac{\partial^2 \psi(x, y, t)}{\partial t^2} \tag{4.20}$$

and that in three dimensions may be written as

$$\frac{\partial^2 \psi(x, y, z, t)}{\partial x^2} + \frac{\partial^2 \psi(x, y, z, t)}{\partial y^2} + \frac{\partial^2 \psi(x, y, z, t)}{\partial z^2} = \frac{1}{v^2} \frac{\partial^2 \psi(x, y, z, t)}{\partial t^2} \tag{4.21}$$

Let us consider the one-dimensional equation (4.19). We see that it concerns second derivatives, implying that the solution will require two integration steps, thereby giving rise to at least two unknowns in the final explicit form for the wave amplitude. This is our boundary value problem. The solution of equation (4.19) will give a general solution, which will require knowledge of the boundary conditions for application to any particular case.

4.5 The solution of the wave equation

In one dimension

We will not perform all the mathematical operations to solve the wave equation directly, but, rather, we will guess the nature of certain solutions and test them to see if they satisfy the wave equation. If this is so, then our guess is an allowed description of wave amplitude. If we take the one-dimensional equation (4.19),

$$\frac{\partial^2 \psi(x, t)}{\partial x^2} = \frac{1}{v^2} \frac{\partial^2 \psi(x, t)}{\partial t^2} \tag{4.19}$$

let us try as a possible solution equation (4.14),

$$\psi(x, t) = \psi_0 \cos(kx - \omega t) \tag{4.14}$$

From equation (4.14), we have

$$\frac{\partial \psi(x, t)}{\partial x} = -k\psi_0 \sin(kx - \omega t)$$

$$\frac{\partial \psi(x, t)}{\partial t} = \omega \psi_0 \sin(kx - \omega t)$$

$$\frac{\partial^2 \psi(x, t)}{\partial x^2} = -k^2 \psi_0 \cos(kx - \omega t)$$

$$\frac{\partial^2 \psi(x, t)}{\partial t^2} = -\omega^2 \psi_0 \cos(kx - \omega t)$$

Substituting in equation (4.19), we have

$$-k^2 \psi_0 \cos(kx - \omega t) = -\frac{\omega^2}{v^2} \psi_0 \cos(kx - \omega t)$$

The equality holds if we identify

$$k^2 = \frac{\omega^2}{v^2}$$

$$\therefore v = \frac{\omega}{k}$$

But we have already demonstrated that the phase velocity v of a wave is ω/k (see equation (4.18)), and so equation (4.14) does indeed represent a valid solution of the wave equation, and therefore is a correct description of a one-dimensional wave.

Equation (4.14), however, is not the only solution, as we shall now demonstrate. Consider the equation

$$\psi(x,t) = \psi_0 \sin(kx - \omega t) \tag{4.16}$$

We have

$$\frac{\partial \psi(x,t)}{\partial x} = k\psi_0 \cos(kx - \omega t)$$

$$\frac{\partial \psi(x,t)}{\partial t} = -\omega \psi_0 \cos(kx - \omega t)$$

$$\frac{\partial^2 \psi(x,t)}{\partial x^2} = -k^2 \psi_0 \sin(kx - \omega t)$$

$$\frac{\partial^2 \psi(x,t)}{\partial t^2} = -\omega^2 \psi_0 \sin(kx - \omega t)$$

and so equation (4.16) also satisfies the wave equation on condition that $v = \omega/k$. Further, let us try the solution

$$\psi(x,t) = \psi_0' \cos\left(k'x - \omega't + \phi\right) \tag{4.22}$$

in which ψ_0', k' and ω' are parameters different from the ψ_0, k and ω we have used previously, and ϕ is some numerical constant called the *phase factor*. Differentiating, we obtain

$$\frac{\partial \psi(x,t)}{\partial x} = -k'\psi_0' \sin\left(k'x - \omega't + \phi\right)$$

$$\frac{\partial \psi(x,t)}{\partial t} = \omega'\psi_0' \sin\left(k'x - \omega't + \phi\right)$$

$$\frac{\partial^2 \psi(x,t)}{\partial x^2} = -(k')^2 \psi_0' \cos\left(k'x - \omega't + \phi\right)$$

$$\frac{\partial^2 \psi(x,t)}{\partial t^2} = -(\omega')^2 \psi_0' \cos\left(k'x - \omega't + \phi\right)$$

and once again we have a solution to the wave equation in which the new wave vector k' and frequency ω' are related by $\omega'/k' = v$.

Finally, let us try the solution

$$\psi(x,t) = \psi_0 \cos\left(kx - \omega t + \phi\right) + \psi_0' \sin\left(k'x - \omega't + \phi'\right) \tag{4.23}$$

We now have

$$\frac{\partial \psi(x,t)}{\partial x} = -k\psi_0 \sin\left(kx - \omega t + \phi\right) + k'\psi_0' \cos\left(k'x - \omega't + \phi'\right)$$

$$\frac{\partial^2 \psi(x,t)}{\partial x^2} = -k^2\psi_0 \cos\left(kx - \omega t + \phi\right) - (k')^2\psi_0' \sin\left(k'x - \omega't + \phi'\right)$$

$$\frac{\partial \psi(x,t)}{\partial t} = \omega\psi_0 \sin\left(kx - \omega t + \phi\right) - \omega'\psi_0' \cos\left(k'x - \omega't + \phi'\right)$$

$$\frac{\partial^2 \psi(x,t)}{\partial t^2} = -\omega^2\psi_0 \cos\left(kx - \omega t + \phi\right) - (\omega')^2\psi_0' \sin\left(k'x - \omega't + \phi'\right)$$

Equation (4.23) is yet another possible solution of the one-dimensional wave equation with the proviso that $v = \omega/k$ and simultaneously $v = \omega'/k'$ as well. We now have four possible solutions of the one-dimensional wave equation:

$$\psi(x,t) = \psi_0 \cos\left(kx - \omega t\right) \tag{4.14}$$

$$\psi(x,t) = \psi_0 \sin(kx - \omega t) \tag{4.16}$$

$$\psi(x,t) = \psi_0' \cos(k'x - \omega't + \phi) \tag{4.22}$$

$$\psi(x,t) = \psi_0 \cos(kx - \omega t + \phi) + \psi_0' \sin(k'x - \omega't + \phi') \tag{4.23}$$

We may therefore infer that any equation of the form

$$\psi(x,t) = \sum_n \psi_n \cos\left(k_n x - \omega_n t + \phi_n\right) + \sum_m \psi_m \sin\left(k_m x - \omega_m t + \phi_m\right) \tag{4.24}$$

is a possible solution of the wave equation. In equation (4.24), ψ_n is the maximum amplitude for the wave of wave vector k_n and frequency ω_n, and ψ_m is the maximum amplitude for the wave of wave vector k_m and frequency ω_m; both are subject to the condition

$$v = \frac{\omega_n}{k_n} = \frac{\omega_m}{k_m}$$

Each of equations (4.14), (4.16), (4.22) and (4.23) is a special case of equation (4.24), and so equation (4.24) is the general solution of equation (4.19), and any one-dimensional wave whatsoever may be expressed by means of equation (4.24). The values of ψ_n, k_n, ω_n, ϕ_n, ψ_m, k_m, ω_m and ϕ_m are all determined by the boundary solutions appropriate to a particular problem, and the extent to which the summation is necessary, meaning the number of different terms in the final form of equation (4.24), is also dependent on the boundary conditions.

The complicated form of equation (4.24) is indeed awesome. The explanation of just what equation (4.24) means will be given in the next section, but before coming to this, we shall complete our mathematical study of the wave equation.

In two dimensions

We shall now discuss the solution to the two-dimensional equation (4.20),

$$\frac{\partial^2 \psi(x,y,t)}{\partial x^2} + \frac{\partial^2 \psi(x,y,t)}{\partial y^2} = \frac{1}{v^2} \frac{\partial^2 \psi(x,y,t)}{\partial t^2}$$

Since we know that a particular solution of the one-dimensional equation (4.19),

$$\frac{\partial^2 \psi(x,t)}{\partial x^2} = \frac{1}{v^2} \frac{\partial^2 \psi(x,t)}{\partial t^2} \qquad (4.19)$$

is

$$\psi(x,t) = \psi_0 \cos(kx - \omega t) \qquad (4.14)$$

then it is a reasonable guess that there is a particular solution to the two-dimensional case of the form

$$\psi(x,y,t) = \psi_0 \cos(k_x x + k_y y - \omega t) \qquad (4.25)$$

We now are dealing with two spatial parameters x and y, and so we associate two different wave vectors k_x and k_y with x and y, respectively. Differentiating,

$$\frac{\partial \psi(x,y,t)}{\partial x} = -k_x \psi_0 \sin(k_x x + k_y y - \omega t)$$

$$\frac{\partial \psi(x,y,t)}{\partial y} = -k_y \psi_0 \sin(k_x x + k_y y - \omega t)$$

$$\frac{\partial \psi(x,y,t)}{\partial t} = \omega \psi_0 \sin(k_x x + k_y y - \omega t)$$

$$\frac{\partial^2 \psi(x,y,t)}{\partial x^2} = -(k_x)^2 \psi_0 \cos(k_x x + k_y y - \omega t)$$

$$\frac{\partial^2 \psi(x,y,t)}{\partial y^2} = -(k_y)^2 \psi_0 \cos(k_x x + k_y y - \omega t)$$

$$\frac{\partial^2 \psi(x,y,t)}{\partial t^2} = -\omega^2 \psi_0 \cos(k_x x + k_y y - \omega t)$$

$$\therefore \frac{\partial^2 \psi(x,y,t)}{\partial x^2} + \frac{\partial^2 \psi(x,y,t)}{\partial y^2} = -(k_x^2 + k_y^2)\psi_0 \cos(k_x x + k_y y - \omega t)$$

implying that equation (4.20) is satisfied if

$$k_x^2 + k_y^2 = \frac{\omega^2}{v^2} \qquad (4.26)$$

Equation (4.25) is seen to be a particular solution of the two-dimensional wave equation, according to the condition of equation (4.26). We can therefore immediately write down the general solution to equation (4.25) as

$$\psi(x, y, t) = \sum_n \psi_n \cos[(k_x)_n x + (k_y)_n y - \omega_n t + \phi_n]$$

$$+ \sum_m \psi_m \sin[(k_x)_m x + (k_y)_m y - \omega_m t + \phi_m] \tag{4.27}$$

in which the various parameters ψ_n, $(k_x)_n$ and so on are determined by the boundary conditions.

In three dimensions

The situation in three dimensions is exactly analogous. We wish to solve the equation

$$\frac{\partial^2 \psi(x, y, z, t)}{\partial x^2} + \frac{\partial^2 \psi(x, y, z, t)}{\partial y^2} + \frac{\partial^2 \psi(x, y, z, t)}{\partial z^2} = \frac{1}{v^2} \frac{\partial^2 \psi(x, y, z, t)}{\partial t^2} \tag{4.21}$$

and we immediately and intuitively see that there is a particular solution of the form

$$\psi(x, y, z, t) = \psi_0 \cos(k_x x + k_y y + k_z z - \omega t) \tag{4.28}$$

giving the general solution

$$\psi(x, y, z, t) = \sum_n \psi_n \cos[(k_x)_n x + (k_y)_n y + (k_z)_n z - \omega_n t + \phi_n]$$

$$+ \sum_m \psi_m \sin[(k_x)_m x + (k_y)_m y + (k_z)_m z - \omega_m t + \phi_m] \tag{4.29}$$

As usual, there is a condition on k_x, k_y, k_z and ω, namely that

$$k_x^2 + k_y^2 + k_z^2 = \frac{\omega^2}{v^2} \tag{4.30}$$

We shall now discuss the significance of the variables k_x, k_y and k_z introduced in equations (4.25) and (4.28). In fact, the form of the expression $(k_x x + k_y y + k_z z - \omega t)$ which occurs in equation (4.28) gives us a strong hint as to the meaning of k_x, k_y and k_z. Let us consider a three-dimensional system of Cartesian coordinates, in which any point may in general be represented by a vector $\mathbf{r}$ whose components are (x, y, z):

$$\mathbf{r} = (x, y, z)$$

Let us now define a vector k with components $(k_x, k_y$ and $k_z)$ as

$$\mathbf{k} = (k_x, k_y, k_z)$$

On forming the scalar product $\mathbf{k} \cdot \mathbf{r}$ we obtain

$$\mathbf{k} \cdot \mathbf{r} = k_x x + k_y y + k_z z$$

and so equation (4.28) becomes

$$\psi(\mathbf{r}, t) = \psi_0 \cos(\mathbf{k} \cdot \mathbf{r} - \omega t)$$

Hence k_x, k_y and k_z are the components of a vector $\mathbf{k}$ which is the three-dimensional analogue of the k we met in the discussion of equation (4.12). It is for this reason that we originally called k the 'wave vector', for the one-dimensional example is but a special case of a general three-dimensional phenomenon.

Since $\mathbf{k}$ is a vector, it has both magnitude and direction. To find the magnitude, we consider the square of the modulus of $\mathbf{k}$, namely

$$|\mathbf{k}|^2 = k^2 = k_x^2 + k_y^2 + k_z^2$$

On comparison of this equation with equation (4.30), we see that

$$k^2 = \frac{\omega^2}{v^2}$$

But from equation (4.15) we see that

$$\frac{\omega^2}{v^2} = \frac{4\pi^2 \nu^2}{v^2} = \frac{4\pi^2 \nu^2}{\lambda^2 \nu^2} = \frac{4\pi^2}{\lambda^2} \tag{4.31}$$

$$\therefore k^2 = \frac{4\pi^2}{\lambda^2}$$

$$\therefore |\mathbf{k}| = k = \frac{2\pi}{\lambda} \tag{4.32}$$

Hence $\mathbf{k}$ is a vector of magnitude $2\pi/\lambda$, which agrees with the one-dimensional definition given in equation (4.12). The direction associated with $\mathbf{k}$ may be determined by remembering that in the one-dimensional case in which a wave travels in the x direction only, the solution of the wave equation is equation (4.14),

$$\psi(x,\,t) = \psi_0 \cos\left(kx - \omega t\right) = \psi_0 \cos\left(k_x x - \omega t\right) \tag{4.14}$$

in which we emphasise that the wave vector has an x component k_x only. Consequently, a wave travelling in the x direction is associated with an x component only, and so we may generalise this to say that when a wave travels in three dimensions, the direction of the $\mathbf{k}$ vector is that of wave propagation. The full definition of the *wave vector* $\mathbf{k}$ is therefore: The wave vector $\mathbf{k}$ is a vector in the direction of wave propagation and whose magnitude is given by $k = 2\pi/\lambda$.

Clearly, in two dimensions, the wave vector has two components (k_x, k_y) and the magnitude of the wave vector is, as usual, $2\pi/\lambda$, as in equations (4.25) and (4.26).

4.6 The principle of superposition

We have now found the general solution to the three-dimensional wave equation as

$$\psi(\mathbf{r}, t) = \sum_n \psi_n \cos[\mathbf{k}_n \cdot \mathbf{r} - \omega_n t + \phi_n] + \sum_m \psi_m \sin[\mathbf{k}_m \cdot \mathbf{r} - \omega_m t + \phi_m] \tag{4.29}$$

Conversely, any function of the type defined by equation (4.29) solves the wave equation, and so represents a wave. The waveform which corresponds to a general amplitude function $\psi(\mathbf{r}, t)$, as in equation (4.29), is, however, very complicated.

In this section, we will discuss the significance of the form of equation (4.29) and its one- and two-dimensional analogues, equations (4.24) and (4.27).

If we return to our original example of plane waves generated by a rule dipping periodically into a tank of water, we know that the wave has a very simple sinusoidal form given by the one-dimensional expression

$$\psi(x, t) = \psi_0 \cos (k_x x - \omega t) \tag{4.14}$$

which we now recognise as a special case of the general equation (4.29). Furthermore, we know that the general solution, equation (4.29), can be applied to any system which obeys the wave equation if we know the relevant boundary conditions. For the case of plane waves in a tank, the boundary conditions themselves are particularly simple. There is a single source of waves (the rule), which dips with constant frequency ω into the water, and so the waves generated have the elementary mathematical description of equation (4.14). But what happens if we take a case in which the boundary conditions are more complicated? Consider, for instance, the wave pattern formed on the surface of the sea in a harbour. There are many ships moving through the water, all of which create disturbances independently of one another, and all of which act as a source of water waves. What we see on the surface of the water is a highly complex pattern formed by all the waves generated by the many different sources. Since we wish to describe the waves on the water's surface, we must consider the solutions of the two-dimensional equation. We know intuitively that the wave pattern cannot be represented solely by the particular function of equation (4.25),

$$\psi(x, y, t) = \psi_0 \cos (k_x x + k_y y - \omega t) \tag{4.25}$$

But now we are in a position to suggest a form for the mathematical expression for the waves on the water in the harbour. The wave amplitude at any point will be a fairly complicated function like that of equation (4.27),

$$\psi(x, y, t) = \sum_n \psi_n \cos[(k_x)_n x + (k_y)_n y - \omega_n t + \phi_n]$$

$$+ \sum_m \psi_m \sin[(k_x)_m x + (k_y)_m y - \omega_m t + \phi_m] \tag{4.27}$$

Because the boundary conditions appropriate to the problem are so complicated, there will be many different terms, each with particular maximum amplitudes ψ_n and ψ_m, and their own wave vectors and frequencies corresponding to each source of waves. We now see that equations (4.24), (4.27) and (4.29) are not quite as formidable as their forms suggest. Clearly, the summation terms look menacing, but they are really very meaningful. Any plane wave of frequency ω and wave vector $\mathbf{k}$ has a mathematical description of the type $\psi_0 \cos(\mathbf{k} \cdot \mathbf{r} - \omega t)$ or $\psi_0 \sin(\mathbf{k} \cdot \mathbf{r} - \omega t)$. When we have a difficult problem with many different sources giving rise to waves of maximum amplitudes ψ_1, ψ_2, ψ_3, ..., ψ_n, wave vectors $\mathbf{k}_1$, $\mathbf{k}_2$, $\mathbf{k}_3$, ..., $\mathbf{k}_n$, and frequencies ω_1, ω_2, ω_3, ..., ω_n, and so on (all subject to the condition $v = \omega_n/k_n$), then equation (4.29) tells us that the total disturbance at any point $\mathbf{r}$ at time t due to all the sources is simply the sum of the disturbances which would be present if each source were acting alone. This is an elegant result.

Given two or more independent sources of waves, we might have thought that the total disturbance propagated would be that due to each source independently, plus a possible effect due to the interaction of one wave with another. Equation (4.29), however, tells us that this is not so. Because the general solution to the wave equation is a linear summation of pure wave-like terms in which each term is of the form $\psi_0 \cos(\mathbf{k}_n \cdot \mathbf{r} - \omega_n t)$ or $\psi_0 \sin(\mathbf{k}_m \cdot \mathbf{r} - \omega_m t)$, the total disturbance is simply the sum of the waves due to each source independently. This is known as the *principle of superposition*, which, formally stated, is: The total wave disturbance due to an array of sources is the sum of the individual disturbances due to each source.

The principle of superposition therefore tells us that the waves due to an array of sources propagate quite independently of one another. This should be familiar to anyone who has ever watched water waves on a large expanse of water such as a lake. On the lake are many different waveforms, of varying maximum amplitude, wavelength and frequency. Yet it is quite easy to choose a particular wave and to follow its progress across the surface of the water. It will meet other waves, but it will pass on unchanged. This is exactly what the principle of superposition and equation (4.29) tell us. In fact, the principle of superposition is one of the most important results in the theory of wave motion. Often, we will be faced with the problem of calculating the total wave amplitude at any point due to an array of sources, which may be quite complicated in form. The way we tackle this problem is to calculate the effect of one single source at the point in which we are interested, and then we add up the contributions due to each source. If the sources are infinitesimally small, then the addition becomes an integration, and we may use the techniques of integral calculus. The problem is then greatly reduced in complexity.

This completes our discussion of the general solutions to the wave equation. For the remainder of the chapter, we will, for the most part, investigate some further properties of the particular solutions

$$\psi(x, t) = \psi_0 \cos(kx - \omega t) \tag{4.14}$$

$$\psi(x, y, t) = \psi_0 \cos(k_x x + k_y y - \omega t) \tag{4.25}$$

and

$$\psi(\mathbf{r}, t) = \psi_0 \cos(\mathbf{k} \cdot \mathbf{r} - \omega t) \tag{4.28}$$

corresponding to a plane wave of frequency ω and wave vector $\mathbf{k}$. All results for the special case may be carried over as necessary when we have to deal with more complicated situations.

4.7 Phase

Reference to the general solution of the three-dimensional wave equation shows that each term is of the type

$$\psi_n(\mathbf{r}, t) = \psi_n \cos(\mathbf{k}_n \cdot \mathbf{r} - \omega_n t + \phi_n)$$

The general amplitude is $\psi_n(\mathbf{r}, t)$, and the associated maximum amplitude is ψ_n. The expression $(\mathbf{k}_n \cdot \mathbf{r} - \omega_n t + \phi_n)$ is called the *phase* of the wave, and it may assume values between 0 and 2π. If the numerical value of $(\mathbf{k}_n \cdot \mathbf{r} - \omega_n t + \phi_n)$ ever exceeds 2π, then when we take the cosine, we subtract integral multiples of 2π until the value is reduced to the range between 0 and 2π. Basically, the phase of the wave tells us the whereabouts in a wave cycle of a particular point on the wave. If

$(\mathbf{k}_n \cdot \mathbf{r} - \omega_n t + \phi_n) = 0$, there is a maximum amplitude;

$(\mathbf{k}_n \cdot \mathbf{r} - \omega_n t + \phi_n) = \dfrac{\pi}{2}$, the amplitude is zero;

$(\mathbf{k}_n \cdot \mathbf{r} - \omega_n t + \phi_n) = \pi$, there is a minimum amplitude;

$(\mathbf{k}_n \cdot \mathbf{r} - \omega_n t + \phi_n) = \dfrac{3\pi}{2}$, the amplitude is again zero;

$(\mathbf{k}_n \cdot \mathbf{r} - \omega_n t + \phi_n) = 2\pi$, we are at the next amplitude maximum.

Figure 4.6 shows the appearance of three waves with the same frequency ω and wavelength λ being propagated side by side, as, for instance, on three adjacent strings. Although the waves are identical in general form, they are seen to be 'out of step', as if the waves in each string had set out at different times.

Since the waves are perfectly regular, the relationship between the wave crests of each wave is constant, and so we are specifically interested in the distance between corresponding wave crests for each wave. Referring to the axes shown in Fig. 4.6, we see that wave 1 has a maximum at the origin at time $t = 0$, implying that the equation of wave 1 is

$$\psi_1(\mathbf{r}, t) = \psi_0 \cos(\mathbf{k} \cdot \mathbf{r} - \omega t) \tag{4.33}$$

whereas wave 2 has zero amplitude at the origin at time $t = 0$, suggesting that the equation appropriate to wave 2 is

$$\psi_2(\mathbf{r}, t) = \psi_0 \sin(\mathbf{k} \cdot \mathbf{r} - \omega t)$$

But

$$\sin \theta = \cos\left(\theta - \frac{\pi}{2}\right)$$

$$\therefore \psi_2(\mathbf{r}, t) = \psi_0 \cos\left(\mathbf{k} \cdot \mathbf{r} - \omega t - \frac{\pi}{2}\right) \tag{4.34}$$

Comparison of equations (4.33) and (4.34) shows that wave 1 is associated with a phase factor $\phi = 0$, and wave 2 with a phase factor $\phi = -\pi/2$. Note, $\pi/2 = 1/4 \times 2\pi$, and that a complete wave cycle corresponds to a change in phase of 2π. The significance

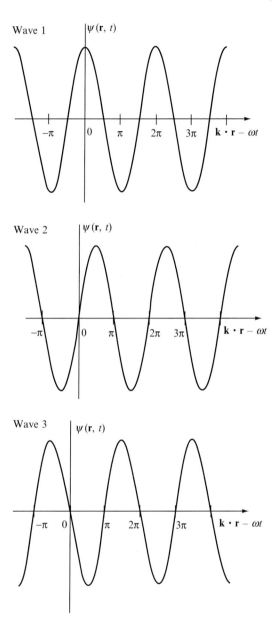

Fig. 4.6 *Phase relationships.* The equations of the three waves are $\psi_1(r, t) = \psi_0 \cos(\mathbf{k} \cdot \mathbf{r} - \omega t)$, $\psi_2(r, t) = \psi_0 \cos(\mathbf{k} \cdot \mathbf{r} - \omega t - \pi/2)$ and $\psi_3(r, t) = \psi_0 \cos(\mathbf{k} \cdot \mathbf{r} - \omega t + \pi/2)$. The three waves differ in their phase factors, and this specifies the degree to which the waves are out of step.

of the value of the phase factor of wave 2 is that the wave crests of wave 2 are exactly one-quarter of a wave cycle, or wavelength, out of step with the wave crests of wave 1. Also, the negative sign associated with the phase factor of wave 2 implies that wave 2 is 'lagging behind' wave 1, for reference to Fig. 4.6 shows that wave 1 is already at a positive maximum when wave 2 is passing through zero. Wave 1 is therefore ahead of wave 2 or, alternatively, wave 2 lags behind wave 1.

The equation of wave 3 is

$$\psi_3(\mathbf{r}, t) = -\psi_0 \sin(\mathbf{k} \cdot \mathbf{r} - \omega t)$$

but since

$$\cos\left(\theta + \frac{\pi}{2}\right) = -\sin\theta$$

we may write

$$\psi_3(\mathbf{r}, t) = \psi_0 \cos\left(\mathbf{k} \cdot \mathbf{r} - \omega t + \frac{\pi}{2}\right)$$

This time the phase factor ϕ is $+\pi/2 = +\frac{1}{4} \times 2\pi$. This tells us that wave 3 is one-quarter of a cycle, or wavelength, out of step with wave 1, and the positive sign implies that wave 3 leads wave (1). Once again, this is because wave 3 reaches a positive maximum before wave 1.

In general, we use phase measurements when we wish to compare waves of the same frequency ω and wave vector $\mathbf{k}$ at a given point $\mathbf{r}$ at time t. Let us consider three general waves whose amplitudes are given by

$$\psi_1(\mathbf{r}, t) = \psi_1 \cos(\mathbf{k} \cdot \mathbf{r} - \omega t + \phi_1)$$
$$\psi_2(\mathbf{r}, t) = \psi_2 \cos(\mathbf{k} \cdot \mathbf{r} - \omega t + \phi_2)$$
$$\psi_3(\mathbf{r}, t) = \psi_3 \cos(\mathbf{k} \cdot \mathbf{r} - \omega t + \phi_3)$$

If we choose wave 1 as a reference wave, we may evaluate the phase difference $\Delta\phi_{12}$ between waves 1 and 2, and $\Delta\phi_{13}$ between waves 1 and 3:

$$\Delta\phi_{12} = \phi_2 - \phi_1$$
$$\Delta\phi_{13} = \phi_3 - \phi_1$$

The sign of $\Delta\phi$ states whether the wave in question leads or lags behind the reference wave, and the degree to which the waves are out of step is given by the ratio $\Delta\phi/2\pi$, which in general is a fraction corresponding to discrepancy between the waves, expressed as a fraction of a cycle, or wavelength.

Two special cases, depicted in Fig. 4.7, are as follows. When $\Delta\phi = 0$, the two waves are 'in phase'; when $\Delta\phi = \pi$, the two waves are 'out of phase'.

It is interesting to consider the total disturbance due to two sources which emit waves either in phase or exactly out of phase. According to the principle of superposition, this is easily done by adding the effect of each source. If the two sources propagate waves in phase, then the disturbances at any point $\mathbf{r}$ at time t due to each source are

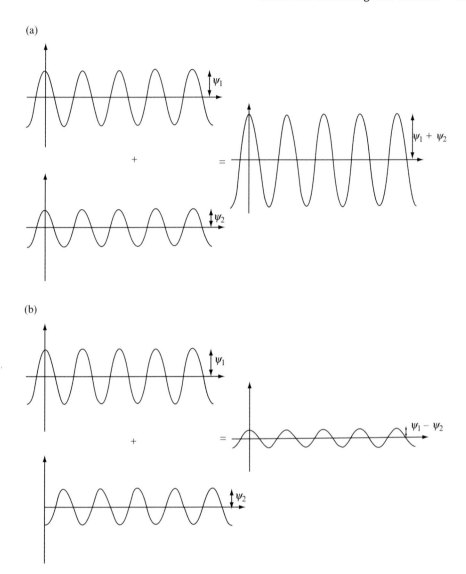

Fig. 4.7 *Interference.* The waves of (a) are exactly in phase and interfere constructively. The waves in (b) are exactly out of phase and interfere destructively. Note that the frequency and wavelength of the resultant wave are the same as those of each of the initial waves, but the amplitude maxima are different.

$$\psi_1(\mathbf{r}, t) = \psi_1 \cos(\mathbf{k} \cdot \mathbf{r} - \omega t)$$
$$\psi_2(\mathbf{r}, t) = \psi_2 \cos(\mathbf{k} \cdot \mathbf{r} - \omega t)$$

We assume that each source propagates waves of the same frequency ω and wave vector $\mathbf{k}$. Also, we may choose our origin of time and space so that the phase factor for each wave is zero. This may be done because the significant quantity is the *phase difference* between the two waves, as opposed to the absolute value of the phase. The total disturbance $\psi_t(\mathbf{r}, t)$ is therefore

$$\psi_t(\mathbf{r}, t) = \psi_1(\mathbf{r}, t) + \psi_2(\mathbf{r}, t) = (\psi_1 + \psi_2) \cos(\mathbf{k} \cdot \mathbf{r} - \omega t)$$

The total disturbance is a wave of the same wavelength and frequency as the original waves, but the total maximum amplitude is just the sum of the individual maximum amplitudes.

In an exactly similar way, we may combine the effect of two sources which propagate waves exactly out of phase:

$$\psi_1(\mathbf{r}, t) = \psi_1 \cos(\mathbf{k} \cdot \mathbf{r} - \omega t)$$
$$\psi_2(\mathbf{r}, t) = \psi_2 \cos(\mathbf{k} \cdot \mathbf{r} - \omega t + \pi)$$

But

$$\cos(\theta + \pi) = -\cos\theta$$
$$\therefore \psi_2(\mathbf{r}, t) = -\psi_2 \cos(\mathbf{k} \cdot \mathbf{r} - \omega t)$$
$$\therefore \psi_t(\mathbf{r}, t) = \psi_1(\mathbf{r}, t) + \psi_2(\mathbf{r}, t) = (\psi_1 - \psi_2) \cos(\mathbf{k} \cdot \mathbf{r} - \omega t)$$

Once again, we have a wave of the same frequency ω and wavelength λ as the original wave, but the maximum amplitude is now the difference of the original maximum amplitudes. If the two sources are such that the maximum amplitudes are the same, that is, if $\psi_1 = \psi_2$, then $\psi_t(\mathbf{r}, t) = 0$ for all $\mathbf{r}$ and t. Hence the two waves exactly cancel each other out. This is known as *total destructive interference*. In general, the mixing of the effects of the waves due to several sources according to the principle of superposition is known as *interference*. The interference is *constructive* if the waves reinforce each other, and *destructive* if the waves tend to cancel each other. The degree to which they affect each other may be described using the word *partial* or *total*, and this depends on the relative magnitudes of the amplitude maxima of each wave. If ψ_1 and ψ_2 are different, as in the example cited immediately above, then we speak of partial destructive interference.

4.8   Waves and complex exponentials

The previous section has explained that the two most significant aspects of a wave motion are its maximum amplitude and its phase. A mathematical description of a wave motion must therefore contain information on both maximum amplitude and phase, as, for instance, does equation (4.28),

$$\psi(\mathbf{r}, t) = \psi_0 \cos(\mathbf{k} \cdot \mathbf{r} - \omega t) \tag{4.28}$$

As we saw in Chapter 2, complex numbers of the form $|z|e^{i\delta}$ also contain information on both maximum amplitude and phase if we identify

$$|z| = \psi_0$$

and

$$\delta = (\mathbf{k} \cdot \mathbf{r} - \omega t)$$

Let us suppose that an alternative representation of a wave amplitude is

$$\psi(\mathbf{r}, t) = \psi_0 e^{i(\mathbf{k} \cdot \mathbf{r} - \omega t)} \tag{4.35}$$

and to see if this does indeed signify an allowed solution of the wave equation, we may investigate whether or not equation (4.35) satisfies the wave equation. For the purposes of demonstration, we shall choose a one-dimensional example and try

$$\psi(x, t) = \psi_0 e^{i(k_x x - \omega t)} \tag{4.36}$$

as a solution of the one-dimensional equation

$$\frac{\partial^2 \psi(x, t)}{\partial x^2} = \frac{1}{v^2} \frac{\partial^2 \psi(x, t)}{\partial t^2} \tag{4.19}$$

On differentiation, we obtain

$$\frac{\partial \psi(x, t)}{\partial x} = ik_x \psi_0 e^{i(k_x x - \omega t)}$$

$$\frac{\partial \psi(x, t)}{\partial t} = -i\omega \psi_0 e^{i(k_x x - \omega t)}$$

$$\frac{\partial^2 \psi(x, t)}{\partial x^2} = -(k_x)^2 \psi_0 e^{i(k_x x - \omega t)}$$

$$\frac{\partial^2 \psi(x, t)}{\partial t^2} = -\omega^2 \psi_0 e^{i(k_x x - \omega t)}$$

Hence equation (4.36) does satisfy the wave equation if, as usual, we identify $v = \omega / k_x$. By analogy, equation (4.35) is a solution of the three-dimensional wave equation, and the general solution may be expressed as

$$\psi(\mathbf{r}, t) = \sum_n \psi_n e^{i(\mathbf{k}_n \cdot \mathbf{r} - \omega_n t + \phi_n)} \tag{4.37}$$

in which the parameters ψ_n, k_n, ω_n, ϕ_n and the extent to which the summation is necessary are determined by the boundary conditions.

We now have two exactly equivalent expressions for plane waves in three dimensions, namely equations (4.28) and (4.35):

$$\psi(\mathbf{r}, t) = \psi_0 \cos(\mathbf{k} \cdot \mathbf{r} - \omega t) \tag{4.28}$$

$$\psi(\mathbf{r}, t) = \psi_0 e^{i(\mathbf{k} \cdot \mathbf{r} - \omega t)} \tag{4.35}$$

Although the forms of these equations are somewhat different, they both solve the wave equation and hence are equally valid mathematical descriptions of a wave. Since the cosine and complex exponential functions have rather different mathematical properties, when we wish to manipulate the expression for wave amplitude mathematically, it may so happen that one of the equations (4.28) and (4.35) is more amenable to the particular process which we are carrying out. In general, the choice of which of these two equations we actually use rests on whichever is the more convenient. It turns out that for the majority of cases, the complex exponential form is the more useful, and therefore, whenever we wish to write a general expression for a plane wave, we shall usually write an equation of the form of equation (4.35). It is for this reason that Chapter 2 included a discussion of complex algebra. It was pointed out that complex numbers are useful in handling the description of phenomena associated with two numbers. Waves are of this general type, since they are associated with both a maximum amplitude and a phase. The exponential form of a complex number is particularly appropriate to this case, and will be used hereafter.

4.9   Intensity

The *intensity* I of a wave is defined as the square of the magnitude of the amplitude of the wave:

$$I = |\psi(\mathbf{r}, t)|^2 \tag{4.38}$$

We take the square of the magnitude of the amplitude for the following reason. Physically, the intensity of a wave is a measure of the energy flowing through unit area of the wavefront:

$$\text{total energy of wavefront} = \text{intensity} \times \text{area}$$

Since both the energy flow and the area are directly measurable quantities, then the intensity must also be measurable, and so intensity must be represented by a real number. In Chapter 2 we saw that the product of any complex number z and its complex conjugate z^* is always a real number $|z|^2$:

$$|z|^2 = zz^* \tag{2.22}$$

If we choose to describe a plane wave by equation (4.35),

$$\psi(\mathbf{r}, t) = \psi_0 e^{i(\mathbf{k} \cdot \mathbf{r} - \omega t)} \tag{4.35}$$

then

$$\psi^*(\mathbf{r}, t) = \psi_0 e^{-i(\mathbf{k} \cdot \mathbf{r} - \omega t)}$$

and

$$\psi(\mathbf{r}, t)\psi^*(\mathbf{r}, t) = \psi_0^2 \tag{4.39}$$

This tells us that the intensity of a plane wave is a constant, equal to ψ_0^2. Hence the energy flow through any unit area of the wave is constant, but this is a special case for plane waves—in a more complicated system, the intensity will not be a constant. To anticipate, when we perform an X-ray diffraction experiment, our experimentally observed quantity is the intensity of a wave, and so the results of the analysis presented in this section are important.

4.10 Waves which are not plane

The entire discussion up to this point has been concerned with the description of plane waves. Although not stated explicitly, the assumption of an analysis based on plane waves was made as soon as we wrote the wave equation (4.19), (4.20) and (4.21) in the form shown, in which partial derivatives are taken with respect to the three Cartesian coordinates x, y and z. Cartesian coordinates are applicable to systems which have the geometry of planes and right angles, and so any equation couched in terms of Cartesian coordinates will automatically generate solutions appropriate to rectangular and planar symmetry. If we are interested in waves which are not plane, then it is more useful to express the wave equation in terms of a coordinate system appropriate to the symmetry under consideration, for then the solution will also have that symmetry. For instance, if we have a point source of sound or light waves, we intuitively know that the waves are propagated as concentric spheres, as shown in Fig. 4.8. Spherical polar

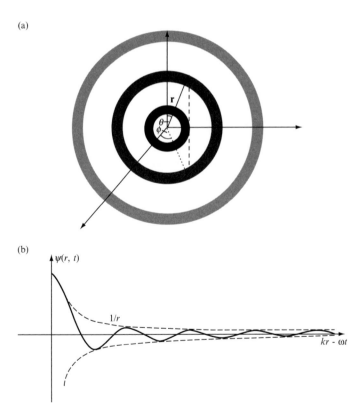

Fig. 4.8 *Spherical waves.* (a) represents a spherical wave from a point source; the spherical polar coordinates r, θ and ϕ are shown. (b) is a graph of wave amplitude against radial distance r. Note that the wave amplitude is independent of the angles θ and ϕ and falls off as $1/r$.

coordinates are the obvious choice of coordinate system, and the three-dimensional wave equation in this case takes the form

$$\frac{1}{r}\frac{\partial^2}{\partial r^2}(r\psi) + \frac{1}{r^2 \sin\theta}\frac{\partial}{\partial\theta}\left(\sin\theta\frac{\partial\psi}{\partial\theta}\right) + \frac{1}{r^2\sin^2\theta}\frac{\partial^2\psi}{\partial\phi^2} = \frac{1}{v^2}\frac{\partial^2\psi}{\partial t^2} \qquad (4.40)$$

where we write ψ for $\psi(r,\theta,\phi,t)$.

A particular solution of equation (4.40) is

$$\psi(\mathbf{r},t) = \frac{\psi_0}{r}e^{i(kr-\omega_n t)} \qquad (4.41)$$

as may be verified by direct substitution where, as usual, $v = \omega/k$. We see that the wave amplitude $\psi(\mathbf{r},t)$ depends solely on the radial distance r and is independent of the angles θ and ϕ, which is just what we expect for a spherically symmetrical system. Further, the intensity of the waves is just $|\psi(\mathbf{r},t)|^2$, given by

$$|\psi(\mathbf{r},t)|^2 = \frac{\psi_0^2}{r^2}$$

The intensity therefore decreases as r^2, which the reader may recognise as the inverse square law. Since the total area of the wavefront increases as r^2, the total energy in each wavefront is constant:

$$\text{total energy of wavefront} = \text{intensity} \times \text{area} \propto \psi_0^2$$

4.11   Electromagnetic waves

One of the greatest of all theoretical physicists was James Clerk Maxwell, who was born in Edinburgh, Scotland, in 1831. Maxwell's researches in the kinetic theory of gases and thermodynamics were quite sufficient to earn him an immortal place in the history of science, but his works on the theory of electromagnetism place him in the ranks of Newton and Einstein. Maxwell studied electricity, magnetism and their interrelated effects, and derived four equations, known as Maxwell's equations, which relate the time and space derivatives of a general electric field $\mathbf{E}$ and a magnetic field $\mathbf{H}$, both of which may be represented by vectors. In brief, the first equation is a statement of Coulomb's inverse square law of electrostatics; the second concerns Faraday's law of electromagnetic induction; the third describes the inverse square law of magnetostatics; and the fourth is an extension of Ampère's current law. With respect to the fourth equation, Maxwell used great insight to demonstrate that the then accepted form of Ampère's law was inconsistent with the other three equations. By invoking a new concept called 'the displacement current', Maxwell corrected Ampère's law to give his fourth equation. These four equations are the cornerstones of electromagnetism.

In principle, any problem whatsoever in electromagnetism may be solved starting solely from these equations, if the boundary conditions are known. Maxwell, however, did not stop at the discovery of these equations. He manipulated them mathematically, and found that, if we consider the electric field $\mathbf{E}$ and the magnetic field $\mathbf{H}$ to have components (E_x, E_y, E_z) and (H_x, H_y, H_z), referred to Cartesian coordinates, then each component, for instance E_x, satisfies a differential equation of the type

$$\frac{\partial^2 E_x}{\partial x^2} + \frac{\partial^2 E_x}{\partial y^2} + \frac{\partial^2 E_x}{\partial z^2} = \frac{1}{c^2} \frac{\partial^2 E_x}{\partial t^2}$$

This is none other than the wave equation! On the basis of this result, Maxwell predicted that wave-like motion of electric and magnetic fields may exist. From his derivation of the wave equation, he was able to express the velocity c of the wave motion in terms of fundamental constants, and he calculated that c was about $3 \times 10^8 \, \mathrm{m\,s^{-1}}$. Maxwell's paper containing this result was published in 1864, and some years previously, in 1862, Foucault had found a reliable value for the velocity of light, which he stated was about $3 \times 10^8 \, \mathrm{ms^{-1}}$. The agreement between Maxwell's calculated value and Foucault's experimental value led Maxwell to suggest that light itself was a manifestation of the *electromagnetic waves* which he had predicted. This was truly a wonderful moment in the history of physics. Electricity, magnetism and optics had been unified in one elegant theory.

We now know that Maxwell was entirely correct, and that light waves are just one form of electromagnetic radiation. In 1888, Heinrich Hertz (in whose honour the unit of frequency is named) discovered another form of electromagnetic radiation, namely radio waves. As we have seen in previous sections, a wave motion is characterised by a frequency ω and a wave vector $\mathbf{k}$, which are related to the phase velocity v of wave propagation, the wavelength λ and the frequency ν, as in equation (4.18),

$$v = \frac{\omega}{k} = \lambda \nu \tag{4.18}$$

The phase velocity v is a constant in any given medium, and so equation (4.18) permits many different wavelengths λ and frequencies ω or ν such that the value for the phase velocity v in a given medium is constant. For light, the phase velocity v is conventionally denoted c, where $c = \omega/k = \lambda\nu$.

Light waves have wavelengths between about $4 \times 10^{-7} \, \mathrm{m}$ (blue) and $7 \times 10^{-7} \, \mathrm{m}$ (red), whereas radio waves have much longer wavelengths, typically about $100 \, \mathrm{m}$. At about the turn of the century, several more phenomena were shown to be electromagnetic waves. γ-rays, discovered by Paul Villard in 1900 as a natural product of radioactive decay, have very short wavelengths, less than $10^{-12} \, \mathrm{m}$. Röntgen discovered X-rays in 1895, and their wavelength is in the range 10^{-11}–$10^{-8} \, \mathrm{m}$. Gradually, a complete range of electromagnetic waves of wavelength from $10^{-12} \, \mathrm{m}$ to many hundreds of metres were identified, and these may be shown on a chart in which the wavelength λ, wave vector magnitude k, frequency ν and frequency ω corresponding to each type of wave are indicated. This diagram is called the electromagnetic spectrum, as shown in Fig. 4.9.

The final important numerical aspect of electromagnetic waves concerns the energy associated with a wave of frequency ω or frequency ν. After quantum theory had been invoked by Max Planck and then validated in 1905 by Albert Einstein in his theory of

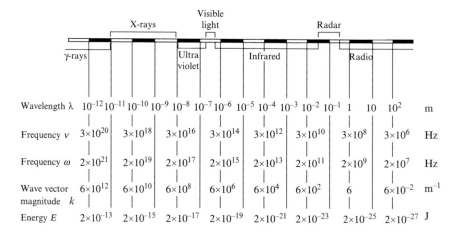

Fig. 4.9 *The electromagnetic spectrum.*

the photoelectric effect, it was established that the energy E associated with a wave is given by

$$E = h\nu = \frac{h\omega}{2\pi}$$

Here, h is a fundamental constant known as Planck's constant, which has the value of 6.62607×10^{-34} J s. Waves with very high frequencies are therefore relatively energetic. The electromagnetic waves of the highest frequencies are the γ-rays and X-rays, and this explains why these forms of radiation are sometimes harmful to living matter.

4.12 The form of electromagnetic waves

As stated above, for the electric field $\mathbf{E} = (E_x, E_y, E_z)$ and the magnetic field $\mathbf{H} = (H_x, H_y, H_z)$, each component of each field obeys the wave equation. An electromagnetic wave therefore consists of an electric field which oscillates in a wave-like manner in a plane perpendicular to a magnetic field, which also executes a wave-like motion. This is depicted in Fig. 4.10, in which is shown an electromagnetic wave propagating in the y direction, associated with an oscillating electric field E_x and an oscillating magnetic field H_z. Reference to Fig. 4.10 shows that the local motion, namely the oscillation of both the electric and the magnetic field, is transverse to the direction of propagation, and so all electromagnetic waves are transverse.

In general, an electromagnetic wave consists of a combination of waves of the type shown in Fig. 4.10, but with superposition of the waves due to all three components of each of the two fields. In certain cases, however, it is possible to produce electromagnetic waves in which there is solely one electric field component, say E_x, and only one magnetic field component, which in this case would be H_z. The electric field component and the magnetic field component therefore define two mutually perpendicular planes in space, and this type of wave is called a *plane-polarised*, or simply *polarised*, wave.

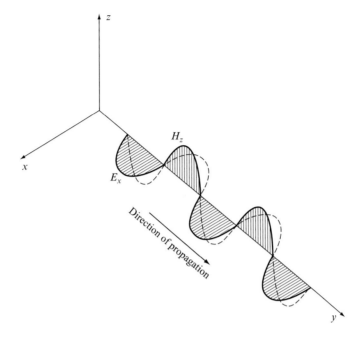

Fig. 4.10 *An electromagnetic wave.* The **E** vector oscillates in the xy plane, and the **H** vector in the yz plane. The wave propagates in the y direction, and is plane polarised. The plane of polarisation is defined as that of the **E** vector, namely the xy plane.

To define the polarisation, we specify the plane of the electric field component, which we call the *plane of polarisation*. Figure 4.10 therefore represents a plane-polarised electromagnetic wave, whose plane of polarisation is the xy plane, and whose direction of propagation is the y direction.

It should be noted that the velocity c of electromagnetic radiation is a constant only for propagation in a particular medium. If electromagnetic waves travel in a different medium, then the velocity will be different, but the frequency–wave-vector relationship $\omega = kc$ still holds. As a consequence of the conservation of energy, it may be proved that when electromagnetic radiation passes from one medium to another, the frequency ω stays constant. This is because the energy E of a wave is given by $E = h\omega/2\pi$, where h is Planck's constant, and the energy E is independent of the medium in which the wave propagates. Hence, if we have two media in which the velocities of electromagnetic radiation are c_1 and c_2 and in which the wave vectors are $\mathbf{k}_1$ and $\mathbf{k}_2$, then we have

$$\omega = k_1 c_1 = k_2 c_2$$

But since

$$k = \frac{2\pi}{\lambda}$$

then

$$\frac{c_1}{c_2} = \frac{\lambda_1}{\lambda_2}$$

The ratio of the velocities of propagation is therefore equal to the ratio of the wavelengths of the radiation in the two media. The velocity of propagation of electromagnetic waves in free space (*in vacuo*) is currently held to be $2.997925 \times 10^8 \, \text{ms}^{-1}$, and this medium is used as a reference medium for other media such as air and water. The *refractive index* n of a medium is defined as the ratio of the velocity of electromagnetic radiation in free space c_0 to that in the medium c_{med}:

$$n = \frac{c_0}{c_{med}} = \frac{\lambda_0}{\lambda_{med}} \tag{4.42}$$

4.13   The interaction of electromagnetic radiation with matter

When an electromagnetic wave passes through matter, what is really happening is that an electric field and a magnetic field of very high frequency are propagated through the matter. The effect of the wave is consequently due to the interaction of these fields with the molecules contained in the matter. In general, the effect of the magnetic field is very small compared with that of the electric field, and so let us see what happens when an electric field $\mathbf{E}$ interacts with a particle, which we shall suppose has a charge q and is free to respond to the electric field. The validity of the latter assumption is discussed at length in Chapters 6 and 9, but for the moment we will not question it. All we require is an order-of-magnitude feeling for the interaction of high-frequency electric fields with particles.

Whenever a particle of electric charge q is placed in an electric field $\mathbf{E}$, it experiences a force $\mathbf{F}$ given by

$$\mathbf{F} = q\mathbf{E}$$

If the particle has a mass m, the force $\mathbf{F}$ will give rise to an acceleration $\mathbf{a}$, as

$$\mathbf{F} = m\mathbf{a}$$
$$\therefore q\mathbf{E} = m\mathbf{a}$$
$$\therefore \mathbf{a} = \frac{q}{m}\mathbf{E}$$

This equation tells us that the particle will accelerate proportionally to its charge-to-mass ratio. But we must remember that the electric field $\mathbf{E}$ is that due to an electromagnetic wave, and so it oscillates with a very high frequency, implying that the particle itself will oscillate with a very high frequency. Now, another result of the theory of electromagnetism is that any charged particle may act as a source of electromagnetic radiation if it is caused to oscillate. We see that the effect of the incident wave is to cause the particle to oscillate, thereby making it a source of radiation. Electromagnetic

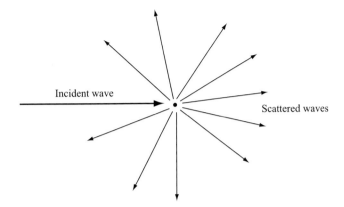

Fig. 4.11 *Scattering.* An incident electromagnetic wave interacts with a charged particle, causing it to oscillate. As a result of electromagnetic theory, an oscillating particle acts as a source of radiation. Waves, called scattered waves, are therefore emitted by the particle, which is said to act as a scattering centre.

waves will therefore be propagated from the particle; this phenomenon is known as *scattering,* and the particle is said to act as a *scattering centre,* as shown in Fig. 4.11.

Further, the intensity of the scattered waves may be proved to be proportional to the square of the acceleration of the particle:

$$I \propto |\mathbf{a}|^2 \propto \left(\frac{q}{m}\right)^2 |\mathbf{E}|^2 \tag{4.43}$$

This states that the intensity of the radiation scattered by a particle of mass m and charge q depends on the ratio $(q/m)^2$.

These results are highly significant. The purpose of this book is principally to describe the interaction of X-rays with crystals. X-rays are a form of electromagnetic radiation, and they interact with particles within the crystal as described above. Specifically, any crystal will contain several types of particle: electrons, protons, neutrons, nuclei, ions, atoms and molecules. Let us see what the effect is of the X-rays on each of these particles.

To do this, we must be careful to consider the size of the particle which interacts with the wave. If, for instance, the particle is large compared with the wavelength of the radiation, as shown in Fig. 4.12(a), there will be several cycles of the wave contained in the same volume of space as that occupied by the particle. The particle will experience the average electric field, which will be close to zero in this instance. On the other hand, if the particle is small compared with the wavelength of the radiation, as in Fig. 4.12(b), it will 'see' only a very small portion of a cycle of the wave, and will experience a well-defined electric field **E**.

This argument, however, depends on what we choose to call a 'particle'. For instance, matter contains 'particles' of several sizes, among which are electrons, nuclei,

(a)

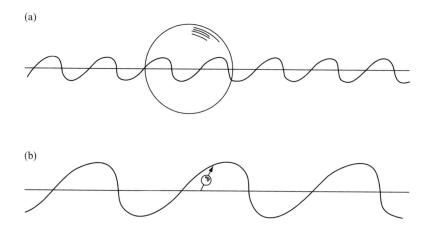

(b)

Fig. 4.12 *The interaction of electromagnetic waves with particles of different sizes.* (a) shows the **E** vector wave interacting with a particle which is large compared with the wavelength of the wave. The particle experiences an average electric field which is close to zero. (b) shows the **E** vector of a wave interacting with a particle small compared with the wavelength of the wave. The particle sees a well-defined electric field, and will respond accordingly.

atoms and molecules. The critical particle size referred to above concerns which particles in general may be considered to be small as compared with the wavelength of the radiation in which we are interested. For example, a molecule may be thought of as an entity in itself or as an aggregate of smaller particles, namely atoms. Since a molecule is in general much larger than the wavelength of an X-ray, which is typically of the order of 10^{-10} m, it will not interact with X-rays as a single particle. In contrast, the atoms, which are of the same size as the X-ray wavelength, will interact with the X-rays, and the effect of the X-rays on the molecule as a whole will be the aggregate of the effects of the X-rays on each atom. Similarly, since electrons and nuclei are also much smaller than an atom, they will also interact with X-rays, so that each atomic effect is really due to the net effect of the electrons, protons and neutrons contained within the atom. The fundamental phenomenon is therefore an interaction of the X-rays with the electrons and nuclei. The total effect due to the components of each atom will then give rise to an effect attributable to the atom as a whole, and the net effect of all the atoms accounts for the effect of the molecules. Then, we may see how the effects of each molecule interact, and establish the total effect on the X-rays due to the crystal as a whole.

We now investigate what the fundamental effect of X-rays is on matter by finding out the interaction between X-rays and neutrons, protons and electrons.

Since neutrons have no charge, they will not scatter X-rays. Protons have the elemental charge e and a mass m_p, and electrons also have the elemental charge e, but a mass m_e. A proton, however, is 1837 times heavier than an electron:

$$m_p = 1837 m_e$$

The intensities of the radiation scattered by an electron I_e and by a proton I_p are given by equation (4.43):

$$I_p \propto \left(\frac{e}{m_p}\right)^2$$

$$I_e \propto \left(\frac{e}{m_e}\right)^2$$

$$\therefore \frac{I_e}{I_p} \propto \left(\frac{e}{m_e}\right)^2 \left(\frac{m_p}{e}\right)^2$$

$$= 1837^2$$

Hence electrons scatter X-rays some one million times more effectively than protons. An X-ray diffraction experiment has the effect of measuring the waves scattered by a crystal, and this result tells us that the scattering is overwhelmingly caused by the electrons. X-ray diffraction therefore looks at the electron distribution in a crystal, and not directly at the positions of the nuclei of atoms. It is only since we know that electrons tend to be close to nuclei that we can infer nuclear positions.

Equation (4.43) has another interesting result if we now consider the total effect of all the electrons in an atom in order to find out the net atomic scattering effect on a beam of X-rays. For an atom of atomic number Z, the total electronic charge is Ze. If for the moment we assume that all the electrons in the atom 'see' the same electric field, then they will all oscillate together, and the waves scattered by each electron will be in phase with those waves scattered by all the other electrons in the same atom. The waves therefore interfere constructively, giving the effect that the waves are due to a single particle of charge Ze. The intensity of the total scattering from a single atom is therefore proportional to $(Ze/m_e)^2$, and so increases as Z^2. In fact, the assumption we have made is not strictly true, as will be explained in more detail in Chapter 9, but to an order of magnitude, we do obtain the physically correct result. Since the scattering by an atom increases as Z^2, heavier atoms of high Z will be much more effective scattering centres than light atoms of low Z. The higher the atomic number of the atom, the greater the scattering, and so heavy atoms such as mercury will give much larger effects than light ones such as hydrogen, carbon, nitrogen and oxygen.

We now have some feeling for the interpretation of an X-ray diffraction experiment. What we measure are the intensities of scattered X-rays, which enable us to determine the location of the electrons in the unit cell of the crystal. This is portrayed on an *electron density map*, which looks somewhat similar to a geographical map in which the contour lines show the heights of mountains. An electron density map for the molecule anthracene is shown in Fig. 4.13.

The contour lines in Fig. 4.13 show the regions where the electron density, or the number of electrons per unit volume, takes various values. Where the electron density is large, we may infer the presence of a nucleus. Since we know that heavy atoms scatter X-rays more effectively than light ones, the positions where the electron density is largest will correspond to the positions of the nuclei of the heaviest atoms. In this way, we are able to locate and identify the positions of the atoms and the

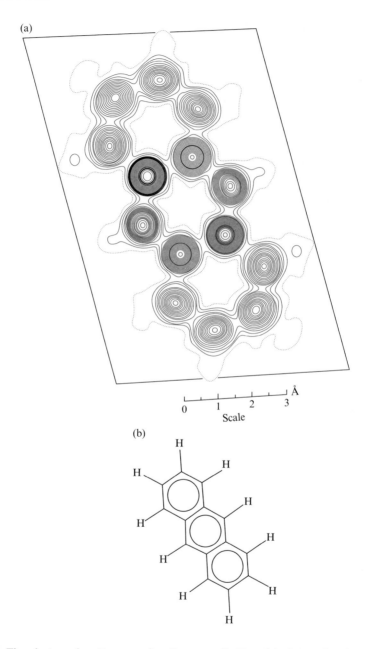

Fig. 4.13 *The electron density map of anthracene,* $C_{14}H_{10}$. (a) shows the electron density map as derived by X-ray diffraction; (b) shows the structural formula, in total agreement with the electron density map. This is a beautiful demonstration of what can be achieved in the crystal structure analysis of small molecules. Proteins do not usually yield electron density maps of this resolution, but in some cases they do. From V. C. Sinclair, J. M. Robertson and A. McL. Mathieson, *Acta Crystallogr.* **3**, 251–256 (1950) (reproduced with permission).

molecules within a unit cell of the crystal, and since the unit cell is repeated regularly throughout all space, the crystal and molecular structures have been determined.

Summary

Any phenomenon which may be described by an amplitude function $\psi(\mathbf{r}, t)$ which satisfies the three-dimensional partial differential equation

$$\frac{\partial^2 \psi(\mathbf{r}, t)}{\partial x^2} + \frac{\partial^2 \psi(\mathbf{r}, t)}{\partial y^2} + \frac{\partial^2 \psi(\mathbf{r}, t)}{\partial z^2} = \frac{1}{v^2}\frac{\partial^2 \psi(\mathbf{r}, t)}{\partial t^2} \tag{4.21}$$

(or its one- or two-dimensional analogue) is known as a *wave*. This equation has the particular solution

$$\psi(\mathbf{r}, t) = \psi_0 e^{i(\mathbf{k}\cdot\mathbf{r}-\omega t)} \tag{4.35}$$

corresponding to a plane wave, and the general solution

$$\psi(\mathbf{r}, t) = \sum_n \psi_n e^{i(\mathbf{k}_n\cdot\mathbf{r}-\omega_n t+\phi_n)} \tag{4.37}$$

in which the parameters ψ_n, k_n, ω_n, ϕ_n and the extent of the summation are determined by the appropriate boundary conditions.

The *wave vector* $\mathbf{k}$ is a vector in the direction of wave propagation and has a magnitude $2\pi/\lambda$, where λ is the wavelength of the wave.

The *frequency* ω of the wave is defined in terms of the periodic time τ of the wave as $\omega = 2\pi/\tau$.

The *phase velocity* v of the wave is the velocity with which the wave crests of the wave travel, and is related to $\mathbf{k}$ and ω as $v = \omega/k$.

The general form of equation (4.37) is a mathematical statement of the *principle of superposition*, a verbal statement of which is: The total wave disturbance due to an array of sources is the sum of the individual disturbances due to each source.

The *phase* of a wave is the quantity $(\mathbf{k} \cdot \mathbf{r} - \omega t + \phi)$.

The *intensity* I of a wave is the square of the magnitude of the wave amplitude:

$$I = |\psi(\mathbf{r}, t)|^2.$$

An *electromagnetic wave* is a wave in which an electric field $\mathbf{E}$ and a magnetic field $\mathbf{H}$ oscillate in a wave-like form in two mutually perpendicular planes. Electromagnetic waves may be classified according to their wavelength: for example, waves of wavelength between 4×10^{-7} m and 7×10^{-7} m constitute visible light; X-rays have wavelengths typically of the order of 10^{-10} m.

When an electromagnetic wave passes through matter, the electric field causes charged particles to oscillate, with the result that the charged particles themselves act as sources of radiation. This is known as *scattering*. The fundamental event occurring when X-rays pass through matter is scattering by electrons and protons. Electrons are much more effective at scattering X-rays than are protons, and so the total scattering due to a crystal is essentially that due to the electrons. The aim of an X-ray diffraction experiment is to determine the intensity of the scattered waves, and from this to deduce the electron density within a unit cell of the crystal. Peaks of electron density may be associated with the positions of atomic nuclei, and the more dense the peak, the higher the atomic number of the atom at that site.

Bibliography

Braddick, H. J. J. *Vibrations, Waves and Diffraction.* McGraw-Hill, London, 1965.

Pain, H. J. *The Physics of Vibrations and Waves*, 6th edn. Wiley, New York, 2005.

Both these books discuss the general physics of wave phenomena, and both deal with Fourier techniques. Pain is stronger on the treatment of electromagnetic wave theory, while Braddick discusses diffraction and associated Fourier techniques more thoroughly.

Bleaney, B. I. and Bleaney, B. *Electricity and Magnetism*, 3rd edn. Oxford University Press, Oxford, 1989. A very clear treatment of all aspects of electromagnetic theory.

5

Fourier transforms and convolutions

This chapter will present the remaining mathematical techniques necessary for the understanding of the theory of diffraction. The discussion will follow the approach of Chapter 2, in that emphasis will be placed on the results of the mathematical formalism, and the way in which these results are used. All examples will be explained in detail, and the result of each one will be used elsewhere in this volume.

5.1 Integrals

Much of this chapter will be concerned with integrals of various types, and so we will first investigate some properties of integrals in general.

There are two possible ways of writing an integral. The first does not have the limits of integration specified, and is called an *indefinite integral*, whereas the second does have defined limits of integration, and is called a *definite integral*. If, as an example, we take the integral of an arbitrary function of x, $f(x)$, an indefinite integral has the form

$$\int f(x)\, dx$$

and a definite integral is written as

$$\int_{x=a}^{x=b} f(x)\, dx \quad \text{or} \quad \int_a^b f(x)\, dx$$

The expression contained within the $\int \ldots dx$ notation, which in this case is just $f(x)$, is referred to as the *integrand*.

As we shall now demonstrate, the information contained in an integral depends on whether the integral is definite or indefinite. Defining $f(x)$ as

$$f(x) = x^2$$

the indefinite integral $\int f(x)\, dx$ may be determined as

$$\int f(x)\, dx = \int x^2\, dx = \tfrac{1}{3}x^3 + c$$

in which c is a constant of integration.

For a definite integral with two arbitrary limits a and b, which are both assumed to be numerical constants, we have

$$\int_a^b f(x)\,dx = \int_a^b x^2\,dx = \tfrac{1}{3}(b^3 - a^3)$$

Since a and b are pure numbers, the result of the definite integral is also a pure number. This is to be contrasted with the result of the indefinite integral, which is a function of the integrand variable, x.

This is a general result which may be stated verbally as follows: For a function $f(x)$ of a single variable x, a definite integral results in a pure number, and an indefinite integral results in another function of x, say, $g(x)$. Mathematically, this becomes

$$\int f(x)\,dx = g(x) \tag{5.1}$$

$$\int_a^b f(x)\,dx = \text{a pure number} \tag{5.2}$$

These statements are important, since they are quite general and may be applied to the integration of any function of a single variable. No matter how complicated the function $f(x)$ may be, it is useful to know whether the result of an integration operation will give rise to another function of x, or simply some number. This information often serves to point to the significance or meaning of an otherwise very complicated looking-expression.

Let us now consider the integration of an arbitrary function $f(x, y)$ which is now a function of two variables x and y. The integration may be performed over either of the variables x or y, and may be definite or indefinite. Four types of integral are possible:

$$\int f(x, y)\,dx \quad \int f(x, y)\,dy$$

$$\int_{x=a}^{x=b} f(x, y)\,dx \quad \int_{y=u}^{y=v} f(x, y)\,dy$$

in which a, b, u and v are pure numbers.

If we take $f(x, y)$ as

$$f(x, y) = x^2 y$$

then the results of the above four integrals are

$$\int f(x, y)\,dx = \tfrac{1}{3}x^3 y + g(y) \tag{5.3}$$

$$\int f(x,y)\, dy = \tfrac{1}{2}x^2 y^2 + h(x) \tag{5.4}$$

$$\int_{x=a}^{x=b} f(x,y)\, dx = \tfrac{1}{3}(b^3 - a^3)y + g(y) \tag{5.5}$$

$$\int_{y=u}^{y=v} f(x,y)\, dy = \tfrac{1}{2}(v^2 - u^2)x^2 + h(x) \tag{5.6}$$

Because the integrand $f(x,y)$ is a function of two variables, instead of there being a constant of integration, we have two arbitrary functions of integration $g(y)$ and $h(x)$. Note that the functions of integration are functions of the variable over which the integral is *not* taken.

The indefinite integrals, equations (5.3) and (5.4), are seen to be functions of both variables, x and y, whereas the definite integrals come out as functions of the variable over which the integrals were not taken. Symbolically, we may express this as

$$\int f(x,y)\, dx = g(x,y) \tag{5.7}$$

$$\int f(x,y)\, dy = g'(x,y) \tag{5.8}$$

$$\int_{x=a}^{x=b} f(x,y)\, dx = h(y) \tag{5.9}$$

$$\int_{y=u}^{y=v} f(x,y)\, dy = h'(x) \tag{5.10}$$

in which $g(x,y)$, $g'(x,y)$, $h(y)$ and $h'(x)$ are those functions which are obtained as a result of the integrations.

Once again, the generalisations expressed by equations (5.7)–(5.10) are very important. Many integrals met in this chapter will be definite integrals, taken over one variable, of functions of two variables, such as that represented by equation (5.9). Although the integration processes themselves may be quite involved algebraically, it is of great use for us to be able to look at an integral and to tell at a glance what sort of function will be derived as a result of the integration procedure. Knowing that an integral will turn out to be a function of one variable in particular enables us to concentrate on the properties of the relevant variable in order to establish the significance of the integral.

5.2 Curve sketching

Most of the integrals to be met during this chapter will be formed from integrands of the general type $f(x,y)g(x)$, in which we have a function of two variables multiplied by a function of one variable. In particular, the function of two variables will often be the one-dimensional expression e^{ikx}, in which the variables are k and x, or its three-dimensional analogue $e^{i\mathbf{k}\cdot\mathbf{r}}$, which has the vector variables $\mathbf{k}$ and $\mathbf{r}$.

Our interest centres on integrands of the general form $f(x)e^{ikx}$ or $f(\mathbf{r})e^{i\mathbf{k}\cdot\mathbf{r}}$. In order to obtain some feeling for these functions, it is useful if we are readily able to draw sketch graphs which indicate the salient behaviour of functions of this type.

Let us consider the graph of $f(x)e^{ikx}$, where $f(x)$ is any arbitrary function of x, which we shall assume is known in any particular instance. Our first task is to draw the graphs of e^{ikx} and $f(x)$ separately, and then to form the product function $f(x)e^{ikx}$. The function e^{ikx}, as we saw in the previous chapter, may be considered as representing the spatial variation of a one-dimensional wave with maximum amplitude unity and of wave vector $k = 2\pi/\lambda$. In one dimension, of course, the wave vector $\mathbf{k}$ is reduced to a scalar. In the following paragraphs, for simplicity we will sketch only the real, or cosine, component of e^{ikx}, which is shown in Fig. 5.1(a), with maximum amplitude unity, and values of zero whenever

$$kx = \pm\frac{\pi}{2}, \pm\frac{3\pi}{2}, \pm\frac{5\pi}{2}, \dots$$

But since

$$k = \frac{2\pi}{\lambda}$$

the corresponding values of x are

$$x = \pm\frac{\lambda}{4}, \pm\frac{3\lambda}{4}, \pm\frac{5\lambda}{4}, \dots$$

Also, the real component of e^{ikx} has maxima and minima, of values ± 1, when

$$kx = 0, \pm\pi, \pm2\pi, \pm3\pi, \dots$$

with values of x of

$$x = 0, \pm\frac{\lambda}{2}, \pm\lambda, \pm\frac{3\lambda}{2}, \dots$$

The values of x which specify maxima, minima and zeros are sufficient to allow us to sketch the graph of e^{ikx}. In general, the 'width' of the waves will be determined by the value of $k = 2\pi/\lambda$ chosen.

The graph of $f(x)$ depends on the nature of $f(x)$ itself, but in general it will have a fairly simple algebraic form. However, for the purposes of this discussion, let us suppose that we know that a particular function $f(x)$ varies with x as depicted in Fig. 5.1(b).

Consideration of the product $f(x)e^{ikx}$ leads to the following discussion. When either $f(x)$ or e^{ikx} is individually zero, then the product function $f(x)e^{ikx}$ must be

(a)

(b)

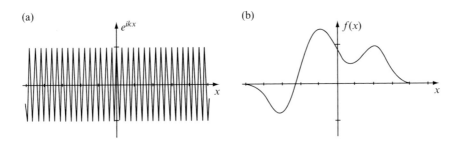

Fig. 5.1 *We desire the product $f(x)e^{ikx}$. The real component of e^{ikx} is a wave as shown in* (a). *$f(x)$ is an arbitrary function of x as in* (b).

zero. We know that the real component of e^{ikx} is zero when

$$x = \pm\frac{\lambda}{4}, \pm\frac{3\lambda}{4}, \pm\frac{5\lambda}{4}, \cdots$$

and the graph of $f(x)$ will allow us to find those values of x for which $f(x)$ is zero. We will then have all those values of x for which the product function is zero.

Another series of points of interest are those for which the real component of e^{ikx} is ±1, which we know will occur when

$$x = 0, \pm\frac{\lambda}{2}, \pm\lambda, \pm\frac{3\lambda}{2}, \cdots$$

Consider, for example, the point $x = 2\lambda$. We know that e^{ikx} has a value of $+1$, and from the graph of $f(x)$ we may determine the value of this function at the point $x = 2\lambda$, which we denote $f(2\lambda)$. The product function is therefore $1 \times f(2\lambda) = f(2\lambda)$, which is just the value of the function $f(x)$ at the point in question. Similarly, when $x = \lambda/2$, e^{ikx} is -1, and the value of the product function $f(x)e^{ikx}$ at this point is simply $-f(\lambda/2)$. At the points

$$x = 0, \pm\frac{\lambda}{2}, \pm\lambda, \pm\frac{3\lambda}{2}, \cdots$$

the values of the product function $f(x)e^{ikx}$ are therefore alternately plus and minus the value of $f(x)$ at the appropriate point. We already have the graph of $f(x)$, and so we may immediately determine the values of $f(x)$ at

$$x = 0, \pm\frac{\lambda}{2}, \pm\lambda, \pm\frac{3\lambda}{2}, \cdots$$

enabling these points to be plotted on the graph of the product function. A significant point to notice is that e^{ikx} can never take a value greater than $+1$ or less than -1. Consequently, on forming the product $f(x)e^{ikx}$, the product function can never be greater in value than $f(x)$ itself, or less in value than $-f(x)$. The product function must therefore be 'enclosed' or 'enveloped' by the function $f(x)$.

With these thoughts in mind, we may set down the rules by which any graph of the form $f(x)e^{ikx}$ may be drawn:

(1) Draw the graphs of $f(x)$ and e^{ikx} separately.
(2) Reflect the graph of $f(x)$ in the x axis to obtain the graph of $-f(x)$.
(3) On a graph which will ultimately be that of the product function $f(x)e^{ikx}$, sketch lightly the forms of $f(x)$ and $-f(x)$.
(4) Mark the points

$$x = \pm\frac{\lambda}{4}, \pm\frac{3\lambda}{4}, \pm\frac{5\lambda}{4}, \ldots$$

at which we know the product function must be zero. Fill in any other zero values known from the graph of $f(x)$.
(5) At the points

$$x = 0, \pm\frac{\lambda}{2}, \pm\lambda, \pm\frac{3\lambda}{2}, \ldots$$

mark in the values of $f(x)$ or $-f(x)$, taking alternate positive and negative values corresponding to e^{ikx} being equal to ±1.
(6) The graph of the product function now has marked on it those values where the product function is zero, and where it has maxima and minima.
(7) Fill in intermediate values with a general sinusoid.

These steps are shown in Fig. 5.2, for the function $f(x)$ of Fig. 5.1. As can be seen, the graph of the product function is a general sinusoid completely enclosed by the function $f(x)$, which is said to act as an *envelope*. Any function of the general type $f(x)e^{ikx}$ will be a general sinusoid enveloped by $f(x)$, and so this allows us to obtain sketch graphs of functions of this type with great ease, according to the rules above.

In this discussion we have only considered the real (or cosine) component of e^{ikx}, since in many crystallographic applications e^{ikx} reduces to the cosine form, as we shall see later. However, functions of the types of both $f(x)\cos kx$ and $f(x)\sin kx$ are general sinusoids enveloped by the function $f(x)$. The only difference between the forms occurs in those values of x for which the product function is zero or has maxima and minima. This is determined by the behaviour of the trigonometric functions $\sin kx$ and $\cos kx$. In general,

$$\cos kx = 0 \text{ when } x = \pm\frac{\lambda}{4}, \pm\frac{3\lambda}{4}, \pm\frac{5\lambda}{4}, \ldots$$

$$= \pm1 \text{ when } x = 0, \pm\frac{\lambda}{2}, \pm\lambda, \pm\frac{3\lambda}{2}, \ldots$$

and

$$\sin kx = 0 \text{ when } x = 0, \pm\frac{\lambda}{2}, \pm\lambda, \pm\frac{3\lambda}{2}, \ldots$$

$$= \pm1 \text{ when } x = \pm\frac{\lambda}{4}, \pm\frac{3\lambda}{4}, \pm\frac{5\lambda}{4}, \ldots$$

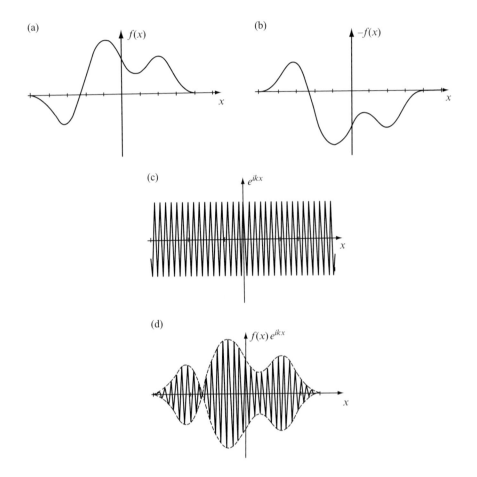

Fig. 5.2 *The formation of the product $f(x)e^{ikx}$.* We first draw $f(x)$ as in (a). On reflection in the x axis, we obtain $-f(x)$ as in (b). The real component of e^{ikx} is shown in (c). The product $f(x)e^{ikx}$ is depicted in (d). Having sketched in the form of the envelope of $f(x)$ and $-f(x)$, we plot the zeros and the maxima and minima which lie on the envelope. The remainder of the curve is completed with a sinusoid.

If we are careful to ensure that we have plotted the zeros in the correct places, then the maxima and minima occur exactly halfway between two successive zeros.

As an example, we shall draw the graph of the function $(\sin x)/x$, which will be obtained as a result of an integral in a later section. This function may be written as $(1/x)(\sin x)$, so that we identify $f(x) = 1/x$ and $k = 1$, so that $\lambda = 2\pi$. The graphs of $1/x$ and $\sin x$ as a function of x are shown in Figs 5.3(a) and 5.3(b), and the product function is depicted in Fig. 5.3(c).

All steps are carried out as in the rules above, by ensuring that the zeros are in the correct places. We find, however, that when $x = 0$ the product function $(\sin x)/x$

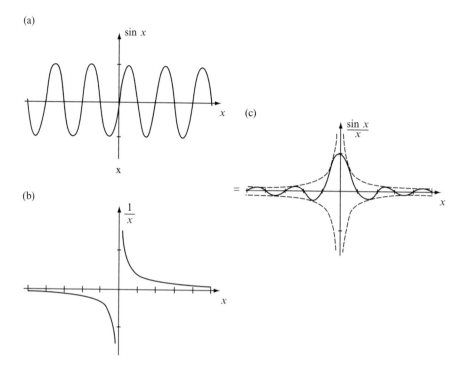

Fig. 5.3 *The function* $(\sin x)/x$. $\sin x$ is shown in (a), and $1/x$ in (b). The product function $\sin x)/x$ is a sinusoid enveloped by $1/x$ and has the value of unity at $x = 0$. The central peak has a width of π, and all other peaks a width of π.

is $0/0$, which is indeterminate. This difficulty is resolved when we consider the power series expansion of $\sin x$, which is valid for small values of x:

$$\sin x = x - \frac{x^3}{3!} + \frac{x^5}{5!} - \frac{x^7}{7!} + \cdots$$

where, for example, $5! = 5 \cdot 4 \cdot 3 \cdot 2 \cdot 1$.

Dividing throughout by x, we have

$$\frac{\sin x}{x} = 1 - \frac{x^2}{3!} + \frac{x^4}{5!} - \frac{x^6}{7!} + \cdots$$

so that when x is zero,

$$\left(\frac{\sin x}{x} \right)_{x=0} = 1$$

The value of $(\sin x)/x$ at the origin is therefore unity. The general form of this graph is a sinusoid which is sharply peaked at the origin. Notice that the width of the central peak is 2π, whereas the widths of all the other subsidiary peaks are just one-half of this, namely π.

5.3   Fourier transforms

There are many examples of mathematical expressions which are encountered in similar form throughout the different branches of science. An instance which we have already met is the wave equation, which is used to describe all forms of wave motion. Another very common mathematical form is the one-dimensional integral

$$\int_{-\infty}^{\infty} f(x)e^{ikx}\, dx$$

which is particularly relevant to our purposes, since it is very important in the mathematical theory of diffraction. Our discussion of integrals earlier in this chapter has shown that, since the integrand is a function of two variables, namely k and x, and that the integration is over x, then the result of the integration is a function of k, which we shall write as $F(k)$:

$$F(k) = \int_{-\infty}^{\infty} f(x)e^{ikx}\, dx \tag{5.11}$$

Integrals of this type are so common that they are given a special name. The function of k defined by equation (5.11) is called the *Fourier transform* of $f(x)$, or simply the *transform* of $f(x)$. The use of the abbreviated title 'transform' may be used only when it is known that we are referring to the Fourier transform, for there are other types of integrals, also called transforms, for example the Laplace transform. In this volume, however, the Fourier transform is the only such integral we shall meet, and so the term 'transform' will always imply the Fourier transform.

When we speak of 'taking the Fourier transform of a function of x' we mean that we wish to compute the function $F(k)$ defined by equation (5.11). As we shall see, in any diffraction situation we shall always be required to compute the Fourier transform of some function, and since the operation in which we compute the transform is carried out so often, we may symbolise equation (5.11) in a rather more compact form as

$$F(k) = Tf(x)$$

where the symbol T stands for 'is the Fourier transform of'.

Equation (5.11) has defined the Fourier transform of a one-dimensional function, but the concept may easily be extended to three dimensions as

$$F(\mathbf{k}) = \int_{\text{all } \mathbf{r}} f(\mathbf{r})e^{i\mathbf{k}\cdot\mathbf{r}}\, d\mathbf{r} = Tf(\mathbf{r}) \tag{5.12}$$

in which the integrand is a function of the vector variables $\mathbf{r}$ and $\mathbf{k}$. Since the integration is over the variable $\mathbf{r}$, the result of the integration is a function of $\mathbf{k}$, which we represent as $F(\mathbf{k})$. With regard to the three-dimensional integral equation (5.12), we must be careful about the meaning of the symbol $d\mathbf{r}$. Since the integral is in three dimensions, the appropriate differential is (in Cartesian coordinates) the volume element $dV = dx\ dy\ dz$ with the limits being plus and minus infinity for x, y and z so that the integral is computed over all space. We would then write equation (5.12) as

$$F(\mathbf{k}) = \int_{\text{all space}} f(\mathbf{r})e^{i\mathbf{k}\cdot\mathbf{r}}\,dV$$

This form, however, obscures to a certain extent the relationship between the variable of integration and the variable $\mathbf{r}$ which occurs in the integrand. For this reason, this book will continue to express three-dimensional integrals using the vector differential $d\mathbf{r}$, which in this situation signifies the volume element $dx\ dy\ dz$. The integral is then taken over all values of the vector $\mathbf{r}$, which will cover all space. Although this device is not mathematically strictly correct, it does have the advantage of emphasising the relevant relationships between the variables in the integrand and the variable of integration.

Let us now consider the meaning of equations (5.11) and (5.12). Concentrating firstly on the integrand of equation (5.11), namely $f(x)e^{ikx}$, we see that this is a product function of the type we discussed in the previous section. If we know the function $f(x)$, then we may sketch the product function $f(x)e^{ikx}$ for any specific value of k we might choose. This will turn out to be a general sinusoid, enveloped by the function $f(x)$, with a wavelength determined by the value of k. Having sketched the graph of the product function, we may then evaluate the integral

$$F(k) = \int_{-\infty}^{\infty} f(x)e^{ikx}\,dx$$

by finding the total area contained between the product graph and the x axis. As long as the product function does not tend to infinity anywhere, and converges as x tends to plus and minus infinity, then this area will be measurable, and so the integral will have a well-defined value. We may then choose a new value of k, sketch the new product function, and find the new area. This process may be continued for all values of k: for each value of k, the area will have a different value, and we may then tabulate the values of k and the corresponding areas. Since each area corresponds to the integral

$$F(k) = \int_{-\infty}^{\infty} f(x)e^{ikx}\,dx$$

for a chosen value of k, we then see that the value of this integral is indeed a function of k. This is the function we call $F(k)$. As an example, some graphs of $f(x)e^{ikx}$ for a given function $f(x)$, for various k values, are shown in Fig. 5.4.

Exactly similar considerations apply to the three-dimensional integral equation (5.12). Although equations (5.11) and (5.12) are possibly somewhat strange at first sight, their mathematical meaning is quite straightforward. Their physical significance will be discussed in depth in Chapter 6.

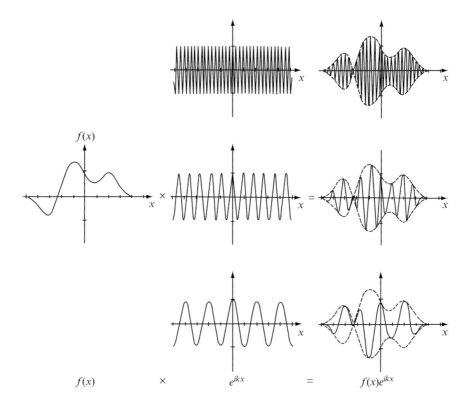

Fig. 5.4 *The function $\int_{-\infty}^{\infty} f(x)e^{ikx}\,dx$.* On the left is an arbitrary function $f(x)$. The centre shows three different functions of the type e^{ikx}. Since $k = 2\pi/\lambda$, the smaller λ, the larger the value of k. The product functions $f(x)e^{ikx}$ for each value of k are shown in the third column. The integral $\int_{-\infty}^{\infty} f(x)e^{ikx}\,dx$ represents the area under the curve of the product function and, as can be seen, this varies with the k value of the waveform. For any value of k, we may plot the product function and find the appropriate area.

In general, we may usually compute the Fourier transform $F(k)$ of any function $f(x)$ by ordinary analytical means. If, however, $f(x)$ is a particularly bizarre function which defies analytical integration then a computer may be programmed to generate values of the function $F(k)$ over a defined range of values of k. The computer does this by calculating the area of the graph of the product function $f(x)e^{ikx}$ for those values of k we require, just as we described above.

During the remainder of this volume, we shall meet many examples of Fourier transforms. In order to denote those functions which are the Fourier transforms of other functions, we shall use lower-case letters, for example $f(x)$, to denote our initial functions, and upper-case letters, for example $F(k)$, to symbolise those functions which are the Fourier transforms of the original functions $f(x)$:

$$F(k) = Tf(x)$$

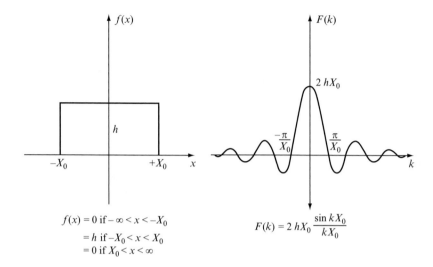

$$f(x) = 0 \text{ if } -\infty < x < -X_0$$
$$= h \text{ if } -X_0 < x < X_0$$
$$= 0 \text{ if } X_0 < x < \infty$$

$$F(k) = 2\,hX_0 \, \frac{\sin kX_0}{kX_0}$$

Fig. 5.5 *The Top Hat and its Fourier transform.* Note that the width of the Top Hat function is $2X_0$, and the width of the central peak of the Fourier transform of the Top Hat is $2\pi/X_0$. Note also that the height of the peak in the Fourier transform is the area under the Top Hat.

Our first example of the actual computation of a Fourier transform is the calculation of the so-called 'Top Hat' function, which is defined as follows. If

$$-\infty < x < -X_0, \qquad f(x) = 0$$
$$-X_0 \le x \le +X_0, \qquad f(x) = h$$
$$X_0 < x < \infty, \qquad f(x) = 0$$

The function is a 'top hat' of height h and width $2X_0$ symmetrically placed about the origin, as shown in Fig. 5.5.

The Fourier transform of $f(x)$ is defined as

$$F(k) = \int_{-\infty}^{\infty} f(x)e^{ikx} \, dx \tag{5.11}$$

Since $f(x)$ vanishes if x is greater than X_0 or less than $-X_0$, the integrand will also vanish in these ranges. The integrand is non-zero between $-X_0$ and $+X_0$ only, for which the value of $f(x)$ is a constant h which may be factored outside of the integral. The Fourier transform of the Top Hat therefore reduces to

$$F(k) = h \int_{-X_0}^{X_0} e^{ikx} \, dx$$

$$= h \left[\frac{e^{ikx}}{ik} \right]_{-X_0}^{X_0}$$

$$= h \frac{e^{ikX_0} - e^{-ikX_0}}{ik}$$

But equation (2.34) of Chapter 2 may be stated as

$$\sin \theta = \frac{e^{i\theta} - e^{-i\theta}}{2i}$$

Identifying

$$\theta = kX_0$$

we have

$$F(k) = 2h \frac{\sin kX_0}{k}$$

Multiplying above and below by X_0, we derive

$$F(k) = 2hX_0 \frac{\sin kX_0}{kX_0} \tag{5.13}$$

Equation (5.13) is the Fourier transform of the Top Hat function. The form of equation (5.13) emphasises two points. Firstly, there is a constant term $2hX_0$. Since the original Top Hat function is a rectangle of height h and width $2X_0$, this is simply the area of the Top Hat. Secondly, there is a function of k, namely $(\sin kX_0)/kX_0$ (remember X_0 is a constant, the half-width of the Top Hat). This function is of the type $(\sin x)/x$, which we have already investigated.

Reference to Fig. 5.3 shows that this is a sinusoid sharply peaked at the origin. Of particular interest is the width of the central peak. The first zeros on each side of the origin occur when

$$\sin kX_0 = 0$$

corresponding to

$$kX_0 = \pm \pi$$

and

$$k = \pm \frac{\pi}{X_0} \tag{5.14}$$

The total width of the central peak is therefore $2\pi/X_0$. The full form of the function $F(k)$ is shown in Fig. 5.5.

With regard to the width of the central peak of the transform function, we notice that it is inversely proportional to the width of the original Top Hat. Whereas the Top Hat had a width of $2X_0$, the Fourier transform has a central-peak width of $2\pi/X_0$. From this we infer that the wider the Top Hat, the narrower the Fourier transform, and, conversely, the narrower the Top Hat, the wider the Fourier transform. This is an example of an extremely important general result:

The more extended the function $f(x)$ as a function of x, the narrower the Fourier transform $F(k)$ as a function of k.
The narrower the function $f(x)$ as a function of x, the wider the Fourier transform $F(k)$ as a function of k.

The importance of this cannot be overemphasised, and it should be borne in mind whenever Fourier transforms are calculated. The physical significance will be discussed when we study diffraction theory in Chapter 6.

5.4   Mathematical conventions and physical reality

At this stage, it is pertinent to make a slight detour away from the main mathematical development of this chapter, and to consider some aspects of the way in which scientists use mathematics to interpret the natural world. When a scientist wishes to derive a mathematical formula, he or she desires a succinct and relevant description of some natural event. We are not strictly concerned with the mathematical niceties of the formula used, but, rather, our attention is directed to using a formula which gives the correct physical answers. This is not to say that mathematical formalism is unnecessary. On the contrary, it is vital that the abstract mathematical properties of important functions be well known so that any scientist may apply the mathematics correctly and with confidence, but in practice, mathematical formalism takes a subsidiary role.

One instance of the way in which we use mathematics relates to certain conventions which are accepted universally. For example, whenever we draw the x axis of Cartesian coordinates, we conventionally take values to the right of the origin as positive, and those to the left as negative. There is no particular reason why we should measure coordinates in this manner, but it is useful if everyone accepts the same convention so that there is a universally accepted mathematical language.

The relevance of this preamble to our immediate purpose is that the definition of the Fourier transform as given in equations (5.11) and (5.12) presupposes a mixture of both mathematical formalism and convention.

The first point to be appreciated is that the sign of the exponential in equations (5.11) and (5.12) is a matter of convention. Fourier transforms may equally as well be defined as

$$F(k) = \int_{-\infty}^{\infty} f(x) e^{-ikx} \, dx$$

or

$$F(\mathbf{k}) = \int_{\text{all } \mathbf{r}} f(\mathbf{r}) e^{-i\mathbf{k}\cdot\mathbf{r}} \, d\mathbf{r}$$

In fact, whenever we apply the Fourier transform to a real case, the answer obtained with either the positive or the negative exponential will in the end give rise to the same result. This book will take the convention of defining the Fourier transform as in equations (5.11) and (5.12) using positive exponentials:

$$F(k) = \int_{-\infty}^{\infty} f(x)e^{ikx}\,dx \tag{5.11}$$

$$F(\mathbf{k}) = \int_{\text{all }\mathbf{r}} f(\mathbf{r})e^{i\mathbf{k}\cdot\mathbf{r}}\,d\mathbf{r} \tag{5.12}$$

The second point concerns the fact that often, especially in mathematics texts, the Fourier transform is defined as

$$F(k) = \frac{1}{2\pi} \int_{-\infty}^{\infty} f(x)e^{ikx}\,dx \tag{5.11a}$$

or

$$F(k) = \frac{1}{\sqrt{2\pi}} \int_{-\infty}^{\infty} f(x)e^{ikx}\,dx \tag{5.11b}$$

in which factors of $1/2\pi$ or $1/\sqrt{(2\pi)}$ are present as shown. The reason for this will be discussed in the next section, but we introduce these alternative definitions here since it is relevant to our discussion of the use of mathematics to scientists in practice. It is true that equations (5.11a) and, especially, (5.11b) are appropriate definitions of the Fourier transform. But our main priority is mathematical relevance and not mathematical punctiliousness. We shall, during the course of this book, write the equation for a Fourier transform very many times, and it becomes very tedious to write factors of $1/2\pi$ and $1/\sqrt{(2\pi)}$ continually. What is of importance to us is the functional relationship between significant variables, and not an exact numerical correspondence. The definitions of the Fourier transform which express most concisely the information we require are equations (5.11) and (5.12), and these are the definitions this book will choose. Whenever we require an exact numerical answer, then we will adopt a suitable procedure to correct for any error we have introduced by omitting factors of $1/2\pi$ or $1/\sqrt{(2\pi)}$. In general, this is easily done in the last few lines of any calculation. For the most part, however, we lose no useful information by choosing the definitions given in equations (5.11) and (5.12), and we gain succinctness of expression.

In conclusion, we may now appreciate that the definition of the Fourier transform

$$F(k) = \int_{-\infty}^{\infty} f(x)e^{ikx}\,dx \tag{5.11}$$

$$F(\mathbf{k}) = \int_{\text{all }\mathbf{r}} f(\mathbf{r})e^{i\mathbf{k}\cdot\mathbf{r}}\,d\mathbf{r} \tag{5.12}$$

is a combination of convention, in the sign of the exponential, and of usefulness in the description of physical reality, in that we include no unnecessary numerical factors.

5.5   The inverse transform

We have seen that the Fourier transform $F(k)$ of any one-dimensional function $f(x)$ may be defined by equation (5.11),

$$F(k) = \int_{-\infty}^{\infty} f(x)e^{ikx}\, dx \tag{5.11}$$

It is a remarkable property of the Fourier transform that the integral written as

$$\int_{-\infty}^{\infty} F(k)e^{-ikx}\, dk$$

is related to the original $f(x)$ by

$$f(x) = \int_{-\infty}^{\infty} F(k)e^{-ikx}\, dk \tag{5.15}$$

This result is proved in Appendix I of this chapter, but what is important is not the proof, but the result itself. From the form of the integral in equation (5.15), we see that the integrand is a function of k and x, and that the integral is taken over k, implying that the result of the integration is a function of x. But it is not just *any* function of x, it is that very *special* function of x from which we derived the function $F(k)$ according to equation (5.11). This is indeed a noteworthy result. Having computed $F(k)$ from $f(x)$ by performing the operation of taking the Fourier transform, we may then 'undo' this operation by regenerating the original $f(x)$ from $F(k)$, by computing the integral defined in equation (5.15). For this reason, equation (5.15) is known as the *inverse Fourier transform*, or simply *inverse transform*, of $F(k)$.

In three dimensions, the situation is exactly analogous, for the transform of $f(\mathbf{r})$ is

$$F(\mathbf{k}) = \int_{\text{all } \mathbf{r}} f(\mathbf{r})e^{i\mathbf{k}\cdot\mathbf{r}}\, d\mathbf{r} \tag{5.12}$$

and the inverse transform is

$$f(\mathbf{r}) = \int_{\text{all } \mathbf{k}} F(\mathbf{k})e^{-i\mathbf{k}\cdot\mathbf{r}}\, d\mathbf{k} \tag{5.16}$$

Just as we symbolised the operation of taking the Fourier transform by the symbol T,

$$F(k) = Tf(x)$$

then the operation of taking the inverse transform may be symbolised by T^{-1},

$$f(x) = T^{-1}F(k)$$

This is internally consistent, for if we write

$$F(k) = Tf(x)$$

we may then take the inverse transform of both sides, obtaining

$$T^{-1}F(k) = T^{-1}Tf(x)$$

But the operation T followed by the operation T^{-1}, symbolised as the combination $T^{-1}T$, takes us back to our original function, and so

$$T^{-1}F(k) = f(x)$$

We may now clarify some aspects of the discussion of the previous section. Mathematically speaking, if we choose to define the Fourier transform of $f(x)$ as

$$F(k) = \int_{-\infty}^{\infty} f(x)e^{ikx}\, dx \tag{5.11}$$

then to obtain numerical correspondence, the inverse transform is not

$$f(x) = \int_{-\infty}^{\infty} F(k)e^{-ikx}\, dk \tag{5.15}$$

but rather

$$f(x) = \frac{1}{2\pi} \int_{-\infty}^{\infty} F(k)e^{-ikx}\, dk \tag{5.11'}$$

which follows from the fact that $k = 2\pi/\lambda$. Alternatively, if we adopt the definition of equation (5.15),

$$F(k) = \frac{1}{2\pi} \int_{-\infty}^{\infty} f(x)e^{ikx}\, dx \tag{5.11a}$$

then the inverse transform is strictly

$$f(x) = \int_{-\infty}^{\infty} F(k)e^{-ikx}\, dk \tag{5.11a'}$$

The most symmetrical definition of the Fourier transform and its inverse is therefore to take

$$F(k) = \frac{1}{\sqrt{2\pi}} \int_{-\infty}^{\infty} f(x)e^{ikx}\, dx \tag{5.11b}$$

and

$$f(x) = \frac{1}{\sqrt{2\pi}} \int_{-\infty}^{\infty} F(k)e^{-ikx}\, dk \tag{5.11b'}$$

Any of the pairs of equations (5.11) and (5.11′), (5.11a) and (5.11a′), and (5.11b) and (5.11b′) is correct mathematically. As explained previously, however, writing the factors of $1/2\pi$ or $1/\sqrt{(2\pi)}$ is tedious and provides no useful information. We shall therefore use the pairs of equations

$$F(k) = \int_{-\infty}^{\infty} f(x)e^{ikx}\, dx \tag{5.11}$$

and

$$f(x) = \int_{-\infty}^{\infty} F(k)e^{-ikx}\, dk \tag{5.15}$$

as our definitions of the Fourier transform and its inverse. Whenever numerical results are required, we shall have to remember that somewhere along the line we have lost a factor of $1/2\pi$.

Another interesting aspect of equations (5.11) and (5.15) concerns the sign appearing in the exponentials. We have already met the fact that the choice of the sign in the exponential of the Fourier transform equation (5.11) is a matter of convention. We might equally as well define the Fourier transform as

$$F(k) = \int_{-\infty}^{\infty} f(x)e^{-ikx}\, dx$$

in which case the inverse transform becomes

$$f(x) = \int_{-\infty}^{\infty} F(k)e^{ikx}\, dk$$

It therefore makes no difference which sign we use for the transform operation as long as we make sure to use the opposite sign when we wish to take the inverse. We now see that equations (5.11) and (5.15) are quite symmetrical in the variables k and x. If we know $f(x)$ then we may compute $F(k)$ by equation (5.11); and if we know $F(k)$ then we may obtain $f(x)$ by means of equation (5.15). The significance of this is that both $f(x)$ and $F(k)$ contain the same amount of *information*. Since the functions $f(x)$ and $F(k)$ are quite symmetrical, and it is quite straightforward to obtain either one from the other, they are essentially expressing the same sort of relationship, but in terms of different variables. The physical relevance of the relationship between the transform of any function and its inverse will be made clear in the next chapter.

In view of the fact that the functions $f(x)$ and $F(k)$ are so symmetrically related as defined by equations (5.11) and (5.15), and that either one may be regarded as the transform of the other, we call any pair of functions $f(x)$ and $F(k)$ *Fourier mates*, and the Fourier mate relationship is symbolised using the double-headed arrow $\leftrightarrow$:

$$f(x) \leftrightarrow F(k)$$

An example of the Fourier mate relationship is that between the Top Hat function and the function $(\sin kX_0)/kX_0$; each is the transform of the other:

$$\text{Top Hat} \leftrightarrow 2hX_0 \frac{\sin kX_0}{kX_0}$$

5.6 Real space and Fourier space

The Fourier transform mates $f(x)$ and $F(k)$ are two functions with the same information content expressed in terms of different variables. The same is true for the three-dimensional functions $f(\mathbf{r})$ and $F(\mathbf{k})$ defined by equations (5.12) and (5.16):

$$F(\mathbf{k}) = \int_{\text{all } \mathbf{r}} f(\mathbf{r}) e^{i\mathbf{k} \cdot \mathbf{r}} \, d\mathbf{r} \tag{5.12}$$

$$f(\mathbf{r}) = \int_{\text{all } \mathbf{k}} F(\mathbf{k}) e^{-i\mathbf{k} \cdot \mathbf{r}} \, d\mathbf{k} \tag{5.16}$$

In general, the variable $\mathbf{r}$ will represent a three-dimensional spatial coordinate corresponding to some point in space. This point may be defined in terms of three-dimensional Cartesian coordinates as the arbitrary point $\mathbf{r} = (x, y, z)$. Any value of $\mathbf{r}$ will correspond to a particular point in space, and the complete range of values of $\mathbf{r}$ will serve to define all space. If we were to measure a distance in this space, the units in which we would express $\mathbf{r}$, and also the components x, y, and z, would be the normal units of length such as metres or inches. In this way, we may think of the variable $\mathbf{r}$ as defining a set of points which may be measured using ordinary units of length. This is quite familiar to us, and the space defined by the variable $\mathbf{r}$ is therefore known as *real space*.

Mathematically, however, any variable may be thought of as defining a 'space' in a manner exactly analogous to the way in which the variable $\mathbf{r}$ defines real space. It is only because the units of $\mathbf{r}$ are familiar to us that we readily accept that $\mathbf{r}$ does indeed define real space.

Let us consider the 'space' defined by the variable $\mathbf{k}$. In equation (4.32) in Chapter 4, $\mathbf{k}$ was defined as a vector of magnitude $2\pi/\lambda$. Since λ has the dimensions of length, $\mathbf{k}$ has the dimensions of $[\text{length}]^{-1}$. $\mathbf{k}$ is a vector which has components (k_x, k_y, k_z), all of which have dimensions of $[\text{length}]^{-1}$, and so we may set up an analogue of three-dimensional Cartesian coordinates, each axis of which represents each component of the wave vector $\mathbf{k}$, as in Fig. 5.6.

As $\mathbf{k}$ takes different values, we may identify points relative to the new coordinate system corresponding to any value of $\mathbf{k}$. Our new coordinate system defines a new 'space' in which distances are measured in units of $[\text{length}]^{-1}$. Although this is somewhat strange, it is the exact analogue of the three-dimensional coordinate system based on the variable $\mathbf{r}$ in which distances are measured in ordinary length units. Just as the r-system was called real space, the space defined by the variable $\mathbf{k}$ is commonly referred to as $\mathbf{k}$ *space, reciprocal space* or *Fourier space*. These names are

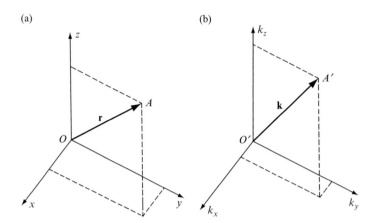

Fig. 5.6 *Real space and Fourier space.*(a) shows a three-dimensional Cartesian coordinate system, in which a vector **r** may be represented in terms of three components (x, y, z). Distances in this system are measured in ordinary units of length, and the system is said to define real space. In (b) is an analogous representation in which a wave vector **k** is expressed in terms of components (k_x, k_y, k_z). Distances in this system are measured in units of $[\text{length}]^{-1}$, and we speak of Fourier space.

self-explanatory in view of the use of the units of reciprocal length, $[\text{length}]^{-1}$, to define distances in the new space, and also when we consider the Fourier transform relationship between the variables **r** and **k**.

We now have two spaces: real space, defined by the variable **r**, and Fourier space, defined by the variable **k**. The mathematical significance of the Fourier transform and its inverse is that the integrals defined by equations (5.12) and (5.16), and their one-dimensional analogues, equations (5.11) and (5.15), are the recipes by which we may transform any function $f(\mathbf{r})$ in real space into the corresponding function $F(\mathbf{k})$ in Fourier space, and back again. Sometimes it is more convenient to express a given formula in terms of the variable **k**. If this is done, then we may always find the corresponding equation in terms of the variable **r** by using the inverse transform.

5.7 Delta functions

A function of considerable importance is the *Dirac delta function*, or δ *function*, which may be defined as

$$\int_{-\infty}^{\infty} \delta(x - x_0)\, dx = 1$$

A δ function at the point $x = x_0$, written as $\delta(x - x_0)$, has a value of infinity at x_0, is zero for all other values of x and integrates to unity. A δ function is depicted in

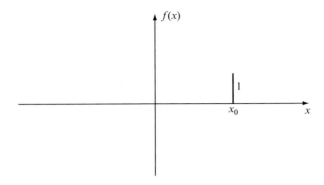

Fig. 5.7 *The δ function.* A pulse of unit magnitude at $x = x_0$ is represented by a δ function, written as $\delta(x - x_0)$. The delta function actually has infinite height when $x = x_0$ but the area under its peak is unity and it is conventional to indicate the area rather than the height. Pulses of other magnitudes may be represented by multiplying the area under the delta function by the appropriate number.

Fig. 5.7. Note that the height of the δ function in this figure is set to 1.0, whereas in the previous sentence we stated that the height was infinite. The reason for this discrepancy is that the height of the line in Fig. 5.7 represents the area under the δ function—the integral—rather than the actual value of the δ function, and this is a very commonly used representation.

The δ function is therefore the mathematical description of a point, or of a phenomenon at a point. For instance, a point charge of magnitude q coulombs at the position $x = -x_0$ is described by the mathematical function $q * \delta(x + x_0)$, in which we essentially multiply a δ function at $x = -x_0$ by q in order to obtain the correct magnitude. Any situation in which we are dealing with the properties of points is described by the use of δ functions. In three dimensions, a point $\mathbf{r} = \mathbf{r}_0$ is represented by a δ function written as $\delta(\mathbf{r} - \mathbf{r}_0)$.

The particular relevance of δ functions to this book is that a crystal lattice is represented by a three-dimensional array of points in space, and may therefore be described using the formalism of the δ function. For a lattice defined by three unit vectors $\mathbf{a}$, $\mathbf{b}$ and $\mathbf{c}$, we learnt in Chapter 3 that a lattice point $\mathbf{r}$ may be represented by

$$\mathbf{r} = p\mathbf{a} + q\mathbf{b} + r\mathbf{c} \tag{3.3}$$

in which p, q and r are positive or negative integers if the unit cell is primitive, or may be fractional for a non-primitive unit cell. Consequently, a lattice may be expressed by an array of δ functions of the type $\delta(\mathbf{r} - [p\mathbf{a} + q\mathbf{b} + r\mathbf{c}])$, in which p, q, and r take on all allowed values corresponding to each lattice point. We may therefore define the *lattice function* $l(\mathbf{r})$ as

$$l(\mathbf{r}) = \sum_{\text{all } p,\, q,\, r} \delta(\mathbf{r} - [p\mathbf{a} + q\mathbf{b} + r\mathbf{c}]) \tag{5.17}$$

in which the summation signifies that we generate all lattice points for every allowed value of p, q and r.

An important mathematical result concerning a one-dimensional δ function $\delta(x - x_0)$ and any arbitrary one-dimensional function $f(x)$ is that

$$\int_{-\infty}^{\infty} f(x)\delta(x - x_0)\, dx = f(x_0) \tag{5.18}$$

This result is intuitively obvious when we consider the form of the product function $f(x)\delta(x - x_0)$. Since the δ function has a value of zero for all x except $x = x_0$, then the product $f(x)\delta(x - x_0)$ is zero throughout the total range of x except for when $x = x_0$. At this special point, the delta function itself has a value of infinity but its integral has the value of unity. Hence

$$\int_{-\infty}^{\infty} f(x)\delta(x - x_0)\, dx = f(x_0) \int_{-\infty}^{\infty} \delta(x - x_0)\, dx = f(x_0)$$

In three dimensions, the analogous theorem is

$$\int_{-\infty}^{\infty} f(\mathbf{r})\delta(\mathbf{r} - \mathbf{r}_0)\, d\mathbf{r} = f(\mathbf{r}_0) \tag{5.19}$$

5.8   Fourier transforms and delta functions

In this section, we shall derive the Fourier transforms of one, two, three, an arbitrary number N and an infinite number of one-dimensional δ functions. We shall require these results in Chapter 7, and this also serves as an introduction to the problem of finding the Fourier transform of N three-dimensional δ functions, which, as discussed in Chapter 8, is required to find the Fourier transform of the lattice function $l(\mathbf{r})$ defined above.

One delta function

Consider a single function at the origin $x = 0$, as shown in Fig. 5.8, represented by $\delta(x)$. The Fourier transform of $f(x) = \delta(x)$ is defined by equation (5.11):

$$F(k) = \int_{-\infty}^{\infty} f(x)e^{ikx}\, dx \tag{5.11}$$

$$= \int_{-\infty}^{\infty} \delta(x)e^{ikx}\, dx \tag{5.20}$$

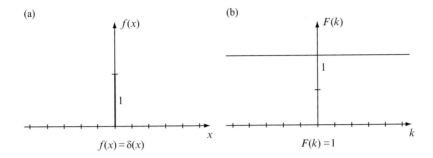

(a) $f(x)$

1

$f(x) = \delta(x)$

x

(b) $F(k)$

1

$F(k) = 1$

k

Fig. 5.8 *A single δ function and its Fourier transform.* The transform of the infinitely narrow δ function is a constant, which is infinitely broad.

We now use the result of equation (5.18), which tells us that the integral defined by equation (5.20) is just e^{ikx} evaluated at $x = 0$:

$$F(k) = \int_{-\infty}^{\infty} \delta(x)e^{ikx} \, dx = [e^{ikx}]_{x=0}$$

$$= e^0 = 1$$

Hence the Fourier transform of a δ function at the origin is a constant, namely, unity. This is shown in Fig. 5.8(b). A δ function at the origin and a constant are therefore Fourier transform mates, implying that the Fourier transform of a constant is a δ function.

Two delta functions

Our second example is an array of two δ functions symmetrically placed about the origin at $x = -x_0$ and $x = +x_0$ as shown in Fig. 5.9,

$$f(x) = \delta(x + x_0) + \delta(x - x_0)$$

From equation (5.11), we have

$$F(k) = \int_{-\infty}^{\infty} f(x)e^{ikx} \, dx \qquad (5.11)$$

$$= \int_{-\infty}^{\infty} \delta(x + x_0)e^{ikx} \, dx + \int_{-\infty}^{\infty} \delta(x - x_0)e^{ikx} \, dx$$

$$= e^{-ikx_0} + e^{ikx_0} \qquad (5.21)$$

(a)

(b)

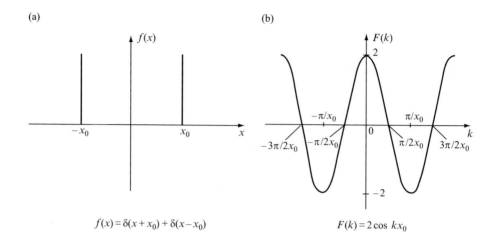

$$f(x) = \delta(x + x_0) + \delta(x - x_0)$$

$$F(k) = 2 \cos kx_0$$

Fig. 5.9 *Two δ functions, and the Fourier transform.*

Equation (5.21) is obtained by using equation (5.18) for each integral. The sum of two complex exponentials may be simplified using equation (2.33), which we derived in Chapter 2, and which may be stated as

$$\cos \theta = \frac{e^{i\theta} + e^{-i\theta}}{2}$$

Identifying

$$\theta = kx_0$$

we obtain

$$F(k) = 2 \cos kx_0 \qquad (5.22)$$

The Fourier transform of two δ functions is therefore a cosine function, as shown in Fig. 5.9(b). Using the Fourier mate relationship, we may state that the Fourier transform of a cosine function is two δ functions.

It is interesting to note that $F(k)$ is first zero on each side of the origin of k when

$$\cos kx_0 = 0$$

$$\therefore kx_0 = \pm \frac{\pi}{2}$$

$$\therefore k = \pm \frac{\pi}{2x_0}$$

The total distance between the first two zeros on each side of the origin is therefore π/x_0. The distance between the original two δ functions was $2x_0$, and once again we see that the narrower the original function, the broader the Fourier transform, and vice versa. This is especially true when we refer to the Fourier transform of a single

δ function. Here the original function is infinitely thin, and the Fourier transform is, appropriately, infinitely broad.

Three delta functions

This time we shall find the Fourier transform of

$$f(x) = \delta(x + x_0) + \delta(x) + \delta(x - x_0)$$

which is the mathematical representation of a δ function at the origin with two other δ functions symmetrically placed on either side, as in Fig. 5.10(a):

$$F(k) = \int_{-\infty}^{\infty} f(x)e^{ikx} \, dx \tag{5.11}$$

$$= \int_{-\infty}^{\infty} \delta(x + x_0)e^{ikx} \, dx + \int_{-\infty}^{\infty} \delta(x)e^{ikx} \, dx + \int_{-\infty}^{\infty} \delta(x - x_0)e^{ikx} \, dx$$

$$= e^{-ikx_0} + 1 + e^{ikx_0}$$

But

$$e^{-ikx_0} + e^{ikx_0} = 2\cos kx_0$$

$$\therefore F(k) = 1 + 2\cos kx_0$$

This function has the appearance shown in Fig. 5.10(b).

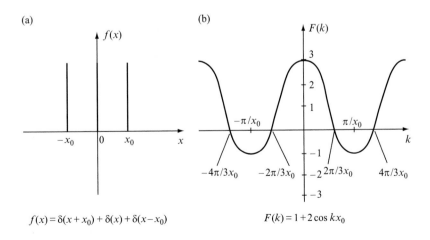

(a)   (b)

$$f(x) = \delta(x + x_0) + \delta(x) + \delta(x - x_0)$$ $$F(k) = 1 + 2\cos kx_0$$

Fig. 5.10 *An array of three δ functions, and the Fourier transform.*

N delta functions

Generalising, we shall now derive the Fourier transform of $N\delta$ functions. For convenience, we shall choose N to be an odd number equal to $2p + 1$, where p is an integer. This representation implies that there is a δ function at the origin, with p other δ functions symmetrically placed on either side of the origin, as shown in Fig. 5.11(a):

$$f(x) = \delta(x + px_0) + \delta(x + [p-1]x_0) + \ldots + \delta(x) + \ldots + \delta(x - [p-1]x_0)$$
$$+ \delta(x - px_0)$$

$$= \sum_{n=-p}^{n=+p} \delta(x - nx_0)$$

$$F(k) = \int_{-\infty}^{\infty} f(x)e^{ikx}\, dx \tag{5.11}$$

$$= \int_{-\infty}^{\infty} \sum_{n=-p}^{n=+p} \delta(x - nx_0)e^{ikx}\, dx$$

$$= e^{-ikpx_0} + e^{-ik(p-1)x_0} + \ldots + 1 + \ldots + e^{ik(p-1)x_0} + e^{ikpx_0}$$

$$= e^{-ikpx_0}\left(1 + e^{ikx_0} + e^{2ikx_0} + e^{3ikx_0} \ldots + e^{(2p)ikx_0}\right) \tag{5.23}$$

(a)

(b)

$$f(x) = \sum_{n=-4}^{n=4} \delta(x - nx_0)$$

$$F(k) = \frac{\sin\frac{Nkx_0}{2}}{\sin\frac{kx_0}{2}} \quad (N=9)$$

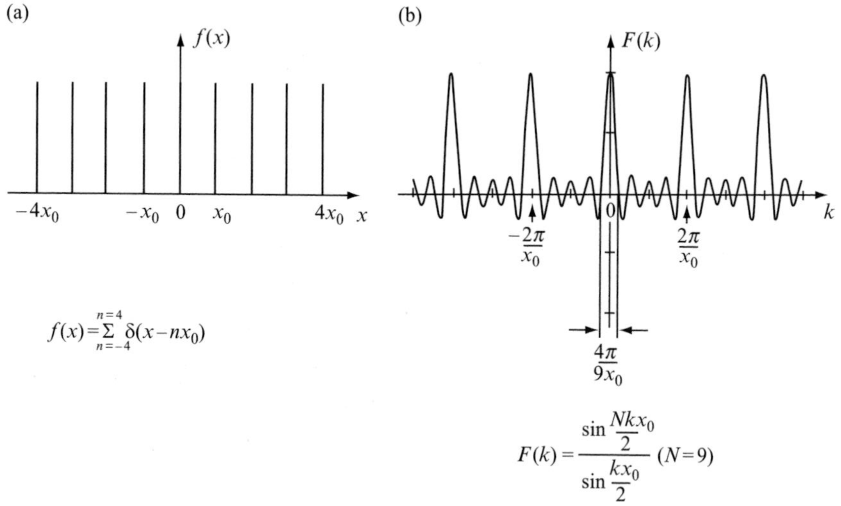

Fig. 5.11 *An array of N δ functions, and the Fourier transform.* (a) is an array of $N = (2p + 1)$ δ functions equally spaced by a distance of x_0, and (b) is its Fourier transform. This is a sharply peaked periodic function, with main peaks separated by $2\pi/x_0$. The width of each main peak is $4\pi/Nx_0$, and there are $N - 2$ subsidiary peaks between each main pair.

This sum of complex exponentials is a geometric progression of $2p + 1$ terms of common ratio e^{ikx_0}, which sums to

$$1 + e^{ikx_0} + e^{2ikx_0} + e^{3ikx_0} \ldots + e^{(2p)ikx_0} = \frac{1 - e^{(2p+1)ikx_0}}{1 - e^{ikx_0}}$$

But $2p + 1$ is just N, the number of δ functions in our array. Equation (5.23) now becomes

$$F(k) = e^{-ikpx_0} \left(\frac{1 - e^{iNkx_0}}{1 - e^{ikx_0}} \right) \tag{5.24}$$

Equation (5.24) is rather complicated, but it may be simplified if we adopt the following procedure. Firstly we note that

$$\frac{1 - e^{-ikx_0}}{1 - e^{-ikx_0}} = 1$$

and so

$$F(k) = e^{-ikpx_0} \left(\frac{1 - e^{iNkx_0}}{1 - e^{ikx_0}} \right) = e^{-ikpx_0} \left(\frac{1 - e^{iNkx_0}}{1 - e^{ikx_0}} \right) \left(\frac{1 - e^{-ikx_0}}{1 - e^{-ikx_0}} \right)$$

$$= \frac{e^{-ikpx_0} - e^{-ik(p+1)x_0} - e^{ik(N-p)x_0} + e^{ik(N-1-p)x_0}}{1 - e^{ikx_0} - e^{-ikx_0} + 1}$$

But $N = 2p + 1$
and so

$$p = \frac{N-1}{2},$$

$$p + 1 = \frac{N+1}{2},$$

$$N - p = \frac{N+1}{2},$$

$$N - 1 - p = \frac{N-1}{2},$$

$$F(k) = \frac{e^{ik[(N-1)/2]x_0} + e^{-ik[(N-1)/2]x_0} - e^{ik[(N+1)/2]x_0} - e^{-ik[(N+1)/2]x_0}}{2 - \left(e^{ikx_0} + e^{-ikx_0} \right)}$$

Using the trigonometric identity

$$e^{i\theta} + e^{-i\theta} = 2 \cos \theta \tag{2.33}$$

we have

$$F(k) = \frac{2 \cos \frac{(N-1)}{2} kx_0 - 2 \cos \frac{(N+1)}{2} kx_0}{2 - 2 \cos kx_0}$$

Two other useful relationships are

$$\cos A - \cos B = 2 \sin \frac{(A+B)}{2} \sin \frac{(B-A)}{2} \quad (B > A)$$

and

$$1 - \cos C = 2 \sin^2 \frac{C}{2}$$

Identifying

$$A = \frac{(N-1)}{2} kx_0$$

$$B = \frac{(N+1)}{2} kx_0$$

$$C = kx_0$$

we obtain

$$F(k) = \frac{4 \sin \frac{Nkx_0}{2} \sin \frac{kx_0}{2}}{4 \sin^2 \frac{kx_0}{2}}$$

$$\therefore F(k) = \frac{\sin \frac{Nkx_0}{2}}{\sin \frac{kx_0}{2}} \tag{5.25}$$

Equation (5.25) is the result we require, and is the Fourier transform of $N\delta$ functions. The form of this function may be sketched if we write $F(k)$ as

$$F(k) = \sin \frac{Nkx_0}{2} \operatorname{cosec} \frac{kx_0}{2}$$

from which we see that $F(k)$ is the product of a sine function with a cosec function. These functions are shown in Figs 5.12(a) and (b), where it can be seen that $\operatorname{cosec}(kx_0/2)$ varies much more slowly than $\sin(Nkx_0/2)$; in fact, the larger the value of N, the more rapidly does the sine function oscillate.

$F(k)$ has peaks whenever

$$\operatorname{cosec} \frac{kx_0}{2} = \infty$$

i.e. when

$$\frac{kx_0}{2} = 0, \pm\pi, \pm 2\pi, \ldots$$

$$k = 0, \pm \frac{2\pi}{x_0}, \pm \frac{4\pi}{x_0}, \ldots$$

Despite the fact that $\operatorname{cosec}(kx_0/2)$ tends to infinity, $\sin(Nkx_0/2)$ is simultaneously zero, and it is easy to show that the limit of the product is finite, and equal to N. $F(k)$ takes the appearance shown in Fig. 5.12(c), and also Fig. 5.11(b). It is remarkable for the extremely sharp peaks which occur whenever k is of the form $\pm 2n\pi/x_0$, where n

(a)   (b)   (c)

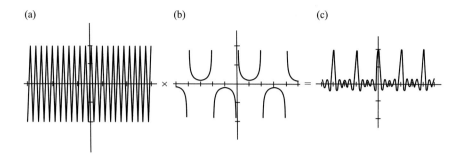

Fig. 5.12 *The function* $\sin N\theta \operatorname{cosec}\theta$. The function $\sin N\theta$ is the rapidly oscillating function shown in (a), which has its first zeros on either side of the origin at $\pm\pi/N$; $\operatorname{cosec}\theta$ is a periodic U-shaped function which tends to infinity at integral multiples of π. The product function is the sharply peaked function shown in (c) with main peaks at integral multiples of π, each of which has a width of $2\pi/N$. There are $N-2$ subsidiary peaks between each main pair.

is any integer, positive, negative or zero. If we concentrate our attention on the peak at the origin, the first zeros on either side of the origin occur when

$$\sin \frac{Nkx_0}{2} = 0$$

$$\therefore \frac{Nkx_0}{2} = \pm\pi$$

$$\therefore k = \pm\frac{2\pi}{Nx_0}$$

The interpeak separation is therefore $2\pi/x_0$, whereas the width of the main peaks is $4\pi/Nx_0$. Since N is a large positive integer, the main peaks are very much narrower than the interpeak distance. Note also that as N increases (that is, when we have a progressively larger number of δ functions in our array), then the main peaks become sharper. Further, there are $(N-2)$ subsidiary peaks between each pair of main peaks.

An infinite number of delta functions

Our last example is that of an infinite array of δ functions equally separated by a distance x_0. The Fourier transform of this array may be determined by considering the behaviour of the Fourier transform of $N\delta$ functions as N tends to infinity. If we refer to Fig. 5.11(b), we see that as N increases, the main peaks become progressively sharper, until in the limit when N approaches infinity, they are themselves reduced to δ functions. The positions of these δ functions are at the same positions as the peaks for the Fourier transform of $N\delta$ functions and they are separated by $2\pi/x_0$. We therefore see that the Fourier transform of an infinite array of δ functions spaced by a distance x_0 is another array of δ functions, spaced by $2\pi/x_0$, as shown in Fig. 5.13:

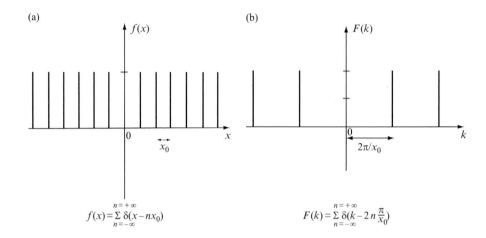

(a)

$f(x)$

0

x_0

x

$$n=+\infty$$
$$f(x)=\Sigma\;\delta(x-nx_0)$$
$$n=-\infty$$

(b)

$F(k)$

0

$2\pi/x_0$

k

$$n=+\infty$$
$$F(k)=\Sigma\;\delta(k-2n\tfrac{\pi}{x_0})$$
$$n=-\infty$$

Fig. 5.13 *An array of an infinite number of δ functions, and the Fourier transform.* (a) *is an infinite array of δ functions spaced by* x_0, *and* (b) *is the Fourier transform, which constitutes a reciprocally spaced infinite array of δ functions. This is the limiting case of Fig. 5.11, in which the main peaks become infinitely narrow as N tends to infinity.*

once again, we notice the reciprocal relationship between the spacing of a function and its Fourier transform. Analytically, this result may be stated as: If

$$f(x) = \sum_{n=-\infty}^{n=\infty} \delta(x - nx_0)$$

then

$$F(k) = \int_{-\infty}^{\infty} \sum_{n=-\infty}^{n=\infty} \delta(x - nx_0)e^{ikx}\,dx$$

$$= \sum_{n=-\infty}^{n=\infty} e^{iknx_0}$$

$$= 1 + \sum_{n=1}^{n=\infty} (e^{iknx_0} + e^{-iknx_0}) = 1 + \sum_{n=1}^{n=\infty} 2\cos(knx_0)$$

But since $\cos(knx_0) = \cos(-knx_0)$, we may again change the lower limit of the summation:

$$F(k) = \sum_{n=-\infty}^{n=\infty} \cos(knx_0)$$

The positive and negative terms in the cosine wave will tend to cancel out unless $k = 2\pi/x_0$. Hence $F(k)$ will be zero at all values of k except when $k = 2\pi/x_0$, and at these

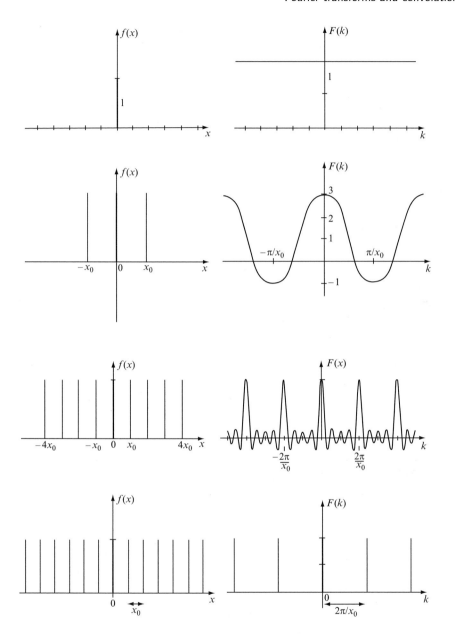

Fig. 5.14 δ *functions and Fourier transforms.* Shown are arrays of one, three, $N = 2p + 1$ and an infinite number of δ functions, and their corresponding Fourier transforms. Note that the positions of the main peaks are determined by the spacing of the δ functions, whereas the width of each main peak is inversely proportional to the total number of δ functions in the array. (The example of two δ functions is omitted, since it constitutes a special case, in that this array contains an even number of δ functions, whereas the other arrays are all of an odd number of δ functions.)

points, $F(k) = \infty$. Hence, $F(k)$ is another δ function with peaks separated by $2\pi/x_0$. This is an extraordinarily elegant result, which underpins much of crystallography, and may be represented by the following equation:

$$F(k) = \sum_{n=-\infty}^{n=\infty} \delta\left(k - \frac{2n\pi}{x_0}\right) \tag{5.26}$$

General discussion

It is useful to correlate the results obtained from our analysis of the Fourier transforms of arrays of δ functions. In Fig. 5.14 are our arrays of δ functions and their Fourier transforms. There are two important points which are made clear in Fig. 5.14:

(1) The positions of the main peaks of the Fourier transform are determined by the spacing of the individual δ functions in the original array.
(2) The width of each main peak and the number of subsidiary peaks are determined by the total number of δ functions in the original array.

These statements will be referred to very often in Chapters 7 and 8.

5.9 Symmetrical and antisymmetrical functions

Any one-dimensional function $f(x)$ is said to be *symmetrical* if $f(x) = f(-x)$. On a graph, a symmetrical function may be identified by the fact that the form of the graph on one side of the y axis is the mirror image of that on the other side. All functions of which we have so far found the Fourier transform, namely the Top Hat and the various arrays of δ functions, have been symmetrical functions.

If $f(x) = -f(-x)$, then the function is said to be *antisymmetrical*. In one dimension, the graph of an antisymmetrical function has a twofold rotation axis at the origin, perpendicular to the plane of the diagram.

A function which is neither symmetrical nor antisymmetrical is called *non-symmetrical*. Examples of symmetrical, antisymmetrical and non-symmetrical functions are shown in Fig. 5.15.

In three dimensions, the mathematical definitions of symmetrical and antisymmetrical functions are the same, but the geometrical interpretations are rather different. A symmetrical function is such that $f(\mathbf{r}) = f(-\mathbf{r})$ and may be identified as having a centre of symmetry at the origin, and so any three-dimensional symmetrical function appears the same under the operation of inversion, which we defined in Chapter 3. To emphasise that such a function has an inversion centre, or centre of symmetry, at the origin, we often speak of a *centrosymmetric function*. A three-dimensional antisymmetric function is defined such that $f(\mathbf{r}) = -f(-\mathbf{r})$, but there is no readily visualised geometrical interpretation.

Of interest is the nature of the Fourier transform of symmetrical and antisymmetrical functions. As an example, let us consider the Fourier transform of the one-dimensional antisymmetrical function defined by

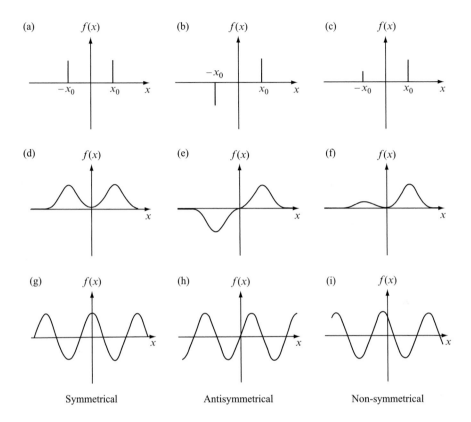

Symmetrical   Antisymmetrical   Non-symmetrical

Fig. 5.15 *Symmetrical, antisymmetrical and non-symmetrical functions.* In one dimension, symmetrical functions are such that the y axis is a mirror line. Antisymmetrical functions have a twofold rotation axis through the origin, perpendicular to the plane defined by the x and y axes. All functions which are neither symmetrical nor antisymmetrical are non-symmetrical.

$$f(x) = -\delta(x + x_0) + \delta(x - x_0)$$

This is a negative δ function at $x = -x_0$ and a positive δ function at $x = +x_0$, as shown in Fig. 5.15(b):

$$F(k) = \int_{-\infty}^{\infty} f(x)e^{ikx} \, dx \tag{5.11}$$

$$= -\int_{-\infty}^{\infty} \delta(x + x_0)e^{ikx} \, dx + \int_{-\infty}^{\infty} \delta(x - x_0)e^{ikx} \, dx$$

$$= -e^{-ikx_0} + e^{ikx_0} \tag{5.21}$$

But equation (2.34) states

$$\sin \theta = \frac{e^{i\theta} - e^{-i\theta}}{2i}$$

Identifying

$$\theta = kx_0$$

we have

$$F(k) = 2i \sin kx_0$$

This result is to be contrasted with the analogous result that we obtained when we found the Fourier transform of a symmetrical pair of δ functions, which, according to equation (5.22), is $2 \cos kx_0$. The significant difference between the answers is that the Fourier transform of the symmetrical function is a real function ($2 \cos kx_0$), whereas that obtained for the antisymmetrical case is an imaginary function ($2i \sin kx_0$), since it contains the imaginary number i. This result carries over to three dimensions, and we may state:

> *The Fourier transform of a symmetrical function is real.*
> *The Fourier transform of an antisymmetrical function is imaginary.*
> *The Fourier transform of a non-symmetrical function is complex, in that it contains both a real and an imaginary part.*

5.10 Convolutions

Another integral form of major importance is

$$c(\mathbf{u}) = \int_{\text{all } \mathbf{r}} f(\mathbf{r})g(\mathbf{u} - \mathbf{r}) \, d\mathbf{r} \tag{5.27}$$

This is known as the *convolution* of $f(\mathbf{r})$ and $g(\mathbf{r})$, and may be written as $f(\mathbf{r}) * g(\mathbf{r})$:

$$c(\mathbf{u}) = f(\mathbf{r}) * g(\mathbf{r}) = \int_{\text{all } \mathbf{r}} f(\mathbf{r})g(\mathbf{u} - \mathbf{r}) \, d\mathbf{r}$$

This integral is symmetrical with respect to the variables $\mathbf{u}$ and $\mathbf{r}$, in that

$$f(\mathbf{r}) * g(\mathbf{r}) = \int_{\text{all } \mathbf{r}} f(\mathbf{r})g(\mathbf{u} - \mathbf{r}) \, d\mathbf{r} = \int_{\text{all } \mathbf{r}} f(\mathbf{u} - \mathbf{r})g(\mathbf{r}) \, d\mathbf{r}$$

This is easily proved using the substitution

$$\mathbf{R} = \mathbf{u} - \mathbf{r}$$

$$\therefore d\mathbf{R} = -d\mathbf{r}$$

so that equation (5.27) becomes

$$\int\limits_{\text{all } \mathbf{r}} f(\mathbf{r})g(\mathbf{u}-\mathbf{r})\,d\mathbf{r} = -\int\limits_{\text{all } \mathbf{R}} f(\mathbf{u}-\mathbf{R})g(\mathbf{R})\,d\mathbf{R}$$

$$= \int\limits_{\text{all } \mathbf{R}} f(\mathbf{u}-\mathbf{R})g(\mathbf{R})\,d\mathbf{R}$$

since $\mathbf{R}$ can take on the same values as $\mathbf{r}$, and the sign of the integral will change on exchanging the upper and lower limits of integration.

But it makes no difference whether we write the variable of integration as $\mathbf{r}$ or $\mathbf{R}$, so, replacing $\mathbf{R}$ by $\mathbf{r}$, we have

$$\int\limits_{\text{all } \mathbf{r}} f(\mathbf{r})g(\mathbf{u}-\mathbf{r})\,d\mathbf{r} = \int\limits_{\text{all } \mathbf{r}} f(\mathbf{u}-\mathbf{r})g(\mathbf{r})\,d\mathbf{r}$$

The evaluation of the convolution integral in equation (5.27) for any two functions $f(\mathbf{r})$ and $g(\mathbf{r})$ may be carried out analytically by the usual rules of integration. In practice, however, many of the functions with which we shall be dealing are quite difficult to handle analytically. In view of this, it is useful to determine the form of the convolution integral by the inspection of pertinent graphs. This geometrical interpretation also serves to indicate the meaning of the convolution integral.

We first notice that the integrand of equation (5.27) is a function of the variables $\mathbf{u}$ and $\mathbf{r}$, but that the integration is over $\mathbf{r}$, implying that the result of the integration is a function of $\mathbf{u}$. The significance of the variable $\mathbf{u}$ may be appreciated when we consider the relationship between the function $g(\mathbf{r})$ and the functions $g(\mathbf{u}-\mathbf{r})$ for different values of $\mathbf{u}$. For ease in drawing, let us investigate the behaviour of the two one-dimensional functions $g(x)$ and $g(u-x)$. An arbitrary function $g(x)$ is shown in Fig. 5.16(a).

The function $g(-x)$ is derived from $g(x)$ by replacing the value of the function at $+x$ by its value at $-x$. This gives the graph of Fig. 5.16(b): in one dimension, the operation of replacing x by $-x$ appears as a reflection in the y axis. To form $g(u-x)$ from $g(-x)$, we displace the function $g(-x)$ by u units along the positive x axis, as shown in Fig. 5.16(c). The form of the function $g(-x)$ is now centred about the point $x = u$. In three dimensions, to derive $g(\mathbf{u}-\mathbf{r})$ from $g(\mathbf{r})$, we invert through the origin to replace $\mathbf{r}$ by $-\mathbf{r}$, and then displace by $\mathbf{u}$. The significance of the variable $\mathbf{u}$ is that it determines the position of $g(\mathbf{u}-\mathbf{r})$ from the origin. The relevance of this will become clear when we calculate an example. This is most clearly done by considering a method for forming the convolution function.

Method for finding convolutions $\int\limits_{\text{all } \mathbf{r}} f(\mathbf{r})g(\mathbf{u}-\mathbf{r})\,d\mathbf{r}$:

(1) $f(\mathbf{r})$ and $g(\mathbf{r})$ are given functions.
(2) Form $g(-\mathbf{r})$ by inversion through the origin. In one dimension, this becomes a reflection in the y axis.

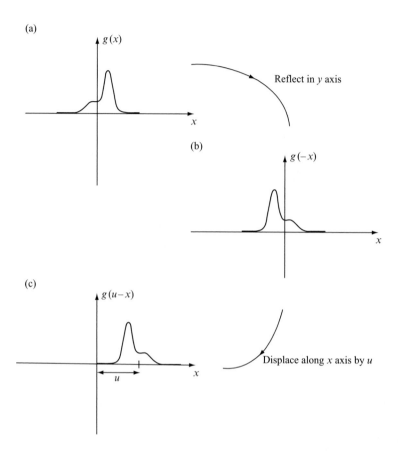

Fig. 5.16 *The formation of* $g(u - x)$ *from* $g(x)$. An arbitrary function $g(x)$ is shown in (a). We first form $g(-x)$, which, in one dimension, corresponds to a reflection in the y axis. $g(u - x)$ is formed by displacing the function $g(-x)$ by u units along the x axis.

(3) Displace $g(-\mathbf{r})$ by any chosen $\mathbf{u}$ to form $g(\mathbf{u} - \mathbf{r})$.

(4) Multiply $f(\mathbf{r})$ by the newly formed $g(\mathbf{u} - \mathbf{r})$ to generate the product function $f(\mathbf{r})g(\mathbf{u} - \mathbf{r})$.

(5) Integrate over all values of $\mathbf{r}$, which in one dimension corresponds to finding the area under the graph of the product function $f(x)g(u - x)$ for the chosen value of u.

(6) Return to step (3), but choose a different value of $\mathbf{u}$. This will, in turn, give a different product function and a different value for the integral.

(7) Repeat for all values of $\mathbf{u}$. This will generate the entire function

$$c(\mathbf{u}) = \int_{\text{all } \mathbf{r}} f(\mathbf{r})g(\mathbf{u} - \mathbf{r}) \, d\mathbf{r}$$

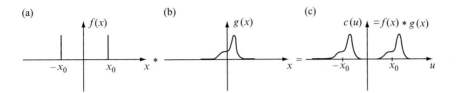

Fig. 5.17 *The convolution integral* $c(u) = \int_{-\infty}^{\infty} f(x)g(u-x)\,dx$. *If we choose* $f(x)$ *to be two* δ *functions at* $x = \pm x_0$, *then the operation of convoluting the array with an arbitrary function* $g(x)$ *results in centring* $g(x)$ *over each function. The steps by which we achieve this are shown in Fig. 5.18.*

As an example, let us consider the convolution of the two functions $f(x)$ and $g(x)$ shown in Fig. 5.17: $f(x)$ is in fact two δ functions, one at $x = -x_0$ and the other at $x = +x_0$, and $g(x)$ is the arbitrary function shown.

The steps by which we form the convolution are shown in Fig. 5.18. The first operation is to form $g(-x)$ from $g(x)$, as in Fig. 5.18(b). We then displace by some value of u, which in the first instance we choose as some large negative value. This implies that the general shape of $g(-x)$ is positioned far along the negative x axis. When we form the product $f(x)g(u-x)$ for this value of u, we see that $f(x)$ is zero whenever $g(u-x)$ is non-zero, and also that $g(u-x)$ is zero at those points when $f(x)$ is non-zero. The product function for this value of u is zero for all values of x, and hence the value of the convolution function is also zero for this particular value of u. We now choose a different value of u, which this time will be a somewhat smaller negative value. This time the product function has a small non-zero value, and so does the value of the convolution for this value of u. The product with the δ function yields a value which we plot on a graph, which will represent the convolution. Our third value of u in Fig. 5.18 is such that the two functions overlap to a greater extent, yielding a higher value for the convolution product.

It should now be apparent that as we choose different values of u, we are 'sliding' the function $g(-x)$ along the x axis from minus infinity to plus infinity. As we do this, we multiply by the stationary function $f(x)$, and record the value of the integral of the product as a function of the relative separation of the functions $f(x)$ and $g(u-x)$. The value of the convolution is therefore largest when the two functions overlap maximally, and zero when they are quite distant. As the function $g(u-x)$ slides over the δ function at $x = -x_0$, the δ function will select values of the g function, and since the δ function 'sees' the point A' before the point B', the effect of the function is to centre the function $g(x)$ (rather than $g(-x)$) over the position at which the δ function is located, as shown in the final diagram of Fig. 5.18. A similar effect occurs as $g(u-x)$ slides over the δ function at $x = x_0$.

This may be proved mathematically when we consider that the value of the integral is

$$c(u) = \int_{-\infty}^{\infty} f(x)g(u-x)\,dx$$

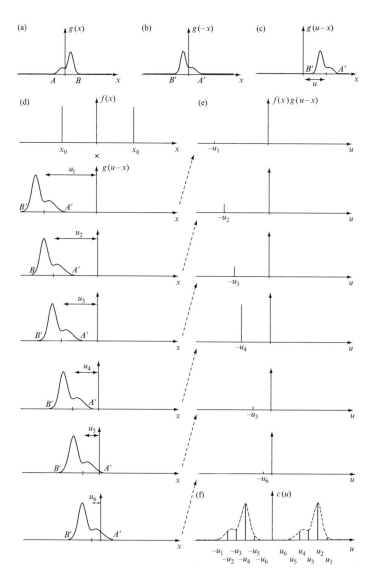

Fig. 5.18 *The formation of the convolution integral.* The arbitrary function $g(x)$ is shown in (a), from which we derive $g(-x)$ as in (b) and $g(u-x)$ as in (c) for an arbitrary value of u. The first diagram in the column (d) shows $f(x)$, which we choose to be a pair of δ functions. The six other diagrams in column (d) represent the function $g(u-x)$ for six different values of u. Starting with u as large and negative, we change u systematically, and for each value of u we plot $g(u-x)$ and form the product function $f(x)g(u-x)$, as in the column (e). If $f(x)$ and $g(u-x)$ do not overlap, then the value of the product function for those values of u is zero. When the value of u is such that $f(x)$ and $g(u-x)$ do overlap on forming the product function, the δ function 'samples' the g function, and we may plot a value as shown. If we collect the values in column (e) onto one graph, we obtain (f), which represents the convolution of the g function with the pair of δ functions.

and if

$$f(x) = \delta(x + x_0) + \delta(x - x_0)$$

then

$$c(u) = \int_{-\infty}^{\infty} \delta(x + x_0)g(u - x)\, dx + \int_{-\infty}^{\infty} \delta(x - x_0)g(u - x)\, dx$$

$$= g(u + x_0) + g(u - x_0)$$

The result of the convolution is therefore to centre the g function at the positions $-x_0$ and $+x_0$, previously occupied by the two δ functions. This process is somewhat difficult to follow verbally, but reference to the various stages of Fig. 5.18, which follows the rules given immediately above, should demonstrate what is happening.

In summary, we may state that the effect of convoluting any function $g(x)$ with an array of δ functions is to set the function $g(x)$ so that it is centred on each of the δ functions, as shown in Fig. 5.19. This will always be the case if the δ functions are separated by a distance greater than the width of the function $g(x)$, so that this function never overlaps more than one δ function at a time.

This result may be carried over to three dimensions, in that the result of convoluting any function $g(\mathbf{r})$ with a three-dimensional array of δ functions is to set the function $g(\mathbf{r})$ at all those points in space at which a δ function is defined.

The relevance of this discussion of the convolution integral may be appreciated when we consider the structure of a crystal. In Section 5.7, we learnt that a lattice may be represented by a three-dimensional array of δ functions as defined by the lattice function $l(\mathbf{r})$,

$$l(\mathbf{r}) = \sum_{\text{all } p,\, q,\, r} \delta\left(\mathbf{r} - [p\mathbf{a} + q\mathbf{b} + r\mathbf{c}]\right) \tag{5.17}$$

in which p, q and r correspond to allowed points defined by the lattice of crystallographic unit vectors $\mathbf{a}$, $\mathbf{b}$ and $\mathbf{c}$. Our study of crystals in Chapter 3 told us that associated with each lattice point in a crystal is an identical unit cell. Suppose we

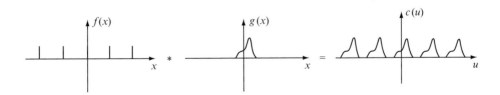

Fig. 5.19 *Convolutions with an array of δ functions. If $f(x)$ is an array of δ functions, and $g(x)$ the arbitrary function shown, then the result of the convolution $f(x) * g(x)$ is to associate the g function with each δ function. This is always true on condition that $g(x)$ is narrower than the spacing of the δ functions, so that $g(x)$ never overlaps two δ functions simultaneously.*

represent the structure of the unit cell by some mathematical function of position $u(\mathbf{r})$, which we shall define as the *unit cell function*. This function will, in general, specify the positions of all the atoms within the unit cell. It will in general be a complicated function, since it contains much information, but all that matters for the moment is that such a function exists, that it is a function of the spatial coordinates $\mathbf{r}$ only and that we may associate this function with each lattice point. The convolution of the lattice function $l(\mathbf{r})$ with the unit cell function $u(\mathbf{r})$ may be written as $l(\mathbf{r}) * u(\mathbf{r})$. What does this represent?

According to our analysis above, whenever we form the convolution of some function $g(\mathbf{r})$ with an array of δ functions represented by $f(\mathbf{r})$, the result is to associate $g(\mathbf{r})$ with each δ function. But the lattice function $l(\mathbf{r})$ is a three-dimensional array of δ functions corresponding to each lattice point. By forming the convolution of $l(\mathbf{r})$ with the unit cell function $u(\mathbf{r})$, we associate the unit cell function with each lattice point as specified by the lattice function $l(\mathbf{r})$, and in this way we build up the entire crystal structure! This is the great relevance of the convolution integral, for now we have a complete mathematical representation of a crystal structure. If we assume that the complete crystal may be represented by some mathematical function $c(\mathbf{u})$, which we shall call the *crystal structure function*, then we may write

$$\text{crystal structure function} = \text{lattice function} * \text{unit cell function}$$

$$c(\mathbf{u}) = l(\mathbf{r}) * u(\mathbf{r})$$

This result is very important, for it gives us a mathematical way of representing a crystal as a whole. It is also significant in that it separates conceptually the contribution to the crystal structure of the lattice from that of the contents of the unit cell.

5.11 The Fourier transform of a convolution

This section will discuss the Fourier transform of the convolution of a function $f(\mathbf{r})$ with a function $g(\mathbf{r})$. What we require is the function of $\mathbf{k}$ defined by

$$F(\mathbf{k}) = T[f(\mathbf{r}) * g(\mathbf{r})]$$

The mathematical investigation of this problem is rather complicated, and is done in Appendix II of this chapter. What is of immediate and extreme importance is the result: The Fourier transform of the convolution of $f(\mathbf{r})$ with $g(\mathbf{r})$ is the product of the individual transforms of $f(\mathbf{r})$ and $g(\mathbf{r})$:

$$T[f(\mathbf{r}) * g(\mathbf{r})] = [Tf(\mathbf{r})].[Tg(\mathbf{r})] \tag{5.28}$$

Conversely, the Fourier transform of the product of two functions $f(\mathbf{r})$ and $g(\mathbf{r})$ is the convolution of the individual transforms of $f(\mathbf{r})$ and $g(\mathbf{r})$:

$$T[f(\mathbf{r}).g(\mathbf{r})] = [Tf(\mathbf{r})] * [Tg(\mathbf{r})] \tag{5.29}$$

These two statements are collectively known as the *convolution theorem*.

The importance of the convolution theorem cannot be overemphasised. It will be used often. In succeeding chapters we shall be faced with the problem of calculating the Fourier transform of a function which will be either the convolution of two functions or the product of two functions. Equations (5.28) and (5.29) tell us how to go about solving these problems. Specifically, one of our main tasks will be to find the Fourier transform of the crystal structure function, which, as we saw in the previous section, is simply the convolution of the lattice function with the unit cell function. Equation (5.28) states that the transform of the crystal structure function is the product of the transform of the lattice function with that of the unit cell function:

$$Tc(\mathbf{u}) = T[l(\mathbf{r}) \, * \, u(\mathbf{r})]$$
$$= [Tl(\mathbf{r})].[Tu(\mathbf{r})] \tag{5.30}$$

In Chapter 8, we shall derive the equations for the Fourier transform of the lattice function, and in Chapter 9 we shall calculate the Fourier transform of the unit cell function.

5.12   The Patterson function

The final section of this chapter will deal with the properties of another integral which will be discussed in Chapter 12. A treatment of the mathematical behaviour of this integral is included here, since it is very similar to the convolution integral. We define the *Patterson integral* as

$$p(\mathbf{u}) = \int_{\text{all } \mathbf{r}} f(\mathbf{r})g(\mathbf{u} + \mathbf{r}) \, d\mathbf{r} \tag{5.31}$$

Comparison of the Patterson integral with the convolution integral

$$c(\mathbf{u}) = \int_{\text{all } \mathbf{r}} f(\mathbf{r})g(\mathbf{u} - \mathbf{r}) \, d\mathbf{r} \tag{5.27}$$

shows that the only difference between them is the fact that the Patterson integral has $g(\mathbf{u} + \mathbf{r})$ where the convolution integral has $g(\mathbf{u} - \mathbf{r})$. The general interpretation of the Patterson integral is identical to that of the convolution integral, with the exception that in forming the Patterson integral, we do not invert $g(\mathbf{r})$ through the origin, as we did in step (2) of the method of forming the convolution integral. Other than that, the significance of the variable $\mathbf{u}$ is the same, in that it specifies the separation of the two functions as one of them slides past the other. The rules for forming the Patterson integral are given here.

Method for determining Patterson integrals $\int_{\text{all } \mathbf{r}} f(\mathbf{r})g(\mathbf{u} + \mathbf{r}) \, d\mathbf{r}$:

(1) $f(\mathbf{r})$ and $g(\mathbf{r})$ are given functions.
(2) Displace $g(\mathbf{r})$ by any chosen $\mathbf{u}$ to form $g(\mathbf{u} + \mathbf{r})$.
(3) Multiply $f(\mathbf{r})$ by the newly formed $g(\mathbf{u} + \mathbf{r})$ to generate the product function $f(\mathbf{r})g(\mathbf{u} + \mathbf{r})$.

(4) Integrate over all values of $\mathbf{r}$, which in one dimension corresponds to finding the area under the graph of the product function $f(x)g(u+x)$ for the chosen value of u.

(5) Return to step (2), but choose a different value of $\mathbf{u}$. This will, in turn, give a different product function and a different value for the integral.

(6) Repeat for all values of $\mathbf{u}$. This will generate the entire function

$$p(\mathbf{u}) = \int_{\text{all } \mathbf{r}} f(\mathbf{r})g(\mathbf{u}+\mathbf{r})\,d\mathbf{r}$$

Of particular interest is the case in which both $f(\mathbf{r})$ and $g(\mathbf{r})$ happen to be the same function, in which instance we speak of the *self-Patterson integral* $P(\mathbf{u})$, or simply the *Patterson function* of $f(\mathbf{r})$:

$$P(\mathbf{u}) = \int_{\text{all } \mathbf{r}} f(\mathbf{r})f(\mathbf{u}+\mathbf{r})\,d\mathbf{r} \tag{5.32}$$

A noteworthy example is the Patterson function formed by the one-dimensional function $f(x)$, shown in Fig. 5.20(a). This is an array of three δ functions which were deliberately chosen to be of different strengths and placed in a non-symmetric manner about the origin. To generate the Patterson function, we slide the function over itself, form the product function and integrate. The Patterson function is non-zero only when any two δ functions overlap, and all the interesting positions are depicted in Fig. 5.21.

The graph of the Patterson function is an array of seven δ functions, as shown in Fig. 5.20(b). On comparing the original function $f(x)$ with the Patterson function $P(\mathbf{u})$, we may make the following generalisations:

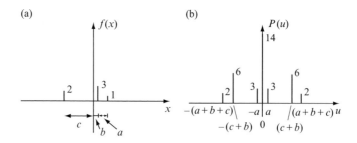

Fig. 5.20 *The Patterson function.* (a) shows a non-symmetrical array of three δ functions of different strengths. The Patterson function of (a) is the diagram shown in (b), which is an array of seven δ functions. Note that the Patterson function is symmetrical, and that the peaks in the Patterson function occur only at values of u corresponding to the distances between pairs of δ functions in $f(x)$. Further, the strength of any δ function in the Patterson function is the sum of the products of each pair of δ functions in the original $f(x)$ separated by the appropriate value of u. The steps by which the Patterson function is derived are shown in Fig. 5.21.

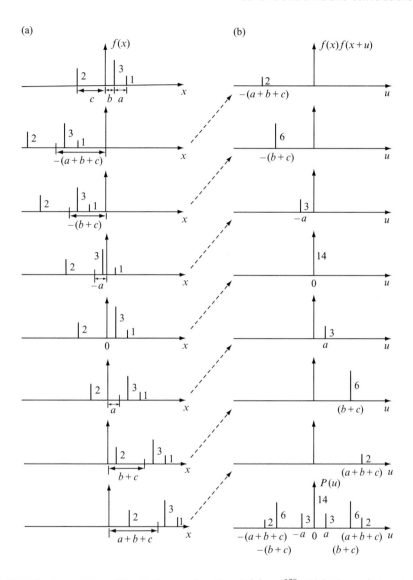

Fig. 5.21 *The formation of the Patterson function* $P(u) = \int_{-\infty}^{\infty} f(x)f(u+x)\,dx$. *At the top of column (a) is a function* $f(x)$. *To form the Patterson function* $P(u)$, *we displace* $f(x)$ *by some distance* u *to form* $f(u+x)$, *and for each value of* u, *we multiply the original* $f(x)$ *by the new* $f(u+x)$ *and integrate over all* x. *Since* $f(x)$ *is an array of* δ *functions, the product function* $f(x)f(u+x)$ *is zero unless* u *corresponds to a value such that two* δ *functions overlap. There are seven such values of* u, *all of which are shown in column (a). Column (b) shows the values of the product functions corresponding to these values of* u. *On integration, we obtain only the value of the* δ *function of the product function, and collecting these together on one graph, we derive the Patterson function, as shown at the bottom of column (b).*

(1) Despite the fact that $f(\mathbf{r})$ is non-centrosymmetric, the Patterson function $P(\mathbf{u})$ is always centrosymmetric.

(2) If $f(\mathbf{r})$ has N peaks, then the Patterson function $P(\mathbf{u})$ has a large peak at the origin plus $N(N-1)$ peaks at points other than the origin, some of which might overlap.

(3) Each peak in the Patterson function $P(\mathbf{u})$ occurs at values of $\mathbf{u}$ corresponding to plus and minus the separation of a pair of peaks in the original function $f(\mathbf{r})$.

(4) The strength of each Patterson peak is the sum of the products of the strengths of each pair of peaks in the original function $f(\mathbf{r})$ which are separated by the corresponding value of $\mathbf{u}$.

The relevance of these results to crystallographers will be discussed in Chapter 12.

Summary

We choose to define the *Fourier transform* $F(\mathbf{k})$ of a function $f(\mathbf{r})$ as

$$F(\mathbf{k}) = \int_{\text{all } \mathbf{r}} f(\mathbf{r}) e^{i\mathbf{k}\cdot\mathbf{r}} \, d\mathbf{r} \qquad (5.12)$$

The *inverse transform* is given by

$$f(\mathbf{r}) = \int_{\text{all } \mathbf{k}} F(\mathbf{k}) e^{-i\mathbf{k}\cdot\mathbf{r}} \, d\mathbf{k} \qquad (5.16)$$

In one dimension, the analogous equations are

$$F(k) = \int_{-\infty}^{\infty} f(x) e^{ikx} \, dx \qquad (5.11)$$

$$f(x) = \int_{-\infty}^{\infty} F(k) e^{-ikx} \, dk \qquad (5.15)$$

$F(\mathbf{k})$ and $f(\mathbf{r})$, and $F(k)$ and $f(x)$ are referred to as *Fourier mates*.

A table of Fourier mates met with in this chapter is given in Table 5.1.

The more extended a function $f(\mathbf{r})$ as a function of $\mathbf{r}$, then the narrower the Fourier transform $F(\mathbf{k})$ as a function of $\mathbf{k}$.

The narrower a function $f(\mathbf{r})$ as a function of $\mathbf{r}$, then the wider the Fourier transform $F(\mathbf{k})$ as a function of $\mathbf{k}$.

The *Dirac δ function*, written as $\delta(\mathbf{r} - \mathbf{r}_0)$, corresponds to a peak of infinite height at the point $\mathbf{r} = \mathbf{r}_0$. The integral of the δ function is unity.

The *convolution integral* is defined as

$$c(\mathbf{u}) = \int_{\text{all } \mathbf{r}} f(\mathbf{r}) g(\mathbf{u} - \mathbf{r}) \, d\mathbf{r} = f(\mathbf{r}) * g(\mathbf{r}) \qquad (5.27)$$

Table 5.1 Fourier mates.

<table>
<tr><th colspan="2" align="center">*Function f(x)*</th></tr>
<tr><th align="center">Equation</th><th align="center">Graph</th></tr>
<tr>
<td align="center">

$-\infty < x < -X_0, f(x) = 0$
$-X_0 < x < X_0, f(x) = h$
$X_0 < x < \infty, f(x) = 0$

</td>
<td align="center">

(a)

</td>
</tr>
<tr>
<td align="center">$\delta(x)$</td>
<td align="center">

(b)

</td>
</tr>
<tr>
<td align="center">$\delta(x+x_0) + \delta(x-x_0)$</td>
<td align="center">

(c)

</td>
</tr>
<tr>
<td align="center">$\delta(x+x_0) + \delta(x) + \delta(x-x_0)$</td>
<td align="center">

(d)

</td>
</tr>
<tr>
<td align="center">

$$\sum_{n=-p}^{n=+p} \delta(x-nx_0)$$

</td>
<td align="center">

(e)

</td>
</tr>
<tr>
<td align="center">

$$\sum_{n=-\infty}^{n=\infty} \delta(x-nx_0)$$

</td>
<td align="center">

(f)

</td>
</tr>
</table>

(*continued*)

Table 5.1 Continued

Function F(k)

Graph	Equation
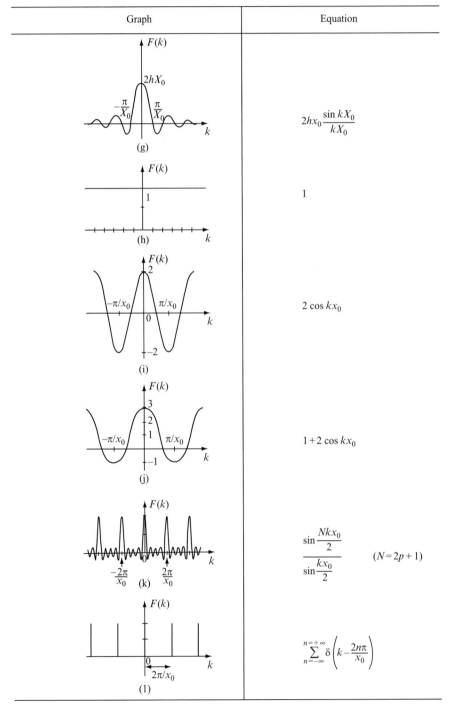 (g)	$2hx_0 \dfrac{\sin kX_0}{kX_0}$
(h)	1
(i)	$2 \cos kx_0$
(j)	$1 + 2 \cos kx_0$
(k)	$\dfrac{\sin \dfrac{Nkx_0}{2}}{\sin \dfrac{kx_0}{2}} \qquad (N = 2p + 1)$
(l)	$\displaystyle\sum_{n=-\infty}^{n=+\infty} \delta\left(k - \dfrac{2n\pi}{x_0}\right)$

The *Patterson integral* is defined as

$$p(\mathbf{u}) = \int_{\text{all } \mathbf{r}} f(\mathbf{r})g(\mathbf{u}+\mathbf{r})\,d\mathbf{r} \qquad (5.31)$$

For any function $f(\mathbf{r})$, the *Patterson function* is defined as

$$P(\mathbf{u}) = \int_{\text{all } \mathbf{r}} f(\mathbf{r})f(\mathbf{u}+\mathbf{r})\,d\mathbf{r}$$

The *convolution theorem* states that:

- The Fourier transform of the convolution of $f(\mathbf{r})$ with $g(\mathbf{r})$ is the product of the individual transforms of $f(\mathbf{r})$ and $g(\mathbf{r})$:

$$T[f(\mathbf{r})^*\,g(\mathbf{r})] = [Tf(\mathbf{r})].[Tg(\mathbf{r})] \qquad (5.28)$$

- The Fourier transform of the product of two functions $f(\mathbf{r})$ and $g(\mathbf{r})$ is the convolution of the individual transforms of $f(\mathbf{r})$ and $g(\mathbf{r})$:

$$T[f(\mathbf{r}).g(\mathbf{r})] = [Tf(\mathbf{r})]^*\,[Tg(\mathbf{r})] \qquad (5.29)$$

A lattice is a three-dimensional array of δ functions described by the *lattice function* $l(\mathbf{r})$,

$$l(\mathbf{r}) = \sum_{\text{all } p,\,q,\,r} \delta\left(\mathbf{r} - [p\mathbf{a} + q\mathbf{b} + r\mathbf{c}]\right) \qquad (5.17)$$

in which the summation extends over all lattice points. The contents of the unit cell may be described by the *unit cell function* $u(\mathbf{r})$. The *crystal structure function* $c(\mathbf{u})$ is the convolution of the lattice function with the unit cell function:

$$c(\mathbf{u}) = l(\mathbf{r})^*u(\mathbf{r})$$

All structural information concerning the crystal is contained within the crystal structure function $c(\mathbf{u})$.

Appendix I

Proof of Fourier's theorem

To prove that if

$$F(k) = \frac{1}{\sqrt{2\pi}} \int_{-\infty}^{\infty} f(x)e^{ikx}\,dx \qquad (5.11b)$$

then

$$f(x) = \frac{1}{\sqrt{2\pi}} \int_{-\infty}^{\infty} F(k)e^{-ikx}\,dk \qquad (5.11b')$$

Since this is a strict mathematical proof, we include the correct factors of $1/\sqrt{2\pi}$.

The proof presented requires a definition of the δ function. There are several possible mathematical definitions of the δ function, but that of use here is

$$\delta(x - x_0) = \frac{1}{2\pi} \int\limits_{-\infty}^{\infty} e^{ik(x-x_0)} \, dk \qquad (5.33)$$

This definition will be used without proof—the interested reader is referred to *The Fourier Integral and Its Applications*, by A. Papoulis, McGraw-Hill, New York, 1962. Substituting for $F(k)$ in equation (5.11b′), we obtain

$$g(x) = \frac{1}{2\pi} \int\limits_{-\infty}^{\infty} \left(\int\limits_{-\infty}^{\infty} f(x')e^{ikx'} \, dx' \right) e^{-ikx} \, dk \qquad (5.34)$$

We have made two alterations here. Firstly, we have called the function defined by equation (5.11b′) $g(x)$, since at this stage we know only that the integral represents a function of x, but we do not yet know that it is the specific function $f(x)$ used in equation (5.11b). Secondly, in order not to confuse the x's occurring in equations (5.11b) and (5.11b′), we write the variable of equation (5.11b) as x', and that of equation (5.11b′) as x. Continuing from equation (5.34), we may reverse the order of integration, obtaining

$$g(x) = \frac{1}{2\pi} \int\limits_{-\infty}^{\infty} f(x') \left(\int\limits_{-\infty}^{\infty} e^{ik(x'-x)} \, dk \right) dx'$$

But the inner integral, according to our definition, equation (5.33), is a δ function $\delta(x' - x)$.

$$g(x) = \int\limits_{-\infty}^{\infty} f(x')\delta(x' - x) \, dx'$$

But from equation (5.18),

$$\int\limits_{-\infty}^{\infty} f(x')\delta(x' - x) \, dx' = f(x)$$

$$\therefore g(x) = f(x) = \frac{1}{\sqrt{2\pi}} \int\limits_{-\infty}^{\infty} F(k)e^{-ikx} \, dk$$

The theorem is therefore proved. The proof in three dimensions is exactly analogous.

Appendix II

Proof of convolution theorem

To prove the convolution theorem,

$$T[f(x) * g(x)] = [Tf(x)].[Tg(x)]$$

and the converse,

$$T[f(x).g(x)] = [Tf(x)] * [Tg(x)]$$

In fact, the form of the convolution theorem as stated above is not the correct mathematical statement, which is

$$T[f(x) * g(x)] = \sqrt{2\pi}[Tf(x)].[Tg(x)]$$

and

$$T[f(x).g(x)] = \frac{1}{\sqrt{2\pi}}[Tf(x)] * [Tg(x)]$$

in which the now familiar factors of $\sqrt{2\pi}$ are included. Since this is to be a rigorous mathematical proof, we shall use the correct definitions.

We require

$$T[f(x) * g(x)] = \frac{1}{\sqrt{2\pi}} \int\limits_{-\infty}^{\infty} \left(\int\limits_{-\infty}^{\infty} f(x)g(u-x)\,dx \right) e^{iku}\,du$$

Since the convolution is a function of u, to form the Fourier transform of the convolution we multiply by e^{iku} and integrate over u. We may multiply and divide by e^{ikx} and, changing the order of integration, we obtain

$$\frac{1}{\sqrt{2\pi}} \int\limits_{-\infty}^{\infty} \left(\int\limits_{-\infty}^{\infty} f(x)g(u-x)\,dx \right) e^{iku}\,du = \frac{1}{\sqrt{2\pi}} \int\limits_{-\infty}^{\infty} f(x)e^{ikx} \left(\int\limits_{-\infty}^{\infty} g(u-x)e^{ik(u-x)}\,du \right) dx$$

For any value of x, we may change the variable of the second integral as

$$U = u - x$$

$$\therefore dU = du$$

$$\therefore T[f(x) * g(x)] = \frac{1}{\sqrt{2\pi}} \int\limits_{-\infty}^{\infty} f(x)e^{ikx} \left(\int\limits_{-\infty}^{\infty} g(U)e^{ikU}\,dU \right) dx$$

$$= \frac{1}{\sqrt{2\pi}} \int\limits_{-\infty}^{\infty} f(x)e^{ikx}\,dx \int\limits_{-\infty}^{\infty} g(U)e^{ikU}\,dU$$

But it does not matter what name we give to the variable of integration of the second integral, and so

$$T[f(x) * g(x)] = \frac{1}{\sqrt{2\pi}} \int\limits_{-\infty}^{\infty} f(x)e^{ikx}\,dx. \int\limits_{-\infty}^{\infty} g(x)e^{ikx}\,dx$$

$$\therefore T[f(x) * g(x)] = \sqrt{2\pi}[Tf(x)].[Tg(x)]$$

thereby proving the convolution theorem.

The proof of the converse theorem,

$$T[f(x).g(x)] = \frac{1}{\sqrt{2\pi}}[Tf(x)] * [Tg(x)]$$

is as follows. Writing the familiar definitions

$$F(k) = \frac{1}{\sqrt{2\pi}} \int_{-\infty}^{\infty} f(x)e^{ikx} \, dx = Tf(x)$$

$$G(k) = \frac{1}{\sqrt{2\pi}} \int_{-\infty}^{\infty} g(x)e^{ikx} \, dx = Tg(x)$$

$$f(x) = \frac{1}{\sqrt{2\pi}} \int_{-\infty}^{\infty} F(k)e^{-ikx} \, dk$$

$$g(x) = \frac{1}{\sqrt{2\pi}} \int_{-\infty}^{\infty} G(k)e^{-ikx} \, dk$$

we have

$$T[f(x).g(x)] = \frac{1}{\sqrt{2\pi}} \int_{-\infty}^{\infty} f(x)g(x)e^{ikx} \, dx$$

$$= \frac{1}{\sqrt{2\pi}} \int_{-\infty}^{\infty} \left(\frac{1}{\sqrt{2\pi}} \int_{-\infty}^{\infty} F(k')e^{-ik'x} \, dk' \right) g(x)e^{ikx} \, dx$$

$$= \frac{1}{\sqrt{2\pi}} \int_{-\infty}^{\infty} F(k') \left(\frac{1}{\sqrt{2\pi}} \int_{-\infty}^{\infty} g(x)e^{i(k-k')x} \, dx \right) dk'$$

$$= \frac{1}{\sqrt{2\pi}} \int_{-\infty}^{\infty} F(k')G(k - k') \, dk'$$

This has the general form of the convolution integral, equation (5.27), and so represents the convolution of the transform of $f(x)$ with the transform of $g(x)$, and the theorem is proved.

The proofs in three dimensions are exactly similar.

Of course, from the point of view of practical applications, we may use the convolution theorem as first stated, without the factors of $\sqrt{2\pi}$.

Bibliography

Bracewell, R. M. *The Fourier Transform and Its Applications*, 3rd edn. McGraw-Hill, New York, 1999.

Papoulis, A. *The Fourier Integral and its Applications*. McGraw-Hill, New York, 1962.

Both these texts discuss Fourier theory with particular reference to applications to engineering and electronics.

Braddick, H. J. J. *Vibrations, Waves and Diffraction*. McGraw-Hill, London, 1965. Later chapters discuss the application of Fourier techniques to optical diffraction.

Lipson, H. and Taylor, C. A. *Fourier Transforms and X-ray Diffraction*. Bell, London, 1958. A discussion of the relevance of Fourier methods to X-ray crystallography.

Stuart, R. D. *An Introduction to Fourier Analysis*. Methuen, London, 1966. A useful monograph on the mathematics of Fourier analysis and Fourier transforms.

6
Diffraction

6.1 The interaction of waves with obstacles

This chapter introduces and discusses a most important aspect of the behaviour of waves, namely, what happens when a wave motion interacts with an obstacle placed in its path. In Chapter 4 we learnt that a wave may be described mathematically by the expression

$$\psi = \psi_0 \, e^{i(\mathbf{k}\cdot\mathbf{r}-\omega t)} \tag{4.35}$$

but this represents only a very specific type of wave. Equation (4.35) is the equation of an infinite plane wave with wave vector $\mathbf{k}$ and frequency ω. When an infinite plane wave interacts with an obstacle, one of the effects of the obstacle is to perturb the wave motion in some way such that equation (4.35) is no longer the correct description of the motion. During the course of this chapter, we will investigate how equation (4.35) is modified to take account of the interaction of the wave with the obstacle.

Physicists use two words to describe the interaction of waves with obstacles. The first of these is *scattering*, which was mentioned in Chapter 4. In general, we talk of scattering when the dimensions of the perturbing obstacle are comparable to the wavelength of the appropriate wave motion. If the obstacle is much larger than the wavelength of the wave motion, we talk of *diffraction*. Thus we say that light waves are scattered by dust particles in the atmosphere, whereas light is diffracted by an obstacle such as a small ball bearing. This latter event is noteworthy historically and will be discussed below.

There is a subtlety in nomenclature here which does require explanation. Consider a crystal composed of a large number of atoms or molecules, whose size will be of the order of 1 nm (10 Å). The crystal size will be of the order of 1 mm. When X-rays of wavelength 0.1 nm (1 Å) interact with the crystal, each atom will scatter the waves, and is said to act as a *scattering centre*. The combination of scattering events due to all the molecules in the crystal will give rise to a total net effect which may be ascribed to the crystal as a whole. Since the crystal is very much larger than the X-ray wavelength, we say that the crystal has diffracted the X-rays. We now see that scattering is a fundamental property of the atoms, and it is a combination of these scattering events which gives rise to the macroscopic phenomenon of diffraction. Strictly speaking, diffraction is an essentially macroscopic phenomenon which is the result of many microscopic scattering events.

It should be mentioned, however, that the distinction in meaning between the words 'scattering' and 'diffraction' is often neglected. In that they both describe the same overall physical phenomenon, it is not uncommon to find them used interchangeably.

6.2 The diffraction of water waves

Our discussion of diffraction is facilitated if we examine the specific example of the interaction of various obstacles with plane water waves as generated on the surface of water in a shallow tank. Consider a rectangular tank, known as a *ripple tank*, containing water which is subject to a disturbance which gives rise to plane waves, as, for instance, that due to the periodic dipping into the water of a rule whose length is just smaller than the width of the tank. If the tank is long, then reflection from the far end is negligible. Also, it is assumed that the plane waves are propagated accurately parallel to the side walls of the tank so that no stray reflections occur. Initially, we observe a series of plane waves as shown in Fig. 6.1.

If the tank has a transparent base, it is possible to shine light from beneath upwards through the tank so that the wave crests cast shadows on a screen parallel to, and above, the tank. With this arrangement, a moving pattern of waves is seen on the screen. Now consider the effect of switching the light source on and off so that the light is on only when each wave crest has advanced exactly one wavelength. In this case, the light will illuminate a pattern of waves which is identical to that which was present on the previous occasion that the light was on. The image seen on the screen will therefore be the same every time the light is on, and in this way, the image of the wave motion is 'frozen in'. By synchronising the illumination of the wave pattern with the frequency of the motion, we observe a stationary image. This is known as the *stroboscopic effect*, and its use is that it is much easier to see what the wave motion is doing when we observe a stationary pattern, as shown in Fig. 6.2. In practice, the frequency of illumination of the light source is continuously variable so that we may adjust the light to the frequency of the water waves. This apparatus is known as a *stroboscopic ripple tank*.

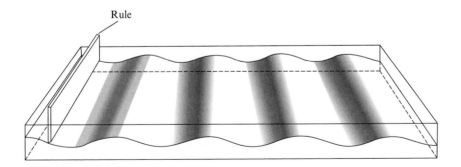

Fig. 6.1 *The ripple tank.* Water waves are propagated in a shallow tank by the action of a mechanically driven rule which oscillates in a plane perpendicular to the water surface, and parallel to the end walls of the tank. A moving pattern of waves is seen on the surface of the water.

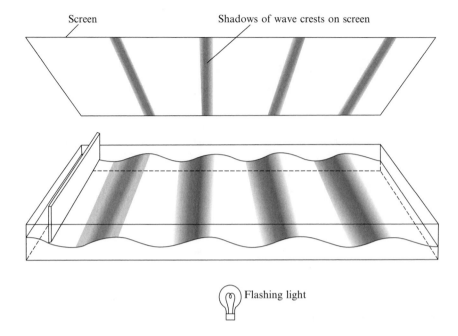

Fig. 6.2 *The stroboscopic ripple tank.* The tank is illuminated from below, and the wave crests throw shadows on the screen. If the illumination is constant, a moving pattern is observed on the screen. If the light is synchronised with the wave motion such that the light is on only when the wave pattern has advanced exactly one wavelength, then the light 'sees' the same pattern on each illumination, implying that a stationary pattern is seen on the screen.

We now place, perpendicular to the side walls of the tank and about halfway along the tank's length, two slats with a gap between them which is considerably less than the wavelength of the water waves. The slats project above the surface of the water and constitute an obstacle in the path of the waves, implying that diffraction occurs. To be truly accurate, the diffraction occurs at the edge of each slat, but we may think of the system as showing diffraction from a slit which is narrow compared with the wavelength of the wave motion. With our stroboscopic light, we 'freeze' the wave crests and observe that, on the source side of the obstacle, the waves are still plane waves whose wavefronts are parallel to the obstacle. But the waves on the further side of the obstacle have the appearance shown in Fig. 6.3.

This experiment shows that the effect of the obstacle is therefore to transform plane waves into curved waves. The action of the obstacle is to 'bend' the waves around its edges so that the disturbance is propagated into the geometrical shadow of the slit. The term 'geometrical shadow' means the following. If we imagine two lines drawn perpendicular to the plane wavefronts at the edges of the slit, then the zone within these lines is known as the geometrical image of the slit, and the zone outside the two lines is the geometrical shadow.

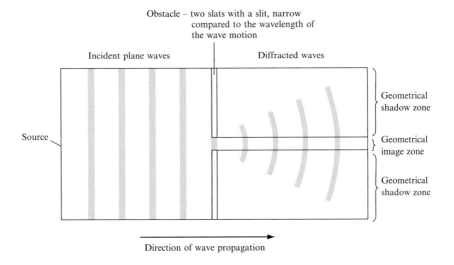

Obstacle – two slats with a slit, narrow
compared to the wavelength of
the wave motion

Incident plane waves   Diffracted waves

Geometrical
shadow zone

Source

Geometrical
image zone

Geometrical
shadow zone

Direction of wave propagation

Fig. 6.3 *Diffraction by a narrow slit.* The diagram shows the pattern observed on the screen of a stroboscopic ripple tank into which an obstacle is placed. The diffracted waves travel into the area of geometrical shadow. If the wave–obstacle interaction were bullet-like, we would expect no disturbance in the geometrical shadow. Note that the wavelength of the diffracted waves is equal to that of the incident waves.

As a counterexample, consider the effect of firing bullets exactly perpendicular to the obstacle. Because bullets travel in straight lines, all those bullets passing through the slit would continue on undeflected into the zone of geometrical image, but those bullets which hit the slats would, for the purposes of this example, be stopped. No bullets would travel into the zone of geometrical shadow, and so no disturbance would be detected there. This gives us a test as to whether a phenomenon is wave-like or 'bullet-like'. When the motion interacts with an obstacle, we observe the zone of geometrical shadow. If a disturbance is present, we may correctly infer that we are dealing with a wave motion. If there is a disturbance only in the zone of geometrical image, then it is possible that the motion is bullet-like. Notice the use of the word 'possible' in the last sentence. As will become clear later in this section, although observation of a disturbance in the zone of geometrical shadow is a necessary and sufficient condition for wave motion, observation in the geometrical image zone only is not a specific criterion of bullet-like motion.

The difference in the behaviour of a wave motion and a bullet-like motion on interaction with an obstacle such as a slit is of great historical importance and will be discussed when we review the diffraction of light.

Our first example shows that one of the effects of the diffraction of waves is to cause a 'bending around corners' such that plane waves are transformed into waves of a more complicated nature.

We will now investigate the effect of making the slit wider, as shown in the sequence of diagrams in Fig. 6.4. In Fig. 6.4(a), the slit is comparable in width to

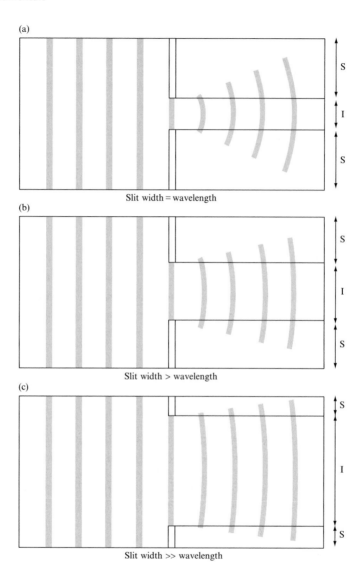

Fig. 6.4 *The variation in the diffraction pattern with obstacle size.* The diagrams show the patterns observed on the screen of a stroboscopic ripple tank as the width of the slit obstacle is increased. Diffraction still occurs, but its effect becomes increasingly masked as the slit is widened as compared with the wavelength of the wave motion. Wave propagation is from left to right. S corresponds to zones of geometrical shadow; I to zones of geometrical image.

the wavelength of the waves; in Fig. 6.4(b), it is rather wider than the wavelength; and in Fig. 6.4(c), the slit is very much wider than the wavelength. In each case, diffraction still occurs, and the wave is propagated into the zone of geometrical shadow. Although the waves are still 'bent around the corners', the overall effect has changed.

Compare Fig. 6.3 with Fig. 6.4(c). In the former case, the diffracted waves are quite semicircular, whereas in the latter case, a more adequate description of the diffracted waves would be 'plane with curved edges'. As the slit width is increased, as compared with the wavelength of the wave motion, the specific effects of diffraction become less observable. Unless we are quite careful in our observation of the system depicted in Fig. 6.4(c), we might not notice the slight curvature at the periphery of the wave, and conclude (erroneously) that the effect of the obstacle was exactly as would be observed if we had used the bullets of our previous example. Thus when the slit width is wide compared with the wavelength of the wave motion, the diffraction effects are masked, and so the distinction between bullet-like motion and wave motion is less apparent. This explains why, as mentioned above, the observation of a disturbance only in the zone of geometrical image is not an unambiguous criterion of bullet-like motion.

6.3   Diffraction and information

Let us now re-examine the examples of the diffraction of water waves by a slit discussed in Section 6.2. Reference to Fig. 6.3 shows that the waves on the source side of the obstacle are plane, whereas those beyond the obstacle are almost semicircular. The plane waves are a result of the particular disturbance at the source, which in this instance is the periodic dipping of a rule into the water. The semicircular waves are caused by the interaction of the plane waves with the obstacle. Looking at this from a slightly different viewpoint, we may say that the plane waves, before they interact with the obstacle, do not 'know' that the obstacle is present, whereas those waves further from the source beyond the obstacle *do* 'know' that the obstacle is there because they have interacted with it. Because the diffracted waves are different from the incident waves, then the diffracted waves contain some *additional information* which the incident waves do not possess.

This concept is clarified by comparing the effects of different obstacles as in Figs. 6.3 and 6.4. In each case the incident waves are plane waves with the same wavelength and frequency, and although the diffracted waves also have the same frequency and wavelength, the overall wave pattern is different for each obstacle. Now, the *sole* difference between each example is the nature of the obstacle with which the incident waves interact. We can therefore correlate the differences in wave pattern with the fact that the waves were diffracted by different obstacles, implying that the nature of the wave pattern after diffraction is determined by the obstacle causing the diffraction. Hence the overall diffraction pattern depends solely on the nature of the obstacle.

That the diffraction pattern of an obstacle contains information concerning the structure of the obstacle is demonstrated by the following example. Suppose that the obstacle in the water wave experiment shown in Fig. 6.3 were in a 'black box' so that an observer could see only plane waves incident on an unknown obstacle, and the semicircular waves coming away from the obstacle. The observer would compare the incident and the diffracted waves, and on seeing that they were different, would conclude that an obstacle was present. If the observer knew the behaviour exhibited by slit obstacles, he or she would correctly deduce that the obstacle was a slit narrow compared with the wavelength of the wave motion. Had the obstacle been more complicated, then the diffraction pattern would have been more complicated, but if we know the full theory of diffraction, then it should be possible for us to infer the nature of the obstacle from investigation of the diffraction pattern.

Let us now apply these ideas to the case of the diffraction of X-rays by a crystal. The X-rays are incident on the crystal and are diffracted, and the diffraction pattern is recorded on, for example, a photographic plate. The details of the diffraction pattern are dependent solely on the nature of the obstacle, which in this case is a crystal containing certain molecules in a well-ordered, three-dimensional array. The diffracted waves therefore contain information concerning both the nature of the molecules and the way in which they are packed in space. This information is present in a complicated manner in the diffraction pattern, but we know that the information is there. If we know the rules which diffraction phenomena obey, when we look at the diffraction photograph in the correct manner, we will be able to extract the relevant information from the diffraction photograph and, using this information, we can reconstruct the obstacle.

This is just what an X-ray diffraction experiment sets out to do. By looking at the X-ray diffraction pattern, we may derive the pertinent information from which we may deduce the crystal parameters, as well as the structure of the crystal's molecules.

The argument so far may be summarised as follows.

When a wave interacts with an obstacle, diffraction occurs. The detailed behaviour depends solely on the diffracting obstacle, and so the diffracted waves may be regarded as containing information on the structure of the obstacle. It is the purpose of X-ray crystallography to examine the X-rays diffracted by a crystal so that the crystal structure may be determined.

6.4   The diffraction of light

Our example of the diffraction of water waves has given us the criteria for an experimental arrangement with which we may observe the diffraction of light. Firstly, we require a source of plane light waves, which may be approximated as follows. If a light source is shielded by a screen containing a pinhole, we have effectively a point source which gives rise to spherical light waves. If we observe these waves at a long distance from the pinhole, over any small area, the curvature of the wavefront will be small, and, locally, the wave will be approximately plane. (In fact it is not necessary to use plane waves, in that diffraction occurs when any wave motion interacts with an obstacle. Plane waves are chosen for this example merely to make the analysis simpler, and so that the system is analogous to the water wave case discussed in the previous section.) If a narrow slit is placed in the path of the waves, diffraction effects should be observed. Clearly, our criterion of narrowness is that the slit should be as close as possible to the wavelength of light. If yellow (for instance, sodium) light is used, the wavelength is some 600 nm, which means that the slit, practically speaking, has to be very narrow indeed.

It is interesting at this juncture to review some aspects of the history of optics. In the middle of the seventeenth century, there arose two schools of thought as to the physical nature of light. That led by Isaac Newton (1642–1727) suggested that light was a bullet-like motion, implying that luminous objects emitted a stream of bullet-like particles. On the other hand, Christiaan Huygens (1629–1695) proposed that light was a wave motion analogous to water waves. Such was the prestige of Newton at this time that the 'bullet' theory was accepted for almost a century, merely because Newton had said so.

In 1803, Thomas Young (1773–1829) published the results of an experiment in which diffraction effects were observed. Now, on the basis of our examples using water waves above, the distinction between the 'bullet' and wave theories will become apparent in a diffraction-type experiment only if the width of the obstacle is not too much greater than the wavelength of the wave motion. Because we now know the wavelength of light to be so small, we see that the crucial experiment to decide between the theories will work only if a very small obstacle is used. Prior to Young's experiment, the obstacles used were very much larger than the wavelength of light and so any diffraction effects were masked, as in the water wave example of Fig. 6.4(c). This was one of the reasons that the wave–particle controversy lingered on for so long. In fact, Young's experiment was rather more subtle in that instead of using a single slit, he used a pair of narrow slits, and in this way (see Chapter 7, Section 7.5, for an analysis of Young's experiment), he made the effect of diffraction considerably more noticeable.

Despite the apparent conclusiveness of Young's experiment, the wave–particle argument went on for several more years. One of the more famous anecdotes connected with this concerns the French physicist Poisson (1781–1840), who vigorously supported the particle theory. At a meeting of the French Academy of Sciences in 1818, he presented a mathematical argument which predicted that on the basis of the wave theory, if a perfectly round obstacle (what criteria would you use to define 'perfectly round' in this case?) such as a small ball bearing were placed in a light beam, then, instead of there being a perfect geometrical shadow, the effect of diffraction would be to produce a bright spot exactly at the centre of the geometrical shadow. He stated that since no such effect had ever been observed, then the wave theory was incorrect. Arago (1786–1853), who was present at the meeting, decided to try the experiment, and having obtained a sufficiently perfectly spherical obstacle, observed that the bright spot was indeed present at the centre of the geometrical shadow! Thus the wave theory was vindicated, and the spot at the centre of the shadow is now known as 'Poisson's bright spot'. The criterion of roundness for the obstacle is, of course, that the radius of the obstacle should be constant to within the wavelength of light. If the deepest pit or highest bump on the surface is no more than a wavelength of light below or above the mean radius of the obstacle, then the light waves effectively 'see' a perfect sphere. Poisson had been misled because no obstacle spherical to that degree of precision had been investigated. (In fact, to be historically accurate, it is now known that the existence of Poisson's bright spot at the centre of the shadow of a spherical obstacle was first recorded by Miraldi in 1723. This observation had been forgotten, and Poisson and Arago were unaware of it.)

In conclusion, we may say that diffraction effects are observed only if the obstacle used is not too much larger than the wavelength of light. This gives us a test for the applicability of the ray as opposed to the wave treatment of optics. Ray optics deals with the propagation of rays, which are normal to wavefronts. In this respect, the propagation of rays is analogous to the 'bullet' theory, and so is applicable only for interactions of light with relatively large obstacles. Wave optics, on the other hand, must be used when light interacts with obstacles that are comparable in size with the wavelength of the light used.

6.5   X-ray diffraction

The extension of the ideas of diffraction to the interaction of X-rays with crystals is straightforward. The wavelength of X-rays is typically of the order of 0.1 nm (1 Å), and so they will be diffracted or scattered by objects not too much larger than this. Atoms and molecules in general will satisfy this requirement, and so will diffract X-rays. There is, however, an important subtlety concerned with the fact that we are dealing with a crystal. The crystalline nature of the diffracting obstacle implies that the molecular scattering centres are arranged in a well-defined, three-dimensional, ordered array. The significance of this is that there is a specific distance between any two molecules, implying that there will be a well-defined phase relationship between the waves scattered from any pair of scattering centres. The relevance of this is that the waves scattered from the complete set of molecules within the crystal will combine together in a definite manner, giving rise to a clear and precise diffraction pattern.

Let us compare this situation with what we would expect from the diffraction of X-rays by a liquid. Each molecule still acts as a scattering centre, and diffracts the X-rays. But because the structure of the liquid phase is such that there are no well-defined intermolecular distances, there will be no definite phase relationship between waves scattered from different molecules, and on the average, the diffraction pattern from the liquid as a whole will tend to be rather diffuse and imprecise.

It is the three-dimensional ordering of the crystalline state which allows us to obtain clear diffraction data, giving us readily understandable information on the nature of the crystal and the molecules it contains.

The discussion of diffraction given in the last few pages is a more detailed explanation of the effects described in Chapter 1 in the section 'Why do we use X-rays?' and confirms the conclusions reached there.

6.6   The mathematics of diffraction

In this section, we will investigate how to obtain a mathematical description of diffraction. For the purposes of the following argument, it will be assumed that the structure and all physical properties of the diffracting obstacle are known. Using this information, we will derive a mathematical expression for the diffraction pattern, thereby giving us a formula by which we may predict the diffraction pattern of any given obstacle. Later on, we will investigate the inverse process, in which we know the diffraction pattern and wish to deduce the nature of the obstacle. One further assumption is made. It will be assumed that plane electromagnetic waves of a single frequency ω and wavelength λ are incident normally on the obstacle. The case of non-normal incidence will be treated in Chapter 8. Note that since the direction of the incident waves is specified, as well as the wavelength λ, then the wave vector **k** of the incident wave is determined in both magnitude and direction.

Frequency and wavelength considerations

Our obstacle is in the path of electromagnetic waves of a single frequency ω. The microscopic way in which the molecules interact with the waves is that the electric field vector **E** of a wave causes the electrons in the molecules to oscillate. Each electron is therefore being accelerated, and so gives rise to electromagnetic radiation as discussed in Chapter 4. These waves constitute the diffracted waves and, from fundamental electromagnetic theory, the frequency of the diffracted waves is equal to the frequency of oscillation of the electrons. This, in turn, is determined by the frequency of the **E** vector of the incident wave, and the way in which the electrons respond to this vector. Now, in general, the effect of the **E** vector on any electron is a purely passive process, in that the electron will respond exactly as the **E** vector forces it. Under the assumption that the electron may be regarded as free (this is discussed below), then the electron cannot behave in any way other than to oscillate exactly in phase with the **E** vector of the incident wave. Since we are using incident waves of frequency ω, the electrons will oscillate with frequency ω, implying that the diffracted waves will also have frequency ω. The incident and diffracted waves therefore have the same frequency ω.

The velocity c, frequency ω and wavelength λ of the wave are related by equation (4.18),

$$c = \frac{\lambda \omega}{2\pi} \tag{4.18}$$

In any medium, the velocity of electromagnetic radiation c is constant, and since we know that the frequency ω of the diffracted waves is equal to that of the incident waves, equation (4.18) tells us that the wavelength λ of the waves is also unchanged. One effect, therefore, of the diffracting obstacle is that the diffracted waves have the same frequency and wavelength as the incident waves.

This argument depends on the assumption that the electrons in the molecules may be regarded as free so that they do indeed oscillate in phase with the incident **E** vector. But we know that electrons in molecules are bound to atomic nuclei and so are not, in fact, free. What is needed is a criterion by which we may decide whether an electron behaves as though it were effectively free. In the Bohr theory of the hydrogen atom, an electron is considered as orbiting about the nucleus, and it is a straightforward calculation to show that the frequency of orbital motion of the lowest electronic Bohr orbit in the hydrogen atom is of the order of 10^{16} Hz. The frequency of X-rays of wavelength 0.1 nm (1 Å) is about 10^{19} Hz, some three orders of magnitude greater than the 'natural' frequency of an electron in hydrogen. The physical significance of this is that since the X-rays have a frequency considerably greater than that of the orbital motion of the electron, then during any one cycle of the X-radiation, the electron has hardly changed its position relative to the nucleus. In other words, the electron's instantaneous behaviour is determined by the incident X-ray beam and not by the attraction to the nucleus. In this respect, the electron has 'forgotten' that the nucleus is present, and reacts to the **E** vector as if the nucleus were not there. The electron therefore behaves as if it were free.

The above argument is very much an order-of-magnitude discussion based on a physical model known to be inaccurate. To perform the calculation properly, one would have to take account of quantum mechanical effects. The strength of the argument, though, is that, physically speaking, it tells us the significant answers. Any electron does orbit a nucleus with a characteristic frequency, and if that frequency is very much less than that of the incident radiation, then to all intents and purposes the electron is free and oscillates in phase with the **E** vector of the incident wave. If, however, the orbital frequency is comparable to the frequency of the X-radiation, the electron cannot be regarded as free, and so will not oscillate in phase with the **E** vector. This is known as *anomalous scattering*, and will be treated in Chapter 9.

For the purposes of the following argument, we will assume that the electrons can be regarded as free and so it is valid to state that the frequency and wavelength of the incident and diffracted waves are the same.

Spatial considerations

If the diffracting obstacle leaves the frequency and the wavelength of the motion unchanged, then it must have some other effect. To determine this, let us go back to our model of the **E** vector of the incident wave interacting with an electron. The oscillation induced in the electron will cause it to emit electromagnetic radiation in essentially all directions. Every electron in the obstacle is excited and acts as a source of radiation. We now invoke the principle of superposition, which tells us that the net effect of an array of sources is the sum of the effects of the individual sources. In that each electron in the crystal emits waves in all directions, the overall effect of diffraction by the obstacle as a whole is to propagate waves in all directions in space.

Whereas the incident wave is propagated in a fixed direction (in our example, this direction is normal to the obstacle), the diffracted waves emitted as a result of electronic excitations are in all directions.

The conclusion we have reached is that the effect of a diffracting obstacle is to transform plane waves into a series of waves going in many different directions. Obviously, the more complicated the obstacle, the more involved are the overall diffraction effects. We have now justified the experimental observations discussed in Section 6.2. Reference to Fig. 6.3 will show that the effect of the diffracting obstacle (a slit which is narrow compared with the wavelength of the wave motion) on a series of plane waves is to leave the frequency and the wavelength unchanged, but to spread the waves in space, which in this particular instance gives rise to the semicircular pattern shown.

Thus the effect of the diffracting obstacle is to give rise to spatially different waves, implying that the information in the waves is somehow contained in the way in which they are spread out in space. This tells us that to interpret a diffraction pattern, we must sample the pattern in many different spatial locations, for it is only by this means that we will discern the information we require.

The significance of the wave vector k

Let us now review the properties of the wave vector **k**. This quantity is a vector **k** whose magnitude k is given by the equation

$$k = |\mathbf{k}| = \frac{2\pi}{\lambda} \tag{4.32}$$

and whose direction is normal to the advancing wavefront. The only other parameters relevant in the description of a wave are its maximum amplitude ψ_0 and its frequency ω, both of which are scalar quantities and represent magnitudes only. Thus the sole physical parameter of a wave which contains spatial information is the wave vector **k**.

For the incident wave, the **k** vector is determined in both magnitude and direction. We have just demonstrated that the diffracted waves are spatially diverse, so that they all have different directions. But we also know that they have the same wavelength λ as the incident wave. We see, therefore, that any diffracted wave may be represented by a wave vector **k** whose magnitude is the same as that of the incident wave, but whose direction is different. Consequently, the total set of diffracted waves may be represented by a set of wave vectors all of which have the same magnitude, equal to that of the incident wave, but different directions.

We may now predict that the mathematical description of the diffraction event may be expressed in terms of a set of wave vectors, each of which have the same magnitude but are associated with different directions in space. Thus the, spatial information is contained in the variation of the directional part of the wave vectors, and the fact that the wavelengths of the incident and diffracted waves are the same is represented by the constancy of the magnitude of the wave vectors of the diffracted waves.

The mathematical description of the diffraction pattern

This section will derive a mathematical expression for the waves diffracted by an obstacle. For the present discussion, we shall assume that all aspects of the structure and physical behaviour of the obstacle are known. As the incident beam passes through the obstacle, two phenomena occur:

(a) Waves are propagated through the obstacle.
(b) The obstacle perturbs these waves.

Our mathematical description of the overall effect of diffraction is a combination of the above two events, summed for all points in the obstacle.

If we choose an arbitrary origin within the obstacle, then any point in the obstacle may be defined by a position vector **r**, which is a vector from the origin to our given location. The diffraction event at any place in the crystal is determined by the behaviour of a local volume of space, and we may express an infinitesimal volume element centred on **r** by the vector differential $d\mathbf{r}$.

We will now discuss the mathematical description of the two effects mentioned above. The volume element $d\mathbf{r}_1$ centred on $\mathbf{r}_1$ gives rise to a particular wave with wave vector $\mathbf{k}$ and frequency ω, which have the same magnitudes as those parameters for the incident wave. Since the mathematical description of diffraction must contain information about waves, we expect a function such as

$$e^{i(\mathbf{k}\cdot\mathbf{r}_1-\omega t)}$$

to occur.

To determine the perturbing effect of the obstacle, we argue as follows. Each volume element within the obstacle perturbs the waves in a particular manner. Since we have a complete physical description of the obstacle, then we may ascribe a number to the effect that a particular location within the obstacle has on the waves, and the value of the number used will vary from place to place within the obstacle. In view of the variation of the perturbing effect of the obstacle with location, this effect may be described by a function of position $f(\mathbf{r})$. For any given value of $\mathbf{r}$ corresponding to any position within the obstacle, $f(\mathbf{r})$ may be evaluated, and the perturbing effect specified. For the moment, it does not matter what $f(\mathbf{r})$ actually is; all that is important is that we appreciate that a function $f(\mathbf{r})$ exists, that it is a function of position only and that the information contained within the function concerns the mathematical description of the scattering event at any position $\mathbf{r}$. The complexity of $f(\mathbf{r})$ will depend on the complexity of the obstacle it describes. In subsequent chapters we will investigate various forms of $f(\mathbf{r})$ corresponding to different obstacles, but for the moment we will assume that $f(\mathbf{r})$ is known. Consider a volume element $d\mathbf{r}_1$ centred on $\mathbf{r}_1$ as shown in Fig. 6.5. The effect of this element on the wave motion may be described mathematically by the expression

$$f(\mathbf{r}_1)d\mathbf{r}_1$$

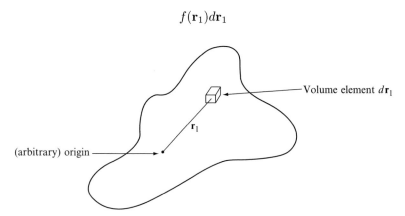

Fig. 6.5 *Diffraction by a volume element.* The volume element $d\mathbf{r}_1$ centred on $\mathbf{r}_1$ affects the waves in a manner which may be represented by the function $f(\mathbf{r})\,d\mathbf{r}_1$. A wave $e^{i(\mathbf{k}\cdot\mathbf{r}_1-\omega t)}$ is propagated, and the mathematical expression for the contribution of the volume element $d\mathbf{r}_1$ to the overall diffraction pattern must be some mathematical combination of $f(\mathbf{r})\,d\mathbf{r}_1$ and $e^{i(\mathbf{k}\cdot\mathbf{r}_1-\omega t)}$.

The multiplication by $d\mathbf{r}_1$ is necessary since $f(\mathbf{r})$ is defined at a point, whereas we know that physically, the diffraction event occurs over a volume element.

We have now demonstrated that in a volume element $d\mathbf{r}_1$ centred on $\mathbf{r}_1$, there is a wave $e^{i(\mathbf{k}\cdot\mathbf{r}_1 - \omega t)}$ and the effect of the volume element on this wave may be described by the expression $f(\mathbf{r}_1)\,d\mathbf{r}_1$. The mathematical description of the outgoing diffracted wave from the volume element must combine the effect of the element on the wave and the description of the wave itself, implying that the diffracted wave must be represented by some mathematical combination of these two functions. Now, there are really very few ways in which two mathematical functions may be combined. The fundamental arithmetical operations are addition, subtraction, multiplication and division, but since subtraction is the inverse of addition, and division is the inverse of multiplication, we see that the only two significantly different ways of combining these two functions are as in equations (6.1) and (6.2):

$$\text{diffracted wave from volume element } d\mathbf{r}_1 = f(\mathbf{r}_1)\,d\mathbf{r}_1 + e^{i(\mathbf{k}\cdot\mathbf{r}_1 - \omega t)} \qquad (6.1)$$

or

$$\text{diffracted wave from volume element } d\mathbf{r}_1 = f(\mathbf{r}_1)e^{i(\mathbf{k}\cdot\mathbf{r}_1 - \omega t)}\,d\mathbf{r}_1 \qquad (6.2)$$

To determine which of the equations (6.1) and (6.2) is in fact the correct description of the diffracted wave, let us consider a system of a wave incident on a perfectly absorbing obstacle, as, for instance, light waves falling on an ideal black body. Since the incident waves are totally absorbed, the effect of the obstacle is to reduce the incident waves to zero, implying that $f(\mathbf{r})$ is zero for all points within the obstacle. Also, we know that there are no diffracted waves. Substituting $f(\mathbf{r})$ as zero in equation (6.1) leaves a finite, non-zero expression for the diffracted wave, whereas substitution in equation (6.2) implies that there is no diffracted wave. Physically, we know that the latter is the correct case. Since equation (6.1) has been shown to give an erroneous result, it cannot represent the diffracted waves. On the other hand, the mathematical form of equation (6.2) is consistent with our physical observations and so we conclude that equation (6.2) may indeed represent the wave diffracted from a volume element $d\mathbf{r}_1$:

$$\text{wave diffracted from volume element } d\mathbf{r}_1 = f(\mathbf{r}_1)e^{i(\mathbf{k}\cdot\mathbf{r}_1 - \omega t)}\,d\mathbf{r}_1 \qquad (6.2)$$

We have now found a possible mathematical description of the act of diffraction. That equation (6.2) has the correct form and gives the right answers will be justified in the succeeding sections, but at present we will develop the equation a little further.

We now consider another volume element $d\mathbf{r}_2$ centred on the position vector $\mathbf{r}_2$. From equation (6.2), we may write

$$\text{wave diffracted from volume element } d\mathbf{r}_2 = f(\mathbf{r}_2)e^{i(\mathbf{k}\cdot\mathbf{r}_2 - \omega t)}\,d\mathbf{r}_2$$

and similarly for other volume elements $d\mathbf{r}_3$, $d\mathbf{r}_4$ and so on. The expressions for the waves diffracted from each volume element are all equations of waves, and we know that waves obey the principle of superposition, which states that the total effect from an array of sources is just the sum of the effects of the individual sources. This tells us that if we wish to find the expression for the resultant diffraction pattern of the obstacle as a whole, then all we have to do is sum the waves diffracted from each volume element in the obstacle:

$$\text{diffraction pattern} = f(\mathbf{r}_1)e^{i(\mathbf{k}\cdot\mathbf{r}_1 - \omega t)}\, d\mathbf{r}_1 + f(\mathbf{r}_2)e^{i(\mathbf{k}\cdot\mathbf{r}_2 - \omega t)}\, d\mathbf{r}_2$$

$$+ f(\mathbf{r}_3)e^{i(\mathbf{k}\cdot\mathbf{r}_3 - \omega t)}\, d\mathbf{r}_3 + \dots \tag{6.3}$$

Equation (6.3) is the sum of a large number of infinitesimal quantities and so may be expressed as an integral. The summation extends over all volume elements in the obstacle, and so the integral will be a definite integral whose limits are determined by the spatial extent of the obstacle. To represent this, the subscript V will be used to remind the reader that the integral is evaluated over the volume of the obstacle:

$$\text{diffraction pattern} = \int_V f(\mathbf{r})e^{i(\mathbf{k}\cdot\mathbf{r} - \omega t)}\, d\mathbf{r} \tag{6.4}$$

Equation (6.4) is what we have been looking for. It tells us the mathematical form describing the waves diffracted by an obstacle, and therefore represents the diffraction pattern of that obstacle.

6.7 Diffraction and Fourier transforms

Rewriting equation (6.4),

$$\text{diffraction pattern} = \int_V f(\mathbf{r})e^{i(\mathbf{k}\cdot\mathbf{r} - \omega t)}\, d\mathbf{r} \tag{6.4}$$

we see that it contains information on space (because of the variables $\mathbf{k}$ and $\mathbf{r}$) and also time (in view of the variables ω and t). The integration is carried out over the variable $\mathbf{r}$, and so we may factorise out the term $e^{-i\omega t}$, giving

$$\text{diffraction pattern} = e^{-i\omega t}\int_V f(\mathbf{r})e^{i\mathbf{k}\cdot\mathbf{r}}\, d\mathbf{r} \tag{6.5}$$

Let us now consider what happens when we carry out a diffraction experiment and record data. For X-rays of wavelength of order 0.1 nm (1 Å), the corresponding frequency is of order 10^{19} Hz, implying that the $\mathbf{E}$ vector executes one cycle every 10^{-19} s. This has a very important consequence. No recording device exists which is capable of monitoring a phenomenon occurring as fast as 10^{19} times per second. This implies that any diffraction experiment using X-rays must necessarily involve taking a time average over a period of time very much longer than the period of the X-rays. From Chapter 4, we know that any diffraction experiment records the intensity, as opposed to the amplitude, of the waves. The reason for this is concerned with the fact that the mathematical description of wave amplitude involves complex numbers, which cannot be determined in any real experiment. All that we can measure is the appropriate real number, which in this case is the intensity. We now see that any X-ray diffraction experiment measures the time average of the intensity of the diffracted waves. Let us just consider what taking a time average implies. The word 'average' signifies that the value we obtain will be the same no matter at what time we start to record data. The taking of a time average therefore implies that the measurements we make are independent of time. Hence the very manner in which we record X-ray diffraction data eliminates time as a variable. The purpose of this discussion is to demonstrate that time information is, for all practical purposes, irrelevant. This will be seen to agree with the conclusion we reached earlier when we stated that 'the information in the waves is somehow contained in the way in which they are spread out in space', for there is no reference to time in this statement. Since we are interested in information on space, we can neglect in any equations all those terms which contain information on time. Hence we may rewrite equation (6.5) as

$$\text{diffraction pattern} = \int_V f(\mathbf{r})e^{i\mathbf{k}\cdot\mathbf{r}}\,d\mathbf{r} \tag{6.6}$$

The justification for dropping the term $e^{-i\omega t}$ is purely one of convenience. We know that this term contains no relevant information, and that in the final stage, when we compute time averages of intensities, it will drop out anyway. In view of this, it serves no useful purpose to keep writing $e^{-i\omega t}$ in diffraction equations, and so equation (6.6) contains all the useful information we require.

There is one further adjustment which we wish to make to equation (6.6), concerning the limits on the integral and the nature of $f(\mathbf{r})$. $f(\mathbf{r})$ represents a function which determines the effect of the obstacle on the incident beam, and is defined only where the obstacle exists in space. Since the obstacle will have a finite extent, $f(\mathbf{r})$ is defined only for those values of $\mathbf{r}$ which correspond to locations within the obstacle. Also, since we are concerned only with the effect the obstacle has on the incident waves, the integral in equation (6.6) is definite and is taken over the (finite) volume of the obstacle. In order to make equation (6.6) mathematically 'nicer' (the reason for this will immediately become apparent), we adopt the following mathematical trick. Until now, $f(\mathbf{r})$ has been defined only for values of $\mathbf{r}$ corresponding to locations within the obstacle, and so has a finite range. We now choose to define $f(\mathbf{r})$ over an infinite range, dividing the values which $f(\mathbf{r})$ may take into two domains:

If $\mathbf{r}$ does not lie within the obstacle, $f(\mathbf{r})$ is zero.
If $\mathbf{r}$ does lie within the obstacle, $f(\mathbf{r})$ has the appropriate value corresponding to the effect of that position on the waves.

$f(\mathbf{r})$ is now formally defined over an infinite range, and so we may replace the finite limits on the integral by infinite limits, corresponding to all possible values of the variable $\mathbf{r}$:

$$\text{diffraction pattern} = \int_{\text{all }\mathbf{r}} f(\mathbf{r})e^{i\mathbf{k}\cdot\mathbf{r}}\,d\mathbf{r} \tag{6.7}$$

Since $f(\mathbf{r})$ is zero everywhere outside the obstacle, the integrand will vanish identically for those values of $\mathbf{r}$, and so locations outside the obstacle will not contribute to the value of the integral. For values of $\mathbf{r}$ within the obstacle, the result of performing the integration in equation (6.7) is identical to that performed in equation (6.6). The sole reason for this mathematical manipulation is to define $f(\mathbf{r})$ in such a way that we may place infinite limits on the integral. This is a purely formal procedure, and makes no difference in practice. But, mathematically, this has a most significant effect.

Reference back to Chapter 5, Section 5.4, gives the definition of the Fourier transform of any function $f(\mathbf{r})$ as

$$\text{Fourier transform of } f(\mathbf{r}) = \int_{\text{all } \mathbf{r}} f(\mathbf{r})e^{i\mathbf{k}\cdot\mathbf{r}} \, d\mathbf{r} \qquad (5.12)$$

Compare this with equation (6.7). The two equations are identical. This is a most important result, for it tells us how to compute the diffraction pattern of any obstacle for which $f(\mathbf{r})$ is known. All we have to do is compute the Fourier transform of $f(\mathbf{r})$ and there we have the answer! It is for this reason that the mathematics of the Fourier transform was investigated in Chapter 5, and also that the last few pages have discussed the alteration of equation (6.5) to the form of equation (6.7). Reference to equation (6.5) shows that at that point we were nearly at the Fourier transform, apart from the $e^{-i\omega t}$ term and the limits of the integral. We then found that we could formally eliminate the $e^{-i\omega t}$ factor without losing any relevant information, and that by defining $f(\mathbf{r})$ so that we could put infinite limits on the integral, we arrive at equation (6.7), which, mathematically, is a very well-known equation.

If we call $f(\mathbf{r})$ the *amplitude function* of the obstacle, we have reached the conclusion that:

The diffraction pattern of any obstacle is the Fourier transform of the amplitude function.

Once this is understood, the actual process of calculating the diffraction pattern of any obstacle is essentially the 'mechanical' one of turning the appropriate mathematical 'handles' in order to evaluate the integral. If the obstacle is complex in structure, then $f(\mathbf{r})$ will be very complicated, and so the computation of the integral is not easy. But this is a mere technicality. Conceptually, the phenomenon of diffraction is contained in equation (6.7). What is important is that we appreciate what equation (6.7) says, and why it is that the Fourier transform does indeed represent the diffraction pattern of an obstacle.

6.8 The significance of the Fourier transform

This section is designed to point out several aspects of equation (6.7) in order that the reader will achieve some feeling for its significance. During this discussion it will be demonstrated that equation (6.7) is the correct expression for the diffraction pattern of an obstacle, in that it contains, in an appropriate manner, all the information we would require in a description of the diffraction event. As a result of this, we shall have justified our choice of equation (6.2) over equation (6.1) in our earlier discussion.

The first notable point about equation (6.7) is that the integrand is a function of both $\mathbf{k}$ and $\mathbf{r}$, and that the integral is taken over $\mathbf{r}$. In view of the analysis of the meaning of integrals in Chapter 5, we see that this implies that the overall result of the integration is to give us a function of $\mathbf{k}$, which, since it is the Fourier transform of $f(\mathbf{r})$, we shall write as $F(\mathbf{k})$:

$$\text{diffraction pattern} = F(\mathbf{k}) = \int\limits_{\text{all } \mathbf{r}} f(\mathbf{r})e^{i\mathbf{k}\cdot\mathbf{r}}\,d\mathbf{r} \qquad (6.8)$$

$F(k)$ is known as the *diffraction pattern function*.

The fact that the mathematical description $F(\mathbf{k})$ of the diffraction pattern defined by equation (6.8) is a function of the wave vector $\mathbf{k}$ implies that the information contained in the diffraction pattern manifests itself according to different values of $\mathbf{k}$. This is exactly what we predicted earlier in Section 6.6. During the development of diffraction theory, we showed that the information in the diffraction pattern was essentially spatial, and that the only way in which waves may contain spatial information is by means of their wave vectors $\mathbf{k}$. Equation (6.8) is therefore totally in agreement with our physical intuition concerning what is happening during diffraction.

Let us now look at the integrand of equation (6.8). $e^{i\mathbf{k}\cdot\mathbf{r}}$ is our usual representation of a wave, and the fact that this is multiplied by the function $f(\mathbf{r})$ indicates that $f(\mathbf{r})$ is acting as an amplitude-type function. The integrand therefore represents a wave whose amplitude is a rather special function of position as determined by $f(\mathbf{r})$. We also saw earlier that the significance of the integration was to sum the effects of local scattering centres, which were assumed to respond passively under the influence of the $\mathbf{E}$ vector of the incident beam. Consider a location $\mathbf{r}_1$ and a scatterer at that position whose effect is described by $f(\mathbf{r}_1)$. Suppose we removed the passive scatterer and replaced it by an active source of radiation which emitted waves of the same frequency and wavelength as the incident X-ray beam, as in Fig. 6.6. If the amplitude of the waves emitted by this active source were given by $f(\mathbf{r}_1)$, then the source would propagate waves obeying the equation

$$\text{wave emitted by source at } \mathbf{r}_1 = f(\mathbf{r}_1)e^{i\mathbf{k}\cdot\mathbf{r}_1} \qquad (6.9)$$

In this equation, the $e^{-i\omega t}$ is tacitly assumed to be present. In the absence of the incident X-ray beam, the source at $\mathbf{r}_1$ would give rise to exactly the same effect as does the passive scatterer at $\mathbf{r}_1$ in the presence of the incident beam.

Exactly the same argument may be repeated for all other locations inside the obstacle, so that any location n gives rise to a wave as follows:

$$\text{wave emitted by source at } \mathbf{r}_n = f(\mathbf{r}_n)e^{i\mathbf{k}\cdot\mathbf{r}_n} \qquad (6.9)$$

The overall effect of all the sources may therefore be obtained by a summation:

$$\text{wave emitted by all sources} = \sum\limits_{\text{all } n} f(\mathbf{r}_n)e^{i\mathbf{k}\cdot\mathbf{r}_n}$$

For a very large number of infinitesimal sources, this summation becomes the now familiar integral:

$$\text{wave emitted by all sources} = \int\limits_{\text{all } \mathbf{r}} f(\mathbf{r})e^{i\mathbf{k}\cdot\mathbf{r}}\,d\mathbf{r}$$

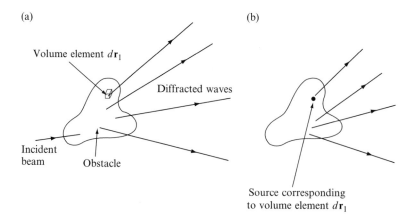

(a)

Volume element $d\mathbf{r}_1$

Diffracted waves

Incident beam

Obstacle

(b)

Source corresponding to volume element $d\mathbf{r}_1$

Fig. 6.6 *These two systems are equivalent.* (a) represents an obstacle which diffracts an incident beam. At any volume element, the obstacle perturbs the beam in a way described by $f(\mathbf{r})$. (b) represents an array of sources such that each source emits waves with an amplitude $f(\mathbf{r})$ corresponding to the appropriate volume element in the original obstacle. The diffraction pattern formed by (a) is identical to the total effect of the array of sources in (b). An observer who samples only the diffraction pattern is unable to determine whether that pattern is due to a passive obstacle which diffracts an incident beam or an array of active sources, implying that the two systems shown are equivalent.

In this way, we may regard the effect of the passive obstacle on the incident X-ray beam as identical to that which would be observed if the X-ray beam were removed and the obstacle were replaced by an array of active sources, each of which propagated a wave with the appropriate amplitude as given by $f(\mathbf{r})$. It is for this reason that $f(\mathbf{r})$ is called the amplitude function. Since $f(\mathbf{r})$ is defined as zero outside the obstacle, we infer that we need place no sources outside the obstacle, which also agrees with our physical intuition.

We now have two physical systems. On the one hand is a passive obstacle which diffracts waves. On the other hand is an array of active sources chosen so that the waves they propagate correspond exactly to the values of $f(\mathbf{r})$. If equation (6.8) is correct, then these two systems must be physically identical: if an observer were to record the diffracted waves only, then he or she would be unable to decide whether they were due to the diffraction of an incident beam by a passive obstacle or due to a set of active sources.

How does this discussion correspond to our knowledge of physical reality?

Firstly, refer back to Fig. 6.3, which shows the diffraction of waves by a narrow slit. In the area away from the slit in the direction of propagation, the semicircular pattern is identical to that which would have been obtained with a point source placed in the slit. In fact, the reader has probably come across this device many times in optics experiments. To obtain a point source of light, we usually use an incandescent filament, and place in front of it a pinhole, which we say acts as a point source. The reason for this is obvious. Waves from the incandescent filament are diffracted by the pinhole, which gives rise to a diffraction pattern of hemispherical waves. These are just what we expect from a point source. Hence the systems of a point source and of a non-point source plus a suitably diffracting obstacle (in this case a pinhole) are indeed physically identical. In that equation (6.8) predicts this, we have further justification that equation (6.8) and the arguments leading to its derivation are correct.

Secondly, let us reconsider the physical model used in Section 6.6, where we discussed the interaction of the **E** vector of the incident X-ray beam with the electrons in the obstacle. During that enquiry, we found that the **E** vector of the incident waves forced the electrons into oscillation, thereby causing them to emit radiation. If the electrons were to oscillate of their own accord, as in an active source, then the two situations of the forced electron and of the electron as its own source of radiation are clearly the same. This also testifies to the veracity of equation (6.8).

This discussion is also relevant in a historical sense. One of the early proponents of the wave theory was Christiaan Huygens. He proposed that light was propagated through matter as a result of what he called 'secondary sources'. Briefly, Huygens envisaged that the interaction of light with matter was to cause the 'particles of matter' (of course, he did not know of atoms and so on) to become 'secondary sources' in that they themselves gave rise to 'secondary wavelets'. Thus any obstacle in the path of light was regarded as an array of secondary sources, each of which was capable of emitting a wavelet, such that the wavelets combined to give the overall diffraction effect. This model was really remarkably perceptive for the middle of the seventeenth century, and clearly parallels our more modern model relating electronic excitation to the incident **E** vector. Using his theory, Huygens demonstrated, among other things, that reflection and refraction (which are really just two special cases of diffraction) would obey the well-known laws. In particular, refraction would follow Snell's law (which relates to the bending of light, for example at an air/glass interface) if the light in the denser medium travelled more slowly than in the rarer medium. Newton, using the particle theory of light, also showed that Snell's law would be obeyed at an interface, but only if the light in the denser medium travelled more quickly than in the rarer medium. The wave and particle theories therefore predict exactly contradictory behaviour for the velocity of light in different media. Accurate measurement of the velocity of light in different media was therefore crucial in deciding between the wave and particle theories. Because light has such a high velocity, accurate experimental data were not obtained until Jean Foucault (1819–1868) succeeded in 1850 in measuring the velocity of light in air and in water. He found that light travelled faster in the air than in the water, and since water is optically denser than air, the wave theory had won the final battle. It is only since the development of quantum theory that the particle aspect of light has re-emerged under the guise of 'photons'. Quantum theory, however, is not within the scope of this book, and will not be discussed.

6.9 Fourier transforms and phase

Further insight into the nature of the Fourier transform may be obtained by considering the phase relationships between the waves scattered by the various scattering centres in an obstacle. Consider a linear array of scattering centres in an obstacle as shown in Fig. 6.7.

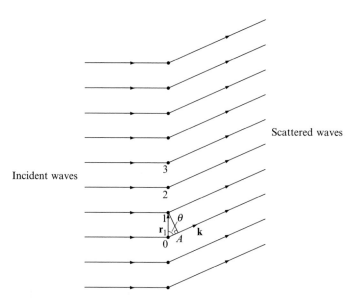

Fig. 6.7 *The phase of the scattered waves.* Waves are incident normally on a linear array, and we are interested in those waves scattered in a direction corresponding to the wave vector **k**. If point 0 is the zero of phase, we have

$$OA = r_1 \cos \theta$$

Number of wavelengths contained in $OA = \frac{r_1 \cos \theta}{\lambda}$

$\therefore$ Phase at point $1 = \frac{2\pi}{\lambda} r_1 \cos \theta = \mathbf{k} \cdot \mathbf{r}_1$

A wave is incident normally on the array, and we wish to find the total wave scattered in a direction corresponding to a particular wave vector **k**. For the wave scattered from point 1, the phase difference between this wave and that scattered from the origin O may be found by determining by how many wavelengths the two waves are out of step. If point 1 corresponds to a vector $\mathbf{r}_1$, then the length OA is $r_1 \cos \theta$, and the number of wavelengths corresponding to this distance is $(r_1 \cos \theta)/\lambda$. Since one wavelength corresponds to a phase difference of 2π, then a distance of $(r_1 \cos \theta)/\lambda$ corresponds to a phase difference of $(2\pi r_1 \cos \theta)/\lambda$. Hence the phase difference between the waves scattered from points 0 and 1 is

$$\text{phase difference} = \frac{2\pi}{\lambda} r_1 \cos \theta$$

But

$$k = \frac{2\pi}{\lambda}$$

and since the angle between the vectors $\mathbf{r}_1$ and **k** is θ, then

$$\text{phase difference} = \mathbf{k} \cdot \mathbf{r}_1$$

If we arbitrarily choose the origin to be the zero of phase, then

$$\text{phase at point } 0 = 0$$
$$\text{phase at point } 1 = \mathbf{k} \cdot \mathbf{r}_1$$

and, similarly,

$$\text{phase at point } 2 = \mathbf{k} \cdot \mathbf{r}_2$$

and so on.

If the scatterer at point n scatters waves with an amplitude $f(\mathbf{r}_n)\, d\mathbf{r}_n$, then we may write

wave scattered from point $0 = f(\mathbf{r}_0)\, d\mathbf{r}_0$
wave scattered from point $1 = f(\mathbf{r}_1)e^{i\mathbf{k}\cdot\mathbf{r}_1}\, d\mathbf{r}_1$
wave scattered from point $2 = f(\mathbf{r}_2)e^{i\mathbf{k}\cdot\mathbf{r}_2}\, d\mathbf{r}_2$

and likewise for all other scattering centres in the array. Using the principle of superposition, the total scattered wave in the direction corresponding to the wave vector $\mathbf{k}$ is

$$\text{total scattered wave} = f(\mathbf{r}_0)d\mathbf{r}_0 + f(\mathbf{r}_1)e^{i\mathbf{k}\cdot\mathbf{r}_1}\, d\mathbf{r}_1 + f(\mathbf{r}_2)e^{i\mathbf{k}\cdot\mathbf{r}_2}\, d\mathbf{r}_2 + \ldots$$

$$= \int_{\text{obstacle}} f(\mathbf{r})e^{i\mathbf{k}\cdot\mathbf{r}}\, d\mathbf{r}$$

in which the integration is over the whole obstacle.

With a suitable definition of $f(\mathbf{r})$, we may replace the finite limits of integration by infinite limits, yielding

$$\text{total scattered wave} = \int_{\text{all }\mathbf{r}} f(\mathbf{r})e^{i\mathbf{k}\cdot\mathbf{r}}\, d\mathbf{r}$$

For all values of $\mathbf{k}$, this corresponds to the entire diffraction pattern. We recognise this as the Fourier transform of the amplitude function $f(\mathbf{r})$, but this derivation shows that the significance of the exponential term is that it specifies the phase relationships between the scattering centres and an arbitrary origin. A similar derivation for the case of non-normal incidence will be discussed in the preamble to Chapter 8.

6.10   Fourier transforms and the wave equation

A highly significant interpretation of the relationship of the Fourier transform to the description of the diffraction pattern is obtained by comparing the Fourier transform with the general solution of the three-dimensional wave equation. Since any diffraction pattern is a generalised wave motion, then the description of the diffraction pattern must be an allowed solution of the wave equation.

Referring to Chapter 4, we recall that the general solution of the three-dimensional wave equation

$$\frac{\partial^2 \psi(x,y,z,t)}{\partial x^2} + \frac{\partial^2 \psi(x,y,z,t)}{\partial y^2} + \frac{\partial^2 \psi(x,y,z,t)}{\partial z^2} = \frac{1}{v^2}\frac{\partial^2 \psi(x,y,z,t)}{\partial t^2} \qquad (4.21)$$

may be expressed in terms of complex exponentials as

$$\psi(\mathbf{r},t) = \sum_n \psi_n e^{i(\mathbf{k}_n \cdot \mathbf{r} - \omega_n t + \phi_n)} \qquad (4.37)$$

The parameters ψ_n, $\mathbf{k}_n$, ω_n, ϕ_n and the extent to which the summation is necessary are determined by the boundary conditions appropriate to any particular problem, which in turn are themselves given by the structure and geometry of the various sources and obstacles involved. In a complicated case, we will require many terms in the summation. If the boundary conditions may be regarded as being caused by a continuous distribution of sources, then the summation becomes an integral, which we may write as

$$\sum_n \psi_n e^{i(\mathbf{k}_n \cdot \mathbf{r} - \omega_n t + \phi_n)} \rightarrow \int \psi(\mathbf{r}) e^{i(\mathbf{k} \cdot \mathbf{r} - \omega t + \phi)}\, d\mathbf{r}$$

in which the amplitude term is now a continuous function of $\mathbf{r}$. As before, we may drop the time-dependent term, and by a suitable definition of $\psi(\mathbf{r})$, which we now write as $f(\mathbf{r})$, we may place infinite limits on the integral. The zero of phase may be chosen as the origin of $\mathbf{r}$, and the general solution of the wave equation becomes

$$F(\mathbf{k}) = \int_{\text{all } \mathbf{r}} f(\mathbf{r}) e^{i\mathbf{k} \cdot \mathbf{r}}\, d\mathbf{r}$$

This is none other than the Fourier transform of the amplitude function $f(\mathbf{r})$! We now see that the Fourier transform does indeed represent an allowed solution of the three-dimensional wave equation which is particularly suited to a case in which the boundary conditions are complicated. In any real diffraction situation, the boundary conditions are indeed very complicated, and so we appreciate that the Fourier transform is an entirely appropriate description of the diffraction event.

6.11 Fourier transforms and information

Let us analyse equation (6.8) from yet another standpoint. The integrand is the product of two functions. One is the amplitude function $f(\mathbf{r})$, and the other is the wave exponential $e^{i\mathbf{k} \cdot \mathbf{r}}$. For a given value of $\mathbf{k}$, $e^{i\mathbf{k} \cdot \mathbf{r}}$ is as a function of $\mathbf{r}$, essentially a sine or cosine wave as shown in Fig. 6.8(a). Thus we see that $e^{i\mathbf{k} \cdot \mathbf{r}}$ is a highly repetitive function. In contrast, $f(\mathbf{r})$ is a function which describes the behaviour of the obstacle, and may be quite complicated. Suppose that $f(\mathbf{r})$ may be represented as in Fig. 6.8(b). The product function $f(\mathbf{r})e^{i\mathbf{k} \cdot \mathbf{r}}$ may be drawn as discussed in Chapter 5, and is shown in Fig. 6.8(c), which shows it to be a wave motion enveloped by the function $f(\mathbf{r})$. Basically, it is a wave motion of constant frequency and wavelength, but very varied in amplitude. This is called an *amplitude-modulated* wave.

(a)   (b)

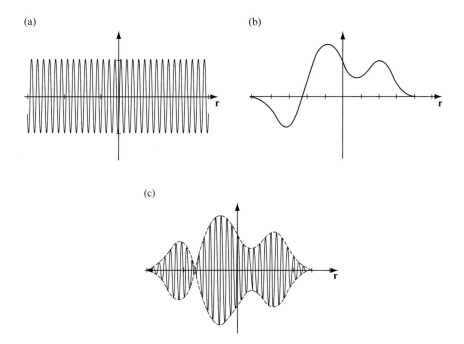

(c)

Fig. 6.8 *The information in* $f(\mathbf{r})e^{i\mathbf{k}\cdot\mathbf{r}}$. (a) Graph of the cosine component of $e^{i\mathbf{k}\cdot\mathbf{r}}$ for a particular value of $\mathbf{k}$. (b) Graph of an arbitrary $f(\mathbf{r})$ which contains information about the obstacle. (c) Graph of the product function $f(\mathbf{r})e^{i\mathbf{k}\cdot\mathbf{r}}$, which also has a similar information content. Hence the product function $f(\mathbf{r})e^{i\mathbf{k}\cdot\mathbf{r}}$ is $f(\mathbf{r})$ enveloping a wave with the same wavelength as the wave $e^{i\mathbf{k}\cdot\mathbf{r}}$.

This is analogous to one of the means by which sound can be transmitted by electromagnetic waves. In a radio station, there is an electronic system that generates radio frequency electromagnetic waves corresponding to the particular wavelength of the given radio station. This 'carrier wave' has a specific wave vector $\mathbf{k}$, and may be represented by a function analogous to $e^{i\mathbf{k}\cdot\mathbf{r}}$. Speech or music in the form of audio frequency waves can then be mixed with the radio frequency carrier wave such that the radio frequency wave is enveloped by the audio frequency wave. This occurs because audio frequencies (of order $500\,\mathrm{Hz}$) are very much lower than radio frequencies (of order $10^5\,\mathrm{Hz}$). The audio wave contains information such as speech or music, and corresponds to $f(\mathbf{r})$. The electronic mixing process is equivalent to the mathematical formation of the product $f(\mathbf{r})e^{i\mathbf{k}\cdot\mathbf{r}}$, and the result is an amplitude-modulated radio wave—hence the term 'AM radio'. In a conventional AM radio, the electronic circuit and the speaker convert the audio frequency information back into sound waves. Even with modern digital radio transmission, the digital signal has to be modulated in essentially the same way. This is a clear demonstration that a mathematical function of the form of equation (6.8) is indeed well equipped to contain information, such as the structure of a diffracting obstacle. We have now justified the description of the diffraction pattern in terms of a Fourier transform on several grounds:

(a) The description is, as required, a function of $\mathbf{k}$.
(b) The description agrees with our physical knowledge of the interaction of waves with obstacles.
(c) The description represents a general solution of the three-dimensional wave equation.
(d) The description is capable of containing the required information.

One further, and highly satisfying, demonstration of the significance of the Fourier transform will be given in the next few sections.

6.12 The inverse transform

We now know that the diffraction pattern function $F(\mathbf{k})$ may be derived from the amplitude function $f(\mathbf{r})$ by the Fourier transform, equation (6.8),

$$F(\mathbf{k}) = \int_{\text{all } \mathbf{r}} f(\mathbf{r}) e^{i\mathbf{k}\cdot\mathbf{r}} \, d\mathbf{r} \tag{6.8}$$

implying that $F(\mathbf{k})$ and $f(\mathbf{r})$ are Fourier transform mates,

$$F(\mathbf{k}) \leftrightarrow f(\mathbf{r})$$

This in turn implies that we may express $f(\mathbf{r})$ as the inverse transform of $F(\mathbf{k})$:

$$f(\mathbf{r}) = \int_{\text{all } \mathbf{k}} F(\mathbf{k}) e^{-i\mathbf{k}\cdot\mathbf{r}} \, d\mathbf{k} \tag{6.10}$$

Let us investigate what equation (6.10) means. $F(\mathbf{k})$ is the diffraction pattern function, and for any value of $\mathbf{k}$ it represents a diffracted wave. Since $\mathbf{k}$ is a vector of constant magnitude but variable direction, $F(\mathbf{k})$ is essentially giving information on the spatial distribution of the diffraction pattern. $e^{-i\mathbf{k}\cdot\mathbf{r}}$ is a wave, and the relevance of the negative sign will be discussed below. The integral in equation (6.10) is over $\mathbf{k}$, which means that we are adding the effects of the diffraction pattern over all directions in space. Since the integrand is a function of both $\mathbf{r}$ and $\mathbf{k}$, and the integration is over $\mathbf{k}$, the result of the integration will be a function of $\mathbf{r}$. Equation (6.10) tells us that the result of the integration is indeed a function of $\mathbf{r}$, and it is a very special function of $\mathbf{r}$. Specifically, it is the exact function which determines the effect of the diffracting obstacle on the incoming waves, and contains information on the structure of the obstacle.

In any diffraction experiment, we have an obstacle of unknown structure, and we observe the diffraction pattern. In fact, what we actually measure is the intensity of the diffraction pattern, which is related to the function $F(\mathbf{k})$ by

$$\text{primary data} = \text{intensity of diffraction pattern} = |F(\mathbf{k})|^2$$

For the moment, we will assume that our measurements are sufficient to determine $F(\mathbf{k})$ completely. (In fact, as we shall see, these are not sufficient data, since it is impossible to determine an unambiguous value of $F(\mathbf{k})$ from $|F(\mathbf{k})|^2$. This, however, is a technical nuisance as opposed to a conceptual difficulty, and we will assume for the moment that our primary data allow us to obtain the complete set of $F(\mathbf{k})$ values for all $\mathbf{k}$.) In practice, this means that we sample the diffraction pattern over all space, corresponding to all possible values of $\mathbf{k}$. For each value of $\mathbf{k}$, we can measure $F(\mathbf{k})$, and so we have the complete function, which we may tabulate or plot on a graph.

Once we have obtained the function $F(\mathbf{k})$ experimentally, it is then solely a question of computation to derive $f(\mathbf{r})$. The calculation of the integral in equation (6.10) may be laborious, but since we know $F(\mathbf{k})$, equation (6.10) is the formula by which we may determine $f(\mathbf{r})$.

Here is the answer to the question of X-ray diffraction. The obstacle is a crystal, which is a three-dimensional array of molecules. In cases of biological interest, the molecules may be very highly complex, such as proteins, but we know that the crystal scatters X-rays, and that this effect can be determined by some function $f(\mathbf{r})$. This function will be very complicated, for it contains information on the structure of the crystal as a whole, and also the structure of the molecules contained within the crystal. A beam of X-rays is directed at the crystal and diffraction occurs. The diffraction pattern corresponds to a complete set of diffracted waves, which propagate from the crystal in all directions. The diffraction pattern is sampled by a sensitive X-ray detector, which yields the values of the intensity $|F(\mathbf{k})|^2$ for all directions in space. At the moment, we assume that these data allow us to find the values of $F(\mathbf{k})$. We now use equation (6.10) to calculate $f(\mathbf{r})$, and the complete crystal structure is determined. The actual effort the total investigation requires can be considerable— indeed, the remainder of this book will study the exact manner in which the various stages of the process are carried out, but, conceptually, the problem is now solved, for we know what is happening in a diffraction experiment.

Information on the crystal structure is contained in the manner in which the crystal perturbs the incident waves into diffracted waves, i.e. in the spatial distribution of the diffraction pattern. Essentially, the diffraction pattern and the crystal structure contain equivalent amounts of information, since their mathematical descriptions are interchangeable. If we know the structure, $f(\mathbf{r})$ is known and we can use equation (6.8) to compute the diffraction pattern $F(\mathbf{k})$. Alternatively, if we know the diffraction pattern, $F(\mathbf{k})$ is known, and so we may compute $f(\mathbf{r})$ from equation (6.10).

Experimentally, it is equation (6.10) which connects our experimental data to our goal, namely the calculation of $f(\mathbf{r})$. For illustrative purposes, however, it is conceptually easier to assume that $f(\mathbf{r})$ is known, and in this way we may use this more meaningful knowledge to determine the diffraction pattern. It is for this reason that we started with the assumption that $f(\mathbf{r})$ was known. Starting from our intuitive feelings for what was happening during diffraction, we developed the concept of the transfer of information from the crystal to the diffracted waves. Using this, we found that equation (6.8) gave the diffraction pattern. As soon as this is established, it is a mathematical fact that equation (6.10) is true, and this allows us to determine $f(\mathbf{r})$ from our experimental diffraction data.

The fact that $f(\mathbf{r})$ and $F(\mathbf{k})$ are Fourier transform mates is indeed an elegant and most beautiful result. Intuitively, we know that the diffraction pattern must contain all the information concerning the crystal structure, and that the information content of $f(\mathbf{r})$ and $F(\mathbf{k})$ must be identical. Equations (6.8) and (6.10) show this to be true, and we are left with no doubt that the arguments used to establish equation (6.8) as the mathematical description of the diffraction phenomenon are correct.

6.13   The significance of the inverse transform

This section will discuss one further aspect of equation (6.10). We firstly consider another principle of optics, known as *the principle of the reversibility of light paths*. This states, quite simply, that if a light wave travels from one place A to another place B, then it is possible for a light wave to travel from B back to A. Although expressed in terms of light waves, the principle holds for all forms of electromagnetic radiation. Let us write equations (6.10) and (6.10) again:

$$F(\mathbf{k}) = \int_{\text{all } \mathbf{r}} f(\mathbf{r})e^{i\mathbf{k}\cdot\mathbf{r}}\, d\mathbf{r} \tag{6.8}$$

$$f(\mathbf{r}) = \int_{\text{all } \mathbf{k}} F(\mathbf{k})e^{-i\mathbf{k}\cdot\mathbf{r}}\, d\mathbf{k} \tag{6.10}$$

Given $f(\mathbf{r})$, we may use equation (6.8) to compute, for any value of $\mathbf{k}$, $F(\mathbf{k})$. $F(\mathbf{k})$ represents a diffracted wave for a particular value of $\mathbf{k}$, that is, in a particular spatial direction. Looking at the integrand of equation (6.10), we see that it is a product of a wave function $e^{-i\mathbf{k}\cdot\mathbf{r}}$ and the function $F(\mathbf{k})$, which acts essentially as an amplitude function. Exactly as we did when we considered the significance of equation (6.8), we shall assume that for a particular direction in space corresponding to a wave vector $\mathbf{k}_1$, we place a source which propagates waves with the correct wavelength, frequency and amplitude as given by $F(\mathbf{k}_1)$. For all positions in space we place suitable sources, and we obtain a spatial distribution of sources as shown in Fig. 6.9. Each source will emit a wave described mathematically by the expression $F(\mathbf{k})e^{-i\mathbf{k}\cdot\mathbf{r}}$. The significance of the negative sign is that we wish the waves to be propagated *back towards* the original position of the obstacle. When we sum the effect of all the sources corresponding to the integration of equation (6.10), we derive a new function $f(\mathbf{r})$.

We now invoke the principle of the reversibility of light paths, as applied to X-rays. Firstly, consider the obstacle as an array of active sources as in Fig. 6.6. In the discussion of equation (6.8) we derived the result that such an array of active sources would give rise to a diffraction pattern described by $F(\mathbf{k})$. Now consider the removal of the obstacle, and the insertion of an array of active sources which propagate waves described by $F(\mathbf{k})$. The effect of these sources is to combine to create an image of the object! This must be true, for if the sources corresponding to $f(\mathbf{r})$ combine to give $F(\mathbf{k})$, then by the principle of reversibility of light paths, the sources $F(\mathbf{k})$ must combine to give $f(\mathbf{r})$. But what about that minus sign in equation (6.10)? We now see its significance. When we consider the sources in the obstacle as given by $f(\mathbf{r})$, we

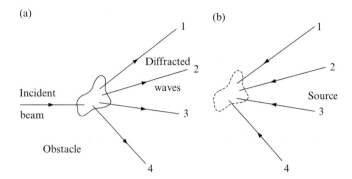

Fig. 6.9 *These two systems are identical.* (a) Shows an obstacle which gives rise to a diffraction pattern. At any point, for instance 2, the diffraction pattern is determined by $F(\mathbf{k})$, where $\mathbf{k}$ is appropriate to position 2. (b) Represents an assembly of X-ray sources placed in locations equivalent to where the diffraction pattern was sampled in (a). Each source emits waves described by $F(\mathbf{k})$; these waves travel back towards the original position of the obstacle. By the principle of reversibility of light waves, a complete set of the sources of the system (b) combines to form an image of the obstacle.

see that they emit waves travelling outwards. If we wish to consider the sources $F(\mathbf{k})$, in order for them to combine to give an image of the obstacle, then the waves must travel in the other direction, namely *back towards* the original position of the obstacle. Hence we need the minus sign.

Looking at it from another standpoint, this discussion has given us the recipe for finding 'the diffraction pattern of a diffraction pattern'. Given an array of sources $f(\mathbf{r})$, they create a diffraction pattern $F(\mathbf{k})$. If the diffraction pattern is regarded as an array of sources, then these sources themselves form a second diffraction pattern, which may be regarded as 'the diffraction pattern of the diffraction pattern' of the original obstacle. Our discussion has shown us that this regenerates the original obstacle. This is entirely in agreement with the mathematical properties of the Fourier transform. The act of diffraction is equivalent to taking the Fourier transform of the obstacle. Given an obstacle described by $f(\mathbf{r})$, its diffraction pattern is the Fourier transform of $f(\mathbf{r})$, namely $F(\mathbf{k})$. If a second diffraction event occurs, this corresponds to forming the Fourier transform of $F(\mathbf{k})$. As seen in Chapter 5, this gives us back the original obstacle function $f(\mathbf{r})$.

In fact, we have been using this very fact for a long time. Consider the schematic diagram of a microscope shown in Fig. 6.10.

The action of a microscope is as follows. Light from the source illuminates the specimen, which is close to the objective lens. Whenever light interacts with an obstacle, diffraction occurs, and so the specimen diffracts light from the source. This diffracted light is caught by the objective lens, which itself constitutes an obstacle, and so a second diffraction event occurs. The nature of this second diffraction event, of course, depends on the nature of the obstacle, which in this case is the very special piece of optical equipment called a lens. What makes a lens special is that it acts as an

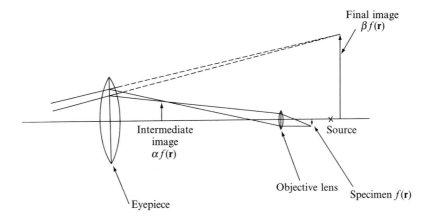

Fig. 6.10 *Schematic representation of a microscope.* The light diffracted by the specimen $f(\mathbf{r})$ is caught by the objective lens, which acts to take the Fourier transform of the diffraction pattern, thereby throwing an intermediate image $\alpha f(\mathbf{r})$, where α is a magnifying factor. The intermediate image now acts as an object for the eyepiece, which forms the final image $\beta f(\mathbf{r})$, β being the overall magnification. Ideally, all the information is transferred without distortion, and so the images are identical in form to the specimen, and magnified.

analogue computer which physically takes Fourier transforms: the objective lens takes the Fourier transform of the diffraction pattern of the specimen, thereby creating an image of the specimen, the intermediate image, in the space between the objective and the eyepiece. The familiar image-forming property of a lens is therefore due to the fact that a lens physically creates Fourier transforms; or, as expressed above, a lens produces the diffraction pattern of a diffraction pattern, thereby forming an image. The eyepiece acts in an exactly similar manner with respect to the intermediate image to give the final image.

As mentioned in Chapter 1, the question 'If you can have a light microscope or an electron microscope, why can't you have an X-ray microscope?' arises. The invention of an X-ray microscope would save all the bother of crystallising proteins and analysing their X-ray diffraction data, but the problem is that no one has yet discovered a means of making suitable X-ray lenses, primarily because no method has been discovered to 'refract' X-rays—as, for example, glass refracts light or magnetic fields can 'refract' a beam of electrons. Since these do not exist at present, there is no X-ray analogue of the microscope.

The X-ray diffraction experiment can now be seen to be analogous to the first event in the action of a microscope. Our primary data constitute the X-ray diffraction pattern, corresponding to that formed by the diffraction of the source light by the specimen in a microscope. Instead of using a physical analogue computer such as a lens to re-create an image, we have to combine the information in the diffraction pattern mathematically, using a computer program.

6.14 Experimental limitations

At this stage it is relevant to introduce a discussion of certain practical limitations which occur in any diffraction experiment. When a diffraction event occurs, the information is spread throughout all space. Unless all space is monitored, then some information is inevitably lost, and the reconstruction of the obstacle from the diffraction data will be incomplete. This is exemplified by the construction of a microscope. After diffraction by the specimen, the objective lens collects information only from that region of space contained within the solid angle subtended at the lens by the specimen. Any diffraction outside that solid angle is not collected by the lens, and so that information is lost. Consequently, the intermediate image formed between the lenses of the microscope may not be a perfect replica of the original specimen. In that the very fine detail of the specimen will give the broadest diffraction information, it is this information which is most likely to be lost. For this reason, an efficient microscope has the specimen very close to the objective lens, which has as large an aperture as possible, thereby maximising the effective solid angle. It is here that practical difficulties come in, for effects such as spherical aberration and the like increase as the lens aperture increases, and these lens aberrations serve to distort the information passing through the lens. A microscope designer therefore has to compromise between maximising the collection of diffraction information and minimising lens defects. In general, it is true that some information is irrevocably lost, so that what we see through a microscope is never a 'true' image of the specimen.

Similar remarks apply to X-ray diffraction. The information is contained in all space, and it may be physically impossible to scan all space to collect it. Every X-ray detector subtends a finite solid angle at the crystal, and so some information may be lost. Hence the structure computed from equation (6.10) may not be the true structure. However, by rotating the crystal, and sometimes the detector too, it is possible to monitor the diffraction at virtually all angles, so enabling an accurate structure to be determined.

Another experimental limitation lies in the fact that we can measure only the intensity, and not the complex amplitude, of the diffraction pattern. Reference to equation (6.8) shows that $F(\mathbf{k})$ is in general a complex quantity. Since we cannot measure complex numbers in any real experiment, our primary data are the real quantity $|F(\mathbf{k})|^2$. Since $F(\mathbf{k})$ is complex, it can be written in complex exponential form as in equation (2.27):

$$F(\mathbf{k}) = |F(\mathbf{k})|e^{i\phi} \qquad (2.27)$$

Here ϕ is a phase factor, corresponding to an angle on the Argand diagram. For any $F(\mathbf{k})$, some of the information content is expressed in $|F(\mathbf{k})|$, and the rest (usually the majority) is in the phase factor ϕ. Now, our observable is just $|F(\mathbf{k})|^2$, which gives no information on ϕ. Hence we see that much of the potential information in the diffraction pattern is lost because we can record only intensity. This is a far more serious limitation than the purely practical problem of collecting data from all space. The problem caused by the loss of the information contained in ϕ is referred to as *the phase problem*. It was the great and painstaking work of the early crystallographers that led to the ways by which the phase problem may be resolved. These methods will be discussed in detail in Chapters 12–14.

This concludes our discussion of equations (6.8) and (6.10). The physical significance of what is happening, combined with the mathematical properties of the functions which describe the events, surely represents a most aesthetic branch of the natural sciences.

The purpose of this chapter has been to describe what is happening when diffraction takes place, and to give the reasoning which shows that equations (6.8) and (6.10) are the only ones compatible with the physical situation. The arguments presented have been based more on physical insight than on mathematical rigour, and although equation (6.8) was derived in a conceptually satisfactory manner, a true-blooded mathematician might quake at the lack of mathematical exactitude of the discussion. This book, however, is not designed for mathematicians, and so no attempt will be made to present the material in any more punctilious a form. If the reader wishes to see the whole story in greater depth, a number of more detailed texts are listed in the Bibliography.

Summary

Whenever a wave motion interacts with an obstacle, *diffraction* occurs. The diffracted waves form a pattern which is determined solely by the nature and structure of the diffracting obstacle. The diffracted waves have, in general, the same wavelength and frequency as the incident waves, but are propagated in all directions in space. The diffraction pattern contains information of the structure of the diffracting obstacle by virtue of the variation of this pattern with direction. Waves can contain directional information only through their wave vectors $\mathbf{k}$, implying that the diffraction pattern may be represented as a function of $\mathbf{k}$, $F(\mathbf{k})$.

If the structure of the obstacle is represented by an amplitude function $f(\mathbf{r})$, then the diffraction pattern function $F(\mathbf{k})$ and the amplitude function $f(\mathbf{r})$ are Fourier transform mates:

$$F(\mathbf{k}) = \int_{\text{all } \mathbf{r}} f(\mathbf{r})e^{i\mathbf{k}\cdot\mathbf{r}}\, d\mathbf{r} \tag{6.8}$$

$$f(\mathbf{r}) = \int_{\text{all } \mathbf{k}} F(\mathbf{k})e^{-i\mathbf{k}\cdot\mathbf{r}}\, d\mathbf{k} \tag{6.10}$$

Equations (6.8) and (6.10) may be used to calculate whichever of $f(\mathbf{r})$ or $F(\mathbf{k})$ is unknown. Since the two functions are Fourier transform mates and are mathematically interchangeable, the information content of the diffraction pattern is the same as the information content of the corresponding obstacle.

In an X-ray diffraction experiment, we sample the diffraction pattern in space and obtain values of the time average of the intensity $|F(\mathbf{k})|^2$, corresponding to as many values of $\mathbf{k}$ as we can measure. Assuming that this allows us to find the corresponding values of $F(\mathbf{k})$, we may then use equation (6.10) to calculate $f(\mathbf{r})$, and the structure of the crystal is solved. In practice, it is not quite as easy as this, since values of

$|F(\mathbf{k})|^2$ do not provide unambiguous values of $F(\mathbf{k})$. This difficulty is known as the *phase problem.*

Bibliography

On optical physics

Born, M. and Wolf, E. *Principles of Optics: Electromagnetic Theory of Propagation, Interference and Diffraction of Light,* 7th edn. Cambridge University Press, Cambridge, 1999. The standard text on the mathematical theory of optics.

Jenkins, F. A. and White, H. E. *Fundamentals of Optics,* 4th edn. McGraw-Hill, New York, 1976.

Longhurst, R. S. *Geometrical and Physical Optics.* Longman, London, 1986.

Stone, J. M. *Radiation and Optics.* McGraw-Hill, New York, 1963.

These last three books are all at about the same level, and discuss similar topics. Stone is the most mathematical, and Longhurst contains rather more information on topics such as the properties of optical instruments and the optical properties of crystals.

On the application of Fourier transform techniques

Goodman, J. W. *An Introduction to Fourier Optics.* Roberts, Greenwood Village, CO, 2004. Discusses the use of the Fourier transform in diffraction, and contains a full account of optical information processing.

Taylor, C. A. and Lipson, H. *Optical Transforms.* Bell, London, 1964. Discusses the analogy between the diffraction of light and the diffraction of X-rays. Contains 54 superbly illustrative plates.

Review I

The contents of the preceding six chapters of this book form the conceptual foundations of the study of X-ray diffraction. These six chapters fall neatly into three groups: Chapters 1 and 3 deal with crystallography; Chapters 2 and 5 with mathematical techniques; and Chapters 4 and 6 with the theory of wave motion. The present brief interlude is included in order to bring together the main results of these six chapters so that the reader may be reassured that the bases are understood.

Crystals are solids characterised by a long-range three-dimensional order. The geometrical arrangement of the atoms, molecules or ions in space may be represented by the *lattice function* $l(\mathbf{r})$, which is a three-dimensional array of δ functions corresponding to the allowed lattice positions. The *unit cell function* $u(\mathbf{r})$ describes the arrangement of particles within a single unit cell, and it is this function which contains the detailed information on the structure of individual molecules. The *crystal structure function* $c(\mathbf{u})$ is the *convolution* of the lattice function with the unit cell function:

$$c(\mathbf{u}) = l(\mathbf{r}) * u(\mathbf{r})$$

Whenever a wave motion interacts with an obstacle, *diffraction* occurs. The nature of the diffraction pattern is determined solely by the obstacle, and may be considered to contain information on the structure of the obstacle. If the obstacle may be described by an *amplitude function* $f(\mathbf{r})$, then the diffraction pattern amplitude $F(\mathbf{k})$ is given by the *Fourier transform* of $f(\mathbf{r})$:

$$F(\mathbf{k}) = \int_{\text{all } \mathbf{r}} f(\mathbf{r}) e^{i\mathbf{k}\cdot\mathbf{r}} \, d\mathbf{r} \tag{6.8}$$

By the *Fourier inversion theorem*, we may express $f(\mathbf{r})$ in terms of $F(\mathbf{k})$ as

$$f(\mathbf{r}) = \int_{\text{all } \mathbf{k}} F(\mathbf{k}) e^{-i\mathbf{k}\cdot\mathbf{r}} \, d\mathbf{k} \tag{6.10}$$

If we know the amplitude function $f(\mathbf{r})$, we may use equation (6.8) to derive the diffraction pattern $F(\mathbf{k})$. Conversely, the use of equation (6.10) permits us to infer the amplitude function $f(\mathbf{r})$ from the diffraction pattern function $F(\mathbf{k})$. In any practical case, our primary data will give us information concerning $F(\mathbf{k})$, and it is by calculating the integral of equation (6.10) that we can derive the corresponding $f(\mathbf{r})$. For a crystal, the amplitude function is simply the crystal structure function, and so we may obtain information on the crystal structure.

A major problem is that the primary data are the *intensities* of the diffraction pattern, which are mathematically represented by $|F(\mathbf{k})|^2$. This is a real quantity, and unfortunately it is impossible to derive an unambiguous value of the complex quantity

$F(\mathbf{k})$ directly from a single observation of $|F(\mathbf{k})|^2$. However, we need the correct value of $F(\mathbf{k})$ for equation (6.10). This predicament is known as the *phase problem*. Once we have a method for solving the phase problem, then we may calculate the requisite values of $F(\mathbf{k})$, allowing us to solve the crystal structure.

The next two parts of this book deal with the methods by which crystal structures may be solved. Few new concepts will be introduced—basically, at this stage, the reader should know what is happening, for the succeeding chapters merely contain a somewhat sophisticated list of recipes by which certain useful quantities may be calculated.

Part II, 'Diffraction theory', comprises Chapters 7, 8 and 9. In order for the reader to become familiar with diffraction patterns and the manner in which information is contained within them, Chapter 7 will describe one-dimensional obstacles and their corresponding diffraction patterns. Since the examples taken are one-dimensional, the mathematics is relatively straightforward, and the results of the various operations can be readily represented graphically on a two-dimensional page. Various generalisations concerning diffraction patterns will be derived, and these will carry over to the three-dimensional case, discussed in Chapters 8 and 9. The lattice function $l(\mathbf{r})$ and its diffraction pattern form the subject of Chapter 8, which introduces the very important concept of the Ewald sphere—a clear understanding of this geometrical construction is vital for the interpretation of the diffraction patterns of real crystals. In Chapter 9, the effect of the contents of the unit cell will be discussed. This chapter is the key for chemical and biological applications of X-ray crystallography, for it is here that we discuss the formulae which allow detailed molecular information to be derived.

Part III, 'Structure solution', describes how the theory discussed in Part II is used in practice.

Part II
Diffraction theory

7
Diffraction by one-dimensional obstacles

7.1 The geometrical arrangement

In the last chapter, we derived the very important result that the diffraction pattern function $F(\mathbf{k})$ defining the diffraction of waves by an obstacle described by an appropriate amplitude function $f(\mathbf{r})$ can be computed using the Fourier transform equation, equation (6.8):

$$F(\mathbf{k}) = \int_{\text{all } \mathbf{r}} f(\mathbf{r})e^{i\mathbf{k}\cdot\mathbf{r}}\,d\mathbf{r} \qquad (6.8)$$

The amplitude function $f(\mathbf{r})$ is zero outside the obstacle, but inside the obstacle it specifies the interaction between a point $\mathbf{r}$ and the incident wave. In general, the obstacle will be three-dimensional, implying that $f(\mathbf{r})$ is three-dimensional. In order that the reader may become acquainted with the way in which information is presented in a diffraction pattern, this chapter will investigate the diffraction patterns of a number of one-dimensional objects, allowing us to represent $f(\mathbf{r})$ as a one-dimensional function $f(x)$—we will return to three dimensions in Chapters 8 and 9.

All the examples will be based on the geometrical arrangement depicted in Fig. 7.1.

The plane of the diagram corresponds to the xz plane, with the y direction perpendicular to the plane of the paper and directed towards the reader. The obstacle whose diffraction pattern we are to determine will always be along the x axis, centred on the origin. Plane waves are assumed to be incident normally onto the obstacle from the left, travelling in the direction of the positive z axis. As a result of diffraction by the obstacle, the diffracted waves will spread out in the xz plane to the right of the obstacle. The parameter $\mathbf{r}$ in equation (6.8) corresponds to any point within the obstacle. For a one-dimensional obstacle along the x axis, $\mathbf{r}$ has solely an x component, and relative to the usual three base vectors $\mathbf{i}$, $\mathbf{j}$ and $\mathbf{k}$ of a right-handed Cartesian coordinate system, we may express an arbitrary point in the obstacle as a vector $\mathbf{r}$ given by

$$\mathbf{r} = (x, 0, 0)$$

The amplitude function $f(\mathbf{r})$ now becomes the one-dimensional function $f(x)$.

We now evaluate the scalar product $\mathbf{k}\cdot\mathbf{r}$, which appears in the exponential term $e^{i\mathbf{k}\cdot\mathbf{r}}$. Here, $\mathbf{k}$ represents the wave vector of a scattered wave, and this is a vector of magnitude $k = 2\pi/\lambda$, whose direction is parallel to that of propagation of the wave. One of the effects of the diffracting obstacle is to transform the incident plane waves

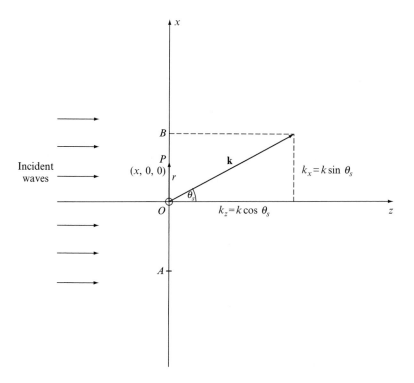

Fig. 7.1 *Diffraction in one dimension.* Waves are incident normally onto an obstacle which is along the x axis, centred on the origin. The y axis is perpendicular to the paper, directed towards the reader. A and B denote the limits of the obstacle along the x axis. Any point P within the obstacle is represented by a vector $\mathbf{r} = (x, 0, 0)$, which has only an x component. Diffracted waves spread out in the xz plane in different directions. The wave vector of one particular diffracted wave is shown, and this has both an x and a z component, so that $\mathbf{k} = (k_x, 0, k_z)$. The scalar product $\mathbf{k} \cdot \mathbf{r}$ becomes $(k_x, 0, k_z) \cdot (x, 0, 0) = k_x x = kx \sin \theta_s . \theta_s$ is the scattering angle.

into a series of waves propagating in different directions. In order to evaluate the scalar product, let us take a scattered wave travelling in an arbitrary direction as in Fig. 7.1. The wave vector of this wave is contained solely in the xz plane, and so this vector has only an x component, k_x, and a z component, k_z:

$$\mathbf{k} = (k_x, 0, k_z)$$
$$\therefore \mathbf{k} \cdot \mathbf{r} = (k_x, 0, k_z) \cdot (x, 0, 0)$$

which, according to equation (2.9), becomes

$$\mathbf{k} \cdot \mathbf{r} = k_x x$$

But the x component k_x of the wave vector is given by

$$k_x = k \sin \theta_s$$

in which θ_s is the angle between the **k** vector and the *normal* to the obstacle. It is important to note that it is this angle—the angle between the **k** vector and the normal—which is conventionally used in the description of one-dimensional diffraction patterns, and *not* the angle between the **k** vector and the obstacle. The reason for this is that the incident waves were originally normal to the obstacle, but have been deviated by the act of diffraction. θ_s represents the angle through which a particular wave has been deflected, and is known as the *scattering angle*. The scalar product **k·r** can now be written as

$$\mathbf{k \cdot r} = kx \sin \theta_s$$

Our expression for the diffraction pattern function $F(\mathbf{k})$ now becomes

$$F(\mathbf{k}) = \int_{-\infty}^{\infty} f(x)e^{ikx \sin \theta_s}\, dx \tag{7.1}$$

Inspection of the right-hand side of equation (7.1) shows that, in view of the fact that the magnitude k is a constant (remember that the wavelength of a wave is unchanged on diffraction), the only two variables are x and the scattering angle θ_s. Since the integration is over x, the result of the integration is a function of the angle θ_s, or, alternatively, of $\sin \theta_s$. The diffraction pattern function of a one-dimensional obstacle may therefore be written as $F(\sin \theta_s)$:

$$F(\sin \theta_s) = \int_{-\infty}^{\infty} f(x)e^{ikx \sin \theta_s}\, dx \tag{7.2}$$

Equation (7.2) is the form that equation (6.8) takes in the one-dimensional case, and it is this equation which we shall use in this chapter. For a variety of obstacles, we shall derive the appropriate form of the amplitude function $f(x)$, perform the integration of equation (7.2) and so generate the amplitude of the diffraction pattern as a function of $\sin \theta_s$. In general, $F(\sin \theta_s)$ is complex, and so we shall obtain the intensity of the diffraction pattern from this as $|F(\sin \theta_s)|^2$. This is our experimentally observed quantity. It should be noted, however, that all the examples to be used in this chapter are centrosymmetric, and, as we saw in Section 5.9, all centrosymmetric obstacles give rise to real Fourier transforms.

It is useful to pause for a moment to consider the relevance of the fact that the diffraction pattern function, which we originally identified as a function $F(\mathbf{k})$ of **k**, has now become a function $F(\sin \theta_s)$ of the scattering angle θ_s. In Chapter 6, we find the following two sentences: 'Thus the effect of the diffracting obstacle is to give rise to spatially different waves, implying that the information in the waves is somehow contained in the way in which they are spread out in space. This tells us that to interpret a diffraction pattern, we must sample the pattern in many different spatial locations, for it is only by this means that we will discern the information we require.'

The variation in the amplitude of a diffraction pattern is therefore ultimately a function of a spatial coordinate. One such coordinate is the angular displacement about an origin with respect to a defined axis. In our example in Fig. 7.1, the origin is shown, and the reference axis which we use is the z axis, defined as the direction of the incident X-rays. In this way, we should really expect that the diffraction pattern function will indeed turn out to be expressed in terms of an angle such as θ_s.

All of the examples to be discussed may be considered also in terms of obstacles to light waves. The diffraction pattern intensities we shall derive can actually be observed if a suitable optical experiment is performed.

7.2   One narrow slit

For our first example, we shall examine the diffraction pattern due to an infinite opaque sheet along the x axis containing a narrow slit at the origin. Our criterion of narrowness is that the slit is very narrow compared with the wavelength of the wave motion we are using. When the waves impinge normally on the obstacle, the only place through which they may pass is at the narrow slit. Elsewhere, since the sheet is perfectly opaque, the waves are absorbed. We wish to find a suitable amplitude function $f(x)$ which describes mathematically the fact that the obstacle will absorb the waves everywhere except at the slit, through which the waves may pass unhindered. In order to do this, we require that where the waves are absorbed, the amplitude function is zero, for this ensures that the product $f(x)e^{ikx \sin \theta_s}$ is zero, implying that there are no diffracted waves. At the slit, however, the waves passing through the slit do not 'know' that the obstacle is there and they 'think' that they are in free space. The waves pass on quite unhindered, and the slit has no effect whatsoever on the wave behaviour within the slit. Mathematically, we may express this by defining the amplitude function to be unity over the slit, for then the product $f(x)e^{ikx \sin \theta_s}$ becomes $e^{ikx \sin \theta_s}$ which is simply the free-space expression for a wave. The amplitude function is therefore a function which is zero everywhere over the screen, and unity at the slit. Since the slit is narrow, the complete amplitude function for a narrow slit may be represented by a δ function placed at the origin, corresponding to the position of the slit, as in Fig. 7.2(a).

We now have the required mathematical description of the amplitude function $f(x)$ of a narrow slit, namely

$$f(x) = \delta(x)$$

Substituting for $f(x)$ in equation (7.2),

$$F(\sin \theta_s) = \int_{-\infty}^{\infty} f(x)e^{ikx \sin \theta_s} \, dx \tag{7.2}$$

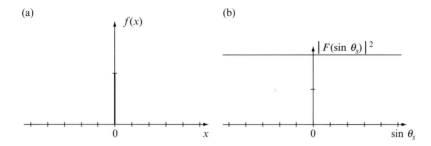

Fig. 7.2 *Diffraction by one narrow slit.* The amplitude function $f(x)$ for a single narrow slit is a δ function at the origin, as in (a). The diffraction pattern intensity as a function of $\sin \theta_s$ is shown in (b). The intensity is constant, independent of the angle θ_s, and corresponds to the intensity we would expect from a point source.

we obtain

$$F(\sin \theta_s) = \int_{-\infty}^{\infty} \delta(x)e^{ikx \sin \theta_s}\, dx \qquad (7.3)$$

Equation (7.3) requires us to take the Fourier transform of a δ function. Reference to Chapter 5 reminds us that the value of the δ function $\delta(x)$ is zero for all values of x other than $x = 0$, and that when $x = 0$, $e^{ikx \sin \theta_s} = e^0 = 1$. The integral of equation (7.3) therefore reduces to the integral of the δ function itself, which is defined as unity:

$$F(\sin \theta_s) = \int_{-\infty}^{\infty} \delta(x)e^{ikx \sin \theta_s}\, dx = \int_{-\infty}^{\infty} \delta(x)\, dx = 1$$

From this result, we can determine the intensity of the diffraction pattern as

$$|F(\sin \theta_s)|^2 = 1 \qquad (7.4)$$

It is of course the intensity $|F(\sin \theta_s)|^2$ that is measured, and equation (7.4) implies that the intensity of the diffraction pattern of a single narrow slit is a constant, and is independent of the scattering angle θ_s, as shown in Fig. 7.2(b).

What is the significance of this? This becomes clear when we remember that we express the diffraction pattern intensity as a function of the scattering angle θ_s between the z axis and the direction of the corresponding diffracted wave. The fact that we have calculated that the diffraction pattern intensity is constant implies that no matter what angle we are at relative to the z axis, we see the same intensity: the intensity is uniform at all angles. This is exactly what we would expect if there were a point source positioned at the origin, for at all angles from a point source, the intensity is the same.

What we have proved is that the diffraction pattern of a single narrow slit is exactly equivalent to the disturbance produced by an active point source. This validates the use of a pinhole as a point source of light, and also verifies several of the arguments of the preceding chapter.

7.3   One wide slit

Our second example is that of an opaque screen containing a slit which is wide compared with the wavelength of the incident wave motion. It must not, however, be too wide, for then the diffraction effects will be masked, and so we shall choose as a typical width something like 100 times the wavelength. As before, the screen absorbs the waves incident on it, whereas waves falling on the slit pass on unhindered. An appropriate amplitude function is therefore a 'Top Hat' function of unit height and of width $2X_0$, as shown in Fig. 7.3a.

Our amplitude function may be defined by

$$f(x) = 0 \quad \text{if} \quad -\infty < x < -X_0$$

$$f(x) = 1 \quad \text{if} \quad -X_0 \leq x \leq +X_0$$

$$f(x) = 0 \quad \text{if} \quad X_0 < x < \infty$$

We require the Fourier transform of this function, namely

$$F(\sin \theta_s) = \int_{-\infty}^{\infty} f(x) e^{ikx \sin \theta_s} \, dx$$

This integral exists only where the amplitude function is non-zero, and so we may replace the infinite limits on the integral by finite limits corresponding to the width of the slit, between which limits $f(x) = 1$:

$$F(\sin \theta_s) = \int_{-X_0}^{X_0} e^{ikx \sin \theta_s} \, dx$$

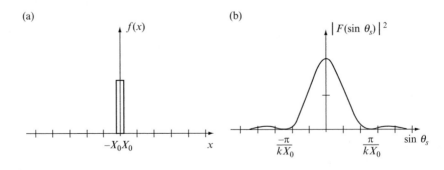

(a)

(b)

Fig. 7.3 *Diffraction by one wide slit.* One wide slit may be represented by an amplitude function in the form of a 'Top Hat', of unit height and of width $2X_0$, as in (a); (b) shows the diffraction pattern intensity, which is proportional to $\sin^2(kX_0 \sin \theta_s)/(kX_0 \sin \theta_s)^2$. This is a broad-peaked function, of characteristic width $2\pi/kX_0$.

Quoting from Chapter 5, equation (5.13), with the substitution of the variable $k \sin \theta_s$ for k, we obtain the result that

$$F(\sin \theta_s) = 2X_0 \frac{\sin(kX_0 \sin \theta_s)}{kX_0 \sin \theta_s}$$

and the intensity becomes

$$|F(\sin \theta_s)|^2 = 4X_0^2 \frac{\sin^2(kX_0 \sin \theta_s)}{(kX_0 \sin \theta_s)^2} \tag{7.5}$$

In contrast to equation (7.4), which showed that the intensity of the diffraction pattern of a single narrow slit does not vary with the scattering angle θ_s but is a constant, equation (7.5) indicates that the intensity of the diffraction pattern of a single wide slit is not constant, but does indeed depend on the scattering angle θ_s in what appears to be quite a complex way. But not too complex: the factor $4X_0^2$ is a constant, and the rest is the square of the $(\sin x)/x$ function depicted in Fig. 5.5(b), where x is $kX_0 \sin \theta_s$. The intensity is therefore the broad-peaked function plotted in Fig. 7.3(b). If light is used as a source of waves and a suitable obstacle is constructed, then this diffraction pattern may be displayed on a screen as a series of light and dark bands corresponding to different values of the scattering angle θ_s.

There are several points to notice about this diffraction pattern. Firstly, the width of the slit is $2X_0$, whereas the width of the central peak corresponds to an angular deviation $\Delta(\sin \theta_s)$ given by

$$\Delta(\sin \theta_s) = \frac{2\pi}{kX_0}$$

If the angle θ_s is small, then to a good approximation

$$\sin \theta_s \approx \theta_s$$

and so the actual angular deviation $\Delta \theta_s$ is $2\pi/kX_0$. This may be regarded as the width of the central peak of the diffraction pattern, and it is reciprocally related to the width of the slit $2X_0$. Hence the wider the slit, the narrower the diffraction pattern, and, conversely, the narrower the slit, the more extended the diffraction pattern. Taking this to its limit, we infer that an infinitely narrow slit will have an infinitely broad diffraction pattern, proving that the diffraction pattern of a narrow slit is a constant regardless of angle, as derived in the previous example.

Secondly, it should be noted that the intensities of the secondary maxima become weaker very rapidly, but the widths of all secondary maxima are the same, equal to half the width of the central maximum.

7.4   Two narrow slits

An appropriate form of the amplitude function for two narrow slits is a pair of δ functions, one at $+x_0$ and the other at $-x_0$, corresponding to the positions of the two slits shown in Fig. 7.4(a). We therefore have

$$f(x) = \delta(x + x_0) + \delta(x - x_0)$$

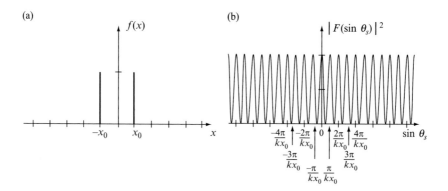

Fig. 7.4 *Diffraction by two narrow slits.* An array of two δ functions (a) is the appropriate amplitude function, giving the diffraction pattern $|F(\sin\theta_s)|^2 \alpha \cos^2(kx_0 \sin\theta_s)$, as shown in (b).

and

$$F(\sin\theta_s) = \int_{-\infty}^{\infty} f(x)e^{ikx\,\sin\theta_s}\,dx \tag{7.2}$$

The Fourier transform of two δ functions was found in Chapter 5, equation (5.22), as

$$F(k) = 2\cos kx_0 \tag{5.22}$$

which for our present case we write as

$$F(\sin\theta_s) = 2\cos(kx_0\sin\theta_s)$$

The intensity of the diffraction pattern is therefore given by

$$|F(\sin\theta_s)|^2 = 4\cos^2(kx_0\sin\theta_s) \tag{7.6}$$

If we observe the diffraction pattern close to the z axis, the angle θ_s is small, and to a good approximation $\sin\theta_s \approx \theta_s$. This implies that the central part of the diffraction pattern is a series of light and dark bands, known as $\cos^2$ *fringes*, as shown in Fig. 7.4(b).

7.5   Young's experiment

In 1803, Thomas Young published the results of an experiment which demonstrated conclusively the wave nature of light. This experiment is a landmark in the history of physics, since it caused the downfall of Newton's corpuscular theory of light. Young's experimental arrangement is depicted in Fig. 7.5.

A narrow slit is placed in front of a light source so as to act as a point source. On the farther side of the single narrow slit is an opaque sheet containing a pair of narrow slits such that each slit in this sheet is equidistant from the single slit source. Diffraction occurs at the pair of slits, and the diffraction pattern is caught on a screen.

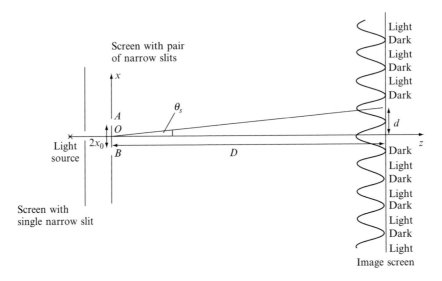

Fig. 7.5 *Young's experiment.* Light from a source passes through a single narrow slit, and then through a pair of narrow slits, A and B, which are equidistant from the source slit. The diffraction pattern is caught on the image screen, and a series of light and dark bands, known as $\cos^2$ fringes, are observed. The scattering angle θ_s corresponding to the first off-axis bright fringe is assumed to be small, which holds if D is large. It is then true that
$\sin \theta_s \approx \tan \theta_s \approx \frac{d}{D}$

This experiment is one to determine the diffraction pattern of a pair of narrow slits, which, as we have shown, gives rise to a series of $\cos^2$ fringes. If the distance between the image screen and the opaque sheet containing the pair of slits is represented by D, then we may calculate the positions of the bright fringes as follows.

Consider the intensity along the image screen at a distance d from the z axis. Equation (7.6) above states that

$$|F(\sin \theta_s)|^2 = 4 \cos^2 (k x_0 \sin \theta_s)$$

For small angles θ_s, we may write

$$\sin \theta_s \approx \theta_s \approx \tan \theta_s$$

$$\therefore \sin \theta_s \approx \frac{d}{D}$$

$$\therefore |F(\theta_s)|^2 = 4 \cos^2 \left(\frac{k x_0 d}{D} \right)$$

At the centre of a bright fringe, the intensity is a maximum, for which

$$\cos^2 \frac{kx_0 d}{D} = 1$$

$$\therefore \frac{kx_0 d}{D} = 0, \pi, 2\pi, 3\pi$$

$$\therefore d = 0, \frac{\pi D}{kx_0}, \frac{2\pi D}{kx_0}, \frac{3\pi D}{kx_0}, \dots$$

Hence there is a bright fringe on the z axis corresponding to $d = 0$, and then a series of bright fringes spaced equally by a distance $\Delta = \pi D/kx_0$. Rearranging this expression, we may write

$$k = \frac{\pi D}{x_0 \Delta}$$

but since

$$k = \frac{2\pi}{\lambda}$$

then

$$\lambda = \frac{2x_0 \Delta}{D}$$

D, x_0 and Δ are all experimentally measurable, and from these distances we may compute the wavelength of the light used.

The reader familiar with basic physics will have met Young's experiment before, but will be used to a rather different method by which the diffraction pattern may be found. In order to show that the Fourier transform method agrees with, and is more elegant than, the 'classical' method, the latter is summarised here.

With reference to Fig. 7.6, we define some symbols. Consider a point P on the image screen at a distance d from the z axis. Let the distance from slit A to this point be r_1. Let the distance from the slit B to this point be r_2. Let the distance from the origin to this point be r.

The wave emanating from slit A is a spherical wave, and at any point r, at time t, the amplitude of this wave is given by equation (4.41):

$$\psi_A = \frac{\psi_0}{r} \cos(kr - \omega t)$$

The wave due to slit A at the point P is therefore

$$\psi_A = \frac{\psi_0}{r_1} \cos(kr_1 - \omega t)$$

Similarly, the wave from slit B arriving at point P at the same time t as that from slit A is given by

$$\psi_B = \frac{\psi_0}{r_2} \cos(kr_2 - \omega t)$$

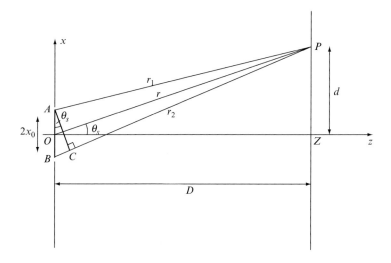

Fig. 7.6 *The geometry of Young's experiment.* For clarity, the diagram is distorted, since in fact $D \gg d$. Using small-angle approximations, the triangles ABC and OPZ are similar, and $r_2 - r_1 \approx BC \approx 2x_0 \sin\theta_s \approx 2x_0 \tan\theta_s \approx 2x_0 d/D$.

Since the light source is the same for both slits, the maximum amplitude ψ_0 is the same for both waves.

Using the principle of superposition, the total disturbance at point P is

$$\psi_{\text{Tot}} = \psi_A + \psi_B$$

$$= \frac{\psi_0}{r_1} \cos(kr_1 - \omega t) + \frac{\psi_0}{r_2} \cos(kr_2 - \omega t)$$

If the image screen is relatively far from the screen containing the two slits, then to a good approximation

$$r_1 \approx r_2 \approx r$$

$$\therefore \psi_{\text{Tot}} = \frac{\psi_0}{r} [\cos(kr_1 - \omega t) + \cos(kr_2 - \omega t)]$$

At this stage we substitute r for r_1 and r_2 in the $1/r_1$ and l/r_2 terms only, since these vary much more slowly than do the cosine terms. Using the fact that

$$\cos A + \cos B = 2\cos\frac{A+B}{2}\cos\frac{A-B}{2}$$

we may write

$$\psi_{\text{Tot}} = 2\frac{\psi_0}{r}\cos\left(\frac{k(r_1+r_2)}{2} - \omega t\right)\cos\frac{k(r_2-r_1)}{2}$$

We may now approximate $(r_1 + r_2)/2$ by r, but since $(r_2 - r_1)/2$ is small, we must leave it as it is for the present. We may now write

$$\psi_{\text{Tot}} = 2\frac{\psi_0}{r} \cos(kr - \omega t) \cos\frac{k(r_2 - r_1)}{2}$$

At any time, the quantity $2(\psi_0/r)\cos(kr - \omega t)$ is a constant, and so

$$\psi_{\text{Tot}} \propto \cos\frac{k(r_2 - r_1)}{2}$$

and the intensity becomes

$$|\psi_{\text{Tot}}|^2 \propto \cos^2\frac{k(r_2 - r_1)}{2}$$

The quantity $(r_2 - r_1)$ may be expressed in terms of the dimensions of the apparatus as follows. For small angles θ_s, $(r_2 - r_1)$ is approximately equal to the distance BC in Fig. 7.6. Also, the triangles ABC and OPZ are similar and

$$BC = AB \sin\theta_s$$

But AB is the separation $2x_0$ of the slits, and for small angles,

$$\sin\theta_s \approx \theta_s \approx \tan\theta_s = \frac{d}{D}$$

$$\therefore r_2 - r_1 = \frac{2x_0 d}{D}$$

$$\therefore |\psi_{\text{Tot}}|^2 \propto \cos^2\frac{kx_0 d}{D}$$

This has the same functional dependence as equation (7.6), demonstrating that the two methods give the same result. It is hoped that comparison of these two methods for solving the problem of the diffraction pattern of two narrow slits will convince the reader that the classical method is useful only in the very simplest of cases, whereas the Fourier transform method, though requiring a more detailed mathematical background, is more direct, more elegant and very much more simple to use even in the rather complicated cases which we shall meet in due course.

7.6   Two wide slits

The example of the diffraction pattern of two wide slits introduces a new and extremely important technique: the use of the concept of convolution. We require a suitable amplitude function for two slits, each of width $2X_0$, centred on $x = -x_0$ and $x = +x_0$ as shown in Fig. 7.7(a).

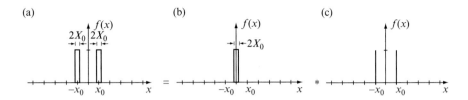

Fig. 7.7 *Two wide slits may be regarded as the convolution of one wide slit and two δ functions.*

One way of representing this function would be to define the amplitude function $f(x)$ according to each domain of x as

$$f(x) = 0 \quad \text{if} \quad -\infty < x < -(x_0 + X_0)$$
$$= 1 \quad -(x_0 + X_0) \le x \le -(x_0 - X_0)$$
$$= 0 \quad -(x_0 - X_0) < x < (x_0 - X_0)$$
$$= 1 \quad (x_0 - X_0) \le x \le (x_0 + X_0)$$
$$= 0 \quad (x_0 + X_0) < x < \infty$$

This list is somewhat indigestible, for it is far more significant to regard the amplitude function $f(x)$ as the *convolution* of the amplitude function for one wide slit with the function for two narrow slits as depicted in Fig. 7.7(b).

$$f(\text{two wide slits}) = f(\text{one wide slit}) * f(\text{two narrow slits})$$

The great relevance of this result becomes apparent when we compute the Fourier transform of the amplitude function of two wide slits. We require

$$Tf(\text{two wide slits}) = T[f(\text{one wide slit}) * f(\text{two narrow slits})]$$

We now invoke the convolution theorem (Section 5.11), which states that *The Fourier transform of a convolution is the product of the individual Fourier transforms.* We can now immediately write

$$Tf(\text{two wide slits}) = Tf(\text{one wide slit}) \cdot Tf(\text{two narrow slits})$$

But we have already found the Fourier transform of one wide slit in Section 7.3 as

$$F(\sin \theta_s) = 2X_0 \frac{\sin(kX_0 \sin \theta_s)}{kX_0 \sin \theta_s}$$

and that for a pair of narrow slits was shown in Section 7.4 to be

$$F(\sin \theta_s) = 2\cos(kx_0 \sin \theta_s)$$

The amplitude of the diffraction pattern for two wide slits becomes

$$F(\sin \theta_s) = 4X_0 \frac{\sin(kX_0 \sin \theta_s)}{kX_0 \sin \theta_s} \cdot \cos(kx_0 \sin \theta_s)$$

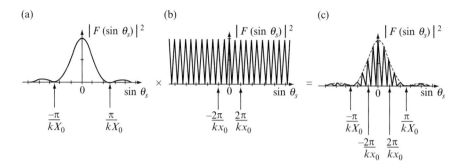

Fig. 7.8   *The diffraction pattern of two wide slits.* (a) shows the diffraction pattern of one wide slit, and (b) that of two δ functions. By the convolution theorem, the diffraction pattern of two wide slits is the product of the patterns shown in (a) and (b), which product is depicted in (c). This corresponds to the cos² fringes due to the pair of δ functions being modulated by the diffraction pattern of the single wide slit. Note that since the single wide slit is narrower than the pair of functions considered together, the peaks of the diffraction pattern of the single wide slit are broader than those of the pair of δ functions.

and the corresponding intensity is given by

$$|F(\sin\theta_s)|^2 = 16X_0^2 \frac{\sin^2(kX_0\sin\theta_s)}{(kX_0\sin\theta_s)^2} \cdot \cos^2(kx_0\sin\theta_s) \tag{7.7}$$

The analytical form of equation (7.7) is hard to digest, but what is important is that the intensity is proportional to the product of a $(\sin^2 x)/x^2$ function with a $\cos^2$ function. This product may be determined by inspection according to the rules elucidated in Chapter 5. The two separate functions and the product function are shown in Fig. 7.8.

Figure 7.8 demonstrates a number of important concepts. The most notable point is that the form of the diffraction pattern is a series of $\cos^2$ fringes modulated by the $(\sin^2 x)/x^2$ function. That the latter function extends over the $\cos^2$ function, and not vice versa, may be demonstrated algebraically by finding those points at which each function is individually first zero. The $\cos^2$ function is first zero at an angle θ_{s1}, given by

$$kx_0\sin\theta_{s1} = \pi/2$$

$$\therefore \sin\theta_{s1} = \frac{\pi}{2kx_0}$$

whereas the $(\sin^2 x)/x^2$ function is first zero at an angle θ_{s2} according to

$$kX_0\sin\theta_{s2} = \pi$$

$$\therefore \sin\theta_{s2} = \frac{\pi}{kX_0}$$

Since

$$x_0 \gg X_0$$

that is, the distance between the slits is greater than the width of either slit individually, then

$$\theta_{s2} > \theta_{s1}$$

and so the $\cos^2$ function has its first zero at a smaller value of $\sin\theta_s$, and hence of θ_s, than does the $(\sin^2 x)/x^2$ function.

The significance of this may be appreciated when we remember that a general feature of Fourier transforms, and hence diffraction patterns, is that the narrower the obstacle, the more extended the diffraction pattern, and vice versa. Our present obstacle is composed of the convolution of two parts: a pair of δ functions separated by a distance $2x_0$, and a slit of width $2X_0$, which although wide compared with the wavelength of the wave motion, is narrow compared with the distance $2x_0$ between the slits. The diffraction pattern will therefore combine the effects of the relatively broad pattern of a single wide slit and the relatively condensed pattern due to the pair of δ functions.

7.7   Three narrow slits

The amplitude function of three narrow slits may be represented by an array of three functions as

$$f(x) = \delta(x + x_0) + \delta(x) + \delta(x - x_0)$$

According to Section 5.8, the Fourier transform of this function is given by $(1 + 2\cos kx_0)$, which, with the substitution of $k\sin\theta_s$ for k, gives us the amplitude of the diffraction pattern as

$$F(\sin\theta_s) = 1 + 2\cos(kx_0 \sin\theta_s)$$

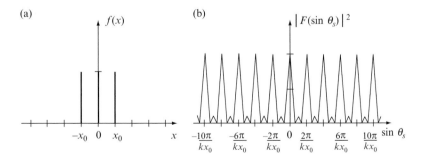

Fig. 7.9 *Diffraction by three narrow slits.* The amplitude function for three narrow slits is an array of three δ functions as in (a), giving the diffraction pattern $|F(\sin\theta_s)|^2 = (1 + 2\cos(kx_0 \sin\theta_s))^2$, as in (b).

and the intensity becomes

$$|F(\sin \theta_s)|^2 = (1 + 2\cos(kx_0 \sin \theta_s))^2 \tag{7.8}$$

The amplitude function and the diffraction pattern intensity are shown in Fig. 7.9.

There are alternating strong and weak peaks, the strong ones being nine times more intense than the weak ones.

7.8   Three wide slits

Proceeding as in the example of the two wide slits, we may invoke the concept of convolution and write

$$f(\text{three wide slits}) = f(\text{one wide slit}) * f(\text{three narrow slits})$$
$$\therefore Tf(\text{three wide slits}) = T[f(\text{one wide slit}) * f(\text{three narrow slits})]$$
$$= Tf(\text{one wide slit}) \cdot Tf(\text{three narrow slits})$$

The intensity of the diffraction pattern is given by

$$|F(\sin \theta_s)|^2 = |Tf(\text{one wide slit})|^2 \cdot |Tf(\text{three narrow slits})|^2$$

(a)

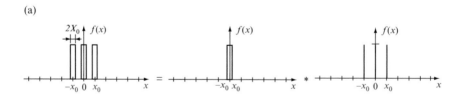

(b)

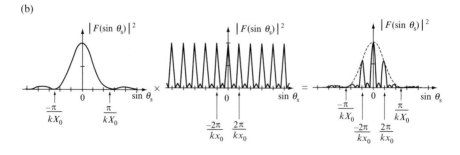

Fig. 7.10 *Diffraction by three wide slits.* (a) depicts the how the amplitude function for three wide slits can be represented as the convolution of one wide slit with an array of three δ functions. (b) shows the diffraction patterns of one wide slit and three δ functions, and how they combine to give the diffraction pattern of three wide slits. Note the envelope effect.

Quoting from equations (7.5) and (7.8), we have

$$|F(\sin\theta_s)|^2 = 4X_0^2 \frac{\sin^2(kX_0\sin\theta_s)}{(kX_0\sin\theta_s)^2} \cdot (1 + 2\cos(kx_0\sin\theta_s))^2 \tag{7.9}$$

As shown in Fig. 7.10(b), the diffraction pattern of three wide slits consists of that of three narrow slits modulated by that due to one wide slit.

7.9   *N* narrow slits

We now generalise the argument to consider the case of N narrow slits, equally spaced by a distance x_0, for which the amplitude function $f(x)$ is an array of N δ functions as represented in Fig. 7.11(a). Following the example of Chapter 5, Section 5.8, for convenience we choose N to be an odd number, equal to $2p+1$, allowing us to write

$$f(x) = \sum_{n=-p}^{n=+p} \delta(x - nx_0)$$

The Fourier transform of this function is given in equation (5.25),

$$F(k) = \frac{\sin\frac{Nkx_0}{2}}{\sin\frac{kx_0}{2}} \tag{5.25}$$

for which the corresponding intensity is

$$|F(k)|^2 = \frac{\sin^2\frac{Nkx_0}{2}}{\sin^2\frac{kx_0}{2}}$$

Replacing k by $k\sin\theta_s$, we obtain the intensity of the diffraction pattern as

$$|F(\sin\theta_s)|^2 = \frac{\sin^2\frac{Nkx_0\sin\theta_s}{2}}{\sin^2\frac{kx_0\sin\theta_s}{2}} \tag{7.10}$$

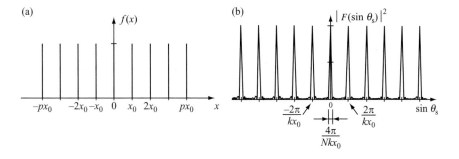

Fig. 7.11 *Diffraction by N narrow slits.* (a) shows an array of $N = (2p+1)$ δ functions, and the corresponding diffraction pattern is as in (b). This is a sharply peaked function, with main peaks spaced equally by $2\pi/kx_0$. The width of each main peak is $4\pi/Nkx_0$, and there are $(N-2)$ very weak subsidiary peaks between each main pair.

This is a periodic, sharply peaked function, depicted in Fig. 5.11(b) and also in Fig. 7.11(b). Using the small-angle approximation that $\sin\theta_s \approx \theta_s$, the angular width of the main peaks is $4\pi/Nkx_0$, and so the greater the number of slits in the array, the sharper are the main peaks. Each pair of main peaks is separated by an angular distance of $2\pi/kx_0$, which is *independent of the total number of slits*. An optical diffraction grating is a screen of N narrow slits, where N can be as great as several thousand lines per inch. The diffraction pattern caused by such a grating may be caught on a screen and is seen to be a series of very bright, fine lines corresponding to the main peaks. The $(N-2)$ subsidiary peaks between each main pair are often too weak to be visible.

7.10 N wide slits

The next example is that of N wide slits. Each slit has a width of $2X_0$, and is centred on a δ function of the array of the previous example. The amplitude function is the convolution of the single-wide-slit function with an array of N δ functions:

$$f(N \text{ wide slits}) = f(\text{one wide slit}) * f(N \text{ narrow slits})$$

Using the convolution theorem, we obtain

$$Tf(N \text{ wide slits}) = T[f(\text{one wide slit}) * f(N \text{ narrow slits})]$$
$$= Tf(\text{one wide slit}) \cdot Tf(N \text{ narrow slits})$$

And the intensity $|F(\sin\theta_s)|^2$ becomes

$$|F(\sin\theta_s)|^2 = |Tf(\text{one wide slit})|^2 \cdot |Tf(N \text{ narrow slits})|^2$$

$$|F(k)|^2 = 4X_0^2 \frac{\sin^2(kX_0\sin\theta_s)}{(kX_0\sin\theta_s)^2} \cdot \frac{\sin^2\frac{Nkx_0\sin\theta_s}{2}}{\sin^2\frac{kx_0\sin\theta_s}{2}} \tag{7.11}$$

The form of the amplitude function is shown in Fig. 7.12(a), and that of the diffraction pattern intensity in Fig. 7.12(b).

The diffraction pattern of the single wide slit has its first zero at an angle given by

$$\sin\theta_s = \frac{\pi}{kX_0}$$

whereas the diffraction pattern of the N δ functions has its first zero at an angle corresponding to

$$\sin\theta_s = \frac{\pi}{Nkx_0}$$

Since N is a large integer and $x_0 > X_0$, the diffraction pattern of the single wide slit modulates that of the array of δ functions as shown. This is what we expect, since the single wide slit is more compact than the array of N narrow slits considered as a whole, implying that the peaks in the diffraction pattern of the wide slit will be the broader of the two diffraction patterns.

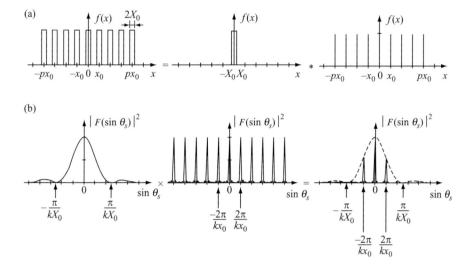

Fig. 7.12 *Diffraction by N wide slits.* The amplitude function of (a) gives rise to the diffraction pattern in (b), which is a sharply peaked function modulated by the pattern of a single wide slit.

We see that the effect of widening the slits is to alter the intensities of the main peaks, but that the positions of the main peaks are the same as those for the array of N narrow slits. We will discuss this more fully in a succeeding section.

7.11 An infinite number of narrow slits

For an infinite number of narrow slits separated by a distance x_0, the amplitude function is an infinite array of δ functions as shown in Fig. 7.13. Following the argument of Section 5.8, the diffraction pattern will be another infinite array of δ functions as in Fig. 7.13(b).

We have

$$f(x) = \sum_{n=-\infty}^{n=\infty} \delta(x - nx_0)$$

and, from equation (5.26),

$$F(\sin\theta_s) = \sum_{n=-\infty}^{n=\infty} \delta\left(\sin\theta_s - \frac{2n\pi}{kx_0}\right)$$

and

$$|F(\sin\theta_s)|^2 = \left|\sum_{n=-\infty}^{n=\infty} \delta\left(\sin\theta_s - \frac{2n\pi}{kx_0}\right)\right|^2 \tag{7.12}$$

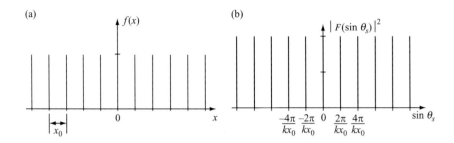

Fig. 7.13 *Diffraction by ∞ narrow slits.* An infinite array of narrow slits (a) results in a diffraction pattern of an infinite array of infinitely sharp peaks (b). This is the limiting case of N narrow slits as $N \to \infty$.

The diffraction pattern consists of an array of infinitely sharp peaks separated by an angular distance given by

$$\Delta(\sin \theta_s) = \frac{2\pi}{kx_0}$$

Note once again the reciprocal relationship between the obstacle and the diffraction pattern: the closer the slits, the more widely separated are the diffraction peaks.

An infinite array of δ functions is, of course, a one-dimensional model of a crystal lattice, and we have just shown that the diffraction pattern of this lattice is another infinite array of δ functions, which define a second lattice. As we saw in Section 5.6, this second lattice exists in Fourier, or reciprocal, space, in contrast to the first lattice, which exists in real space. We therefore refer to the first lattice as a *real lattice* and to the second as a *reciprocal lattice*. This leads to a most important result:

1. *The diffraction pattern of a real lattice is a second lattice, the reciprocal lattice.*
2. *The real lattice and the reciprocal lattice are Fourier mates.*

We shall explore the concept of the reciprocal lattice and the significance of the Fourier mate relationship in much more detail in the next chapter.

7.12   An infinite number of wide slits

Our final example concerns what happens when we widen the slits of our infinite array. Exactly as before, we have

$$f(\infty \text{ wide slits}) = f(\text{one wide slit}) * f(\infty \text{ narrow slits})$$

$$Tf(\infty \text{ wide slits}) = T[f(\text{one wide slit}) * f(\infty \text{ narrow slits})]$$

$$= Tf(\text{one wide slit}) \cdot Tf(\infty \text{ narrow slits})$$

$$\text{Intensity} = |F(\sin \theta_s)|^2 = |Tf(\text{one wide slit})|^2 \cdot |Tf(\infty \text{ narrow slits})|^2$$

$$|F(\sin \theta_s)|^2 = 4X_0 \frac{\sin^2(kX_0 \sin \theta_s)}{(kX_0 \sin \theta_s)^2} \left| \sum_{n=-\infty}^{n=\infty} \delta \left(\sin \theta_s - \frac{2n\pi}{kx_0} \right) \right|^2 \tag{7.13}$$

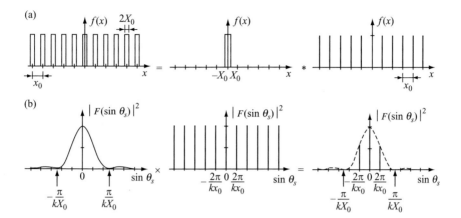

Fig. 7.14 *Diffraction by ∞ wide slits.* An infinite array of wide slits (a) gives the diffraction pattern in (b), which may be thought of as a series of δ functions sampling the diffraction pattern of a single wide slit at places determined by the spacing of the δ functions in the original array.

The diffraction pattern is a series of δ functions enveloped by the effect of a single wide slit, as shown in Fig. 7.14(b).

As before, the positions of the main peaks are unchanged on widening the slits, but whereas the diffraction pattern of an infinite number of narrow slits has all peaks of equal intensity, that of an infinite array of wide slits has peaks of different intensities as determined by the envelope.

7.13 The significance of the diffraction pattern

With the results of the above examples, we may now examine the form of each diffraction pattern in order to determine the manner in which information concerning the structure of the obstacle is presented to us.

Firstly, let us compare the diffraction patterns due to three, N and an infinite number of narrow slits. These patterns are reproduced in Fig. 7.15. As can be seen, increasing the number of slits has two effects:

(a) As the number of slits increases, the main peaks become sharper and narrower, whilst the subsidiary peaks become rapidly less intense.
(b) The positions of and separation between the main peaks stay constant, independent of the number of peaks. The main peaks are separated by an angular deflection given by

$$\Delta(\sin\theta_s) = \frac{2\pi}{kx_0}$$

This distance is determined solely by the wavelength of the wave motion, and the separation x_0 of any two neighbouring narrow slits.

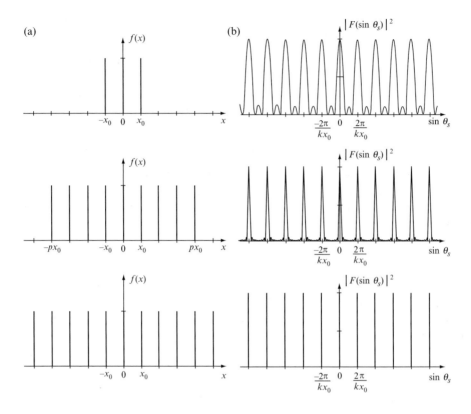

Fig. 7.15 *The significance of the diffraction pattern (1).* (a) shows the amplitude functions for three, N and ∞ narrow slits, and (b) shows the corresponding diffraction patterns. This indicates that the positions of the main peaks are determined by the spacing of the slits in the obstacle, and also that the width, or shape, of each main peak is determined by the overall width of the entire obstacle. Note that the intensity of each main peak is the same.

In the limit of an infinite number of narrow slits, the subsidiary peaks are gone, and we are left with an array of δ functions separated by the characteristic spacing $2\pi/kx_0$.

The example of an infinite number of narrow slits is a one-dimensional analogue of a crystal lattice. In Chapter 5, we learnt that a crystal lattice may be represented by a three-dimensional array of δ functions. The dimensions of a typical crystal are many orders of magnitude greater than the wavelength of X-rays, and so the case of an infinite number of δ functions is of great relevance. Specifically, we see that the distance between the main peaks in the diffraction pattern of a crystal is determined solely by the distance between adjacent δ functions in the obstacle, this being the lattice spacing of the crystal. By measuring the distance between the main peaks in a diffraction pattern, we may therefore obtain information about the lattice spacing in the crystal. Although the mathematics is rather more complicated in three dimensions, this general qualitative result is true:

The position of the main peaks in a diffraction pattern is determined solely by the lattice spacing in an obstacle.

As the number of slits in the above examples increases, each main peak becomes sharper. This implies that the sharpness or, more generally, the shape of each main peak is determined by the total number of slits in the obstacle or, equivalently, by the overall shape of the obstacle. We shall return to this in the next section, but once again the general result carries over into three dimensions and we may state:

The shape of each main peak is determined by the overall shape of the obstacle.

At this stage, we are now able to interpret two features of any diffraction picture whatsoever. By investigating the spacing of the main peaks, we may infer the lattice spacing within the obstacle, whereas the shape of each main peak contains information on the overall shape of the obstacle.

Let us now enquire into the effect of replacing the narrow slits by wide ones. Figure 7.16 shows the diffraction patterns for one, three, N and an infinite number of wide slits.

On comparison with the equivalent diffraction patterns for narrow slits as shown in Fig. 7.15, we see that the positions of the peaks are the same, but whereas an array of narrow slits gives rise to peaks of the same intensity, the pattern for an array of wide slits has peaks of differing intensities as determined by the envelope function. This envelope function is the same no matter how many wide slits we have, and is that due to a single wide slit. The effect of increasing the number of slits may be thought of as a sampling of the diffraction pattern of a single wide slit. In the case of one wide slit, the entire diffraction pattern may be recorded on a screen. But for N wide slits, the effect of the repeating of the lattice is to cause the overall pattern of a single wide slit to be sampled only in those positions determined by the diffraction pattern of the lattice.

The significance of the width of the slit is that our single wide slit is the one-dimensional analogue of the motif of a three-dimensional crystal. Just as a real crystal is the convolution of a motif with a lattice, then the array of N wide slits is the convolution of a motif (one wide slit) with a lattice (N narrow slits). The effect of the motif is therefore one of modifying the intensity of each main peak.

The effect of the motif is to alter the intensity of each main peak, but the positions of the main peaks are unchanged.

In this way, information concerning the motif may be obtained by collecting data on the intensity of every main peak in the diffraction pattern. Suppose we are given a diffraction pattern of N slits of unknown width. From this, at each main peak position, we could draw vertical lines according to the intensity of each main peak. Knowing that these are determined by an envelope, we might guess at the form of the envelope in between the main peaks, thereby generating the entire diffraction pattern of the motif, as in Fig. 7.17.

Once the complete motif diffraction pattern is known, we may compute the inverse Fourier transform so as to derive the complete structure of the motif, and thereby determine the width of each wide slit. In essence, this is how the structure of individual

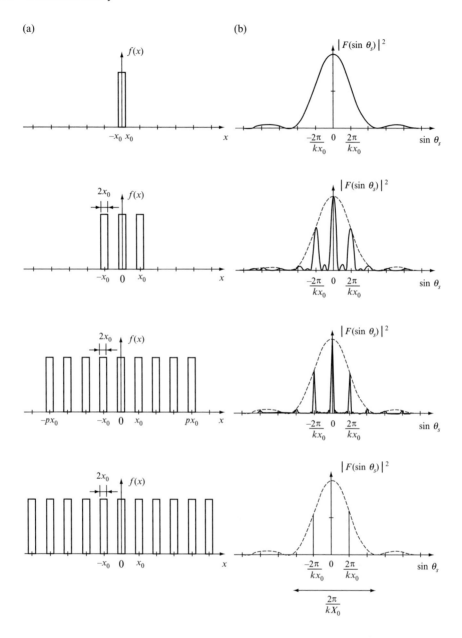

Fig. 7.16 *The significance of the diffraction pattern (2).* The amplitude functions for one, three, N and ∞ wide slits are depicted in (a), and the appropriate diffraction patterns in (b). Introducing a motif has the effect of altering the intensities of the main peaks, but the positions of these peaks remain the same, for this is determined by the lattice. Also, the shape of each main peak remains the same, for this is a function of the overall shape of the entire obstacle.

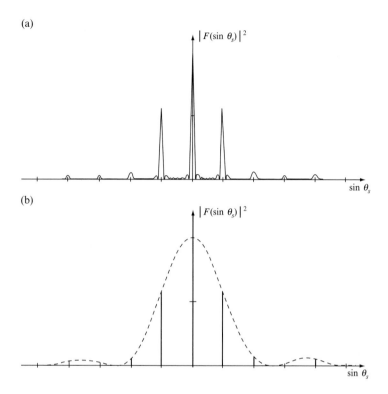

(a)

$|F(\sin\theta_s)|^2$

$\sin\theta_s$

(b)

$|F(\sin\theta_s)|^2$

$\sin\theta_s$

Fig. 7.17 *The derivation of the motif.* Suppose we see a diffraction pattern such as that shown in (a). We may measure the intensity of each peak, and plot the relative intensities on a graph such as (b). We know that the height of each peak is determined by an envelope function, which we may guess as having the form shown. The dotted curve corresponds to the diffraction pattern of the motif, and once this is known, we may determine the motif itself.

molecules is determined, but, of course, guesswork is not used as it was above, and the actual techniques will be met in due course.

We may now state three general rules regarding diffraction patterns:

The positions of the main peaks give information on the lattice.
The shape of each main peak gives information on the overall obstacle shape.
The set of intensities of all the main peaks gives information on the structure of the motif.

The above three rules are perhaps the most important statements in this entire book. For now no longer is the array of spots on an X-ray diffraction picture incomprehensible. Although we do not yet know how to derive quantitative information from the picture, we do have a qualitative understanding of which aspects of the information concerning a crystal structure are to be found in which locations. This also clarifies several observations made in earlier chapters to the effect that the conceptual separation of a crystal structure into a lattice and a motif implies that the information relating to each is to be found in a different manner in the diffraction pattern. We now see that this is so, for the lattice determines the positions of the main peaks, whereas the motif controls their intensities.

7.14   Another way of looking at N wide slits

In the final section of this chapter, we shall discuss another, highly meaningful way of deriving the diffraction pattern of N wide slits. Previously, we had represented this array as the convolution of one wide slit with an array of N narrow slits.

Let us focus our attention on the representation of the lattice. The lattice consists of N δ functions, and corresponds to a finite lattice. An alternative way of representing a finite lattice is to consider it to be part of an infinite lattice which has been limited in some way so that the infinite number of lattice sites is reduced to N. We can do this mathematically by introducing a new function, which we shall call the *shape function*, which has the following properties:

The shape function is zero everywhere outside an obstacle.
The shape function corresponds to the macroscopic shape of the obstacle within the obstacle.

For our present example, the shape function is a large 'Top Hat' function which has a total width of $(N-1)x_0$, and unit height. The effect of multiplying an infinite lattice by this function is to cause all lattice points outside the shape function to go to zero, whereas those lattice sites within it are unaltered. This process is depicted in Fig. 7.18.

Since a finite lattice may now be represented by an infinite lattice multiplied by the appropriate shape function, the amplitude function of a finite lattice may be represented as

$$f(\text{finite lattice}) = f(\text{infinite lattice}) \cdot f(\text{shape function})$$

But since

$$f(\text{obstacle}) = f(\text{motif}) * f(\text{finite lattice})$$

then

$$f(\text{obstacle}) = f(\text{motif}) * [f(\text{infinite lattice}) \cdot f(\text{shape function})] \tag{7.14}$$

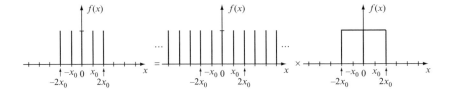

Fig. 7.18 *The shape function.* A 'Top Hat' shape function as on the extreme right has the effect of limiting an infinite lattice to give a finite lattice.

Now, the diffraction pattern function $F(\sin\theta_s)$ is given by the Fourier transform of $f(\text{obstacle})$:

$$F(\sin\theta_s) = Tf(\text{obstacle})$$
$$= T\{f(\text{motif}) * [f(\text{infinite lattice}) \cdot f(\text{shape function})]\}$$

But since the Fourier transform of a convolution is the product of the individual Fourier transforms,

$$F(\sin\theta_s) = Tf(\text{motif}) \cdot T[f(\text{infinite lattice}) \cdot f(\text{shape function})]$$

At this point we apply the convolution theorem once more, but this time in the form that the Fourier transform of a product is the convolution of the individual Fourier transforms. This allows us to write

$$T[f(\text{infinite lattice}) \cdot f(\text{shape function})] = Tf(\text{infinite lattice}) * Tf(\text{shape function})$$

Hence the diffraction pattern function $F(\sin\theta_s)$ becomes

$$F(\sin\theta_s) = Tf(\text{motif}) \cdot [Tf(\text{infinite lattice}) * Tf(\text{shape function})] \qquad (7.15)$$

Equation (7.15) expresses the fact that the diffraction pattern function due to a crystal-like obstacle is made up of three parts. Firstly we have the diffraction pattern due to an infinite lattice. This is then convoluted with the diffraction pattern of the shape function, and finally we multiply the result of this convolution by the diffraction pattern of the motif.

All of this seems very complicated, so let us look at the particular example of the diffraction pattern of N wide slits. Firstly, we break down the structure of the obstacle into the convolution of a motif with a finite lattice:

$$f(N \text{ wide slits}) = f(\text{one wide slit}) * f(N \text{ narrow slits})$$

We then introduce the shape function as

$$f(N \text{ wide slits}) = f(\text{one wide slit}) * [f(\infty \text{ narrow slits}) \cdot f(\text{shape function})]$$

This stage is depicted in Fig. 7.19. We now take the Fourier transform and, using the convolution theorem twice, we obtain

$$Tf(N \text{ wide slits}) = T\{f(\text{one wide slit}) * [f(\infty \text{ narrow slits}) \cdot f(\text{shape function})]\}$$
$$= Tf(\text{one wide slit}) \cdot T[f(\infty \text{ narrow slits}) \cdot f(\text{shape function})]$$
$$= Tf(\text{one wide slit}) \cdot [Tf(\infty \text{ narrow slits}) * Tf(\text{shape function})]$$

The intensity is the square of this last equation.

One wide slit is represented by a 'Top Hat' of width $2X_0$, for which the diffraction pattern intensity is a $(\sin^2 x)/x^2$ function of characteristic angular width $2\pi/kX_0$. The intensity of the diffraction pattern of an infinite lattice of narrow slits separated by x_0 is an infinite array of δ functions spaced by $2\pi/kx_0$. And, in this particular case, the shape function is a wide 'Top Hat', of total width $(N-1)x_0$, and this too will have a diffraction pattern in the form of a $(\sin^2 x)/x^2$ function whose characteristic width is $4\pi/(N-1)kx_0$, which, since N is large, is approximately $4\pi/Nkx_0$.

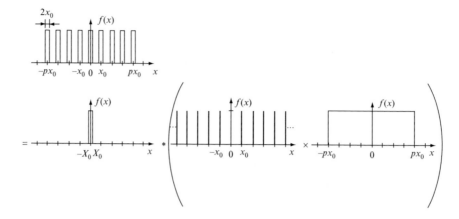

Fig. 7.19 *N wide slits may be represented by the convolution of one wide slit with the product of an infinite lattice and the appropriate shape function.*

We may follow how the diffraction pattern is put together from these three components by referring to Fig. 7.20.

Figure 7.20(a) shows the diffraction pattern of the infinite array of narrow slits. We then convolute this with the diffraction pattern, Fig. 7.20(b), of the shape function. The width of the diffraction pattern of the shape function is $4\pi/Nkx_0$, whereas the

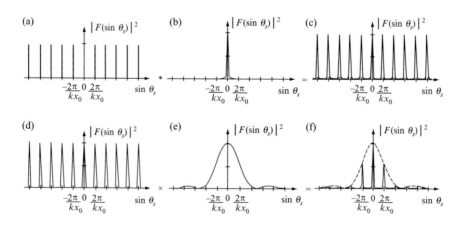

Fig. 7.20 *The diffraction pattern of N wide slits.* As shown in the text, the diffraction pattern of N wide slits may be represented by the square of Tf(one wide slit) $\cdot[Tf(\infty$ narrow slits) $*$ Tf(shape function)]. The diffraction patterns of narrow slits and the shape function are shown in (a) and (b), respectively, and the operation of their convolution gives rise to (c), the diffraction pattern of N narrow slits. On multiplying this, shown in (d), by the diffraction pattern of one wide slit, (e), we derive (f), the diffraction pattern of N wide slits, which agrees with that shown in Fig. 7.12.

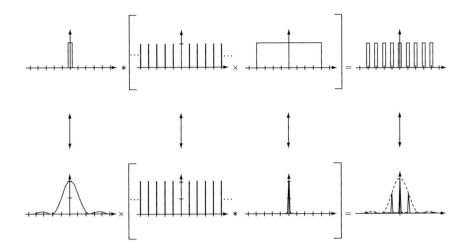

Fig. 7.21 *The diffraction pattern of N wide slits, showing Fourier transform and convolution relationships.*

spacing between the peaks of the diffraction pattern of the infinite lattice is $2\pi/kx_0$. Since N is a large number, the diffraction pattern of the shape function is very much narrower than the interpeak spacing of the lattice diffraction pattern, and so on taking the convolution, the shape function diffraction pattern never overlaps two δ functions at once. Consequently, the form of the convolution is to associate the diffraction pattern of the shape function with each δ function, giving the result of Fig. 7.20(c). This has the effect of 'blurring' the infinitely sharp peaks of the infinite array. Figure 7.20(c), of course, represents the diffraction pattern of N narrow slits, and the reader is advised to check this by comparison with Fig. 7.11.

We now multiply the diffraction pattern of N narrow slits by that of a single wide slit, as shown in Figs 7.20(d) and (e). This modulates the intensities of the peaks, giving the result in Fig. 7.20(f). This is quite identical to the pattern which we obtained in Fig. 7.12. Figure 7.21 summarises these processes.

Note how the smallest structural unit of the obstacle, the motif, has the most extended diffraction pattern, whereas the largest unit, as denoted by the shape function which specifies the macroscopic shape of the obstacle, has the most compact diffraction pattern.

In fact, any crystal has only three types of structural information:

(a) that concerning the lattice;
(b) that concerning the motif;
(c) that concerning the shape of the entire crystal.

Equation (7.14) demonstrates how we may interpret the contributions of these aspects of crystal structure in a mathematical manner. Furthermore, equation (7.15) shows how they individually affect the diffraction pattern. Firstly, an infinite lattice will give rise to an infinite array of δ functions in the diffraction pattern. Secondly, the intensities of the peaks are determined by the motif. Thirdly, the peaks will, in principle, be somewhat smeared out as a result of the fact that the crystal is finite. However, in X-ray diffraction, the vast size of the crystal compared with the wavelength means that this last effect is much less important than the first two. In practice, smearing of the diffraction spots is a very common effect that is caused predominantly by disorder in the crystal, as we shall see later in this book.

Summary

The amplitude functions $f(x)$ and corresponding diffraction pattern intensities $|F(\sin\theta_s)|^2$ have been found for one-dimensional arrays of one, two, three, N and an infinite number of narrow and wide slits. These results are shown in Table 7.1.

Investigation of the nature of the amplitude function of a crystal-like obstacle shows that this function may be represented as the convolution of the amplitude function of a motif with that of a finite lattice:

$$f(\text{obstacle}) = f(\text{motif}) \; * \; f(\text{finite lattice})$$

By introducing the concept of the shape function, we may express the amplitude function of a finite lattice as the product of that for an infinite lattice with that for the shape function:

$$f(\text{finite lattice}) = f(\text{infinite lattice}) \cdot f(\text{shape function})$$

We now have

$$f(\text{obstacle}) = f(\text{motif}) \; * \; [f(\text{infinite lattice}) \cdot f(\text{shape function})] \qquad (7.14)$$

Equation (7.14) specifies how information on the structures of a crystal lattice, the motif and the crystal shape is contained in the crystal's overall amplitude function.

The diffraction pattern function is the Fourier transform of the amplitude function of the obstacle:

$$\text{diffraction pattern function} = Tf(\text{obstacle})$$

$$= T\{f(\text{motif}) \; * \; [f(\text{infinite lattice}) \cdot f(\text{shape function})]\}$$

Using the convolution theorem twice, this implies

$$\text{diffraction pattern function} = Tf(\text{motif}) \cdot [Tf(\text{infinite lattice}) \; * \; Tf(\text{shape function})] \qquad (7.15)$$

The diffraction pattern intensity may be derived from the diffraction pattern function, equation (7.15), by squaring if the function is real. In general, the diffraction pattern function is complex, and to find the intensity, we must multiply by the complex conjugate.

Equation (7.15) shows how the information concerning the various aspects of crystal structure manifests itself in the diffraction pattern.

Study of the diffraction patterns of the one-dimensional obstacles described in this chapter illustrates the meaning of equation (7.15) in a most lucid manner. From these

Table 7.1 Diffraction patterns of one-dimensional obstacles: arrays of (a) narrow slits and (b) wide slits.

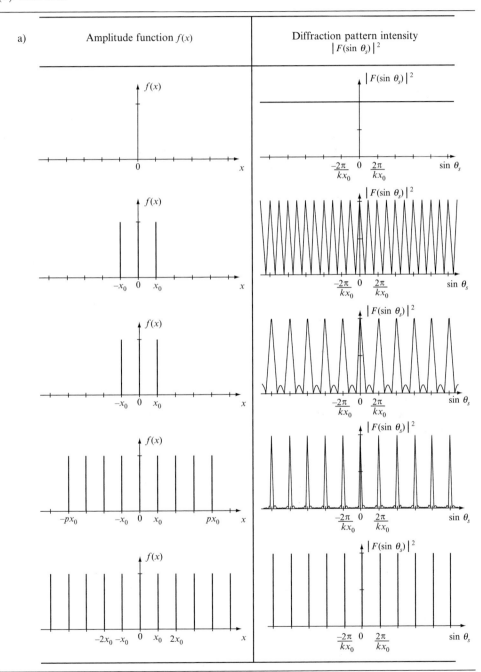

(*continued*)

Table 7.1 Continued

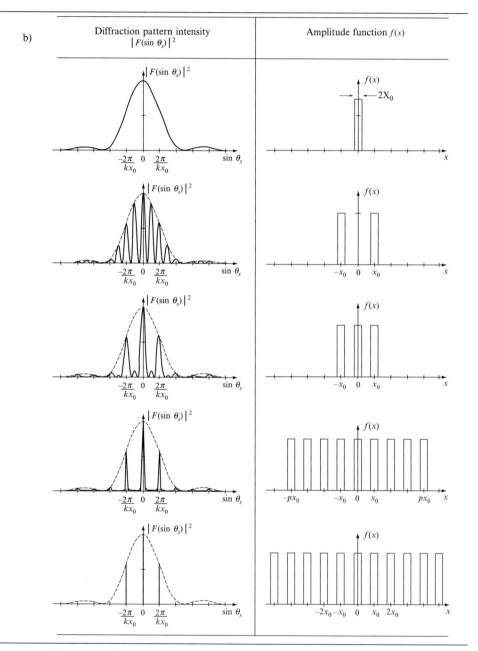

Note. The patterns for two narrow slits and for two wide slits do not correspond numerically to the case $N = 2$ as a particular instance of N slits. The reason for this is that N was assumed to be an odd integer, whereas 2 is even. This does not affect the physical result, but the counting scheme is different for odd and even N.

patterns, we may infer the following rules for the way in which information about the structure of a crystal-like obstacle is presented in a diffraction pattern:

The positions of the main peaks give information on the lattice.
The shape of each main peak gives information on the overall shape of the object.
The set of intensities of all the main peaks gives information on the structure of the motif.

Furthermore:

The diffraction pattern of a real lattice is a second lattice, the reciprocal lattice.
The real lattice and the reciprocal lattice are Fourier mates.

Bibliography

Braddick, H. J. J. *Vibrations, Waves and Diffraction.* McGraw-Hill, London, 1965. Discusses optical diffraction using Fourier techniques.

Francon, M. *Diffraction-Coherence in Optics.* Pergamon, New York, 1966. A lucid discussion of optical diffraction.

Longhurst, R. S. *Geometrical and Physical Optics*, 2nd edn. Longman, London, 1986. A standard text on optics.

8

Diffraction by a three-dimensional lattice

8.1   The diffraction pattern of a crystal

In this chapter and the next, we shall derive the mathematical form of the diffraction pattern of a crystal. Before we start the mathematics, however, it is useful to predict the form of the result that we expect to obtain, so that we can keep this in the back of our minds. According to equation (7.14), we may write the amplitude function of a crystal in the form

$$f(\text{crystal}) = f(\text{motif}) * [f(\text{infinite lattice}) \cdot f(\text{shape function})] \qquad (7.14)$$

and on taking the Fourier transform we obtain

$$Tf(\text{crystal}) = Tf(\text{motif}) \cdot [Tf(\text{infinite lattice}) * Tf(\text{shape function})] \qquad (7.15)$$

The diffraction pattern intensity is therefore given by

$$\text{diffraction pattern intensity} = |Tf(\text{crystal})|^2$$

$$= |Tf(\text{motif})|^2 \cdot |[Tf(\text{infinite lattice}) * Tf(\text{shape function})]|^2$$

In the previous chapter, we found that the diffraction pattern of an infinite, one-dimensional array of δ functions in real space was another infinite one-dimensional array of δ functions in Fourier, or reciprocal, space. Since an infinite crystal lattice may be represented by an infinite three-dimensional array of δ functions, we may, by analogy with the one-dimensional case, suggest that the diffraction pattern of an infinite crystal lattice is a three-dimensional array of δ functions in reciprocal space. The actual crystal lattice is called the *real lattice*, and that array which is the diffraction pattern of the infinite real lattice is termed the *reciprocal lattice*, where we allude to the reciprocal relationship between any function and its Fourier transform.

The intensities of the main peaks of the diffraction pattern are determined by the envelope effect of the Fourier transform of the motif. The amplitude function of the motif contains information concerning the detailed structure of the unit cell and so will be a mathematical description of the structures of the molecules within the unit cell. This function can be enormously complicated, especially if we are considering a crystal of a biological macromolecule such as a protein. The Fourier transform of the motif will be correspondingly complicated, and so the pattern of lighter and darker spots we see on an X-ray photograph is likely to be highly intricate. This has the effect of making the mathematics and arithmetic of unravelling the information we require rather laborious, but conceptually all we have to do is take the inverse transform of the diffraction data to derive information on the amplitude function of the motif, and hence determine the motif's molecular structure.

In this chapter, we will be concerned only with the diffraction patterns of a finite and an infinite lattice. The envelope effect of the diffraction pattern of the motif will be considered in the next chapter.

One of the problems encountered in the discussion of the diffraction pattern of a three-dimensional lattice is that of visualisation. The real lattice is awkward to represent on a two-dimensional page, and since the reciprocal lattice is also three-dimensional, the difficulty is encountered once again. With a representation of the reciprocal lattice, however, the problem is aggravated by the fact that all photographic techniques used in X-ray diffraction necessarily record only a two-dimensional image of the projection of a three-dimensional object. When one wishes to relate the reciprocal lattice to the direction of waves travelling in space, and then deduce how they would affect a two-dimensional film, much imagination is required. The reader is alerted to this difficulty, and is advised to study the diagrams carefully.

Our strategy is as follows. In the previous chapter, we found that the diffraction patterns of one-dimensional objects could be described in terms of the scattering angle θ_s as a function of $\sin \theta_s$. In three dimensions, we must specify three angles, but this leads to an unwieldy representation. In order to specify three-dimensional spatial relationships, we shall describe the diffraction pattern in terms of the wave vectors of the scattered waves. Diffraction patterns will be expressed as a function of the wave vector $\mathbf{k}$, which, since the magnitude k is constant, is equivalent to defining the diffraction pattern in terms of three-dimensional spatial coordinates.

8.2   Non-normally incident waves

Since we are dealing with three-dimensional objects, we return to the Fourier transform equation discussed in Chapter 6, namely equation (6.8):

$$F(\mathbf{k}) = \int_{\text{all } \mathbf{r}} f(\mathbf{r})e^{i\mathbf{k}\cdot\mathbf{r}}\, d\mathbf{r} \qquad (6.8)$$

This equation was derived on the assumption that the incident waves were everywhere normal to the obstacle. We now generalise equation (6.8) to deal with the case where the incident waves impinge on the obstacle from any angle.

To do this, we recall that in Section 6.9 we discussed the relevance of equation (6.8) with reference to the phase difference between a wave scattered from an arbitrary origin and another scattered from an arbitrary point within the obstacle. We showed that the scalar product $\mathbf{k} \cdot \mathbf{r}$ represented this phase difference, but this expression was derived assuming normally incident waves. The significance of the direction of the incident waves is that when the waves are normal to an obstacle, they are all in phase when they reach the obstacle. If the incident waves approach the obstacle at some arbitrary angle α, then there is a phase difference between waves hitting the obstacle, as well

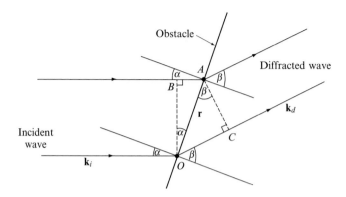

Fig. 8.1 *Non-normally incident waves.* The path difference d between the wave through the origin O and that through A is given by $d = OC - BA = r(\sin\beta - \sin\alpha)$.

∴ Phase difference $= \frac{2\pi}{\lambda} r (\sin\beta - \sin\alpha) = (\mathbf{k}_d - \mathbf{k}_i) \cdot \mathbf{r}$

as that between the diffracted waves. We now modify equation (6.8) to take account of any phase differences between the incident waves.

The geometrical arrangement is depicted in Fig. 8.1. Waves are incident on the obstacle at an arbitrary angle α and are diffracted at an angle β, both angles being relative to a normal to the obstacle. Consider two points within the obstacle, an origin O and an arbitrary point A. We wish to find the total phase difference between the diffracted waves from these two points, and so we must take account of the following two contributions:

(a) Since the incident waves are not normal to the obstacle, they arrive at O and A with a difference in phase.
(b) There is also a phase difference between the diffracted waves.

The phase difference $\Delta\phi$ is related to the geometrical path difference d by the equation

$$\Delta\phi = \frac{2\pi d}{\lambda}$$

which is derived from the fact that a geometrical path difference of one wavelength corresponds to a phase difference of 2π. The geometrical path difference is simply the actual difference in distance between the paths traversed by each wave. This is calculated by dropping perpendiculars AC and OB as shown. Since OB is parallel to the incident wavefront, the phases at O and B are the same, as are the phases at A and C for the diffracted wave. The wave through the origin traces out the path OC and that through the point A traces out BA, giving the path difference d as

$$d = OC - BA$$

Representing the distance OA by r, we have

$$d = r\sin\,\beta - r\sin\,\alpha = r(\sin\,\beta - \sin\,\alpha)$$

Phase difference $\Delta\phi = \frac{2\pi}{\lambda} r \left(\sin\beta - \sin\alpha \right)$

We now introduce a vector notation. Let the incident wave be represented by a wave vector $\mathbf{k}_i$, and the diffracted wave by a vector $\mathbf{k}_d$. The point A now corresponds to the vector distance $\mathbf{r}$, as measured from the origin O. Since the wavelength of a wave motion is unchanged on diffraction, we have

$$k_i = k_d = \frac{2\pi}{\lambda}$$

But, with the angles α and β defined as in Fig. 8.1,

$$\mathbf{k}_i \cdot \mathbf{r} = \frac{2\pi}{\lambda} r \sin\alpha$$

and

$$\mathbf{k}_d \cdot \mathbf{r} = \frac{2\pi}{\lambda} r \sin\beta$$

$$\therefore \text{Phase difference } \Delta\phi = \mathbf{k}_d \cdot \mathbf{r} - \mathbf{k}_i \cdot \mathbf{r} = (\mathbf{k}_d - \mathbf{k}_i) \cdot \mathbf{r}$$

Defining a new vector $\Delta\mathbf{k}$ as

$$\Delta\mathbf{k} = \mathbf{k}_d - \mathbf{k}_i \qquad (8.1)$$

then

$$\text{phase difference } \Delta\phi = \Delta\mathbf{k} \cdot \mathbf{r}$$

The correct expression for the Fourier transform now becomes

$$F(\Delta\mathbf{k}) = \int\limits_{\text{all } \mathbf{r}} f(\mathbf{r}) e^{i\Delta\mathbf{k}\cdot} \, d\mathbf{r} \qquad (8.2)$$

in which we express the Fourier transform as a function of the newly defined vector $\Delta\mathbf{k}$. Note that in the case of normal incidence, the angle α is $0°$, implying that $\sin\alpha$ is zero. $\Delta\mathbf{k} \cdot \mathbf{r}$ then reduces to $\mathbf{k}_d\cdot\mathbf{r}$, which is simply $\mathbf{k} \cdot \mathbf{r}$ where the fact that we are using the diffracted wave vector is tacitly understood. This will then cause equation (8.2) to assume the form of equation (6.8). The relationship between the incident wave vector $\mathbf{k}_i$, the diffracted wave vector $\mathbf{k}_d$ and the new vector $\Delta\mathbf{k}$ is shown in Fig. 8.2.

Both $\mathbf{k}_i$ and $\mathbf{k}_d$ have the same magnitude, and $\Delta\mathbf{k}$ represents a vector describing the change in direction which has occurred as a result of the diffraction. $\Delta\mathbf{k}$ is therefore known as the *scattering vector*, and the angle between $\mathbf{k}_i$ and $\mathbf{k}_d$ is called the *scattering angle*. In the context of diffraction by a three-dimensional lattice, it is convenient to introduce an angular variable θ which is related to the scattering angle θ_s (as defined in Chapter 7) by $\theta_s = 2\theta$. Consequently, the scattering angle is denoted by 2θ as shown in Fig. 8.2. The reasons for this will become clear later in this chapter. Three-dimensional diffraction patterns are usually described at first in terms of the scattering vector $\Delta\mathbf{k}$. Since in any real case the direction of the incident wave is known, this is tantamount to expressing the diffraction pattern as a function of the diffracted wave vector, and hence of the direction of the diffracted waves.

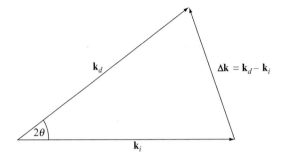

Fig. 8.2 *The scattering vector.* The scattering vector $\mathbf{\Delta k}$ is the vector difference $\mathbf{k}_d - \mathbf{k}_i$. In the context of diffraction by crystals, the scattering angle is denoted by 2θ.

8.3   The diffraction pattern of a finite three-dimensional lattice

We shall now derive the diffraction pattern of a finite three-dimensional lattice. For a lattice defined by the crystallographic unit vectors $\mathbf{a}$, $\mathbf{b}$ and $\mathbf{c}$, any lattice point $\mathbf{r}$ may, as we saw in Chapter 3, be defined by

$$\mathbf{r} = p\mathbf{a} + q\mathbf{b} + r\mathbf{c} \tag{3.3}$$

in which p, q and r are positive or negative integers if the lattice is primitive, or may have certain fractional values for non-primitive lattices.

The amplitude function $f(\mathbf{r})$ for the lattice will be a three-dimensional array of δ functions corresponding to all allowed lattice sites. This is just the lattice function $l(\mathbf{r})$ introduced in Chapter 5,

$$f(\mathbf{r}) = l(\mathbf{r}) = \sum_{\text{all } p,\, q,\, r} \delta\left(\mathbf{r} - [p\mathbf{a} + q\mathbf{b} + r\mathbf{c}]\right) \tag{5.17}$$

in which the summation extends over all the allowed values of p, q and r.

Substituting equation (5.17) in equation (8.2), we have

$$F(\mathbf{\Delta k}) = \int_{\text{all } \mathbf{r}} \sum_{\text{all } p,\, q,\, r} \delta\left(\mathbf{r} - [p\mathbf{a} + q\mathbf{b} + r\mathbf{c}]\right) e^{i\mathbf{\Delta k} \cdot \mathbf{r}} d\mathbf{r} \tag{8.3}$$

This equation is the three-dimensional analogue of equation (5.23). Since the δ functions are non-zero only at points defined by $\mathbf{r} = p\mathbf{a} + q\mathbf{b} + r\mathbf{c}$, the effect of integrating over the array of δ functions is to select out only those points $\mathbf{r}$ at which there is a lattice site, so that equation (8.3) therefore reduces to

$$F(\mathbf{\Delta k}) = \sum_{\text{all } p,\, q,\, r} e^{i\mathbf{\Delta k} \cdot (p\mathbf{a} + q\mathbf{b} + r\mathbf{c})}$$

$$= \sum_{\text{all } p,\, q,\, r} e^{ip\mathbf{\Delta k} \cdot \mathbf{a}} \cdot e^{iq\mathbf{\Delta k} \cdot \mathbf{b}} \cdot e^{ir\mathbf{\Delta k} \cdot \mathbf{c}} \tag{8.4}$$

Equation (8.4) factors out terms dependent on only one of the indices p, q and r at a time. The summation over p, q and r may now be applied to each term separately, giving

$$F(\Delta \mathbf{k}) = \sum_{\text{all } p} e^{ip\Delta \mathbf{k} \cdot \mathbf{a}} \cdot \sum_{\text{all } q} e^{iq\Delta \mathbf{k} \cdot \mathbf{b}} \cdot \sum_{\text{all } r} e^{ir\Delta \mathbf{k} \cdot \mathbf{c}}$$

$F(\Delta \mathbf{k})$ represents the diffraction pattern function, and the intensity is given by $|F(\Delta \mathbf{k})|^2$:

$$|F(\Delta \mathbf{k})|^2 = \left| \sum_{\text{all } p} e^{ip\Delta \mathbf{k} \cdot \mathbf{a}} \right|^2 \cdot \left| \sum_{\text{all } q} e^{iq\Delta \mathbf{k} \cdot \mathbf{b}} \right|^2 \cdot \left| \sum_{\text{all } r} e^{ir\Delta \mathbf{k} \cdot \mathbf{c}} \right|^2 \qquad (8.5)$$

To evaluate this equation, let us consider just one summation, say that over p. If we take our origin of coordinates at a lattice point on the edge of the lattice, then p runs over all positive allowed values from 0 to some number $(P-1)$, giving a total of P lattice points in all:

$$\sum_{p=0}^{p=P-1} e^{ip\Delta \mathbf{k} \cdot \mathbf{a}} = 1 + e^{i\Delta \mathbf{k} \cdot \mathbf{a}} + e^{i2\Delta \mathbf{k} \cdot \mathbf{a}} + \cdots + e^{i(P-1)\Delta \mathbf{k} \cdot \mathbf{a}}$$

This is a geometric progression which sums to

$$\sum_{p=0}^{p=P-1} e^{ip\Delta \mathbf{k} \cdot \mathbf{a}} = \frac{1 - e^{iP\Delta \mathbf{k} \cdot \mathbf{a}}}{1 - e^{i\Delta \mathbf{k} \cdot \mathbf{a}}}$$

We require

$$\left| \sum_{p=0}^{p=P-1} e^{ip\Delta \mathbf{k} \cdot \mathbf{a}} \right|^2 = \left| \frac{1 - e^{iP\Delta \mathbf{k} \cdot \mathbf{a}}}{1 - e^{i\Delta \mathbf{k} \cdot \mathbf{a}}} \right|^2$$

$$= \frac{1 - e^{iP\Delta \mathbf{k} \cdot \mathbf{a}}}{1 - e^{i\Delta \mathbf{k} \cdot \mathbf{a}}} \cdot \frac{1 - e^{-iP\Delta \mathbf{k} \cdot \mathbf{a}}}{1 - e^{-i\Delta \mathbf{k} \cdot \mathbf{a}}}$$

$$= \frac{2 - \left(e^{iP\Delta \mathbf{k} \cdot \mathbf{a}} + e^{-iP\Delta \mathbf{k} \cdot \mathbf{a}} \right)}{2 - \left(e^{i\Delta \mathbf{k} \cdot \mathbf{a}} + e^{-i\Delta \mathbf{k} \cdot \mathbf{a}} \right)}$$

But

$$\cos \theta = \frac{e^{i\theta} + e^{-i\theta}}{2}$$

$$\therefore \left| \sum_{p=0}^{p=P-1} e^{ip\Delta \mathbf{k} \cdot \mathbf{a}} \right|^2 = \frac{2 - 2 \cos P\Delta \mathbf{k} \cdot \mathbf{a}}{2 - 2 \cos \Delta \mathbf{k} \cdot \mathbf{a}}$$

But since $\cos A = 1 - 2\sin^2(A/2)$, we have

$$\left|\sum_{p=0}^{p=P-1} e^{ip\Delta\mathbf{k}\cdot\mathbf{a}}\right|^2 = \frac{\sin^2 \frac{P\Delta\mathbf{k}\cdot\mathbf{a}}{2}}{\sin^2 \frac{\Delta\mathbf{k}\cdot\mathbf{a}}{2}}$$

Similarly,

$$\left|\sum_{q=0}^{q=Q-1} e^{iq\Delta\mathbf{k}\cdot\mathbf{b}}\right|^2 = \frac{\sin^2 \frac{Q\Delta\mathbf{k}\cdot\mathbf{b}}{2}}{\sin^2 \frac{\Delta\mathbf{k}\cdot\mathbf{b}}{2}}$$

and

$$\left|\sum_{r=0}^{r=R-1} e^{ir\Delta\mathbf{k}\cdot\mathbf{c}}\right|^2 = \frac{\sin^2 \frac{R\Delta\mathbf{k}\cdot\mathbf{c}}{2}}{\sin^2 \frac{\Delta\mathbf{k}\cdot\mathbf{c}}{2}}$$

The intensity of the diffraction pattern is therefore given by

$$|F(\Delta\mathbf{k})|^2 = \frac{\sin^2 \frac{P\Delta\mathbf{k}\cdot\mathbf{a}}{2}}{\sin^2 \frac{\Delta\mathbf{k}\cdot\mathbf{a}}{2}} \cdot \frac{\sin^2 \frac{Q\Delta\mathbf{k}\cdot\mathbf{b}}{2}}{\sin^2 \frac{\Delta\mathbf{k}\cdot\mathbf{b}}{2}} \cdot \frac{\sin^2 \frac{R\Delta\mathbf{k}\cdot\mathbf{c}}{2}}{\sin^2 \frac{\Delta\mathbf{k}\cdot\mathbf{c}}{2}} \tag{8.6}$$

8.4   The diffraction pattern of an infinite lattice

Equation (8.6) is the analytical expression for the diffraction pattern of a finite three-dimensional lattice. It is made up of the product of three terms, each one being of the now familiar $(\sin^2 Nx)/(\sin^2 x)$ form. This equation is the three-dimensional analogue of equation (7.10), which describes the diffraction pattern of N narrow slits. Each of the functions in equation (8.6) is a sharply peaked function of the type depicted in Fig. 7.11(b), and once again in Fig. 8.3(a).

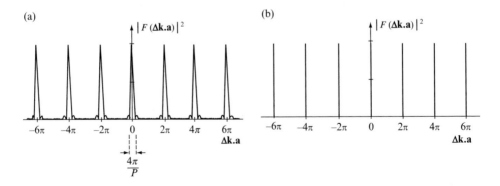

Fig. 8.3 *The function* $\sin^2(P\,\Delta\mathbf{k}\cdot\mathbf{a}/2)/\sin^2(\Delta\mathbf{k}\cdot\mathbf{a}/2)$. The analytical expression for the diffraction pattern of a three-dimensional lattice involves the product of three functions of the type shown in (a). As the lattice becomes infinite, $P \to \infty$, the function becomes an array of δ functions as shown in (b).

In order to describe the form of equation (8.6), we must investigate the behaviour of each term. The first term has maxima whenever

$$\frac{\Delta\mathbf{k}\cdot\mathbf{a}}{2} = 0, \pm\pi, \pm 2\pi, \dots$$

or, in general, when

$$\Delta\mathbf{k}\cdot\mathbf{a} = 2h\pi \tag{8.7}$$

in which h is an integer, which may be positive or negative. The first zeros on each side of the main peaks occur when

$$P\frac{\Delta\mathbf{k}\cdot\mathbf{a}}{2} = \pm\pi$$

and so the peak width may be expressed by

$$\Delta\left(\Delta\mathbf{k}\cdot\mathbf{a}\right) = \frac{4\pi}{P}$$

As P tends to infinity, each peak becomes narrower and narrower until in the limit, the $(\sin^2 Nx)/(\sin^2 x)$ function becomes a series of δ functions spaced at intervals corresponding to

$$\Delta\mathbf{k}\cdot\mathbf{a} = 2h\pi \tag{8.7}$$

as shown in Fig. 8.3(b). This series of functions may be represented as

$$\left[\sum_{\text{all } h} \delta\left(\Delta\mathbf{k}\cdot\mathbf{a} - 2h\pi\right)\right]^2$$

Exactly similar remarks apply to the other two factors in equation (8.6). As Q approaches infinity, the second factor becomes a series of δ functions

$$\left[\sum_{\text{all } k} \delta\left(\Delta\mathbf{k}\cdot\mathbf{b} - 2k\pi\right)\right]^2$$

spaced at intervals determined by

$$\Delta\mathbf{k}\cdot\mathbf{b} = 2k\pi \tag{8.8}$$

in which k is an integer. Note that this k is an index and has no connection with the magnitude of the wave vector.

Likewise, the third term becomes

$$\left[\sum_{\text{all } l} \delta\left(\Delta\mathbf{k}\cdot\mathbf{c} - 2l\pi\right)\right]^2$$

representing another set of δ functions spaced according to

$$\Delta\mathbf{k} \cdot \mathbf{c} = 2l\pi \qquad (8.9)$$

for integral l.

In the limit of an infinite crystal, equation (8.6) becomes

$$|F(\Delta\mathbf{k})|^2 = \left[\sum_{\text{all } h} \delta\left(\Delta\mathbf{k} \cdot \mathbf{a} - 2h\pi\right)\right]^2 \cdot \left[\sum_{\text{all } k} \delta\left(\Delta\mathbf{k} \cdot \mathbf{b} - 2k\pi\right)\right]^2 \cdot \left[\sum_{\text{all } l} \delta\left(\Delta\mathbf{k} \cdot \mathbf{c} - 2l\pi\right)\right]^2$$

$$(8.10)$$

in which h, k and l are positive and negative integers that may be used in all combinations. Equation (8.10) is therefore the diffraction pattern of an infinite three-dimensional lattice.

Much of the remainder of this chapter will be a discussion of equation (8.10) and its consequences. At the end of the chapter we shall investigate the effect of finite crystal size, and see how this modifies the results we shall derive for an infinite lattice.

Our intuitive feeling for diffraction patterns tells us that the diffraction pattern of an infinite three-dimensional lattice, the *real lattice*, is another infinite three-dimensional array, which we call the *reciprocal lattice*. Equation (8.10) must therefore represent this lattice, so let us see exactly how it does so.

The first important point to notice is that equation (8.10) is made up of three terms each of the general type

$$\sum_{\text{all } n} \delta\left(\mathbf{r} \cdot \mathbf{x} - nx_0\right)$$

in which we have a δ function of the scalar product of a vector variable $\mathbf{r}$ and a constant unit vector $\mathbf{x}$ associated with a set of scalars x_0. We know that a δ function of the type $\delta(\mathbf{r} - \mathbf{r}_0)$ refers to the point located at $\mathbf{r} = \mathbf{r}_0$, but we now have to answer the question, 'What does $\delta(\mathbf{r} \cdot \mathbf{x} - x_0)$ represent?'

This function is finite only when $\mathbf{r} \cdot \mathbf{x} = x_0$. If $\mathbf{x}$ is a unit vector, then, as we saw in Section 2.7, this implies that the projection of $\mathbf{r}$ along the direction defined by $\mathbf{x}$ is a constant, equal to x_0. A moment's thought verifies that this condition is fulfilled by any point $\mathbf{r}$ in a plane perpendicular to the direction defined by $\mathbf{x}$, passing through the point $x = x_0$, as shown in Fig. 8.4(a).

Whereas $\delta(\mathbf{r} - \mathbf{r}_0)$ represents a *point*, $\delta(\mathbf{r} \cdot \mathbf{x} - x_0)$ represents a *plane*, implying that

$$\sum_{\text{all } n} \delta\left(\mathbf{r} \cdot \mathbf{x} - nx_0\right)$$

represents a stack of planes, all parallel to the yz plane, and spaced by a distance x_0, as shown in Fig. 8.4(b). Similarly,

$$\sum_{\text{all } m} \delta\left(\mathbf{r} \cdot \mathbf{y} - my_0\right)$$

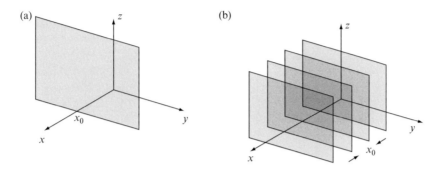

Fig. 8.4 *The function* $\sum_{\text{all } n} \delta\left(\mathbf{r} \cdot \mathbf{x} - nx_0\right)$. *Since* $\mathbf{r} \cdot \mathbf{x}$ *and* x_0 *are scalars, the function* $\delta(\mathbf{r} \cdot \mathbf{x} - x_0)$ *represents a plane parallel to the* yz *plane, cutting the* x *axis at* $x = x_0$ *as in* (a). $\sum_{\text{all } n} \delta(\mathbf{r} \cdot \mathbf{x} - nx_0)$ *corresponds to a stack of parallel planes separated by* x_0 *as in* (b).

represents a stack of planes, this time parallel to the xz plane and spaced by y_0. The product function

$$\sum_{\text{all } n} \delta\left(\mathbf{r} \cdot \mathbf{x} - nx_0\right) \cdot \sum_{\text{all } m} \delta\left(\mathbf{r} \cdot \mathbf{y} - my_0\right)$$

can only be finite only when *both* terms are *simultaneously* non-zero. The first term is non-zero only when $\mathbf{r}$ lies on one of the x_0 series of planes, and the second is non-zero only when $\mathbf{r}$ is on the y_0 set of planes. The product function is therefore non-zero only when $\mathbf{r}$ lies on both sets of planes simultaneously, and so this product function corresponds to a series of lines at the intersections of the two sets of planes as depicted in Fig. 8.5.

We now multiply by the function

$$\sum_{\text{all } s} \delta\left(\mathbf{r} \cdot \mathbf{z} - sz_0\right)$$

which must represent a third stack of planes, parallel in this case to the xy plane, and spaced by z_0. The triple-product function is represented by

$$\sum_{\text{all } n} \delta\left(\mathbf{r} \cdot \mathbf{x} - nx_0\right) \cdot \sum_{\text{all } m} \delta\left(\mathbf{r} \cdot \mathbf{y} - my_0\right) \cdot \sum_{\text{all } s} \delta\left(\mathbf{r} \cdot \mathbf{z} - sz_0\right) \qquad (8.11)$$

and is non-zero only when all three terms are simultaneously non-zero. This will occur only for values of $\mathbf{r}$ which simultaneously lie on all three sets of planes. In general, three sets of planes will intersect at an infinite number of points, which will define a space lattice. This is depicted in Fig. 8.6.

We now see that a function of the general type (8.11) does indeed represent an infinite three-dimensional lattice. If we compare this with the function defined in equation (8.10), we see that the two functions (8.10) and (8.11) are of the same type, but equation (8.10) is a little more complicated, in that we are working in a space defined by the vector $\boldsymbol{\Delta}\mathbf{k}$, which is rather harder to visualise than the real space

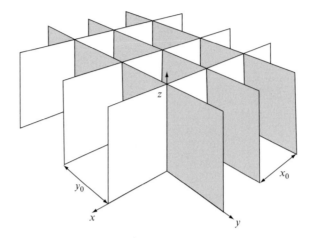

Fig. 8.5 *The function* $\sum\limits_{\text{all } n} \delta(\mathbf{r} \cdot \mathbf{x} - nx_0) \cdot \sum\limits_{\text{all } m} \delta(\mathbf{r} \cdot \mathbf{y} - my_0)$. *Each array of δ functions corresponds to a set of planes, and the product function is finite only at the intersections of the two sets of planes, implying that the product function represents a set of vertical lines as shown.*

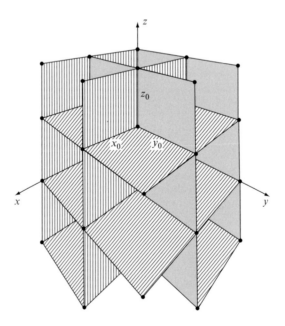

Fig. 8.6 *The function* $\sum\limits_{\text{all } n} \delta(\mathbf{r} \cdot \mathbf{x} - nx_0) \cdot \sum\limits_{\text{all } m} \delta(\mathbf{r} \cdot \mathbf{y} - my_0) \cdot \sum\limits_{\text{all } s} \delta(\mathbf{r} \cdot \mathbf{z} - sz_0)$. *The product of the three arrays of planes is finite only at the simultaneous intersection of the three sets, generating a three-dimensional lattice of points.*

defined by the vector **r**. Nonetheless,

$$\sum_{\text{all } h} \delta\left(\Delta\mathbf{k} \cdot \mathbf{a} - 2h\pi\right)$$

clearly represents a set of planes in $\Delta\mathbf{k}$ space, perpendicular to the direction defined by **a**, and separated by a distance $2\pi/a$, where we divide by the magnitude a of **a** to allow for the fact that **a** is not necessarily a vector of unit magnitude. Equation (8.10) therefore specifies the intersection of three sets of planes, and so defines a space lattice with respect to the vector $\Delta\mathbf{k}$. In Chapter 5, Section 5.6, we found that any function defined with respect to wave vector variables may be thought of as existing in a space we called Fourier space, or reciprocal space. Hence the lattice defined by equation (8.10) is called the *reciprocal lattice*, just as the lattice defined in real space is termed the real lattice.

8.5   The Laue equations

In the previous section, we showed by a geometrical argument that the function defined by equation (8.10) represented a three-dimensional lattice. We shall now justify this analytically.

Our starting point is to write down the conditions such that each term in equation (8.10) is finite. The first is non-zero when

$$\Delta\mathbf{k} \cdot \mathbf{a} = 2h\pi \tag{8.7}$$

where h is an integer. Likewise, the second and third terms are non-zero when

$$\Delta\mathbf{k} \cdot \mathbf{b} = 2k\pi \tag{8.8}$$

and

$$\Delta\mathbf{k} \cdot \mathbf{c} = 2l\pi \tag{8.9}$$

for integral k and l. The three equations

$$\Delta\mathbf{k} \cdot \mathbf{a} = 2h\pi \tag{8.7}$$

$$\Delta\mathbf{k} \cdot \mathbf{b} = 2k\pi \tag{8.8}$$

$$\Delta\mathbf{k} \cdot \mathbf{c} = 2l\pi \tag{8.9}$$

are known as the *Laue equations*, named after Max von Laue (1879–1960), winner of the 1914 Nobel Prize in Physics for the discovery of the diffraction of X-rays by crystals.

What we seek is that value, or those values, of the scattering vector $\Delta\mathbf{k}$ which satisfy all three equations simultaneously. This is the analytical equivalent of finding the values of $\Delta\mathbf{k}$ corresponding to the intersections of three sets of planes. Since **a**, **b** and **c** are constants defined by the real lattice, the only variable is $\Delta\mathbf{k}$, and so the Laue equations represent three simultaneous vector equations which we wish to solve for $\mathbf{\Delta k}$. Once we have done this, we will know which values of $\mathbf{\Delta k}$ correspond to diffraction maxima. Since

$$\mathbf{\Delta k} = \mathbf{k}_d - \mathbf{k}_i$$

then, if we know the wave vector $\mathbf{k}_i$ of the incident beam, we shall have determined the allowed values of the wave vector $\mathbf{k}_d$ for the diffracted waves. This in turn will tell us in which directions relative to the crystal we will find diffraction maxima.

8.6   The solution of the Laue equations

We shall now solve the Laue equations

$$\mathbf{\Delta k} \cdot \mathbf{a} = 2h\pi \tag{8.7}$$

$$\mathbf{\Delta k} \cdot \mathbf{b} = 2k\pi \tag{8.8}$$

$$\mathbf{\Delta k} \cdot \mathbf{c} = 2l\pi \tag{8.9}$$

for the scattering vector $\mathbf{\Delta k}$ which satisfies all three equations simultaneously.

We know from our discussion in Chapter 5, Section 5.6, that $\mathbf{\Delta k}$ exists in Fourier space. Just as any vector $\mathbf{r}$ in real space may be represented in terms of three non-coplanar base vectors $\mathbf{a}$, $\mathbf{b}$ and $\mathbf{c}$ (such as the unit vectors $\mathbf{i}$, $\mathbf{j}$, and $\mathbf{k}$ which define a Cartesian coordinate system) as

$$\mathbf{r} = \alpha\mathbf{a} + \beta\mathbf{b} + \gamma\mathbf{c}$$

in which α, β and γ are scalars, then any vector in Fourier space may be represented in terms of a set of non-coplanar base vectors, which we shall call $\mathbf{a}^*$, $\mathbf{b}^*$ and $\mathbf{c}^*$. In particular, we shall assume that the vector $\mathbf{\Delta k}$ takes the form

$$\mathbf{\Delta k} = \chi(h\,\mathbf{a}^* + k\,\mathbf{b}^* + l\,\mathbf{c}^*) \tag{8.12}$$

in which χ, h, k and l are scalars to be determined.

At the moment, the vectors $\mathbf{a}^*$, $\mathbf{b}^*$ and $\mathbf{c}^*$ are arbitrary. We shall now use equation (8.12) to substitute for $\mathbf{\Delta k}$ in the Laue equations, so deriving the conditions which $\mathbf{a}^*$, $\mathbf{b}^*$ and $\mathbf{c}^*$ must fulfil to satisfy the Laue equations. As a result, we will also determine the scalars χ, h, k and l.

Substituting equation (8.12) for $\mathbf{\Delta k}$ in equation (8.7), we have

$$\mathbf{\Delta k} \cdot \mathbf{a} = \chi(h\,\mathbf{a}^* + k\,\mathbf{b}^* + l\,\mathbf{c}^*) \cdot \mathbf{a} = 2h\pi$$

$$\therefore \mathbf{\Delta k} \cdot \mathbf{a} = \chi h\,\mathbf{a}^* \cdot \mathbf{a} + \chi k\,\mathbf{b}^* \cdot \mathbf{a} + \chi l\,\mathbf{c}^* \cdot \mathbf{a} = 2h\pi \tag{8.13}$$

Similarly, the second and third Laue equations give

$$\mathbf{\Delta k} \cdot \mathbf{b} = \chi h\,\mathbf{a}^* \cdot \mathbf{b} + \chi k\,\mathbf{b}^* \cdot \mathbf{b} + \chi l\,\mathbf{c}^* \cdot \mathbf{b} = 2k\pi \tag{8.14}$$

$$\mathbf{\Delta k} \cdot \mathbf{c} = \chi h\,\mathbf{a}^* \cdot \mathbf{c} + \chi k\,\mathbf{b}^* \cdot \mathbf{c} + \chi l\,\mathbf{c}^* \cdot \mathbf{c} = 2l\pi \tag{8.15}$$

Since $\mathbf{a}^*$, $\mathbf{b}^*$, $\mathbf{c}^*$, χ, h, k and l are as yet undetermined, we may choose such values for them as we find appropriate for the solution of equations (8.13), (8.14) and (8.15).

The first identification we make is to choose

$$\chi = 2\pi$$

We then have

$$\hbar a^* \cdot a + k b^* \cdot a + \ell c^* \cdot a = h \tag{8.16}$$

$$\hbar a^* \cdot b + k b^* \cdot b + \ell c^* \cdot b = k \tag{8.17}$$

$$\hbar a^* \cdot c + k b^* \cdot c + \ell c^* \cdot c = l \tag{8.18}$$

These equations have a neat solution if we select a^*, b^* and c^* such that

$$
\begin{array}{ccc}
a^* \cdot a = 1 & b^* \cdot a = 0 & c^* \cdot a = 0 \\
a^* \cdot b = 0 & b^* \cdot b = 1 & c^* \cdot b = 0 \\
a^* \cdot c = 0 & b^* \cdot c = 0 & c^* \cdot c = 1
\end{array} \tag{8.19}
$$

On choosing a^*, b^* and c^* such that equations (8.19) are true, then equations (8.16), (8.17) and (8.18) are automatically fulfilled with

$$\hbar = h, k = k, \ell = l$$

and all are integers.

We have now shown that the scattering vector

$$\mathbf{\Delta k} = 2\pi(h\mathbf{b^*} + k\mathbf{b^*} + l\mathbf{c^*})$$

satisfies the Laue equations on condition that the equations (8.19) hold. Concentrating on the term in parentheses, we note that since h, k and l are constrained to be integers only, this term will generate a lattice in $\mathbf{\Delta k}$ space, which we call the *reciprocal lattice*.

We now identify how the base vectors a^*, b^* and c^* of the reciprocal lattice relate to the crystallographic unit cell vectors a, b and c of the real lattice. We may do this by studying equations (8.19), in which the relationships are defined.

Firstly, we notice that the equations are symmetrical in that all scalar products of the type $a^* \cdot b$ are zero, whereas those such as $a^* \cdot a$ are unity. Since $a^* \cdot b$ and also $a^* \cdot c$ are both zero, we may conclude that a^* is perpendicular to both b and c simultaneously. We then infer that a^* must be of the form

$$a^* = \xi(b \wedge c)$$

where ξ is some scalar, for only the cross product $b \wedge c$ is known to be perpendicular to both vectors b and c at once. To evaluate the scalar ξ, we remember that

$$a^* \cdot a = \xi a \cdot b \wedge c = 1$$

$$\therefore \xi = \frac{1}{a \cdot b \wedge c} = \frac{1}{V}$$

where V is the volume of the real-lattice cell.

Hence a^* must be given by

$$a^* = \frac{b \wedge c}{a \cdot b \wedge c} = \frac{1}{V} b \wedge c \tag{8.20}$$

But it is also true that the vector of the form

$$\mathbf{a}^* = \frac{\mathbf{c} \wedge \mathbf{b}}{\mathbf{a} \cdot \mathbf{b} \wedge \mathbf{c}} = \frac{1}{V} \mathbf{c} \wedge \mathbf{b} \tag{8.20a}$$

is an equally valid representation of $\mathbf{a}^*$, the only difference being that between $\mathbf{b} \wedge \mathbf{c}$ and $\mathbf{c} \wedge \mathbf{b}$. This concerns the direction in which $\mathbf{a}^*$ is defined relative to $\mathbf{b}$ and $\mathbf{c}$. We shall choose this direction such that $\mathbf{a}^*$, $\mathbf{b}$ and $\mathbf{c}$ form a right-handed set, implying that $\mathbf{a}^*$ is in the same direction as $\mathbf{b} \wedge \mathbf{c}$ rather than $\mathbf{c} \wedge \mathbf{b}$. We shall therefore adopt equation (8.20).

By exactly similar reasoning, we deduce that $\mathbf{b}^*$ and $\mathbf{c}^*$ are given by

$$\mathbf{b}^* = \frac{\mathbf{c} \wedge \mathbf{a}}{\mathbf{a} \cdot \mathbf{b} \wedge \mathbf{c}} = \frac{1}{V} \mathbf{c} \wedge \mathbf{a} \tag{8.21}$$

$$\mathbf{c}^* = \frac{\mathbf{a} \wedge \mathbf{b}}{\mathbf{a} \cdot \mathbf{b} \wedge \mathbf{c}} = \frac{1}{V} \mathbf{a} \wedge \mathbf{b} \tag{8.22}$$

where the vector products are chosen so that $\mathbf{a}^*$, $\mathbf{b}^*$ and $\mathbf{c}^*$ form a right-handed set, like their real-lattice counterparts $\mathbf{a}$, $\mathbf{b}$ and $\mathbf{c}$.

We now have a complete solution of the Laue equations. Any vector $\Delta\mathbf{k}$ of the form

$$\Delta\mathbf{k} = 2\pi(h\mathbf{a}^* + k\mathbf{b}^* + l\mathbf{c}^*) \tag{8.23}$$

represents an allowed solution of the Laue equations such that $\mathbf{a}^*$, $\mathbf{b}^*$ and $\mathbf{c}^*$ are given by equations (8.20), (8.21) and (8.22). If we write

$$\mathbf{S} = h\mathbf{a}^* + k\mathbf{b}^* + l\mathbf{c}^* \tag{8.24}$$

then we have

$$\Delta\mathbf{k} = 2\pi\mathbf{S} \tag{8.25}$$

Let us compare equation (8.24) with the definition of a generalised lattice site $\mathbf{r}$ in the real lattice as

$$\mathbf{r} = p\mathbf{a} + q\mathbf{b} + r\mathbf{c} \tag{3.3}$$

in which p, q and r are positive or negative integers for a primitive lattice. We see that the definition of the vector $\mathbf{S}$ is such that we generate a primitive lattice based on the new vectors $\mathbf{a}^*$, $\mathbf{b}^*$ and $\mathbf{c}^*$, associated with any positive and negative values of the integers h, k and l. For any specific values of h, k and l, which together define a unique point within the reciprocal lattice, $\mathbf{S}_{hkl}$ represents a vector from the origin to the point in the reciprocal lattice indexed by the integers h, k and l. Since this vector is an allowed solution of the Laue equations, it represents the scattering vector corresponding to a diffraction maximum defined by the scattering vector $\Delta\mathbf{k}_{hkl}$,

$$\Delta\mathbf{k}_{hkl} = 2\pi\mathbf{S}_{hkl} \tag{8.25a}$$

Equations (8.25) and (8.25a) are important, and will be referred to frequently. Equation (8.25) is general, and relates any scattering vector $\Delta\mathbf{k}$ to the corresponding *reciprocal-lattice vector* $\mathbf{S}$; equation (8.25a) defines the relationship between the scattering vector $\Delta\mathbf{k}_{hkl}$ associated with a specific diffraction maximum and that particular reciprocal-lattice vector $\mathbf{S}_{hkl}$ defining a unique point within the reciprocal lattice. There is therefore a one-to-one correspondence between diffraction maxima and points within the reciprocal lattice: each diffraction maximum has its own scattering vector $\boldsymbol{\Delta}\mathbf{k}_{hkl}$, which maps onto a specific reciprocal-lattice vector $\mathbf{S}_{hkl}$ defining the unique point within the reciprocal lattice indexed by the three integers h, k and l. The reciprocal lattice is defined by the base vectors $\mathbf{a}^*$, $\mathbf{b}^*$ and $\mathbf{c}^*$, which themselves are determined by the crystallographic unit vectors $\mathbf{a}$, $\mathbf{b}$ and $\mathbf{c}$ of the real lattice according to equations (8.20), (8.21) and (8.22).

Exactly how the reciprocal lattice corresponds to the actual directions in which waves are diffracted from a crystal, and the manner in which the reciprocal lattice is related to the array of spots on an X-ray diffraction photograph, will be discussed in a later section; but for the moment we will investigate the geometry of the reciprocal lattice, and see in some detail how it is related to the real lattice.

8.7 The reciprocal lattice

The definitions of the base vectors $\mathbf{a}^*$, $\mathbf{b}^*$ and $\mathbf{c}^*$ of the reciprocal lattice in terms of those of the real lattice $\mathbf{a}, \mathbf{b}$ and $\mathbf{c}$ are expressed by equations (8.20), (8.21) and (8.22):

$$\mathbf{a}^* = \frac{\mathbf{b} \wedge \mathbf{c}}{\mathbf{a} \cdot \mathbf{b} \wedge \mathbf{c}} = \frac{1}{V}\mathbf{b} \wedge \mathbf{c} \tag{8.20}$$

$$\mathbf{b}^* = \frac{\mathbf{c} \wedge \mathbf{a}}{\mathbf{a} \cdot \mathbf{b} \wedge \mathbf{c}} = \frac{1}{V}\mathbf{c} \wedge \mathbf{a} \tag{8.21}$$

$$\mathbf{c}^* = \frac{\mathbf{a} \wedge \mathbf{b}}{\mathbf{a} \cdot \mathbf{b} \wedge \mathbf{c}} = \frac{1}{V}\mathbf{a} \wedge \mathbf{b} \tag{8.22}$$

where $V = \mathbf{a} \cdot \mathbf{b} \wedge \mathbf{c}$, the volume of the unit cell in the real lattice.

According to equation (8.24), we defined the reciprocal-lattice vector $\mathbf{S}$ as

$$\mathbf{S} = h\mathbf{a}^* + k\mathbf{b}^* + l\mathbf{c}^* \tag{8.24}$$

in which h, k and l are integers only. The convention for representing reciprocal-lattice points is to write the three integers as a triplet (hkl). As in the case of the indexing of a real lattice, if any of h, k or l is negative, this is often indicated by an overbar, as for example $\bar{k}$. We thus refer to reciprocal-lattice points such as $(1\,0\,0)$, $(2\,3\,1)$ and $(2\,\bar{2}\,3)$, and the corresponding reciprocal-lattice vectors may be written as $\mathbf{S}_{100}, \mathbf{S}_{231}$ and $\mathbf{S}_{2\bar{2}3}$.

Examination of equations (8.20), (8.21) and (8.22) shows that the expression $\mathbf{a} \cdot \mathbf{b} \wedge \mathbf{c}$ is present in the denominator of each equation. This is a scalar quantity, and represents the volume V of the unit cell of the real lattice. From the definition of $\mathbf{a}^*$, it is evident that this is a vector which is perpendicular to the plane defined by $\mathbf{b}$ and $\mathbf{c}$. Likewise, $\mathbf{b}^*$ is perpendicular to the plane containing $\mathbf{c}$ and $\mathbf{a}$, and $\mathbf{c}^*$ is perpendicular to that defined by $\mathbf{a}$ and $\mathbf{b}$. Knowing the directions of $\mathbf{a}, \mathbf{b}$ and $\mathbf{c}$, we may represent the spatial relationship between the unit cells of the real and the reciprocal lattice, and Fig. 8.7 shows this relationship for the general case of a triclinic real lattice.

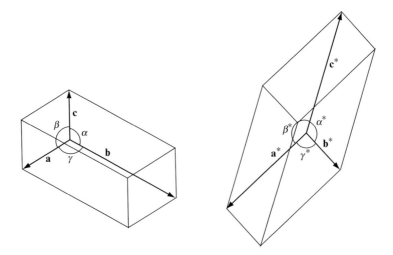

Fig. 8.7 *The unit cells of a real and a reciprocal lattice.* Note that $\mathbf{a}^*$ is perpendicular to the plane of $\mathbf{b}$ and $\mathbf{c}$, and similarly for $\mathbf{b}^*$ and $\mathbf{c}^*$. For clarity, the two figures are not drawn to scale.

The actual calculation of the lengths of the sides of the unit cell of the reciprocal lattice is discussed below; Fig. 8.7 is intended to demonstrate the general relationship between the unit cells of the real and reciprocal lattices, and also to define the crystallographic angles α^*, β^* and γ^* of the reciprocal lattice. Note that whereas the crystallographic angles α, β and γ of the real lattice are conventionally chosen to be obtuse, the corresponding angles α^*, β^* and γ^* of the reciprocal lattice turn out to be acute.

In the general case of a triclinic real lattice, in which the crystallographic axes differ in length and no pair of them are orthogonal, the details of the relationship between the real and reciprocal lattices are complicated, especially if we require algebraic expressions relating a parameter of the real unit cell to those of the unit cell of the reciprocal lattice. But if the real lattice is cubic, tetragonal or orthorhombic, then the three vectors $\mathbf{a}, \mathbf{b}$ and $\mathbf{c}$ are mutually orthogonal, and the geometric and algebraic relationships take on a much simpler form.

As an example, we shall examine the reciprocal lattice of a primitive orthorhombic lattice, which is depicted in Fig. 8.8(a). Since the base vectors are orthogonal, i.e. $\alpha = \beta = \gamma = 90°$, then

$$|\mathbf{a} \wedge \mathbf{b}| = ab$$

$$|\mathbf{b} \wedge \mathbf{c}| = bc$$

$$|\mathbf{c} \wedge \mathbf{a}| = ca$$

and

$$\mathbf{a} \cdot \mathbf{b} \wedge \mathbf{c} = abc$$

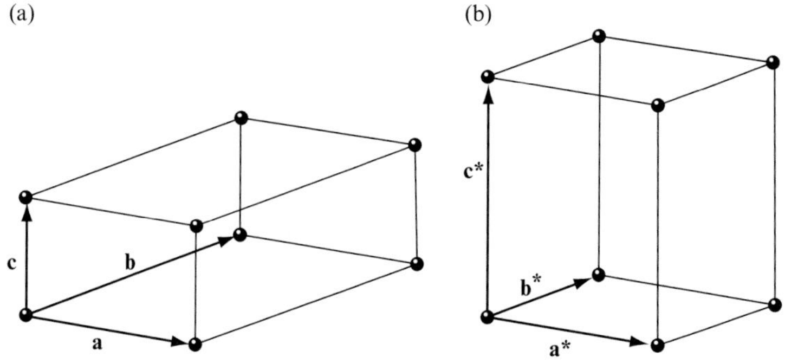

(a)   (b)

Fig. 8.8 *The reciprocal lattice of a primitive orthorhombic real lattice.* The reciprocal lattice is also primitive orthorhombic. For clarity, (b) is drawn on a larger scale than (a).

Hence, equations (8.20)–(8.22) reduce to

$$a^* = \frac{1}{a}$$

$$b^* = \frac{1}{b}$$

$$c^* = \frac{1}{c}$$

From the definitions of $\mathbf{a}^*, \mathbf{b}^*$ and $\mathbf{c}^*$, we see that $\mathbf{a}^*$ is parallel to $\mathbf{a}, \mathbf{b}^*$ to $\mathbf{b}$, and $\mathbf{c}^*$ to $\mathbf{c}$. Furthermore, the relationship between the lengths a, b and c and the lengths a^*, b^* and c^* signifies the relevance of the term 'reciprocal lattice'. The reciprocal lattice was derived as a result of investigating diffraction patterns, and so we expect the familiar reciprocal relationship to arise.

We now know both the magnitudes and the directions of the reciprocal-lattice vectors $\mathbf{a}^*, \mathbf{b}^*$ and $\mathbf{c}^*$, enabling us to plot the reciprocal lattice as in Fig. 8.8(b).

The volume V^* of the unit cell of the reciprocal lattice is given by

$$V^* = \mathbf{a}^* \cdot \mathbf{b}^* \wedge \mathbf{c}^*$$

Since $\mathbf{a}^*, \mathbf{b}^*$ and $\mathbf{c}^*$ are mutually orthogonal, we have

$$V^* = \frac{1}{abc} = \frac{1}{V}$$

This result was derived for the particular case of orthogonal lattice vectors. In fact, the reciprocal relationship between the volumes of the unit cells of the real and reciprocal lattices is true for all crystal classes, and this may be proved in a general analytic manner using the definitions of $\mathbf{a}^*, \mathbf{b}^*$ and $\mathbf{c}^*$. We require

$$V^* = \mathbf{a}^* \cdot \mathbf{b}^* \wedge \mathbf{c}^*$$

$$V^* = \frac{\mathbf{b} \wedge \mathbf{c}}{\mathbf{a} \cdot \mathbf{b} \wedge \mathbf{c}} \cdot \left(\frac{\mathbf{c} \wedge \mathbf{a}}{\mathbf{a} \cdot \mathbf{b} \wedge \mathbf{c}} \wedge \frac{\mathbf{a} \wedge \mathbf{b}}{\mathbf{a} \cdot \mathbf{b} \wedge \mathbf{c}} \right)$$

$$V^* = \frac{1}{V^3} \mathbf{b} \wedge \mathbf{c} \cdot [(\mathbf{c} \wedge \mathbf{a}) \wedge (\mathbf{a} \wedge \mathbf{b})]$$

We need to evaluate the term in square brackets. To do this, we make use of a result from vector analysis which we met in Chapter 2, Section 2.9, namely

$$\mathbf{u} \wedge (\mathbf{v} \wedge \mathbf{w}) = (\mathbf{u} \cdot \mathbf{w}) \mathbf{v} - (\mathbf{u} \cdot \mathbf{v}) \mathbf{w}$$

Identifying

$$\mathbf{u} = \mathbf{c} \wedge \mathbf{a}$$

$$\mathbf{v} = \mathbf{a}$$

$$\mathbf{w} = \mathbf{b}$$

we have

$$(\mathbf{c} \wedge \mathbf{a}) \wedge (\mathbf{a} \wedge \mathbf{b}) = (\mathbf{c} \wedge \mathbf{a} \cdot \mathbf{b}) \mathbf{a} - (\mathbf{c} \wedge \mathbf{a} \cdot \mathbf{a}) \mathbf{b}$$

But

$$\mathbf{c} \wedge \mathbf{a} \cdot \mathbf{b} = \mathbf{a} \cdot \mathbf{b} \wedge \mathbf{c} = V$$

and since $\mathbf{c} \wedge \mathbf{a}$ is by definition orthogonal to $\mathbf{a}$,

$$\mathbf{c} \wedge \mathbf{a} \cdot \mathbf{a} = 0$$

$$\therefore (\mathbf{c} \wedge \mathbf{a}) \wedge (\mathbf{a} \wedge \mathbf{b}) = V \mathbf{a} \qquad (8.26)$$

$$\therefore V^* = \frac{V (\mathbf{a} \cdot \mathbf{b} \wedge \mathbf{c})}{V^3} = \frac{V^2}{V^3}$$

$$\therefore V^* = \frac{1}{V}$$

We shall now prove that for any lattice whatsoever, the reciprocal of the reciprocal lattice is the real lattice, namely $(\mathbf{a}^*)^* = \mathbf{a}, (\mathbf{b}^*)^* = \mathbf{b}$ and $(\mathbf{c}^*)^* = \mathbf{c}$.

We wish to calculate

$$(\mathbf{a}^*)^* = \frac{\mathbf{b}^* \wedge \mathbf{c}^*}{\mathbf{a}^* \cdot \mathbf{b}^* \wedge \mathbf{c}^*} = \frac{1}{V^*} \mathbf{b}^* \wedge \mathbf{c}^* = V \mathbf{b}^* \wedge \mathbf{c}^*$$

Now, from equations (8.21) and (8.22),

$$\mathbf{b}^* \wedge \mathbf{c}^* = \frac{1}{V^2} (\mathbf{c} \wedge \mathbf{a}) \wedge (\mathbf{a} \wedge \mathbf{b})$$

and in equation (8.26) we showed that

$$(\mathbf{c} \wedge \mathbf{a}) \wedge (\mathbf{a} \wedge \mathbf{b}) = V\mathbf{a}$$

$$\mathbf{b}^* \wedge \mathbf{c}^* = \frac{1}{V}\mathbf{a}$$

Therefore, since

$$(\mathbf{a}^*)^* = V\mathbf{b}^* \wedge \mathbf{c}^*$$

we can see that

$$(\mathbf{a}^*)^* = \mathbf{a}$$

and likewise for the other reciprocal-lattice vectors:

$$(\mathbf{b}^*)^* = \mathbf{b}$$

and

$$(\mathbf{c}^*)^* = \mathbf{c}$$

The real lattice and the reciprocal lattice may therefore be derived from each other, and so are equivalent descriptions of the same physical structure.

This situation is entirely analogous to that which we discussed in Chapter 6 concerning the fact that any function and its Fourier transform are equivalent descriptions of the same structure. The above proof is therefore just what we expect, for the reciprocal lattice is merely the Fourier transform of the real lattice.

In general, the direct relationships between the real- and reciprocal-lattice parameters are not as obvious as for the case of the orthorhombic lattice. For completeness, the general formulae for triclinic crystal systems are given below:

$$V = abc\sqrt{1 - \cos^2\alpha - \cos^2\beta - \cos^2\gamma + 2\cos\alpha\cos\beta\cos\gamma}$$

$$a^* = \frac{bc\sin\alpha}{V}$$

$$b^* = \frac{ac\sin\beta}{V}$$

$$c^* = \frac{ab\sin\gamma}{V}$$

$$\cos\alpha^* = \frac{\cos\beta\cos\gamma - \cos\alpha}{\sin\beta\sin\gamma}$$

$$\cos\beta^* = \frac{\cos\alpha\cos\gamma - \cos\beta}{\sin\alpha\sin\gamma}$$

$$\cos\gamma^* = \frac{\cos\alpha\cos\beta - \cos\gamma}{\sin\alpha\sin\beta}$$

Note that for certain crystal systems the formulae can be simplified; for example, for monoclinic cells $V = abc \sin \beta$.

8.8 Reciprocal-lattice vectors and real-lattice planes

There are two theorems concerning the relationship of the reciprocal lattice to the real lattice which are of the utmost importance in describing the way in which the reciprocal lattice is useful in the interpretation and prediction of X-ray diffraction patterns. We shall now derive these theorems, and from them arises a well-known law of crystallography, namely Bragg's law. After that we shall see exactly how we may interpret diffraction patterns with the aid of the reciprocal lattice.

Theorem 1 *The reciprocal-lattice vector $\mathbf{S}_{hkl} = h\mathbf{a}^* + k\mathbf{b}^* + l\mathbf{c}^*$ is perpendicular to the (hkl) set of planes in the real lattice.*

Proof We recall from Chapter 3 that parallel sets of planes in a crystal lattice may be identified by a set of three integers (hkl). Also, the plane of the (hkl) set which is closest to the origin has intercepts on the crystallographic axes of $(1/h, 0, 0)$, $(0, 1/k, 0)$ and $(0, 0, 1/l)$. This plane is shown in Fig. 8.9.

The vector $\overrightarrow{AB}$ lies in the plane, and is given by

$$\overrightarrow{AB} = \left(-\frac{1}{h}, \frac{1}{k}, 0 \right)$$

Consider the reciprocal-lattice vector

$$\mathbf{S}_{\alpha\beta\gamma} = \alpha\mathbf{a}^* + \beta\mathbf{b}^* + \gamma\mathbf{c}^*$$

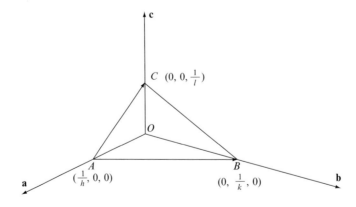

Fig. 8.9 *The plane ABC is the closest to the origin of the (hkl) set of planes. Any vector perpendicular to both AB and AC simultaneously is necessarily perpendicular to the plane ABC.*

This vector is perpendicular to $\overrightarrow{AB}$ if the scalar product $\mathbf{S}_{\alpha\beta\gamma} \cdot \overrightarrow{AB}$ is zero, i.e. if

$$\left(-\frac{1}{h}\mathbf{a} + \frac{1}{k}\mathbf{b}\right) \cdot (\alpha\mathbf{a}^* + \beta\mathbf{b}^* + \gamma\mathbf{c}^*) = 0$$

Since all scalar products of the form $\mathbf{a} \cdot \mathbf{b}^*$ are zero, this implies

$$-\frac{1}{h}\alpha\mathbf{a} \cdot \mathbf{a}^* + \frac{1}{k}\beta\mathbf{b} \cdot \mathbf{b}^* = 0$$

But

$$\mathbf{a} \cdot \mathbf{a}^* = \mathbf{b} \cdot \mathbf{b}^* = 1$$

And so our condition becomes

$$\frac{\alpha}{h} = \frac{\beta}{k} \tag{8.27}$$

Similarly, the vector $\overrightarrow{AC} = (-1/h, 0, 1/l)$ lies in the plane, and this is perpendicular to the vector $\mathbf{S}_{\alpha\beta\gamma}$ defined above if

$$\mathbf{S}_{\alpha\beta\gamma} \cdot \overrightarrow{AC} = \left(-\frac{1}{h}\mathbf{a} + \frac{1}{l}\mathbf{c}\right) \cdot (\alpha\mathbf{a}^* + \beta\mathbf{b}^* + \gamma\mathbf{c}^*) = 0$$

implying that

$$\frac{\alpha}{h} = \frac{\gamma}{l} \tag{8.28}$$

When the vector $\mathbf{S}_{\alpha\beta\gamma}$ is perpendicular to both $\overrightarrow{AB}$ and $\overrightarrow{AC}$ simultaneously, it is then perpendicular to the plane containing both $\overrightarrow{AB}$ and $\overrightarrow{AC}$. Hence the vector $\mathbf{S}_{\alpha\beta\gamma}$ is perpendicular to the (hkl) set of planes if

$$\frac{\alpha}{h} = \frac{\beta}{k} = \frac{\gamma}{l} \tag{8.29}$$

One solution of the equations (8.29) is that

$$\alpha = h, \beta = k, \gamma = l$$

We have now proved that the reciprocal-lattice vector $\mathbf{S}_{hkl}$ given by

$$\mathbf{S}_{hkl} = h\mathbf{a}^* + k\mathbf{b}^* + l\mathbf{c}^*$$

is perpendicular, or normal, to the (hkl) set of planes in the real lattice. Note that to emphasise this, we usually write reciprocal-lattice vectors with the figures h, k and l as subscripts.

Theorem 2 *The magnitude S_{hkl} of the reciprocal-lattice vector $\mathbf{S}_{hkl}$ is related to the spacing d_{hkl} between the (hkl) set of planes in the real lattice by*

$$S_{hkl} = \frac{1}{d_{hkl}}$$

Proof We know that the plane of the (hkl) set which is closest to the origin has intercepts $(1/h, 0, 0)$, $(0, 1/k, 0)$ and $(0, 0, 1/l)$, and for convenience this plane is once again shown in Fig. 8.10.

The plane adjacent to the plane ABC passes through the origin itself, and so the spacing d_{hkl} between the planes in the (hkl) set is equal to the perpendicular distance from the origin O to the plane ABC. As shown in Fig. 8.10, this distance may be calculated by finding the scalar product of the vector $\overrightarrow{OA}$ with a unit vector $\mathbf{u}$ which is normal to the plane ABC:

$$d_{hkl} = \overrightarrow{OA} \cdot \mathbf{u}$$

But we know that the vector

$$\mathbf{S}_{hkl} = h\mathbf{a}^* + k\mathbf{b}^* + l\mathbf{c}^*$$

is perpendicular to the plane ABC, and if the magnitude of this vector is S_{hkl}, then a unit vector normal to the plane is

$$\mathbf{u} = \frac{1}{S_{hkl}} \mathbf{S}_{hkl}$$

$$= \frac{1}{S_{hkl}} (h\mathbf{a}^* + k\mathbf{b}^* + l\mathbf{c}^*)$$

The vector $\overrightarrow{OA}$ can be expressed as

$$\overrightarrow{OA} = \frac{1}{h}\mathbf{a}$$

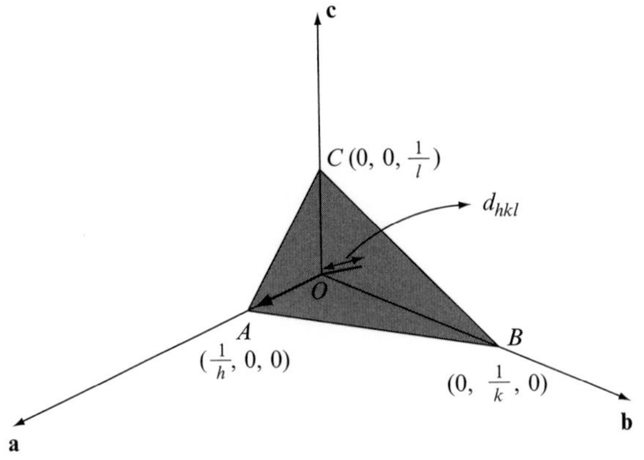

Fig. 8.10 *The spacing of lattice planes. ABC is the closest to the origin of the (hkl) set of planes. An adjacent plane passes through the origin O, and the perpendicular spacing d_{hkl} between the planes is given by $d_{hkl} = \overrightarrow{OA} \cdot \mathbf{u}$, where $\mathbf{u}$ is a unit vector normal to the (hkl) set.*

The distance d_{hkl} can be expanded as follows:

$$d_{hkl} = \overrightarrow{OA} \cdot \mathbf{u}$$

$$= \frac{1}{h}\mathbf{a} \cdot \frac{1}{S_{hkl}}(h\mathbf{a}^* + k\mathbf{b}^* + l\mathbf{c}^*)$$

$$= \frac{1}{hS_{hkl}}(h\mathbf{a} \cdot \mathbf{a}^* + k\mathbf{a} \cdot \mathbf{b}^* + l\mathbf{a} \cdot \mathbf{c}^*)$$

But since

$$\mathbf{a} \cdot \mathbf{a}^* = 1 \quad \mathbf{a} \cdot \mathbf{b}^* = 0 \quad \mathbf{a} \cdot \mathbf{c}^* = 0$$

we have

$$d_{hkl} = \frac{h}{hS_{hkl}}$$

$$\therefore d_{hkl} = \frac{1}{S_{hkl}} \tag{8.30}$$

and our theorem is proved.

8.9 Bragg's law

The theorem proved above, that the magnitude S_{hkl} of the reciprocal-lattice vector $\mathbf{S}_{hkl}$ is the reciprocal of the spacing d_{hkl} between the (hkl) set of planes, has a very important application.

In Fig. 8.11 are shown the incident and diffracted wave vectors $\mathbf{k}_i$ and $\mathbf{k}_d$, and the scattering vector $\boldsymbol{\Delta}\mathbf{k}$. Geometrically, since the magnitudes of the incident and diffracted wave vectors are the same, $\mathbf{k}_i$, $\mathbf{k}_d$ and $\boldsymbol{\Delta}\mathbf{k}$ define an isosceles triangle. Since

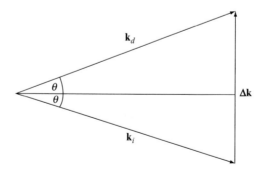

Fig. 8.11 *Bragg's law.* $\mathbf{k}_i$, $\mathbf{k}_d$ and $\boldsymbol{\Delta}\mathbf{k}$ define an isosceles triangle such that
$$\boldsymbol{\Delta}\mathbf{k} = \tfrac{4\pi}{\lambda}\sin\theta$$

where θ is the Bragg angle. Since
$$\boldsymbol{\Delta}\mathbf{k} = 2\pi\mathbf{S}_{hkl} = \tfrac{2\pi}{d_{hkl}}$$

we have
$$\lambda = 2d_{hkl}\sin\theta$$

the angle between $\mathbf{k}_i$ and $\mathbf{k}_d$ is 2θ, we may write that

$$|\Delta\mathbf{k}| = 2k\sin\theta$$

where

$$k = \frac{2\pi}{\lambda}$$

But any scattering vector $\Delta\mathbf{k}$ which corresponds to a diffraction pattern maximum can, from the solution of the Laue equations, be represented in terms of a reciprocal-lattice vector $\mathbf{S}_{hkl}$ as

$$|\Delta\mathbf{k}| = 2\pi S_{hkl}$$

$$\therefore 2\pi S_{hkl} = \frac{2\pi}{\lambda}2\sin\theta$$

$$\therefore S_{hkl} = \frac{2\sin\theta}{\lambda}$$

But equation (8.30) tells us that

$$S_{hkl} = \frac{1}{d_{hkl}}$$

$$\therefore \frac{1}{d_{hkl}} = \frac{2\sin\theta}{\lambda}$$

$$\therefore \lambda = 2d_{hkl}\sin\theta \tag{8.31}$$

Equation (8.31) is known as *Bragg's law*, after Sir Lawrence Bragg (1890–1971), who, jointly with his father Sir William Bragg (1862–1942), won the Nobel Prize in Physics in 1915 for their work on the elucidation of crystal structures using X-ray diffraction. The angle θ, as defined in Fig. 8.11, is called the *Bragg angle* for the (*hkl*) set of planes, which are separated by d_{hkl}. Bragg's law is therefore concerned with the relationship between the positions of diffraction maxima (as determined by the angle θ), the wavelength of the incident radiation λ and the spacing d_{hkl} between the planes of the set (*hkl*). We shall defer a discussion of the relevance of Bragg's law to lattice planes within the crystal until a later section; our first task is to derive and discuss a highly significant geometrical construction.

8.10   The Ewald circle

One of the geometrical interpretations of Bragg's law,

$$\lambda = 2d_{hkl}\sin\theta \tag{8.31}$$

is that originally propounded by the physicist P. P. Ewald (1888–1985), who was a research student at the University of Munich at the time that Laue, Friedrich and Knipping obtained the first X-ray diffraction photographs at that university.

Let us, for simplicity, initially consider a two-dimensional system. The crystal is defined by a two-dimensional real lattice, from which we may derive a two-dimensional reciprocal lattice. The incident wave defined by the wave vector $\mathbf{k}_i$ and the diffracted waves corresponding to various wave vectors $\mathbf{k}_d$ are confined to a plane, implying that the scattering vector $\mathbf{\Delta k}$ also lies in this plane. Bragg's law still holds, except that we should properly refer to the spacing of the lattice planes as d_{hk} since the real lattice is two-dimensional.

The basis of the geometrical construction that we are about to perform concerns the fact that the magnitude of the wavelength of the X-rays is unchanged on diffraction, implying that the magnitudes k_i and k_d are both equal to $2\pi/\lambda$. If we imagine that waves of known wavelength are incident on the two-dimensional crystal from a known direction, then we may draw a vector proportional to the magnitude k_i in the correct direction relative to the crystal. After the diffraction event, diffracted waves described by wave vectors $\mathbf{k}_d$ emerge from the crystal, but in various as yet unknown directions. All we know is the magnitude k_d of the diffracted waves, and so the allowed vectors representing the diffraction event may lie anywhere on a circle of radius proportional to the magnitude k_i.

The stage we have reached so far is depicted in Fig. 8.12. At the centre C of a circle we imagine we have the two-dimensional crystal, with its real lattice in the plane of the paper. The radius r of the circle is proportional to the magnitude of the scattering vector $(2\pi/\lambda)$ and is given by

$$r = \frac{1}{\lambda}$$

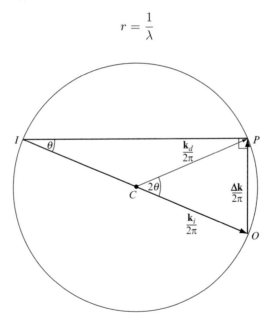

Fig. 8.12 *The Ewald circle.* A crystal at C is in an X-ray beam defined by the vector $\mathbf{k}_i$. The radius r of the circle is $k_i/2\pi = 1/\lambda$. The diffracted beam may be represented by any vector $\mathbf{k}_d$ from the centre C to the circumference of the Ewald circle. The triangle COP is the same triangle as in Fig. 8.11, scaled by $1/2\pi$. ICO is a diameter, defining the right-angled triangle IPO.

Hence

$$r = \frac{\mathbf{k}_i}{2\pi} = \frac{\mathbf{k}_d}{2\pi}$$

For an actual drawing, we choose a convenient scale so that the actual number represented by $1/\lambda$ comes to a useful length. For instance, if the wavelength is 0.1 nm (1Å), then $1/\lambda$ is $10\,\text{nm}^{-1}$, and so a line 100 mm long would be a convenient representation, defining a scale of $10\,\text{mm} \equiv 1\,\text{nm}^{-1}$.

Any vector from the centre of the circle to the circumference represents a possible vector for a scattered wave. The direction of the incident waves is represented by the vector $\overrightarrow{CO}$, and, because the magnitude and direction of the incident wave vector is known, the vector $\overrightarrow{CO}$ may be plotted unambiguously. Since the magnitude of the wavelength is unchanged on diffraction, the representation of the incident wave extends from the centre of the circle to the circumference.

The incident wave direction is shown. Let us suppose that we are interested in waves diffracted in the direction represented by the vector $\overrightarrow{CP}$ relative to the crystal as shown, such that the incident wave vector $\mathbf{k}_i$ makes an angle 2θ with the diffracted wave vector $\mathbf{k}_d$.

The triangle COP represents the diffraction event in both magnitude and direction, implying that the vector $\overrightarrow{OP}$ must represent the scattering vector $\mathbf{\Delta k}$. The triangle COP is in fact a representation of the same physical event as depicted in Fig. 8.11, but we have now enclosed the triangle inside a circle.

We now extend the radius CO back to form a diameter, meeting the circle again at the point I. Note that this construction is carried out for the direction corresponding to the incident waves, for this is the only direction which we know beforehand. It is a fact of the geometry of circles that if the angle $\angle OCP$ is 2θ, then the angle $\angle OIP$ is θ, equal to the Bragg angle as shown. Furthermore, the triangle IPO stands on a diameter of the circle, and so the angle $\angle IPO$ is $90°$.

The radius of the circle is r, and so we may write that the magnitude of the distance OP as

$$OP = 2r \sin \theta$$

But the radius r is just the reciprocal of the wavelength, and so

$$OP = \frac{2}{\lambda} \sin \theta$$

Now the vector $\overrightarrow{OP}$ represents the scattering vector $\mathbf{\Delta k}$ subject to the scaling factor of $1/2\pi$, and so the magnitude of the vector distance OP is related to the magnitude of the scattering vector as

$$OP = \frac{|\mathbf{\Delta k}|}{2\pi}$$

$$\therefore \frac{|\mathbf{\Delta k}|}{2\pi} = \frac{2}{\lambda} \sin \theta = OP$$

We now enquire as to the condition that the vector $\overrightarrow{CP}$ should represent a wave which will give rise to a diffraction maximum. In this case, the scattering vector $\boldsymbol{\Delta k}$ is a vector which satisfies the Laue equations, which will happen whenever $\boldsymbol{\Delta k}$ is related to a reciprocal-lattice vector $\mathbf{S}_{hk}$ as

$$\boldsymbol{\Delta k} = 2\pi \mathbf{S}_{hk} \tag{8.25}$$

$$\therefore \frac{|\boldsymbol{\Delta k}|}{2\pi} = S_{hk}$$

The Bragg angle θ is then related to the magnitude S_{hk} of the reciprocal-lattice vector $\mathbf{S}_{hk}$ if

$$S_{hk} = \frac{2}{\lambda} \sin\theta$$

But equation (8.30) stated that

$$S_{hkl} = \frac{1}{d_{hkl}} \tag{8.30}$$

and so we now have the condition for a diffraction maximum as

$$\frac{1}{d_{hk}} = \frac{2}{\lambda} \sin\theta$$

implying that

$$\lambda = 2d_{hk} \sin\theta$$

This is just the two-dimensional form of Bragg's law!

The geometrical construction depicted in Fig. 8.12 therefore contains Bragg's law inherent in its geometry, subject to the condition that the vector $\overrightarrow{OP}$ corresponds to a reciprocal-lattice vector.

The circle we have drawn is known as the *Ewald circle*, or *reflecting circle*, and it is this construction which holds the key to the interpretation of the reciprocal lattice and its relationship to the physical event of diffraction.

8.11 The reciprocal lattice and diffraction

Let us look at the Ewald circle construction from a slightly different viewpoint. Suppose that we already have some information about the crystal, and that in particular we know the geometry of the reciprocal lattice. Now the reciprocal lattice is an important property of a crystal, and is as valid a concept as the real lattice. Just as any crystal has a real lattice, which is fixed within the macroscopic boundaries of the crystal, then the crystal also possesses a reciprocal lattice, which may also be thought of as being rigidly associated with the crystal, although it is not defined to be within the physical limits of the crystal. When a beam of X-rays is incident on the crystal, it is incident in a well-defined manner onto the real lattice, and since the real lattice and the reciprocal lattice are uniquely defined with respect to one another, then the direction at which the incident waves impinge upon the reciprocal lattice is also well defined. We shall assume that we know the orientation of the incident beam with respect to the real lattice, and hence with respect to the reciprocal lattice.

According to the Ewald circle construction, the vector represented by $\overrightarrow{OP}$, which represents the scattering vector $\mathbf{\Delta k}$ will correspond to a diffraction maximum only if it represents a reciprocal-lattice vector $\mathbf{S}_{hk}$ in both magnitude and direction. Since we know both the orientation of the reciprocal lattice with respect to the incident beam and also the geometry of the reciprocal lattice, then we may draw the reciprocal-lattice points using the point O as the origin.

The scale we choose for plotting the reciprocal-lattice points must be the same as that which was used for the drawing of the Ewald circle. If this is that $10\,\text{mm}$ corresponds to $1\,\text{nm}^{-1}$, then for a square real lattice of side $2\,\text{nm}$, the reciprocal lattice is also square but of side $1/2\,\text{nm}^{-1}$, and so reciprocal-lattice points would be spaced at intervals of $5\,\text{mm}$.

Since the direction of the incident beam corresponds to the vector $\overrightarrow{CO}$, we may now plot the reciprocal-lattice points on the correct scale and in the correct orientation about the point O as origin. With the reciprocal lattice superimposed on the Ewald circle construction, we obtain the diagram of Fig. 8.13.

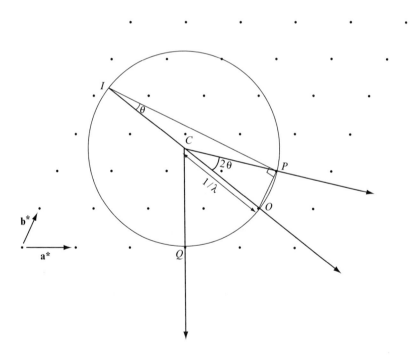

Fig. 8.13 *The Ewald circle and diffraction.* Superposed on the Ewald circle, and to the same scale, is the reciprocal lattice of the crystal. The origin of the reciprocal lattice is O, the point at which the incident beam leaves the Ewald circle. The orientation of the reciprocal lattice is determined by the orientation of the crystal. For a diffraction maximum, a reciprocal-lattice vector from O must lie on the Ewald circle. This condition is satisfied by P and Q, so only two diffraction maxima are seen, one in the direction CP and the other in the direction CQ.

Inspection of Fig. 8.13 shows that in general, only a small number of reciprocal-lattice points will happen to fall on the circumference of the Ewald circle. Such a point is the point P shown, and we may state that the vector $\overrightarrow{OP}$, being both on the Ewald circle and an allowed lattice point, satisfies two conditions simultaneously:

(a) The vector $\overrightarrow{OP}$ is a reciprocal-lattice vector $\mathbf{S}_{hk}$.
(b) Since the vector $\overrightarrow{OP}$ is a chord of the Ewald circle, the Bragg condition is automatically satisfied.

The importance of these two conditions is that we may now identify the vector $\overrightarrow{CP}$ as that corresponding to a diffracted wave which will give rise to a diffraction maximum: diffraction maxima arise *only* for those reciprocal-lattice points that happen to lie on the circumference of the Ewald circle, with diffraction maxima being detected in the directions defined by the vectors from the centre of the Ewald circle to those points.

The great significance of the Ewald construction is therefore that it enables the X-ray crystallographer to determine simultaneously

(a) the direction of the diffracted wave vectors $\mathbf{k}_d$ which correspond to diffraction maxima, and also
(b) the specific reciprocal-lattice vector $\mathbf{S}_{hk}$ to which each diffraction maximum corresponds.

Using these two pieces of information, the crystallographer may then index the diffraction pattern with integers h and k corresponding to the appropriate reciprocal-lattice vectors $\mathbf{S}_{hk}$, and this gives information about the coordinates of the reciprocal lattice and hence the real lattice.

If a photographic film or detector is placed beyond the crystal, then this will record the diffracted waves which satisfy the Bragg condition, and so, for any setting of the crystal, the record on the photograph will be a map of those reciprocal-lattice points which correspond to diffraction maxima. Since there is one reciprocal-lattice vector for each individual diffraction maximum, the photograph is a projection of the reciprocal lattice, with the form of the projection being determined by the geometrical arrangement of the film with respect to the crystal.

As we have seen, for any particular setting of the crystal, only a small proportion of reciprocal-lattice points will lie on the circumference of the Ewald circle, and so only a small number of diffraction maxima will occur. Let us now consider what happens if we rotate the crystal about an axis perpendicular to the plane of Fig. 8.13.

Since the reciprocal lattice may be considered to be rigidly fixed to the crystal, as the crystal rotates about a vertical axis through C by an angle α, the reciprocal lattice will also rotate about a vertical axis through an angle α, but the rotation of the reciprocal lattice is about its own origin O, as opposed to the point C. As the crystal rotates about C, the reciprocal lattice rotates about O, and this has the effect of rotating the reciprocal lattice 'over' the Ewald circle, so as the crystal rotates, different reciprocal-lattice points will come to lie on the Ewald circle and give rise to diffraction maxima, as shown in Fig. 8.14.

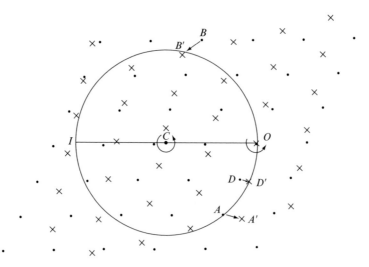

Fig. 8.14 *Rotating the crystal.* If the crystal rotates anticlockwise about an axis through C perpendicular to the plane of the paper, the reciprocal lattice will rotate anticlockwise about a parallel axis passing through the origin O. From an original position indicated by dots, the reciprocal lattice will rotate to a second position indicated by crosses. This brings different reciprocal-lattice points into the diffracting position—we may calculate what rotation is required to bring a particular reciprocal-lattice point into the diffracting position, and in which direction relative to the crystal the intensity maximum may be detected.

By rotating the crystal in the X-ray beam, we may bring the crystal into different orientations such that diffraction peaks corresponding to different reciprocal-lattice vectors may be recorded on a film or electronic detector. A complete rotation of the crystal will allow all those lattice points within a radial distance of $2r$ (where r is the radius of the Ewald circle) of the point O to come into the diffracting position, and so give rise to a spot on the film. Any reciprocal-lattice point outside the radial distance of $2r$ from O can never come into the diffracting position, and so we may define the *limiting circle* as a circle of radius $2r$ about O, outside of which no reciprocal-lattice points may come into the diffracting position.

The limiting circle is another manifestation of the fact we have already met that, although the reciprocal lattice is infinite in extent, not all reciprocal-lattice vectors are detectable: these results are all encapsulated within the Ewald circle's geometry.

The magnitude S_{hk} of a reciprocal-lattice vector $\mathbf{S}_{hk}$ which can give rise to a diffraction effect is therefore limited by

$$S_{hk} \leq 2r$$

If, as is usual, the radius of the Ewald sphere is $1/\lambda$, then this becomes

$$S_{hk} \leq \frac{2}{\lambda} \tag{8.32}$$

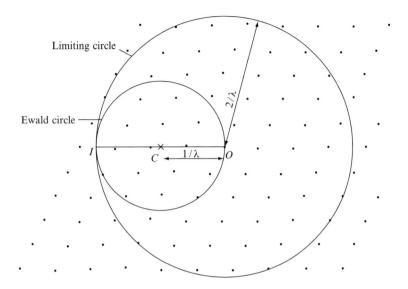

Fig. 8.15 *The limiting circle.* A complete rotation of the crystal about C implies a complete rotation of the reciprocal lattice about O. All points within the limiting circle will intersect the Ewald circle twice, and all points outside the limiting circle can never intersect the Ewald circle. There is thus a limit on the number of available diffraction peaks we may record for a given wavelength of the X-ray beam. If the wavelength is changed, the radius of the Ewald circle will alter and other reciprocal-lattice points may be investigated.

We may prove this relation analytically if we investigate the behaviour of Bragg's law in the form

$$S_{hk} = \frac{2}{\lambda} \sin \theta$$

Since the Bragg angle θ is limited to a maximum of $90°$, $\sin \theta$ must satisfy the condition

$$0 \leq \sin \theta \leq 1$$

We then have

$$0 \leq S_{hk} \leq \frac{2}{\lambda}$$

The relationship of the Ewald circle, the limiting circle and the reciprocal lattice is shown in Fig. 8.15.

8.12 Why X-ray diffraction works

The inequality (8.32),

$$S_{hk} \leq \frac{2}{\lambda} \tag{8.32}$$

is important since it tells us, in effect, why X-ray diffraction works. If we have a crystal lattice such that the side of the unit cell is of the order of 1 nm (10 Å), then the order of magnitude of a reciprocal-lattice vector will be 1 nm^{-1}. Using the above inequality, we see that diffraction effects will be observable if the wavelength of the radiation satisfies the condition that

$$\lambda \leq \frac{2}{S_{hk}}$$

$$\therefore \lambda \leq 2\,\text{nm}$$

Hence the greatest wavelength which will produce a diffraction pattern is of the order of 2 nm (20 Å). X-rays have wavelengths in the 0.1 nm (1 Å) range and so satisfy the inequality, implying that diffraction effects will be observable. Light waves, however, have wavelengths in the 500 nm (5000 Å) range, and so we can never obtain diffraction of light from crystals—in terms of the Ewald circle, the wavelength of light is such that the radius $r = 1/\lambda$ of the Ewald circle is too small to cover any reciprocal-lattice points at all. Conversely, the inequality may be used the other way about, so that if we know that X-rays produce diffraction effects, then we may infer that the typical spacing of lattice sites in crystals is in the nm range. It was the original experiment of Friedrich and Knipping that verified both that X-rays were wave phenomena and that real crystal lattices typically had spacings in the nm range.

8.13   The Ewald sphere

We now extend the discussion from two into three dimensions. We now have to deal with three-dimensional real and reciprocal lattices, and the possibility of diffracted waves emerging from the crystal in all directions. The physical principle that the wavelength of the radiation is unchanged during diffraction is still valid, and so the locus of possible diffracted wave vectors will sweep out a sphere, as opposed to a circle, about the point C which represents the position of the crystal. This is called the *Ewald sphere* or *reflecting sphere*, and is the three-dimensional analogue of the two-dimensional Ewald circle.

   Specifically, the rules for drawing the Ewald sphere construction are listed below. In order to perform this construction, we must know:

(a) The direction of the incident beam relative to the real lattice of the crystal. This will determine the orientation of the reciprocal lattice with respect to the Ewald sphere.
(b) The geometrical properties of the reciprocal lattice.

   The reader may at this stage find it unsatisfactory that we must know the form of the reciprocal lattice before we construct the Ewald sphere. It may perhaps seem that we are putting the cart before the horse, in that we wish to use the Ewald sphere in order to interpret a diffraction pattern so that we may deduce the reciprocal lattice. For the purposes of illustration, however, it is much more useful for us to assume that we know all about the crystal beforehand, for then we may derive the diffraction pattern. Once we see how we may deduce a diffraction pattern if the reciprocal lattice is known, then we shall be able to learn how we may calculate the parameters of an unknown reciprocal lattice from a given diffraction picture.

Rules for the Ewald sphere construction

(1) From a point C, which corresponds to the position of the crystal, draw a sphere of radius r such that

$$r = 1/\lambda$$

(2) Draw a diameter ICO representing the direction of the beam with respect to the crystal.
(3) Choose the point at which the incident beam leaves the Ewald sphere as the origin O of the reciprocal lattice.
(4) With origin O, plot the reciprocal lattice, ensuring that its orientation corresponds to the manner in which the crystal is placed in the incident beam.

The form of the construction is depicted in Fig. 8.16.

When one is carrying out the Ewald sphere construction, there are two important points to bear in mind:

(a) Ensure that the origin of the reciprocal lattice is chosen as that point at which the incident beam *leaves* the Ewald sphere.

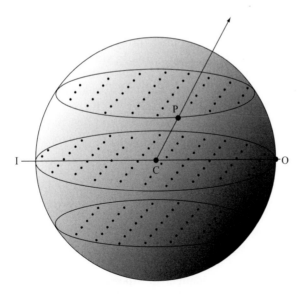

Fig. 8.16 *The Ewald sphere.* In three dimensions, the allowed diffracted wave vectors sweep out a sphere. The crystal is at the centre C, and the origin of the reciprocal lattice is at O, where the incident beam leaves the Ewald sphere. Any lattice point such as P which lies on the surface of the sphere satisfies the diffraction condition, and an intensity maximum may be detected in the direction CP relative to the crystal.

(b) Ensure that the scale used for the Ewald sphere is the same as that used for the reciprocal lattice. Since the sphere has a radius r which represents $1/\lambda$, then the reciprocal-lattice points are plotted directly from knowledge of the magnitudes a^*, b^* and c^* of the base vectors of the reciprocal unit cell. Both $1/\lambda$ and lengths such as a^* have dimensions of nm^{-1} or Å^{-1}, and so a scale should be chosen which is suitable for drawing both the sphere and the reciprocal lattice.

8.14 The Ewald sphere and diffraction

The interpretation of the Ewald sphere in terms of diffraction is of course quite analogous to the previous discussion concerning the two-dimensional Ewald circle. Whenever a reciprocal-lattice point P lies on the surface of the Ewald sphere, the reciprocal-lattice vector $\overrightarrow{OP}$ satisfies Bragg's law. A diffraction maximum will therefore occur in a direction relative to the crystal as determined by the vector $\overrightarrow{CP}$ from the centre of the sphere to the point P, as shown in Fig. 8.17.

In general, for any arbitrary setting of the crystal relative to the incident beam, only a small number of reciprocal-lattice points satisfy the diffraction condition, and so a photographic film or electronic detector will record only a few spots. If we rotate the crystal about an axis through C by an angle α, then the reciprocal lattice will rotate through an equal angle α about a parallel axis through the origin O. This has the effect of rotating the reciprocal lattice 'through' the Ewald sphere so that the rotation causes a different set of reciprocal-lattice points to satisfy the diffraction condition.

Alternatively, we might choose to hold the crystal still, but to alter the direction at which the incident beam strikes the crystal. In this case the reciprocal lattice stays still, but since the origin of the reciprocal lattice is defined as the point at which the incident beam leaves the Ewald sphere, then the Ewald sphere itself rotates about O with respect to the reciprocal lattice. Whether we rotate the crystal or the incident beam, the effect is to cause the Ewald sphere and the reciprocal lattice to rotate relative to each other about O, so that a larger number of reciprocal-lattice points may be brought into the diffracting position.

We now appreciate what is displayed on the photographic image obtained as a result of an X-ray diffraction experiment. Each spot corresponds to a point at which a diffracted beam which satisfies Bragg's law has struck the film. Using the Ewald sphere construction, we see that this beam must have originated at the crystal, and passed 'through' a reciprocal-lattice point positioned on the surface of the Ewald sphere on its way to the film. What is displayed on the film is therefore a projection of the reciprocal lattice, the form of which depends on the geometrical relationship between the film and the crystal, and the manner in which the crystal is moved relative to the incident beam and the film in order to bring other reciprocal-lattice points into the diffraction position. Any X-ray photograph is therefore a two-dimensional projection of a part of the reciprocal lattice of a crystal. This is amply demonstrated by, for example, Fig. 8.18. If we know the geometry of the experimental system, then we know the form of the projection, and so we may deduce information concerning the reciprocal lattice and hence the real lattice.

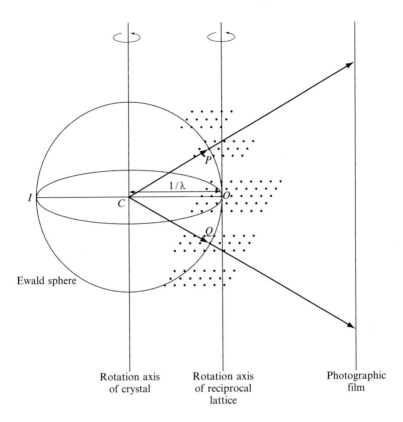

Fig. 8.17 *The Ewald sphere and diffraction.* Any point of the reciprocal lattice which is on the surface of the Ewald sphere gives rise to a diffraction maximum, which may be recorded on a photographic film. If the crystal rotates about C, the reciprocal lattice rotates about a parallel axis through the origin O, thereby bringing more reciprocal-lattice points into the diffracting position. The photographic record is a projection of the reciprocal lattice.

Another feature of an X-ray diffraction picture is that the intensities of the spots are different. This we know to be due to the envelope effect of the Fourier transform of the motif, and the question of the intensities of the spots will be dealt with in the next chapter. The significance of the present chapter is to explain how the positions of the intensity maxima arise.

As in the two-dimensional case, rotation of the crystal can bring into the diffracting position only those reciprocal-lattice points contained within a sphere whose radius is equal to the diameter of the Ewald sphere. This larger sphere is known as the *limiting sphere*, and is the three-dimensional analogue of the two-dimensional limiting circle. It is also the geometric condition corresponding to the mathematical inequality which governs X-ray diffraction, namely

$$S_{hkl} \leq \frac{2}{\lambda} \tag{8.32}$$

Figure 8.19 shows the full geometrical arrangement for the three-dimensional case.

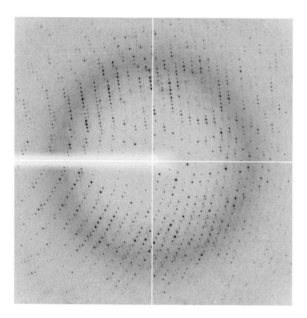

Fig. 8.18 *An X-ray diffraction image of a typical protein.* The image clearly demonstrates a regular lattice structure, corresponding to the layers of the reciprocal-lattice points which intersect the Ewald sphere. The crystal was rotated through 0.5 degrees during the exposure. (Courtesy of Dr S. Kolstoe, UCL.)

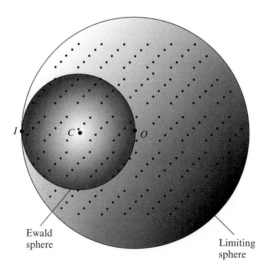

Fig. 8.19 *The limiting sphere.* Relative motion of the Ewald sphere and the reciprocal lattice allows all reciprocal-lattice points within the limiting sphere to intersect the Ewald sphere. A diffraction pattern can therefore contain information on only a finite selection of reciprocal-lattice points.

8.15   Bragg's law and crystal planes

The entire discussion of the Ewald sphere was one way of interpreting Bragg's law, equation (8.31),

$$\lambda = 2d_{hkl} \sin \theta \tag{8.31}$$

We shall now describe a different way of interpreting Bragg's law which is often useful to bear in mind. Bragg's law as stated in equation (8.31) relates the Bragg angle θ to the wavelength λ of the X-rays and the spacing d_{hkl} of the (hkl) set of planes. The geometrical relationship between the wave vectors $\mathbf{k}_i$ and $\mathbf{k}_d$ of the incident and diffracted waves and the scattering vector $\mathbf{k}$ is drawn once again in Fig. 8.20(a), giving the now familiar isosceles triangle.

Now the scattering vector $\mathbf{\Delta k}$ is related to the reciprocal-lattice vector $\mathbf{S}_{hkl}$ by

$$\mathbf{\Delta k} = 2\pi \mathbf{S}_{hkl} \tag{8.25}$$

implying that $\mathbf{\Delta k}$ and $\mathbf{S}_{hkl}$ are parallel. But we also know from Section 8.8 that the vector $\mathbf{S}_{hkl}$ is perpendicular to the (hkl) set of planes, implying that $\mathbf{\Delta k}$ is also perpendicular to this set. From the geometry of Fig. 8.20(a) we see that the line AX, which defines the Bragg angle θ, is perpendicular to the scattering vector $\mathbf{\Delta k}$, and so this line must be parallel to the (hkl) set of planes.

We now redraw Fig. 8.20(a) rather differently. If we project the incident wave vector $\mathbf{k}_i$ back so that it makes the Bragg angle θ with the line AX, then the diffracted wave vector $\mathbf{k}_d$ also makes the Bragg angle θ with AX. This diagram is shown in Fig. 8.20(b) and casts a completely different interpretation on Bragg's law. Since we know that the line AX is parallel to the (hkl) set of planes in the real lattice, we may imagine that the line does in fact represent one member of the set. Figure 8.20(b) suggests that a wave incident at an angle θ emerges at the same angle θ. This is exactly analogous to the event of reflection, and it appears as if the incident wave has been reflected as

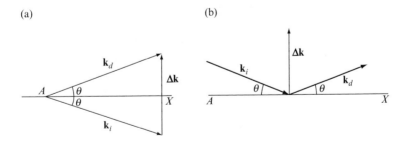

(a)   (b)

Fig. 8.20 *Bragg's law and reflection from lattice planes.* (a) shows the relationship of $\mathbf{k}_i$, $\mathbf{k}_d$ and $\mathbf{\Delta k}$ as in Fig. 8.11. We know that the line AX is perpendicular to $\mathbf{\Delta k}$, which is itself perpendicular to the set of planes (hkl). Hence AX is parallel to the (hkl) set of planes, and may represent one of them. By redrawing (a) we derive (b), in which $\mathbf{k}_i$ and $\mathbf{k}_d$ still make the Bragg angle θ with AX, but this diagram is highly suggestive of a reflection from the (hkl) set of planes.

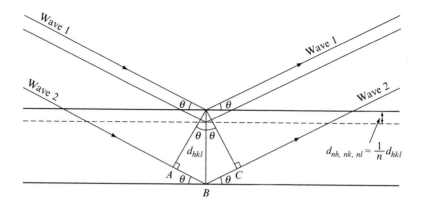

Fig. 8.21 *A physical explanation of Bragg's law.* Two waves are incident on the (hkl) set of planes, and on their reflection from adjacent planes a phase difference occurs. The geometrical path difference is $AB + BC = 2d_{hkl} \sin \theta$. If this is an integral number of wavelengths $n\lambda$, a diffraction maximum will occur. Note that for any n greater than unity, there will always be a 'fictitious' plane (shown dashed) at a distance of $d_{nh\ nk\ nl} = (1/n)d_{hkl}$ such that on reflection there is a path difference of just one wavelength.

if from a plane mirror. Bragg's law may therefore be interpreted as the 'reflection' of X-rays from sets of lattice planes, and this is the origin of the term 'reflecting sphere', which is the synonym of the Ewald sphere.

A physical justification for this effect is demonstrated in Fig. 8.21. Here are drawn two adjacent planes of the (hkl) set, and two waves incident at an angle θ. The waves are assumed to be in phase, with wave 1 'reflected' from the upper plane, and wave 2 from the lower plane. Although the waves were in phase on entering the crystal, wave 2 has travelled a longer distance than wave 1, so that on emerging there is a phase difference between the two waves, and interference effects may occur. The phase difference may be calculated as follows.

If the spacing between the planes is d_{hkl}, then the geometrical path difference d between the paths taken by the waves is given by $2d_{hkl} \sin \theta$. For an intensity maximum, the path difference must be an integral multiple of λ, so we may write

$$n\lambda = 2d_{hkl} \sin \theta$$

where n is an integer.

The integer n introduced above is known as the *order* of the 'reflection', and it specifies by how many wavelengths the two waves are out of step on leaving the crystal. Since the integer n implies that the path difference between waves 1 and 2 is $n\lambda$, we may infer that a *fictitious plane* at a distance d_{hkl}/n below the upper plane would give rise to a wave with a phase difference of λ from wave 1. The fictitious plane does not contain lattice points, and will correspond to the indices $(nh\ nk\ nl)$. In practice, we always assume that the path difference between the incident and diffracted waves is equal to one wavelength, and so diffraction where the path difference is $n\lambda$ is treated as being due to the higher-order fictitious plane and assigned indices $(nh\ nk\ nl)$.

8.16   The effect of finite crystal size

The preceding discussion, as from Section 8.4, has been concerned with the diffraction pattern of a perfect, infinite crystal. In this section we shall consider the modifications to the above theory which we must introduce in order to take account of the finite size of the crystal, which we still assume to be perfect in structure.

The reciprocal lattice of an infinite crystal is an infinite array in which the reciprocal-lattice sites are all ideal mathematical points. As we saw in our discussion of one-dimensional obstacles in the previous chapter, the finite size of a real lattice causes a blurring or smearing-out of these otherwise infinitely sharp reciprocal-lattice points. Hence the diffraction pattern of a finite lattice is an infinite reciprocal lattice formed from points which are not ideal mathematical points but are somewhat fuzzy.

This has two main consequences.

Firstly, when we consider the intersection of the reciprocal lattice with the Ewald sphere, for an infinite real lattice, the reciprocal-lattice points are infinitely sharp, and so the intersection of a reciprocal-lattice point with the Ewald sphere is a well-defined, precise event. For a finite crystal, however, the reciprocal-lattice points are smeared out, and their intersection with the Ewald sphere is less precise.

As shown in Fig. 8.22, the reciprocal-lattice point P is centred on the Ewald sphere and corresponds to the theoretical diffracting position. But because each reciprocal-lattice point is smeared out, the nearby reciprocal-lattice point Q also intersects the Ewald sphere and will give rise to a weak diffraction event. In the ideal case of an infinite perfect crystal, Q would be a perfect point, and would not intersect the Ewald sphere simultaneously with P. What happens in the real case is that the point P correctly gives rise to a diffraction spot, but so does the point Q, even though it really should not. For a given setting of the crystal, a photograph would therefore record a bright spot corresponding to P, and a weaker spot corresponding to the partial intersection of Q. If X-ray crystallography were carried out using stationary crystals, then the presence of these additional 'ghost' spots might be confusing. All X-ray crystallography, however, involves some motion of the crystal, so that sooner or later the point Q would be brought into the exact diffracting position, and the appearance of 'ghost' spots is not a problem.

The second effect of the smearing-out of the reciprocal-lattice points concerns the behaviour of a single spot as it passes through the precise diffracting position. Looking once again at Fig. 8.22, we see that the point P does not intersect with a point on the Ewald sphere, but rather with a small surface area. Instead of there being a single diffracted beam satisfying the diffraction condition at a precise angle, there is instead a cone of allowed diffracted wave vectors as shown. This cone of rays will impinge on the photographic film and leave a slightly blurred or streaked image, depending on the precise shape of the area of intersection of the 'point' P with the Ewald sphere.

The record of the diffraction pattern of a finite crystal is therefore a series of somewhat blurred spots, the blurring being due, at least in part, to the finite size of each reciprocal-lattice point. This in turn is determined by the convolution of the shape function, and so the shape of each diffraction spot is partly determined by the overall shape of the crystal, as we predicted at the end of Chapter 7.

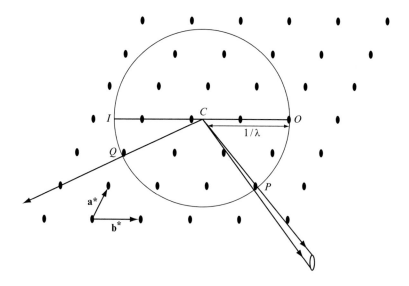

Fig. 8.22 *The effect of finite crystal size.* Shown, for clarity, is an equatorial section through the Ewald sphere construction. The reciprocal lattice of a finite crystal is an infinite lattice of smeared-out points. This has two effects. (i) The point Q is not at the ideal diffracting position, but it does intersect the Ewald sphere, giving rise to a weak diffraction peak. (ii) The point P is at the ideal diffracting position, but the intersection with the Ewald sphere is an element of area, implying that there is a cone of diffracted rays. These will give a streaked or blurred image on the detector or photographic film.

In practice, the finite size of the crystal is not the only factor that affects the shape of the diffraction spots: other factors include disorder within the crystal (which implies that the real and hence the reciprocal lattices are not perfectly spaced) and the possibility that the incident X-ray beam is not perfectly monochromatic (which causes the Ewald sphere to have some 'thickness'). These effects will be discussed in detail later in the book. As it happens, the information contained in the shape of the diffraction spot is never used, and the real practical effect of spot broadening is the nuisance of the possibility of neighbouring closely spaced diffraction peaks obscuring each other as a result of their mutual fuzziness.

Summary

The diffraction pattern of a perfect infinite lattice based on the vectors $\mathbf{a}$, $\mathbf{b}$ and $\mathbf{c}$ is such that the *scattering vector* $\mathbf{\Delta k}$ satisfies the *Laue equations*

$$\mathbf{\Delta k} \cdot \mathbf{a} = 2h\pi \tag{8.7}$$

$$\mathbf{\Delta k} \cdot \mathbf{b} = 2k\pi \tag{8.8}$$

$$\mathbf{\Delta k} \cdot \mathbf{c} = 2l\pi \tag{8.9}$$

in which h, k and l are integers.

The scattering vector $\mathbf{\Delta k}$ is defined by

$$\mathbf{\Delta k} = \mathbf{k}_d - \mathbf{k}_i \tag{8.1}$$

in which $\mathbf{k}_i$ is the wave vector of the incident wave, and $\mathbf{k}_d$ that of the diffracted wave.

Since the incident wave vector $\mathbf{k}_i$ is known in both magnitude and direction, the quantity which is determined by solving the Laue equations is the direction of a diffracted wave vector which gives rise to a diffraction maximum.

Solution of the Laue equations shows that the scattering vector $\mathbf{\Delta k}$ is given by

$$\mathbf{\Delta k} = 2\pi \mathbf{S}_{hkl} \tag{8.25}$$

in which

$$\mathbf{S} = h\mathbf{a}^* + k\mathbf{b}^* + l\mathbf{c}^* \tag{8.24}$$

In these equations h, k and l are integers, and the vectors $\mathbf{a}^*$, $\mathbf{b}^*$ and $\mathbf{c}^*$ define a lattice called the *reciprocal lattice*. The vectors $\mathbf{a}^*$, $\mathbf{b}^*$ and $\mathbf{c}^*$ of the reciprocal lattice are related to the vectors $\mathbf{a}$, $\mathbf{b}$ and $\mathbf{c}$ of the real lattice by

$$\mathbf{a}^* = \frac{\mathbf{b} \wedge \mathbf{c}}{\mathbf{a} \cdot \mathbf{b} \wedge \mathbf{c}} \tag{8.20}$$

$$\mathbf{b}^* = \frac{\mathbf{c} \wedge \mathbf{a}}{\mathbf{a} \cdot \mathbf{b} \wedge \mathbf{c}} \tag{8.21}$$

$$\mathbf{c}^* = \frac{\mathbf{a} \wedge \mathbf{b}}{\mathbf{a} \cdot \mathbf{b} \wedge \mathbf{c}} \tag{8.22}$$

Any vector $\mathbf{S}_{hkl}$ in the reciprocal lattice corresponds (as determined by equation (8.25)) to a value of $\mathbf{\Delta k}$ which will give rise to a diffraction maximum. Two important theorems concerning the reciprocal-lattice vector $\mathbf{S}_{hkl}$ are:

(a) The reciprocal-lattice vector $\mathbf{S}_{hkl}$ is perpendicular to the (hkl) set of planes of the real lattice.
(b) The magnitude S_{hkl} of the vector $\mathbf{S}_{hkl}$ is related to the spacing d_{hkl} between the (hkl) set of planes of the real lattice by

$$d_{hkl} = \frac{1}{S_{hkl}} \tag{8.30}$$

From these two theorems, we derived *Bragg's law*,

$$\lambda = 2d_{hkl} \sin \theta \tag{8.31}$$

in which θ is the Bragg angle, and is equal to the angle which both the incident and the diffracted rays make with the (hkl) set of planes. Bragg's law can be derived directly from the Laue equations and so serves as a condition for diffraction. The law may be interpreted either as the 'reflection' of X-rays from the (hkl) set of planes of the real lattice or in terms of the *Ewald sphere* construction.

The Ewald construction is the most useful aid in interpreting and predicting diffraction patterns. By using this construction, it was established that any X-ray diffraction photograph is a two-dimensional projection of part of the reciprocal lattice of a crystal.

The intensity $|F(\Delta \mathbf{k})|^2$ of the diffraction pattern of a finite perfect crystal lattice is given by

$$|F(\Delta \mathbf{k})|^2 = \frac{\sin^2 \frac{P \, \Delta \mathbf{k} \cdot \mathbf{a}}{2}}{\sin^2 \frac{\Delta \mathbf{k} \cdot \mathbf{a}}{2}} \cdot \frac{\sin^2 \frac{Q \, \Delta \mathbf{k} \cdot \mathbf{b}}{2}}{\sin^2 \frac{\Delta \mathbf{k} \cdot \mathbf{b}}{2}} \cdot \frac{\sin^2 \frac{R \, \Delta \mathbf{k} \cdot \mathbf{c}}{2}}{\sin^2 \frac{\Delta \mathbf{k} \cdot \mathbf{c}}{2}} \tag{8.6}$$

in which P, Q and R signify the extent of the real lattice. The diffraction pattern of a finite perfect lattice may be interpreted in terms of an infinite reciprocal lattice, the points of which are now no longer mathematical points but are somewhat smeared out.

9
The contents of the unit cell

In the previous chapter, we investigated the form of the diffraction pattern of a crystal lattice, and we found that this could be interpreted in terms of the reciprocal lattice. We now turn our attention to the diffraction pattern of the motif, which for a crystal is the contents of the unit cell. Since the diffraction pattern of the crystal as a whole is given by

$$Tf(\text{crystal}) = Tf(\text{motif}) \cdot [Tf(\text{infinite lattice})^* \ Tf(\text{shape function})] \qquad (7.15)$$

the effect of the diffraction pattern of the motif is to envelop that of the finite crystal lattice. Whereas the diffraction pattern of the lattice is an array of peaks of equal intensity, the pattern of a crystal is an array of peaks whose intensities are determined by the diffraction pattern of the motif. In order to derive information concerning the motif, we must look at the intensities of all the diffraction maxima and, by means of the inverse Fourier transform, we may calculate parameters pertaining to the molecules within the unit cell.

Since the motif is composed of molecules arranged in a definite manner within the unit cell, the fundamental diffraction event is the scattering of the incident X-rays by the electrons within the individual atoms. The overall diffraction pattern of the motif may then be interpreted in terms of the distribution of electrons within the unit cell.

Our first task is therefore to study the scattering of X-rays by a single electron. After that, we shall derive equations for the intensity of each diffraction maximum in terms of the overall electron distribution within the unit cell. From these equations, we shall then be able to see how we may use information on the intensities of the diffraction maxima to calculate the atomic and molecular configuration of the unit cell.

9.1   The scattering of X-rays by a single electron

The discussion in this section is a rather more thorough treatment of the interaction of X-rays with an electron than that given in Section 4.13, 'The interaction of electromagnetic radiation with matter'.

Experiment shows that there are two main types of interaction of X-rays with electrons; these are called *Thomson scattering* and *Compton scattering*. We shall discuss each in turn.

Thomson scattering

The physical model which is the basis of Thomson scattering concerns the effect of the electric vector $\mathbf{E}$ of an electromagnetic wave on an electron, as discussed briefly in Chapter 4. This treatment was first developed by Sir J. J. Thomson at the turn of the 20th century.

The assumption made in the Thomson analysis is that the electron is perfectly free and responds passively to the electromagnetic wave in which it is placed. We shall assume that an electron of charge e and mass m experiences an incident X-ray characterised by an electric field vector $\mathbf{E}_{in}$. The force experienced by the electron is therefore $e\mathbf{E}_{in}$, causing an acceleration $\mathbf{a}$ given by

$$\mathbf{a} = \frac{e}{m}\mathbf{E}_{in}$$

This equation implies that the acceleration is related to the incident electric field vector by a well-defined phase relationship. A result of classical electromagnetic theory is that any accelerated charge acts as a source of radiation. The electron is accelerated by the incident field, and so gives rise to scattered waves of the same frequency ω as the incident wave and characterised by an electric field vector $\mathbf{E}_{scat}$. A standard proof from electromagnetic theory is that the amplitude of the electric field E_{scat} of the scattered waves is related to the amplitude of the incident electric field E_{in} of unpolarised waves by

$$\frac{E_{scat}}{E_{in}} = \frac{e^2}{4\pi\varepsilon_0 r mc^2}\sqrt{\frac{(1 + \cos^2 2\theta)}{2}} \tag{9.1}$$

where ε_0 is a fundamental electromagnetic constant, of magnitude 8.854×10^{-12} $F\,m^{-1}$; c is the velocity of electromagnetic radiation; r is the distance from the electron at which the scattered field is sampled; and θ is the appropriate Bragg angle.

We shall return to a discussion of equation (9.1) in the next section, but for the moment we shall discuss the concepts behind the Thomson theory.

An overview of Thomson scattering shows that the basis of the physical process is the interaction of a wave with a particle. This is a purely classical effect, for we need invoke no quantum mechanical concepts in the derivation of equation (9.1). The fact that we correlate the acceleration of an electron to the electric field vector $\mathbf{E}_{in}$ of the incident wave implies that we are considering effects such as the phase relationship between the incident electric field and the motion of the electron. When we study the effect of an incident wave on an assembly of electrons, this correlation automatically ensures that the phase relationship between waves scattered from different electrons is well defined. Any system in which phase relationships are well defined is described as a *coherent* system, and so Thomson scattering is often referred to as *coherent scattering*. This implies that the overall effect of an assembly of electrons may be calculated by using the principle of superposition, in which the total scattered wave is the sum of the individual waves scattered by each electron. Diffraction effects may be observed, and so Thomson scattering is a valid model for the diffraction of X-rays by crystals.

Compton scattering

The Compton effect was discovered in 1923 when Arthur Compton (1892–1962) found that the waves scattered by a crystal placed in a monochromatic X-ray beam were characterised by two wavelengths. The majority of the scattered waves have the same wavelength as the incident wave, but there is also a definite set of waves which have a wavelength rather longer than that of the incident waves. Now, the basis of the Thomson treatment is that an electron responds passively to the electric field vector **E** of the incident wave, so that the wavelength of the scattered waves necessarily equals that of the incident waves. Yet the discovery of the Compton effect, for which Compton received the Nobel Prize in Physics in 1927, produced clear evidence that this was not always the case. This implies that the Thomson model of the interaction of X-rays with electrons does not tell the whole story.

Whereas Thomson scattering may be explained on a purely classical model in which we consider the interaction of a particle with a wave, the Compton effect requires a quantum mechanical explanation.

The model used to describe Compton scattering considers a collision between an electron and a quantum of electromagnetic radiation called a *photon*, which may be considered as a particle. The basis of Compton scattering is therefore a particle–particle collision analogous to a mechanical 'billiard ball' collision. Using this model, we may calculate that the scattered photons will be characterised by two wavelengths just as observed experimentally.

Inherent in the formalism of Thomson scattering is a well-defined phase relationship between the incident X-ray beam and the electronic motion which gives rise to the scattering. The concept of phase is applicable only to wave-like phenomena, and we may contrast this with the model of Compton scattering. The Compton calculation uses a particle–particle collision model, and this is the antithesis of a wave motion analysis. The Compton model, therefore, rejects the concept of phase, and we may infer that well-defined phase relationships in Compton scattering do not exist. This implies that when we consider the combined effect of Compton scattering on an assembly of electrons, there are no phase relationships between the scattering events. Consequently, the principle of superposition cannot be used to assess the overall scattering effect of an assembly of electrons. Compton scattering may also be referred to as *incoherent scattering*.

Now, the entire discussion of diffraction is founded on the principle of superposition and well-defined phase relationships. In that Compton scattering does not give such phase relationships, we may conclude that no discrete diffraction effects result from Compton scattering: the sole effect of Compton scattering is to give rise to an overall background scattering, which has no direct influence on the diffraction pattern as such.

The conclusion we reach after our discussion of Thomson and Compton scattering is that the diffraction pattern of a crystal may be explained solely by the coherent Thomson scattering by the electrons within the molecules. Such Compton scattering as occurs does not affect the diffraction pattern, but produces an overall background scattering.

9.2 The scattering of X-rays by a distribution of electrons

Since the diffraction of X-rays by a crystal may be explained solely in terms of Thomson scattering, we shall investigate equation (9.1) rather more closely, and apply it to the case of a continuous distribution of electrons.

Our first task is to discuss the assumption made by Thomson, namely, that the electron is perfectly free and responds passively to the incident electric field. As we saw in Chapter 4, electrons in crystals are not free in the absolute sense of the word, but are bound to atomic nuclei, about which they move with a characteristic frequency. The Thomson assumption is not as restrictive as stated above, for the physical significance of the assumption is that the electric field of the incident wave should be the most important force acting on an electron at any given instant. This comes down to the condition that the frequency of the incident wave must be very much greater than the characteristic frequency of the electron's motion around the nucleus. In general, this is the case, and the Thomson formalism may safely be applied to the case of electrons bound to atoms. If, however, the energy of the incident X-ray beam is close to an energy which will give rise to a change in the quantum state of the electron, then the details of the Thomson formula are not correct, and this situation is known as *anomalous scattering*, which we shall discuss later in this chapter.

We shall assume that the Thomson treatment is applicable, and, writing equation (9.1) again,

$$\frac{E_{scat}}{E_{in}} = \frac{e^2}{4\pi\varepsilon_0 rmc^2}\sqrt{\frac{(1+\cos^2 2\theta)}{2}} \tag{9.1}$$

we see that the ratio of the scattered to the incident field is determined by the product of two terms. Defining the quantity f_e as

$$f_e = \frac{e^2}{4\pi\varepsilon_0 rmc^2} \tag{9.2}$$

we have

$$\frac{E_{scat}}{E_{in}} = f_e\sqrt{\frac{(1+\cos^2 2\theta)}{2}} \tag{9.3}$$

The Thomson formula as quoted in equations (9.1) and (9.3) was derived assuming that the incident X-rays were unpolarised. If we carry the calculation through for an incident wave in a defined state of polarisation, it turns out that the ratio of the scattered and incident fields is of the form

$$\frac{E_{scat}}{E_{in}} = f_e p(2\theta) \tag{9.4}$$

where $p(2\theta)$ is some other function of the Bragg angle θ. What is important is that no matter what the polarisation of the incident wave is, the ratio E_{scat}/E_{in} is always proportional to f_e, the proportionality being determined by a function related to the polarisation of the incident wave. The function $p(2\theta)$ is called the *polarisation factor*, and if we disregard this, we may represent the field of the wave scattered by an electron as

$$E_{scat} = f_e E_{in} \tag{9.5}$$

bearing in mind that in any real case we must insert the polarisation factor as appropriate.

All further discussion will be based on equation (9.5), and we shall correct for the polarisation factor whenever necessary.

Since the magnitude of the incident field E_{in} is constant for any given wave, equation (9.5) implies that the magnitude E_{scat} of the scattered wave at any point in space is determined by the quantity f_e. From the definition of f_e in equation (9.2), we see that at any distance r, f_e is determined by a variety of fundamental physical constants, and so represents an intrinsic property of the electron. The physical significance of f_e is that it specifies the ability of an electron to scatter radiation, and f_e is known as the *electronic scattering factor*.

We now consider the field scattered by two electrons A and B as depicted in Fig. 9.1.

If the magnitude of the electric field of the incident wave at electron A is represented by E_{in}, then because there is a spatial separation between the two electrons, the electric field of the incident wave acting on electron B will be of the form $E_{in}e^{i\phi}$, where ϕ is a phase factor corresponding to the number of wavelengths, or the fraction of a wavelength, which separate A and B. Each electron scatters the radiation according to the electronic scattering factor f_e, and we may write

$$(E_{scat})_A = f_e E_{in}$$
$$(E_{scat})_B = f_e E_{in} e^{i\phi}$$

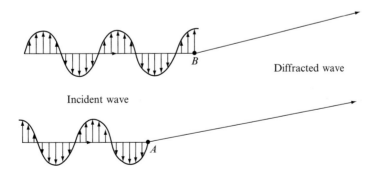

Diffracted wave

Incident wave

Fig. 9.1 *Scattering by two electrons.* A and B are two electrons placed in an incident X-ray beam, the electric field vector $\mathbf{E}_{in}$ of which is shown. There is a phase difference between the incident waves at A and B, and so the total scattered wave $(E_{scat})_{tot}$ is given by $(E_{scat})_{tot} = (f_e + f_e e^{i\phi})E_{in}$, where f_e is the electronic scattering factor as calculated from the Thomson model.

The total scattered field $(E_{scat})_{tot}$ be can found by using the principle of superposition:

$$(E_{scat})_{tot} = (E_{scat})_A + (E_{scat})_B$$

$$\therefore \frac{(E_{scat})_{tot}}{E_{in}} = f_e + f_e e^{i\phi}$$

We may extend this argument to find the total scattered field E_{scat} resulting from an array of n electrons, each with their appropriate phase factor ϕ_n:

$$\frac{(E_{scat})_{tot}}{E_{in}} = \sum_n f_e e^{i\phi_n} \tag{9.6}$$

where the summation goes over all electrons.

It is interesting to compare equation (9.6), which describes the waves scattered by a discrete assembly of electrons, with our general expression for the waves scattered by (or diffracted by) an arbitrary obstacle:

$$F(\mathbf{\Delta k}) = \int_{\text{all } \mathbf{r}} f(\mathbf{r}) e^{i\,\mathbf{\Delta k} \cdot \mathbf{r}}\, d\mathbf{r} \tag{8.2}$$

Equations (9.6) and (8.2) do, of course, describe the same event. In the case of equation (8.2), an obstacle is described by a continuous amplitude function $f(\mathbf{r})$, whereas in equation (9.6), we have a discrete system of electrons characterised by the electronic scattering factor f_e. Similarly, the phase factor ϕ_n is equivalent to the expression $\mathbf{\Delta k} \cdot \mathbf{r}$, for reference to Section 8.2 will show that we derived the term $\mathbf{\Delta k} \cdot \mathbf{r}$ by considering the phase relationship between the waves incident on, and scattered by, two arbitrary points within the obstacle. Finally, our measure of the diffraction pattern function in equation (8.2) is now seen to represent the ratio between the magnitudes of the incident and scattered waves. Since all amplitude and intensity measurements are relative, then it does not really matter in what form a comparison is made, and so the ratio of the scattered to the incident field is as logical as any.

An important consequence of the comparison of equations (8.2) and (9.6) is that we identify the amplitude function of an electron as the electronic scattering factor f_e. As we saw in Chapter 6, the physical significance of the amplitude function $f(\mathbf{r})$ is that it specifies the ability of an obstacle to scatter radiation. Since this is how we defined f_e in equation (9.2), it is gratifying to find that equations (8.2) and (9.6) agree in both a qualitative and a quantitative manner.

We now apply equation (9.6) to the case in which the electrons are distributed continuously. This is a more valid physical model than one of discrete electron positions, since one of the implications of quantum mechanics is that the position of an electron is a most ill-defined quantity. Of more physical significance is a measure of the probability that an electron will be at some point in space.

We may characterise an electronic system by means of the *electron density function* $\rho(\mathbf{r})$. This is a function of position $\mathbf{r}$ only, and its significance is that in any element of volume, centred on the point $\mathbf{r}$, which we represent by $d\mathbf{r}$, the average number of electrons within the volume element $d\mathbf{r}$ is given by $\rho(\mathbf{r})\, d\mathbf{r}$. We use the term 'average number of electrons' in order to comply with the probabilistic interpretation of quantum mechanics, but it is quite acceptable to refer to $\rho(\mathbf{r})\, d\mathbf{r}$ as the number of electrons within the volume element $d\mathbf{r}$. Since a single electron scatters radiation as determined by the electronic scattering factor f_e, $\rho(\mathbf{r})\, d\mathbf{r}$ electrons will scatter as $f_e\rho(\mathbf{r})\, d\mathbf{r}$. But the volume element $d\mathbf{r}$ contains more than one electron, and it appears as if we should include some phase factor for the scattering by the different electrons within $d\mathbf{r}$. As shown in Fig. 9.2, however, if we make the volume element $d\mathbf{r}$ arbitrarily small (which we may do since $d\mathbf{r}$ is an infinitesimal), then all of the electrons within the volume element 'see' the same incident electric field, and so scatter in phase.

Whereas the amplitude function of a single electron is f_e, that of a distribution of electrons is given by $f_e\rho(\mathbf{r})\, d\mathbf{r}$, and so when we wish to find the total wave scattered by the entire distribution of electrons, we may use $f_e\rho(\mathbf{r})$ as the amplitude function in the Fourier transform equation as

$$F(\mathbf{\Delta k}) = \int_{\text{all } \mathbf{r}} f_e\rho(\mathbf{r})e^{i\,\mathbf{\Delta k \cdot r}}\, d\mathbf{r} \tag{9.7}$$

where the integral is over the entire volume in which we are interested.

Equation (9.7) represents the wave scattered by a continuous distribution of electrons, and is clearly the appropriate form of the Fourier transform equation, equation (8.2).

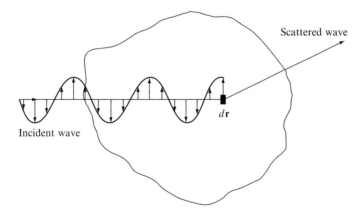

Fig. 9.2 *Scattering by a continuous distribution of electrons.* A continuous distribution of electrons is described by an electron density function $\rho(\mathbf{r})$. If a volume element is represented by $d\mathbf{r}$, then we can make this arbitrarily small so that all the $\rho(\mathbf{r})\, d\mathbf{r}$ electrons within the volume element $d\mathbf{r}$ 'see' the same incident electric field E_{in}. This will happen as long as the linear extent of the volume element $d\mathbf{r}$ is very much smaller than the incident wavelength. In this case, the scattered wave amplitude dE_{scat} is equal to $f_e\rho(\mathbf{r})\, d\mathbf{r}$.

9.3 The diffraction pattern of the motif

Within a single unit cell of a crystal, the atoms are distributed in a certain manner, and so there exists some electron density function $\rho(\mathbf{r})$ which describes the location of the electrons within the atoms of the unit cell. The electron density function will have a large value at positions corresponding to atomic nuclei, and a relatively low value between nuclei. If we can find the electron density function, then we may infer the positions of atoms from the electron density maxima, and derive data such as that pertaining to bond lengths and bond angles. All the fundamental structural information on the contents of the unit cell is contained within the electron density function, and it is this function which we wish to determine from X-ray analysis.

Equation (9.7) states that the diffraction pattern function $F(\mathbf{\Delta k})$ resulting from the scattering of X-rays by a distribution of electrons characterised by an electron density function $\rho(\mathbf{r})$ is

$$F(\mathbf{\Delta k}) = \int_{\text{all } \mathbf{r}} f_e \rho(\mathbf{r}) e^{i\,\mathbf{\Delta k \cdot r}}\, d\mathbf{r} \tag{9.7}$$

For the case in which $\rho(\mathbf{r})$ represents the contents of the unit cell of a crystal, the limits of the integral correspond to those values of $\mathbf{r}$ which define the unit cell, and since f_e is a constant, we have

$$F(\mathbf{\Delta k}) = f_e \int_{\text{unit cell}} \rho(\mathbf{r}) e^{i\,\mathbf{\Delta k \cdot r}}\, d\mathbf{r} \tag{9.8}$$

Reference to equation (9.2) shows that f_e does contain a variable r. However, this r specifies the distance at which we sample the diffracted wave, and is in no way related to the variable $\mathbf{r}$ used to define positions within the unit cell. Hence, for the integration of equation (9.7), f_e is a constant.

The ratio $F(\mathbf{\Delta k})/f_e$ represents the relative scattering ability of the contents of the unit cell as compared with a single electron. This is a useful measure of the scattering ability of the unit cell, and we shall refer to this quantity as $F_{rel}(\mathbf{\Delta k})$:

$$F_{rel}(\mathbf{\Delta k}) = \int_{\text{unit cell}} \rho(\mathbf{r}) e^{i\,\mathbf{\Delta k \cdot r}}\, d\mathbf{r} \tag{9.9}$$

The relative intensity of the diffraction pattern of the unit cell is given by

$$|F_{rel}(\mathbf{\Delta k})|^2.$$

We shall now investigate the integral of equation (9.9), and cast it in a non-vector form. Reference to Fig. 9.3 will show the coordinates which we shall adopt. In Fig. 9.3 is a unit cell of a crystal which, for generality, we assume to be triclinic. The unit cell is defined by three crystallographic vectors $\mathbf{a}$, $\mathbf{b}$ and $\mathbf{c}$. Note that these vectors ($\mathbf{a}$, $\mathbf{b}$ and $\mathbf{c}$) are not necessarily mutually orthogonal.

Any point $\mathbf{r}$ within the unit cell may be defined relative to these axes by the coordinates (X, Y, Z), where X, Y and Z are measured in units of Å. All positions

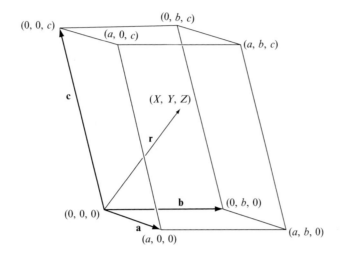

Fig. 9.3 *The geometry of the unit cell.* Shown is the unit cell of a triclinic lattice. From the origin (0, 0, 0) are three unit cell vectors **a**, **b** and **c** which define the crystallographic axes. Any point **r** has absolute coordinates (X, Y, Z) in units of Å. For convenience in calculation, it is more useful to define fractional coordinates (x, y, z), which are dimensionless and are defined with respect to the crystallographic unit vectors **a**, **b** and **c** such that $x = X/a$, $y = Y/b$ and $z = Z/c$. In these coordinates, we have $r = x\mathbf{a} + y\mathbf{b} + z\mathbf{c}$. The volume element $d\mathbf{r}$ centred on **r** is $d\mathbf{r} = V\,dx\,dy\,dz$, where V is the volume of the unit cell.

within the unit cell are such that X, Y and Z are limited by the inequalities

$$0 \le X \le a \quad 0 \le Y \le b \qquad 0 \le Z \le c$$

It is useful to define a new set of coordinates x, y and z such that

$$x = \frac{X}{a} \quad y = \frac{Y}{b} \quad z = \frac{Z}{c}$$

All points within the unit cell are now represented by x, y and z such that

$$0 \le x \le 1 \quad 0 \le y \le 1 \quad 0 \le z \le 1$$

Whereas X, Y and Z represent the *absolute* position of a point within the unit cell, x, y and z represent the *fractional* position with respect to the lengths of the sides of the unit cell, and are necessarily dimensionless fractions. Hence, they are commonly referred to as *fractional coordinates*, and any point **r** becomes

$$\mathbf{r} = x\mathbf{a} + y\mathbf{b} + z\mathbf{c}$$

In order to evaluate the integral of equation (9.9), we need an expression for the volume $d\mathbf{r}$ of an elementary volume. In terms of the *fractional* coordinates (x, y, z), an element of volume is given by

$$d\mathbf{r} = dx\,dy\,dz\,\mathbf{a}\cdot\mathbf{b}\wedge\mathbf{c}$$

But since $\mathbf{a} \cdot \mathbf{b} \wedge \mathbf{c}$ represents the volume V of the unit cell, then

$$d\mathbf{r} = V \, dx \, dy \, dz$$

If we represent the electron density function $\rho(\mathbf{r})$ as a function $\rho(x, y, z)$ of the fractional coordinates x, y and z, equation (9.9) becomes

$$F_{\text{rel}}(\mathbf{\Delta k}) = V \int_{x=0}^{x=1} \int_{y=0}^{y=1} \int_{z=0}^{z=1} \rho(x, y, z) e^{i \, \mathbf{\Delta k} \cdot (x\mathbf{a} + y\mathbf{b} + z\mathbf{c})} \, dx \, dy \, dz \qquad (9.10)$$

in which the variables x, y and z each run from zero to unity, thereby covering the entire unit cell.

Equation (9.10) represents the diffraction pattern function of the contents of the unit cell. At the moment it is a continuous function of the scattering vector $\mathbf{\Delta k}$, and represents the diffraction pattern such as would be detected if the obstacle were just a single unit cell.

Since the unit cell is convoluted with the crystal lattice to form a complete crystal, the diffraction pattern function as given by equation (9.10) envelops the diffraction pattern of the crystal lattice. This latter pattern is a series of sharp peaks as determined by the reciprocal lattice. The overall effect is that the diffraction pattern of the unit cell is sampled at those points where the crystal lattice gives rise to a diffraction maximum. This has the effect of restricting the values of the scattering vector $\mathbf{\Delta k}$ to those which correspond to crystal lattice diffraction maxima. Note that since $\mathbf{\Delta k}$ is proportional to the reciprocal-lattice vector $\mathbf{S}_{hkl}$ as

$$\mathbf{\Delta k} = 2\pi \mathbf{S}_{hkl} \qquad (8.25)$$

$\mathbf{S}_{hkl}$ is also referred to as the Scattering vector. For a particular reciprocal-lattice vector

$$\mathbf{S}_{hkl} = h\mathbf{a}^* + k\mathbf{b}^* + l\mathbf{c}^*$$

the scalar product $\mathbf{\Delta k} \cdot (x\mathbf{a} + y\mathbf{b} + z\mathbf{c})$ in equation (9.10) then becomes

$$\mathbf{\Delta k} \cdot \mathbf{r} = 2\pi \left(h\mathbf{a}^* + k\mathbf{b}^* + l\mathbf{c}^*\right) \cdot (x\mathbf{a} + y\mathbf{b} + z\mathbf{c})$$
$$= 2\pi(hx + ky + lz)$$

in which we have used the relationships (8.19) for the scalar products of the vectors of the real and reciprocal lattices.

For a single particular maximum corresponding to the reciprocal-lattice point hkl, equation (9.10) becomes

$$[F_{\text{rel}}(\mathbf{\Delta k})]_{hkl} = V \int_0^1 \int_0^1 \int_0^1 \rho(x, y, z) e^{2\pi i (hx + ky + lz)} \, dx \, dy \, dz$$

The quantity represented by $[F_{\text{rel}}(\mathbf{\Delta k})]_{hkl}$ specifies the value of the diffraction pattern function at the point indexed as hkl. The multiplicity of subscripts is ugly, and we shall represent this quantity as F_{hkl}:

$$F_{hkl} = V \int_0^1 \int_0^1 \int_0^1 \rho(x,y,z) e^{2\pi i (hx+ky+lz)} \, dx \, dy \, dz \qquad (9.11)$$

F_{hkl} is known as the *structure factor* of the diffraction maximum corresponding to the reciprocal-lattice point *hkl*. In general it is a complex quantity, with a magnitude $|F_{hkl}|$ and a phase represented as ϕ_{hkl}:

$$F_{hkl} = |F_{hkl}| e^{i\phi_{hkl}} \qquad (9.12)$$

The intensity I_{hkl} of the corresponding diffraction peak is related to F_{hkl} as

$$I_{hkl} = |F_{hkl}|^2 \qquad (9.13)$$

Since the intensity I_{hkl} is an experimentally observable quantity, we may relate the electron density function $\rho(x,y,z)$ to the intensity by means of equations (9.11) and (9.13). This implies that if we know the electron density function, then we may compute the integral in equation (9.11) for any particular values of h, k and l, and hence derive the structure factor F_{hkl} of the diffraction maximum indexed as *hkl*. We may then choose a different triplet of integers *hkl*, calculate the appropriate structure factor and, by this procedure, compute both the amplitude and the intensity of the entire set of diffraction maxima for as many reciprocal-lattice points as we care to choose.

9.4 The calculation of the electron density function

Equation (9.11) is useful when we wish to calculate the structure factor F_{hkl} of a diffraction maximum from a known electron density function $\rho(x,y,z)$. Since our experimental data are the set of intensities I_{hkl}, our experimental problem is to deduce the electron density function $\rho(x,y,z)$ from the intensities I_{hkl} or the structure factors F_{hkl}.

In order to find out how this may be done, we return to equation (9.10),

$$F_{\text{rel}}(\boldsymbol{\Delta k}) = V \int_{x=0}^{x=1} \int_{y=0}^{y=1} \int_{z=0}^{z=1} \rho(x,y,z) e^{i \, \boldsymbol{\Delta k} \cdot (x\mathbf{a} + y\mathbf{b} + z\mathbf{c})} \, dx \, dy \, dz \qquad (9.10)$$

in which we emphasise the fact that $F_{\text{rel}}(\boldsymbol{\Delta k})$ is the Fourier transform of $\rho(x,y,z)$. By the Fourier inversion theorem (Section 5.5), we must be able to express the electron density function $\rho(x,y,z)$ in terms of $F_{\text{rel}}(\boldsymbol{\Delta k})$ as

$$\rho(x,y,z) = \frac{1}{V} \int_{\text{all } \boldsymbol{\Delta k}} F_{\text{rel}}(\boldsymbol{\Delta k}) e^{-i \, \boldsymbol{\Delta k} \cdot (x\mathbf{a} + y\mathbf{b} + z\mathbf{c})} \, d(\boldsymbol{\Delta k})$$

in which the integral is taken over all values of the scattering vector $\boldsymbol{\Delta k}$, which simply means over the entire diffraction pattern.

In the previous section we showed that

$$\Delta k \cdot (x\mathbf{a} + y\mathbf{b} + z\mathbf{c}) = 2\pi(hx + ky + lz)$$

$F_{rel}(\Delta k)$ at the reciprocal-lattice site hkl is represented by F_{hkl}, and so we have

$$\rho(x, y, z) = \frac{1}{V} \int\limits_{\text{all } \Delta k} F_{hkl} e^{-2\pi i(hx+ky+lz)} \, d(\Delta k) \qquad (9.14)$$

We know, however, that the diffraction pattern function is not a continuous function of the scattering vector Δk, but a discrete set of values represented by F_{hkl} for each diffraction maximum. When we integrate over all values of Δk, this is equivalent to integrating over the entire diffraction pattern. Since the diffraction pattern is discrete and not continuous, the integral of equation (9.14) degenerates to a summation over all the reciprocal-lattice points indexed by h, k and l. Since h, k and l vary independently, this is a triple summation of the form

$$\rho(x, y, z) = \frac{1}{V} \sum_h \sum_k \sum_l F_{hkl} e^{-2\pi i(hx+ky+lz)} \qquad (9.15)$$

It is equation (9.15) which enables us to calculate the electron density function from the structure factors F_{hkl} of the diffraction maxima, and so equation (9.15) is one of the very cornerstones of the interpretation of X-ray diffraction experiments.

9.5   Fourier synthesis

Equation (9.15) connects the electron density function to the set of structure factors F_{hkl} of all the diffraction maxima. Let us investigate the meaning of this very important equation, for at first sight the triple summation appears to be very complicated.

We shall assume for the moment that our experimental data are the set of structure factors F_{hkl} corresponding to all the diffraction maxima. We shall assume that we have also determined the lattice constants of the real lattice, and so we know the volume V of the unit cell. We suppose that we have a complete map of the reciprocal lattice, each point of which is indexed according to its integers hkl, and the intensities of the diffraction maxima have been measured by, for example, an electronic detector. Assuming that this allows us to calculate the structure factors F_{hkl}, our data are the set of structure factors F_{000}, F_{100}, F_{200}, ... , F_{hkl}, ... and so on for every reciprocal-lattice point.

If we choose values for x, y and z, say x_1, y_1 and z_1, then we may compute for a reciprocal-lattice point, say (2 3 5), the quantity

$$\left(\frac{1}{V}\right) F_{235} e^{-2\pi i(2x_1+3y_1+5z_1)}$$

which will come to some particular value. If we now choose a different reciprocal-lattice point, say (1 1 1), then we may compute

$$\left(\frac{1}{V}\right) F_{111} e^{-2\pi i(x_1+y_1+z_1)}$$

Similar computations for all reciprocal-lattice sites *hkl* for the same values x_1, y_1 and z_1 will generate the complete set of numbers which are to be used in the summation of equation (9.15) for the values of x, y, and z chosen. By summing our calculations for every reciprocal-lattice site, we will find the electron density at the point we have chosen, namely $\rho(x_1, y_1, z_1)$.

We may then choose a different set of coordinates (x_2, y_2, z_2) and compute the complete set of functions

$$\left(\frac{1}{V}\right) F_{hkl} e^{-2\pi i(hx_2 + ky_2 + lz_2)}$$

and add all these together. This will give us the electron density at (x_2, y_2, z_2), which is $\rho(x_2, y_2, z_2)$.

By repeating this procedure for as many values of (x, y, z) as we care to choose, we may compute the electron density function $\rho(x, y, z)$. The entire operation is known as computing a *Fourier synthesis*. Once this has been done, we may find the local maxima of the electron density function, and infer the positions of the atoms. As we shall soon see, the greater the atomic number of the atom, the greater the electron density, and so the places where the electron density is the highest correspond to the locations of the heaviest atoms. If this is combined with the chemical formula of the molecules contained in the crystal (which is usually known prior to an X-ray investigation), then the structure of the molecules may be determined.

As the reader has undoubtedly noticed, the determination of the electron density function $\rho(x, y, z)$ requires an enormous amount of numerical calculation. For a diffraction pattern of 1000 diffraction peaks, the calculation of the electron density function at a single point (x, y, z) within the unit cell requires 1000 calculations. In order to plot the electron density function in a methodical manner, the values of x, y and z are chosen in an ordered, systematic manner. For a very small molecule, we might start with $(0, 0, 0)$, then $(0.1, 0, 0)$ and so on, increasing x, y and z systematically by increments of 0.1. Since x, y and z are fractional and run from 0 to 1, increasing x, y and z in steps of 0.1 requires 10^3 steps to scan the entire unit cell. Since each step needs 10^3 calculations when we sum over the reciprocal-lattice sites, the total structure analysis entails no fewer than 10^6 individual calculations! For larger molecules, where we often have of the order of 100 000 diffraction peaks and the unit cell has to be sampled at around 10^6 positions, the electron density calculation requires something like 10^{11} terms to be evaluated. This was almost a totally impractical proposition for all but the simplest molecules before the technology of modern high-speed computers was developed. Of course, the calculation can be speeded up by sampling the unit cell at coarser intervals, but the resulting electron density function is less clear, and therefore much more difficult to interpret.

The data can be handled by plotting the electron density function for a chosen value of, say, z on a transparent sheet, so that lines are drawn in the xy plane corresponding to points of equal electron density. This represents a 'contour map' of the electron density for that value of z. More sections of electron density can then be drawn on different sheets for different values of z, and these sheets can be stacked over one another to give a three-dimensional contour map of the electron density within a unit cell. The end result is called an *electron density map*. In reality, it is much more convenient for the map to be drawn by computer, which contours the map on x, y and z sections that can be displayed simultaneously. This allows one to inspect a mock three-dimensional representation of the electron density, as shown in Fig. 9.4. From this the atomic positions are more or less apparent, depending on the resolution limit of the diffraction data $(d_{\min})$ and on the increments in x, y and z used during the calculations.

There was, however, the assumption that our primary data were the set of structure factors F_{hkl}. It is unfortunate that this is not the case. Our experimental data are not the set of structure factors F_{hkl} but are the set of intensities $|F_{hkl}|^2$. Whereas F_{hkl} is a complex number, equal to $|F_{hkl}|e^{i\phi_{hkl}}$, the intensity $|F_{hkl}|^2$ is real, all information on the phase ϕ_{hkl} having been lost. But we need F_{hkl} in both magnitude and phase for the Fourier synthesis of equation (9.15). The fact that we can measure only amplitude and not the phase implies that about half the potential information is not available.

In order to perform the Fourier synthesis of equation (9.15), we must find the phase ϕ_{hkl} corresponding to each $|F_{hkl}|$, and the predicament caused by the lack of this information is called the *phase problem*. If it were not for the phase problem, the routine determination of macromolecular structures by X-ray diffraction could be done by a computer program directly from the experimental measurements. The fact that the phase problem has, to a large extent, been solved is a monument to the ingenuity of the X-ray crystallographers of the last few decades. The most commonly used methods for tackling the phase problem in X-ray analysis of biological macromolecules will be discussed later in the book.

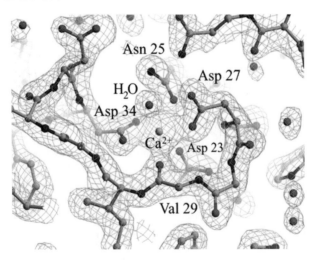

Fig. 9.4 *A typical electron density map for part of a protein.* The electron density function $\rho(x,\ y,\ z)$ is calculated on a fine grid using the Fourier synthesis technique. Contours of equal values of $\rho(x,\ y,\ z)$ are shown as thin, pale lines. In general, peaks of electron density represent the positions of atoms, and the greater the atomic number of an atom, the greater the associated electron density.

9.6   The calculation of structure factors

Equations (9.11) and (9.15),

$$F_{hkl} = V \int_0^1 \int_0^1 \int_0^1 \rho(x, y, z) e^{2\pi i (hx + ky + lz)} \, dx \, dy \, dz \tag{9.11}$$

$$\rho(x, y, z) = \frac{1}{V} \sum_h \sum_k \sum_l F_{hkl} e^{-2\pi i (hx + ky + lz)} \tag{9.15}$$

relate the electron density function $\rho(x, y, z)$ to the structure factors F_{hkl}. Although equation (9.15) is most useful for deriving the electron density function from the structure factors, it so happens that the form of equation (9.11) is not the most appropriate for the calculation of the structure factors from the contents of the unit cell. An equation more suitable than equation (9.11) will now be derived.

The electron density function $\rho(x, y, z)$ as stated in equations (9.11) and (9.15) is the overall electron density function for the unit cell as a whole, and makes no particular reference to the fact that electrons are, for the most part, localised within atoms. We only infer atomic positions from the maxima in the electron density function. An alternative way of representing the distribution of electrons within the unit cell is to consider the electrons as associated with atoms, and then investigate the distribution of atoms within the unit cell.

In order to perform the following calculation, it is convenient at this stage to express equation (9.11) in a vector notation. Defining the vector $\mathbf{r}$ corresponding to the position of any electron as

$$\mathbf{r} = x\mathbf{a} + y\mathbf{b} + z\mathbf{c}$$

the electron density function is written as $\rho(\mathbf{r})$, and equation (9.11) becomes

$$F_{hkl} = \int_{\text{unit cell}} \rho(\mathbf{r}) e^{2\pi i \mathbf{S}_{hkl} \cdot \mathbf{r}} \, d\mathbf{r} \tag{9.16}$$

where $d\mathbf{r}$ is $V \, dx \, dy \, dz$.

With reference to Fig. 9.5, we may express the position $\mathbf{r}$ of any electron with reference to a vector $\mathbf{R}$ from the centre of its associated atom, and a vector $\mathbf{r}_j$ from the origin of the unit cell to the position of the atom centre:

$$\mathbf{r} = \mathbf{r}_j + \mathbf{R}$$

The electron is assumed to be associated with the jth atom within the unit cell. We may describe the distribution of electrons about the jth atom of the unit cell by means of an *atomic electron density function*. As shown in Fig. 9.6(a), we may represent this function in terms of a vector $\mathbf{R}$ from the atom centre to the electron as $\rho_j(\mathbf{R})$.

The form of the function $\rho_j(\mathbf{R})$ is represented in Fig. 9.6(a), where it can be seen that $\rho_j(\mathbf{R})$ has its maximum value when $\mathbf{R} = 0$. This point corresponds to the centre

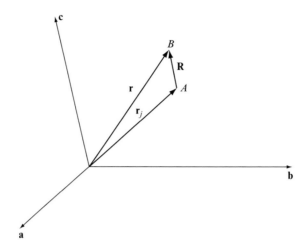

Fig. 9.5 *The position of an electron.* An electron at B is associated with the atom at A. The position of the electron may be defined with respect to the centre of atom A by the vector $\mathbf{R}$ and the position of the atom within the unit cell, $\mathbf{r}_j$, as $\mathbf{r} = \mathbf{r}_j + \mathbf{R}$.

of the jth atom, which is located at the point $\mathbf{r}_j$ within the unit cell, and the electron density function relative to the origin of the unit cell is given by $\rho(\mathbf{r})$, as shown in Fig. 9.6(b). Note that $\rho_j(\mathbf{R})$ represents the electron density function of the jth atom centred around the position $\mathbf{r}_j$ within the unit cell and is equivalent to $\rho_j(\mathbf{r} - \mathbf{r}_j)$.

The overall electron density $\rho(\mathbf{r})$ for the unit cell as a whole is therefore the sum of the individual atomic electron density functions $\rho_j(\mathbf{r} - \mathbf{r}_j)$ for each of the atoms within the unit cell:

$$\rho(\mathbf{r}) = \sum_j \rho_j(\mathbf{r} - \mathbf{r}_j) \tag{9.17}$$

The form of this summation is represented in Fig. 9.6(c), and it is evident that this does represent the overall electron density within the unit cell. Equation (9.17) therefore represents two identical ways of describing the electron density within the unit cell. On the left-hand side is a function describing the overall electron density as a continuum, whereas on the right-hand side the electron density is presented as a summation of the electron densities associated with individual atoms.

Substituting equation (9.17) in equation (9.16), we obtain

$$F_{hkl} = \int_{\text{unit cell}} \sum_j \rho_j(\mathbf{r} - \mathbf{r}_j) e^{2\pi i \mathbf{S}_{hkl} \cdot \mathbf{r}} \, d\mathbf{r} \tag{9.18}$$

But

$$\mathbf{r} = \mathbf{r}_j + \mathbf{R}$$

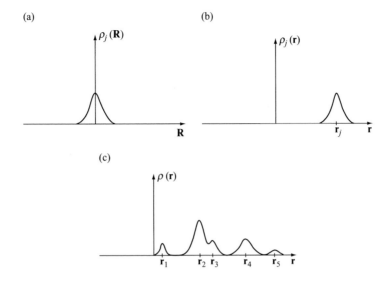

Fig. 9.6 *The electron density function.* In (a) is shown the atomic electron density function $\rho_j(\mathbf{R})$ for the jth atom with respect to its own centre. This is a function which is sharply peaked about the origin, i.e. about the atom centre. If the jth atom is centred on the point $\mathbf{r}_j$ within the unit cell, defined with respect to the origin of the unit cell, the associated electron density function is represented in (b). The function $\rho_j(\mathbf{r} - \mathbf{r}_j)$, which has the shape of (a), represents the contribution from atom j to the electron density at position vector $\mathbf{r}$. (c) shows the superposition of several atomic electron density functions to give the overall electron density function $\rho(\mathbf{r}) = \sum_j \rho_j(\mathbf{r} - \mathbf{r}_j)$.

If we assume that the positions of the atom centres ($\mathbf{r}_j$) are constant,

$$d\mathbf{r} = d\mathbf{R}$$

Strictly speaking, this assumption is valid only for atoms that are not vibrating. In reality, the atoms are oscillating about their equilibrium positions according to the ambient thermal energy. We shall neglect this for the moment, but in Chapter 11 we shall discuss the correction which we eventually have to make.

Substituting for the vector $\mathbf{r}$ in equation (9.18), we have

$$F_{hkl} = \int_{\text{unit cell}} \sum_j \rho_j(\mathbf{R}) e^{2\pi i \mathbf{S}_{hkl} \cdot (\mathbf{r}_j + \mathbf{R})} \, d\mathbf{R}$$

$$F_{hkl} = \int_{\text{unit cell}} \sum_j \rho_j(\mathbf{R}) e^{2\pi i \mathbf{S}_{hkl} \cdot \mathbf{r}_j} e^{2\pi i \mathbf{S}_{hkl} \cdot \mathbf{R}} \, d\mathbf{R}$$

We now make three modifications. Firstly, the order of the summation and the integration is reversed. Secondly, since the term $e^{2\pi i \mathbf{S}_{hkl} \cdot \mathbf{r}_j}$ does not contain the variable of integration, it may be factored out of the integral. Thirdly, since, for any atom j, the function $\rho(\mathbf{R})$ is sharply peaked about the atom centre, the limits of integration can be changed so that the integration is over only the jth atom. If we then sum over all the atoms within the unit cell, this is equivalent to integrating over the unit cell. Our expression for F_{hkl} then becomes

$$F_{hkl} = \sum_j e^{2\pi i \mathbf{S}_{hkl} \cdot \mathbf{r}_j} \int_{atom} \rho_j(\mathbf{R}) e^{2\pi i \mathbf{S}_{hkl} \cdot \mathbf{R}} \, d\mathbf{R} \qquad (9.19)$$

The integral in equation (9.19) depends on the parameters of the jth atom only, so that each different atom has its own value for the integral according to its atomic electron density function $\rho_j(\mathbf{R})$. If we define the *atomic scattering factor* f_j as

$$f_j = \int_{atom} \rho_j(\mathbf{R}) e^{2\pi i \mathbf{S}_{hkl} \cdot \mathbf{R}} \, d\mathbf{R} \qquad (9.20)$$

then the expression for the structure factor F_{hkl} is

$$F_{hkl} = \sum_j f_j e^{2\pi i \mathbf{S}_{hkl} \cdot \mathbf{r}_j}$$

We shall discuss atomic scattering factors in the next section, but for the moment we will develop our expression for the structure factor F_{hkl}.

Returning to the fractional coordinates (x, y, z), we may express the position of the centre of the jth atom by the coordinates (x_j, y_j, z_j):

$$\mathbf{r}_j = x_j \mathbf{a} + y_j \mathbf{b} + z_j \mathbf{c}$$

For the reciprocal-lattice vector $\mathbf{S}_{hkl}$, we have

$$\mathbf{S}_{hkl} = h\mathbf{a}^* + k\mathbf{b}^* + l\mathbf{c}^*$$

$$\therefore \mathbf{S}_{hkl} \cdot \mathbf{r}_j = (h\mathbf{a}^* + k\mathbf{b}^* + l\mathbf{c}^*) \cdot (x_j \mathbf{a} + y_j \mathbf{b} + z_j \mathbf{c})$$

$$= hx_j + ky_j + lz_j$$

by the use of the equations (8.19). The equation for the structure factor F_{hkl} becomes

$$F_{hkl} = \sum_j f_j e^{2\pi i(hx_j + ky_j + lz_j)} \qquad (9.21)$$

The summation in equation (9.21) goes over all the atoms within the unit cell.

We now have two expressions for the structure factor F_{hkl}:

$$F_{hkl} = V \int_0^1 \int_0^1 \int_0^1 \rho(x, y, z) e^{2\pi i(hx + ky + lz)} \, dx \, dy \, dz \qquad (9.11)$$

$$F_{hkl} = \sum_j f_j e^{2\pi i(hx_j + ky_j + lz_j)} \qquad (9.21)$$

The differences between the equations should be noted.

(a) In equation (9.11), the coordinates (x, y, z) refer to any arbitrary position within the unit cell, whereas (x_j, y_j, z_j) in equation (9.21) define the positions of atoms.
(b) $\rho(x, y, z)$ is a continuous function describing the overall electron density, but f_j is a property of each atom.
(c) Equation (9.11) requires an integration over the entire unit cell, but equation (9.21) involves a summation over the positions of the atoms within the unit cell.

As we shall see later, the formalism of equation (9.21) is particularly useful in interpreting the symmetry properties of the unit cell, and also the diffraction pattern. For this reason, it is this equation which is the more useful for the calculation of structure factors from a known unit cell.

In this section, we have established that each atom in the unit cell of the crystal contributes a scattered wave (or wavelet) towards each structure factor F_{hkl}. The wavelet scattered by the jth atom has amplitude f_j and a phase given by $2\pi(hx_j + ky_j + lz_j)$ or $2\pi\mathbf{S}_{hkl} \cdot \mathbf{r}_j$, where $\mathbf{r}_j$ is the position vector of atom j corresponding to its coordinates (x_j, y_j, z_j), and $\mathbf{S}_{hkl}$ is the scattering vector for the diffraction spot of indices h, k and l. The structure factor F_{hkl} can therefore be represented on an Argand diagram as a vector sum of the contributions of all atoms in the unit cell (Fig. 9.7). The fact that the phase angle for the scattering contribution of each atom is given by the dot product $\mathbf{S}_{hkl} \cdot \mathbf{r}_j$ can be demonstrated neatly by considering the Bragg planes that give rise to the diffraction peak. In Fig. 9.8, the position of atom j is shown in relation to the origin O and a set of Bragg planes which are separated by d_{hkl}. From the derivation of Bragg's law, we know that the separation of the Bragg planes d_{hkl} must correspond to a phase difference of 360°, or 2π, in order for constructive interference to occur between the waves reflected by adjacent planes. If atom j produces a scattered

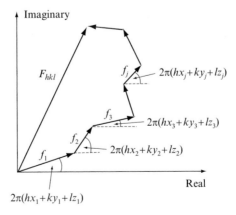

Fig. 9.7 *The contributions of the atoms in the unit cell to the structure factor F_{hkl}. Each atom (j) scatters the incident beam and acts as a source of a wavelet of amplitude f_j and phase $2\pi(hx_j + ky_j + lz_j)$. The vector sum of these scattering contributions is F_{hkl} and can be represented on an Argand diagram as shown.*

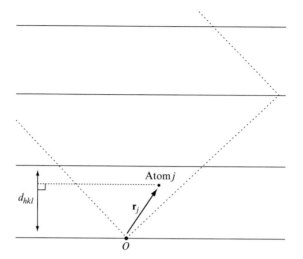

Fig. 9.8 *Phases and Bragg planes.* The position of atom j is shown relative to the origin O of a unit cell (drawn dashed) and reflecting planes *hkl*, which are drawn horizontally. The position vector $\mathbf{r}_j$ of the atom is shown projected onto the normal to the Bragg planes. The phase of the wavelet contributed by atom j to the diffraction from these planes is $2\pi \mathbf{S}_{hkl} \cdot \mathbf{r}_j$, where $\mathbf{S}_{hkl}$ is a vector of magnitude $1/d_{hkl}$ oriented perpendicular to the Bragg planes.

wavelet of phase ϕ_j, then

$$\frac{\phi_j}{2\pi} = \frac{\text{proj}(\mathbf{r}_j)}{d_{hkl}}$$

where $\text{proj}(\mathbf{r}_j)$ is the projection of the vector $\mathbf{r}_j$ onto the interplanar vector, which we know has magnitude d_{hkl}. Previously we showed that

$$\frac{1}{d_{hkl}} = S_{hkl} \tag{8.30}$$

$$\therefore \frac{\phi_j}{2\pi} = S_{hkl} \, \text{proj}(\mathbf{r}_j)$$

and

$$\frac{\phi_j}{2\pi} = \mathbf{S}_{hkl} \cdot \mathbf{r}_j$$

Therefore the component of F_{hkl} scattered by atom j in the direction of the reflected wave has a phase given by $2\pi \mathbf{S}_{hkl} \cdot \mathbf{r}_j$. This demonstrates that the phase of the wavelet that is contributed by an atom to any diffraction spot is directly related to the distance of that atom from the respective Bragg plane passing through the origin.

9.7   Atomic scattering factors

The definition of the atomic scattering factor of the jth atom is

$$f_j = \int_{\text{atom}} \rho_j(\mathbf{R})e^{2\pi i \mathbf{S}_{hkl} \cdot \mathbf{R}} \, d\mathbf{R} \tag{9.20}$$

where $\mathbf{R}$ is a position vector specifying the location of an electron with respect to its associated atom centre. Note that the inclusion of the specific reciprocal-lattice vector $\mathbf{S}_{hkl}$ implies that, for a given $\mathbf{S}_{hkl}$, the value of f_j calculated from equation (9.20) is that which pertains to the diffraction maximum indexed as hkl.

If we compare equation (9.20) with our general expression for the scattering ability of an arbitrary distribution $\rho(\mathbf{r})$ of electrons, equation (9.7),

$$F(\mathbf{\Delta k}) = \int_{\text{all } \mathbf{r}} f_e \rho(\mathbf{r})e^{i\mathbf{\Delta k} \cdot \mathbf{r}} \, d\mathbf{r} \tag{9.7}$$

on identifying $\mathbf{\Delta k}$ as $2\pi \mathbf{S}_{hkl}$ we see that f_j represents the scattering ability of an atom as compared with that of a single electron, which is given by the electronic scattering factor f_e.

Since f_j is a property of a given atom, we may assume some model for the electron density function $\rho_j(\mathbf{R})$ and calculate the corresponding f_j, which represents the relative ability of the atom to scatter radiation. We shall now discuss three possible models for the atomic electron density function $\rho_j(\mathbf{R})$ and compute

$$f_j = \int_{\text{atom}} \rho_j(\mathbf{R})e^{2\pi i \mathbf{S}_{hkl} \cdot \mathbf{R}} \, d\mathbf{R}$$

The electrons are concentrated at the atom centre

In this model, we assume that all the electrons are located at the centre of the atom. If there are Z_j electrons associated with the atom j, then Z_j is the atomic number of that atom, and the atomic electron density function $\rho_j(\mathbf{R})$ is a δ function at the origin of strength Z_j:

$$\rho_j(\mathbf{R}) = Z_j \delta(\mathbf{R})$$

Substituting in equation (9.20), we have

$$f_j = \int_{\text{atom}} Z_j \delta(\mathbf{R})e^{2\pi i \mathbf{S}_{hkl} \cdot \mathbf{R}} \, d\mathbf{R}$$

Z_j is a constant, and the integral through the δ function is unity (see Section 5.8), and so

$$f_j = Z_j$$

If we assume all the electrons in the atom are localised at the central position, i.e. all the electrons are located at a single point, then f_j is simply the total number of electrons within the atom. Since f_j represents the relative scattering ability of the atom as compared with a single electron, then the fact that f_j is equal to the number of electrons within the atom is what we would have predicted. Note that our expression for f_j is independent of the indexing variables h, k and l, implying that in this model, the atomic scattering factor has the same value at every diffraction maximum.

The atomic electron density is spherically symmetric

A rather more accurate representation of the atomic electron density function $\rho_j(\mathbf{R})$ is to assume that it is a spherically symmetric function about the atom centre. In this case, we write the atomic electron density function in terms of the scalar distance R as $\rho_j(R)$. Since the system is spherically symmetric, we use spherical polar coordinates as defined in Fig. 9.9. The axis with respect to which the angle ϕ is defined is known as the *polar axis* (note that the spherical polar angle ϕ is distinct from the phase ϕ_{hkl} of the structure factor). The volume element represented by $d\mathbf{R}$ may be expressed in terms of spherical polar coordinates as

$$d\mathbf{R} = R^2 \sin \phi \; dR \; d\phi \; d\psi$$

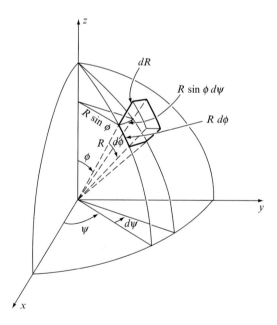

Fig. 9.9 *Spherical polar coordinates.* Any point is defined by the radial distance r from the origin, and two angles ψ and ϕ (z is the polar axis). The volume element $d\mathbf{R}$ is given by $d\mathbf{R} = R^2 \sin \phi \; dR \; d\phi \; d\psi$.

We wish to calculate

$$f_j = \int_{\text{atom}} \rho_j(\mathbf{R})e^{2\pi i \mathbf{S}_{hkl} \cdot \mathbf{R}} \, d\mathbf{R} \tag{9.20}$$

With reference to the geometry of Fig. 9.10, if we choose the direction of the polar axis to be such that it is parallel to the scattering vector $\Delta\mathbf{k}$, then, since for any given reciprocal-lattice point hkl, $\Delta\mathbf{k}$ is equal to $2\pi\mathbf{S}_{hkl}$, the polar axis is also parallel to $\mathbf{S}_{hkl}$. The relevance of this is that we can now express the scalar product $\mathbf{S}_{hkl} \cdot \mathbf{R}$ in terms of the magnitudes S_{hkl} and R, and the angle ϕ:

$$\mathbf{S}_{hkl} \cdot \mathbf{R} = S_{hkl}R\cos\phi$$

Since the system is spherically symmetric, we are free to choose the direction of the polar axis as we wish, and so the above procedure is valid. Equation (9.20) becomes

$$f_j = \int_{\text{atom}} \rho_j(R)e^{2\pi i S_{hkl}R\cos\phi} R^2 \sin\phi \, dR \, d\phi \, d\psi \tag{9.22}$$

Since we are considering only one atom, we may replace the finite limits on the integral in equation (9.22) by infinite ones, although the contribution of the integral outside the actual atom is very small. The variable R therefore runs from zero to infinity, ϕ from 0 to π, and ψ from 0 to 2π. Separating out the variables of integration, equation

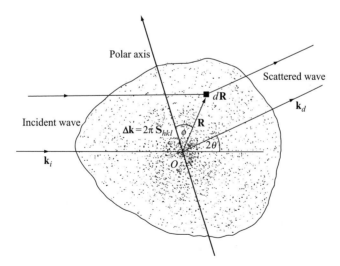

Fig. 9.10 *Scattering by an atom.* The electrons within a volume element $d\mathbf{R}$ at a position vector $\mathbf{R}$, defined with respect to the atom centre, scatter incident waves through an angle 2θ, where θ is the Bragg angle. The scattering vector $\Delta\mathbf{k} = 2\pi\mathbf{S}_{hkl}$ defines the polar axis, and the atomic scattering factor becomes

$$f_j = \int_{\text{atom}} \rho_j(\mathbf{R})e^{2\pi i \mathbf{S}_{hkl} \cdot \mathbf{R}} \, d\mathbf{R}$$

$$= \int_{\psi=0}^{\psi=2\pi} \int_{\phi=0}^{\phi=\pi} \int_{R=0}^{R=\infty} \rho_j(R)e^{2\pi i S_{hkl}R\cos\phi} R^2 \sin\phi \, dR \, d\phi \, d\psi$$

(9.22) becomes

$$
f_j = \int_0^{2\pi} d\psi \int_0^{\infty} R^2 \rho_j(R) \left[\int_0^{\pi} e^{2\pi i S_{hkl} R \cos \phi} \sin \phi \, d\phi \right] dR
$$

$$
= \int_0^{2\pi} d\psi \int_0^{\infty} R^2 \rho_j(R) \left[-\int_0^{\pi} e^{2\pi i S_{hkl} R \cos \phi} \, d(\cos \phi) \right] dR
$$

$$
= 2\pi \int_0^{\infty} R^2 \rho_j(R) \left[-\frac{e^{2\pi i S_{hkl} R \cos \phi}}{2\pi i S_{hkl} R} \right]_0^{\pi} dR
$$

$$
= 2\pi \int_0^{\infty} R^2 \rho_j(R) \left(\frac{e^{2\pi i S_{hkl} R} - e^{-2\pi i S_{hkl} R}}{2\pi i S_{hkl} R} \right) dR
$$

But

$$
\sin \theta = \frac{e^{i\theta} - e^{-i\theta}}{2i} \tag{2.34}
$$

$$
\therefore f_j = 2\pi \int_0^{\infty} R^2 \rho_j(R) \left(\frac{\sin 2\pi S_{hkl} R}{\pi S_{hkl} R} \right) dR
$$

$$
\Rightarrow f_j = 4\pi \int_0^{\infty} R^2 \rho_j(R) \left(\frac{\sin 2\pi S_{hkl} R}{2\pi S_{hkl} R} \right) dR \tag{9.23}
$$

In our discussion of Bragg's law in Section 8.9, we found that for a diffraction maximum,

$$
S_{hkl} = \frac{2}{\lambda} \sin \theta
$$

where θ is the appropriate Bragg angle. Equation (9.23) now gives

$$
f_j = 4\pi \int_0^{\infty} R^2 \rho_j(R) \frac{\sin \left(\frac{4\pi}{\lambda} \sin \theta \right) R}{\left(\frac{4\pi}{\lambda} \sin \theta \right) R} dR
$$

It is conventional to define a new variable s as

$$
s = \frac{4\pi \sin \theta}{\lambda}
$$

$$
\therefore f_j = 4\pi \int_0^{\infty} R^2 \rho_j(R) \left(\frac{\sin sR}{sR} \right) dR \tag{9.24}
$$

The value of the integral after this stage requires some definite knowledge of the form of the atomic electron density function $\rho_j(R)$. However, inspection of the integral in equation (9.24) shows that since the integration is over R, the value of the integral becomes a function of the variable s irrespective of any model we choose for $\rho_j(R)$. Since the variable s is defined as $(4\pi \sin \theta)/\lambda$, the atomic scattering factor f_j is essentially a function of $(\sin \theta)/\lambda$.

Various assumptions as to the form of $\rho_j(R)$ may be made from quantum mechanical studies. The actual form of the calculations does not concern us here, but what is important is the general shape of f_j as a function of $(\sin \theta)/\lambda$, as shown in Fig. 9.11.

The atomic scattering factor f_j is seen to fall off slowly as the Bragg angle increases. Values of f_j for all atoms as a function of $(\sin \theta)/\lambda$ based on various models for $\rho_j(R)$ are to be found in reference works such as the *International Tables for Crystallography*. The way that these are used is that for a given wavelength λ, we may find the value of f_j for any atom at any Bragg angle. If we wish to calculate the structure factor F_{hkl} for a particular diffraction maximum, since

$$F_{hkl} = \sum_j f_j e^{2\pi i(hx_j + ky_j + lz_j)} \tag{9.21}$$

then for any values of h, k and l, we may determine the Bragg angle θ. Having done this, we know both $\sin \theta$ and λ and so we may find the appropriate values of f_j for the diffraction maximum in which we are interested. If we wish to evaluate the structure factor for a different reflection, then the Bragg angle will have changed, and we must

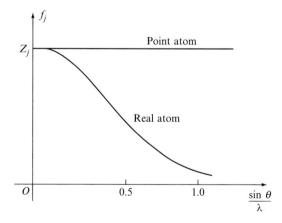

Fig. 9.11 *The atomic scattering factor f_j.* If the atomic electron density function is assumed to be a δ function at the atom centre, the atomic scattering factor is equal to the atomic number Z of the atom for all values of the Bragg angle θ, as in the 'point atom' curve at the top. For all other electron density models, the atomic scattering factor f_j decreases as θ increases, as shown in the 'real atom' curve.

ensure that we use the values of the atomic scattering factors corresponding to the correct value of $(\sin\theta)/\lambda$.

An interesting calculation is to find the value of the integral in equation (9.24) for very small values of the angle θ. We see that the integral contains a $(\sin x)/x$ function and, as $\theta \to 0$, then $s \to 0$ and

$$\frac{\sin sR}{sR} \to 1$$

The integral then becomes

$$f_j = 4\pi \int_0^\infty \rho_j(R) R^2 \, dR$$

Now, $\rho_j(R)$ represents the electron density in electrons per unit volume. This integral must therefore represent the total number of electrons Z_j within the atom, for $\rho_j(R)$ was assumed to be spherically symmetric, and the volume of a sphere is simply $4\pi \int R^2 \, dR$. Hence

$$f_j = Z_j \tag{9.25}$$

Thus for small Bragg angles θ, the value of the atomic scattering factor is the atomic number Z_j of the atom. This implies that heavy atoms are associated with large atomic scattering factors, and light atoms with small ones.

$\rho_j(\mathbf{R})$ is not spherically symmetric

This final stage in sophistication requires a highly accurate model for the electronic structure of each atom. We know that, physically, the assumption of a spherically symmetric electronic distribution is incorrect for all atoms except hydrogen, for elementary chemical quantum mechanics identifies the non-spherically symmetric p, d, f, ... orbitals. Rigorous calculation of atomic scattering factors is therefore a highly complicated problem in quantum mechanics. Various methods have been used, but it is doubtful whether the time spent on calculating accurate values is justified when one considers the overall accuracy of a practical X-ray investigation. For the most part, atomic scattering factors calculated on the assumption of spherical symmetry are quite adequate for most purposes.

9.8   Anomalous scattering

Inspection of equation (9.24) for the atomic scattering factor f_j based on the assumption of a spherically symmetric electron distribution shows that f_j will be a real number. Indeed, this is to be expected, for in our discussion of Fourier transforms in Section 5.9, we saw that any centrosymmetric function necessarily has a real Fourier transform. Since a system with spherical symmetry is necessarily centrosymmetric, f_j is real.

It was mentioned earlier in this chapter that the model used for Thomson scattering is valid unless the energy of the incident wave is close to an energy which will change the quantum state of an electron within an atom. Should this condition not be fulfilled, then the Thomson scattering equation, equation (9.1), does not hold, and we talk of *anomalous scattering*. It is beyond the scope of this volume to prove that when anomalous scattering occurs, it is grossly wrong to consider the electron distribution of an atom to be spherically symmetric. Although a mathematical proof will not be given here, the effect is quite understandable in physical terms. Since the energy of the incident wave is such as to cause electronic transitions within the atom, then we would expect the electron distribution to be highly perturbed, so that it is not unreasonable to expect that such spherical symmetry as may have existed is destroyed. In fact, not only is the spherical symmetry destroyed, but so is the centrosymmetry. Hence when anomalous scattering occurs, the atomic electron density function cannot be assumed to be centrosymmetric. This implies that the Fourier transform of the atomic electron density function, namely the atomic scattering factor, is no longer real. In the case of anomalous scattering, atomic scattering factors must be written in the form

$$f_j = f_{0j} + f'_j + if''_j$$

in which the atomic scattering factor is now a complex number. The quantity f_{0j} which is calculated on the assumption of spherical symmetry is corrected by a real part f'_j and an imaginary part f''_j, both of which may be calculated theoretically.

The most important point to remember from the point of view of the practical crystallographer is that for most purposes, the atomic scattering factor f_j may be considered to be real. In the case of anomalous scattering, the atomic scattering factor f_j is complex.

The fact that the atomic scattering factor f_j of an atom which shows anomalous scattering is complex is highly relevant to one method of tackling the phase problem.

9.9   Crystal symmetry and X-ray diffraction

As we saw in Chapter 3, all crystals possess certain symmetry properties. The fundamental crystal symmetry is that of the Bravais lattice, of which there are 14 types. The next classification is according to the 32 point groups, which correspond to those symmetry operations which do not involve a translation, and so may be considered to act about each lattice point. The symmetry operations included here are proper rotations, improper rotations, reflections and inversions. Finally, every crystal must belong to one of the 230 space groups, which represent the number of different space-filling patterns which may be generated by the combination of the 32 point groups with those symmetry operations which involve a translation. These additional symmetry operations are screw rotations and glides. However, since protein molecules are composed only of L-amino acids, this restricts the crystals that they form to 65 possible space groups.

We have already seen how the structure factors F_{hkl} may be used to generate the electron density function by means of a Fourier synthesis. Before the electron density function is computed, the X-ray diffraction pattern is studied in order to derive information concerning the crystal symmetry. Since the crystal symmetry is determined by the disposition of molecules with respect to the various symmetry elements, the atomic configuration within the unit cell determines the crystal symmetry. According to equation (9.21),

$$F_{hkl} = \sum_j f_j e^{2\pi i (hx_j + ky_j + lz_j)} \tag{9.21}$$

we express the structure factors F_{hkl} in terms of the coordinates (x_j, y_j, z_j) of the atoms within the unit cell. Since these coordinates determine the crystal symmetry, we see that the structure factors F_{hkl} are directly related to the crystal symmetry, and hence contain information concerning the crystal symmetry. It is for this reason that we chose to modify equation (9.11),

$$F_{hkl} = V \int_0^1 \int_0^1 \int_0^1 \rho(x, y, z) e^{2\pi i (hx + ky + lz)} \, dx \, dy \, dz \tag{9.11}$$

into the form of equation (9.21),

$$F_{hkl} = \sum_j f_j e^{2\pi i (hx_j + ky_j + lz_j)} \tag{9.21}$$

In the next few sections we shall see how crystal symmetry shows itself in the diffraction pattern. For the purposes of the following discussion we shall assume that we have measured a complete diffraction pattern, which has been indexed according to the appropriate reciprocal-lattice points *hkl*. The methods by which this may be done will be discussed in later chapters.

9.10 Diffraction pattern symmetry

Cursory examination of any diffraction pattern, such as that in Fig. 9.12, shows that diffraction patterns in general possess symmetry in both the disposition and the intensities of the diffraction maxima. In this section, we shall see how some of the symmetry of the intensities arises.

Let us consider the specific example of a crystal which is assumed to contain a twofold rotation axis along the **c** axis of the unit cell. In this case, for an atom A located at (x_A, y_A, z_A), there is a symmetrically related point $(-x_A, -y_A, z_A)$ which represents the position of an identical atom.

The structure factor F_{hkl} is given by equation (9.21),

$$F_{hkl} = \sum_j f_j e^{2\pi i (hx_j + ky_j + lz_j)} \tag{9.21}$$

We wish to express the contribution to F_{hkl} made by the two symmetrically related atoms located in the positions mentioned above. Denoting the contributions of these two atoms by $(F_{hkl})_A$, we have

$$(F_{hkl})_A = f_A \left(e^{2\pi i (hx_A + ky_A + lz_A)} + e^{2\pi i (-hx_A - ky_A + lz_A)} \right) \tag{9.26}$$

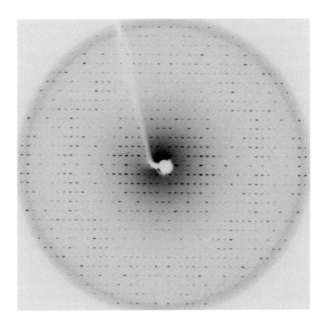

Fig. 9.12 *A typical protein diffraction pattern as photographed using the Buerger precession technique.* This method provides an undistorted view of the reciprocal lattice. Note the systematic absences on the horizontal axis, i.e. spots with odd indices on this axis appear as voids in the otherwise regular pattern.

Similarly, the contribution of these two atoms to the structure factor $F_{\overline{hkl}}$ is given by

$$\left(F_{\overline{hkl}}\right)_A = f_A \left(e^{2\pi i(-hx_A - ky_A + lz_A)} + e^{2\pi i(hx_A + ky_A + lz_A)}\right) \tag{9.27}$$

Comparison of equations (9.26) and (9.27) shows that the contributions of the two atoms related by the twofold axis to the structure factors F_{hkl} and $F_{\overline{hkl}}$ are the same. Since a crystallographic twofold axis will operate on all atoms in the unit cell, the structure factors F_{hkl} and $F_{\overline{hkl}}$ are equal. Since our experimentally observed quantity is the intensity I_{hkl}, a twofold axis in the real crystal implies that the diffraction maxima are related such that

$$I_{hkl} = I_{\overline{hkl}}$$

Hence the reciprocal-lattice points hkl and $\overline{hkl}$ are symmetrically related by a twofold axis so that a twofold axis in the real lattice implies the presence of a twofold axis in the reciprocal lattice, and this result will be indicated in the diffraction pattern.

In an exactly similar manner, we may demonstrate that any proper rotation axis within the real crystal will result in an equivalent proper rotation axis in the reciprocal lattice and therefore in the diffraction pattern. This is displayed in a particularly beautiful manner in X-ray photographs taken with an instrument called the Buerger precession camera. This photographs the reciprocal lattice in an undistorted manner, and the photograph reproduced in Fig. 9.12 was taken by this technique. This method of recording diffraction data has largely been superseded by electronic detectors, which allow 'pseudo-precession' images to be generated from the data, and symmetry information can be deduced from these images.

Examination of the symmetry of the intensities of the diffraction maxima in an X-ray diffraction pattern therefore enables us to identify proper rotation axes in the real crystal. It is left as an exercise for the reader to demonstrate that mirror planes in the real lattice manifest themselves as mirror planes in the reciprocal lattice, although we would not expect to observe such symmetry with protein crystals.

9.11   Friedel's law

An interesting property of equation (9.21),

$$F_{hkl} = \sum_j f_j e^{2\pi i\left(hx_j + ky_j + lz_j\right)} \tag{9.21}$$

appears when we consider the structure factors corresponding to an arbitrary reciprocal-lattice point hkl and its centrosymmetrically related point $\overline{hkl}$. Using equation (9.21), we have

$$F_{hkl} = \sum_j f_j e^{2\pi i(hx_j + ky_j + lz_j)}$$

$$F_{\overline{hkl}} = \sum_j f_j e^{2\pi i\left(\overline{h}x_j + \overline{k}y_j + \overline{l}z_j\right)} = \sum_j f_j e^{-2\pi i\left(hx_j + ky_j + lz_j\right)}\Big|$$

The only difference between the values of F_{hkl} and $F_{\overline{hkl}}$ is in the sign of the complex exponential. In general, we may assume that the atomic scattering factors f_j are real, and, in this case, $F_{\overline{hkl}}$ is simply the complex conjugate of F_{hkl}:

$$F_{hkl}^* = F_{\overline{hkl}}$$

The structure factors F_{hkl} and $F_{\overline{hkl}}$ may be represented on an Argand diagram as in Fig. 9.13.

Since $F_{\overline{hkl}}$ and F_{hkl} are complex conjugates, they may be represented by mirror reflections in the real axis. Of particular importance is the fact that the magnitudes $|F_{hkl}|$ and $|F_{\overline{hkl}}|$ are equal:

$$|F_{hkl}| = |F_{\overline{hkl}}| \tag{9.28}$$

Equation (9.28) is known as Friedel's law, of which one possible verbal statement is

The magnitudes of the structure factors of centrosymmetrically related reciprocal-lattice points are equal.

Since we observe intensities I_{hkl} and $I_{\overline{hkl}}$, Friedel's law implies

$$I_{hkl} = I_{\overline{hkl}} = |F_{hkl}|^2 = |F_{\overline{hkl}}|^2$$

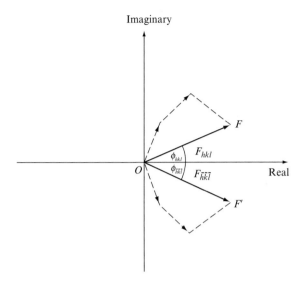

Fig. 9.13 *Friedel's law.* Since $F_{hkl} = \sum_j f_j e^{2\pi i \left(hx_j + ky_j + lz_j \right)}$, the contribution of each atom to F_{hkl} is the product of its atomic scattering factor with a complex exponential depending solely on the coordinates (x_j, y_j, z_j). If f_j is real, this product may be represented on an Argand diagram by a vector of magnitude f_j and phase angle $2\pi(hx_j + ky_j + lz_j)$. The contributions of three atoms are shown. On calculating $F_{\overline{hkl}}$, the magnitudes of the contributions of each atom remain the same, but the phase angle changes sign. Thus $F_{\overline{hkl}}$ is formed from contributions which are the mirror images in the real axis of those which form F_{hkl}. Hence F_{hkl} and $F_{\overline{hkl}}$ are complex conjugates, with equal magnitude and opposite phase angle.

giving an alternative verbal statement of the law as

The intensities of the diffraction maxima corresponding to centrosymmetrically related reciprocal-lattice points are equal.

This effect was first noted in 1913 by the French crystallographer Georges Friedel (1865–1933). A most important aspect of equation (9.21), which can be seen in Fig. 9.13, is that the phases of the two reflections forming a *Friedel pair* (also known as *Friedel mates*) are related in the following way:

$$\phi_{\overline{hkl}} = -\phi_{hkl} \tag{9.29}$$

which follows naturally from the fact that $F_{\overline{hkl}}$ and F_{hkl} are complex conjugates.

In this discussion, we have made no initial assumptions as to the presence or absence of centrosymmetry within the real crystal. Friedel's law was proved on a perfectly general basis, and its significance is that X-ray diffraction pictures in general are centrosymmetric whether or not the crystal has an inversion centre. This is known as *Friedel symmetry*, and is inevitably present in the diffraction patterns of all crystals.

Whereas the previous section demonstrated that mirror planes and rotation axes in the diffraction pattern implied the presence of equivalent symmetry elements in the real crystal, Friedel's law implies that diffraction patterns are inevitably centrosymmetric, and hence we cannot infer the presence of an inversion centre from an X-ray diffraction photograph. This is connected to the fact that our experimental observable is intensity, and implies that there is always an ambiguity as to the presence of an inversion centre within the crystal, although, of course, with proteins we can be sure that inversion and mirror symmetry will not be present.

To summarise the last two sections, we may state that for all crystals (not just proteins), observation of the symmetry of the diffraction pattern allows us to identify proper rotation axes and mirror planes, but not inversion centres and improper rotation axes.

9.12   The breakdown of Friedel's law

The proof of Friedel's law as stated above assumed that the atomic scattering factors f_j were all real. But in Section 9.8 we learnt that any atom, say atom A, which shows anomalous scattering has a complex atomic scattering factor, and we may represent this in the general form $\alpha + i\beta$ on an Argand diagram (Fig. 9.14).

If we suppose that one atom A within the unit cell is an anomalous scatterer, then its anomalous scattering factor may, like any complex number, be represented in exponential form as $|f_A|e^{i\phi_A}$. Hence, we may write the structure factor F_{hkl} as

$$F_{hkl} = \sum_{j \neq A} f_j e^{2\pi i\left(hx_j + ky_j + lz_j\right)} + |f_A|e^{2\pi i(hx_A + ky_A + lz_A)}e^{i\phi_A}$$

in which the summation goes over all the atoms within the unit cell except the anomalous scatterer. The centrosymmetrically related structure factor $F_{\overline{hkl}}$ becomes

$$F_{\overline{hkl}} = \sum_j f_j e^{-2\pi i\left(hx_j + ky_j + lz_j\right)} + |f_A|e^{-2\pi i(hx_A + ky_A + lz_A)}e^{i\phi_A}$$

Comparison of our expressions for F_{hkl} and $F_{\overline{hkl}}$ shows that because of the presence of the factor $e^{i\phi_A}$, which does not change sign, F_{hkl} and $F_{\overline{hkl}}$ are no longer complex conjugates. This is shown in the Argand diagram of Fig. 9.14. The presence of an anomalous scatterer within the unit cell therefore causes a breakdown of Friedel's law, and it is no longer true that the intensities of centrosymmetrically related reciprocal-lattice points are the same.

The above proof states that when an anomalously scattering atom is present, Friedel's law is broken and

$$I_{hkl} \neq I_{\overline{hkl}}$$

A special case does arise, however, for a centrosymmetric crystal containing an anomalous scatterer. Since the crystal is centrosymmetric, the anomalously scattering atoms A must be present in centrosymmetrically related pairs, for example at (x_A, y_A, z_A) and $(-x_A, -y_A, -z_A)$. If we write (F_{hkl}) for the contribution to the structure factor F_{hkl} due to the atoms A only, then

$$(F_{hkl})_A = |f_A|e^{2\pi i(hx_A + ky_A + lz_A)}e^{i\phi_A} + |f_A|e^{-2\pi i(hx_A + ky_A + lz_A)}e^{i\phi_A}$$

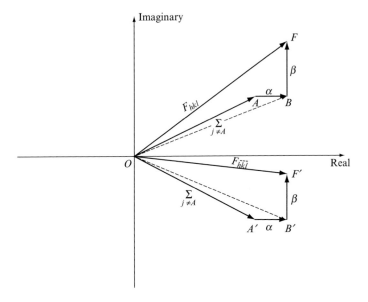

Fig. 9.14 *The breakdown of Friedel's law.* For clarity, we assume that the unit cell has a single anomalous scatterer A at the origin $(0, 0, 0)$, i.e. its phase is 0. This atom has a complex atomic scattering factor $f_A = \alpha + i\beta$. The structure factor F_{hkl} is given by

$$F_{hkl} = \sum_{j \neq A} f_j e^{2\pi i \left(hx_j + ky_j + lz_j \right)} + \alpha + i\beta$$

The summation is represented by the vector $\overrightarrow{OA}$, the real part α by $\overrightarrow{AB}$ and the imaginary part β by $\overrightarrow{BF}$. F_{hkl} is represented by $\overrightarrow{OF}$. The centrosymmetrically related structure factor $F_{\overline{hkl}}$ is given by

$$F_{\overline{hkl}} = \sum_{j \neq A} f_j e^{-2\pi i \left(hx_j + ky_j + lz_j \right)} + \alpha + i\beta$$

The summation is represented by $\overrightarrow{OA'}$, and this is the mirror image of $\overrightarrow{OA}$ in the real axis. The contributions of the anomalous scatterer are represented by $\overrightarrow{A'B'}$ and $\overrightarrow{B'F'}$ as shown, and the structure factor $F_{\overline{hkl}}$ is represented by $\overrightarrow{OF'}$. F_{hkl} and $F_{\overline{hkl}}$ are now no longer mirror images in the real axis, and are no longer complex conjugates. Hence Friedel's law breaks down. Note that if the atom at the origin ceases to scatter anomalously, β goes to zero, and $\overrightarrow{OF}$ approaches $\overrightarrow{OB}$ and $\overrightarrow{OF'}$ approaches $\overrightarrow{OB'}$.

Similarly, the contribution to the structure factor $F_{\overline{hkl}}$ is given by

$$(F_{\overline{hkl}})_A = |f_A| e^{-2\pi i (hx_A + ky_A + lz_A)} e^{i\phi_A} + |f_A| e^{2\pi i (hx_A + ky_A + lz_A)} e^{i\phi_A}$$

Comparison of the expressions for $(F_{hkl})_A$ and $(F_{\overline{hkl}})A$ shows that

$$(F_{hkl})_A = (F_{\overline{hkl}})_A$$

implying that the anomalously scattering atoms contribute equally to F_{hkl} and $F_{\overline{hkl}}$.

Since the contributions to F_{hkl} and $F_{\overline{hkl}}$ due to the atoms other than A are necessarily real, it follows that

$$|F_{hkl}| = |F_{\overline{hkl}}|$$

and so

$$I_{hkl} = I_{\overline{hkl}}$$

Thus, Friedel's law is upheld even though an anomalous scatterer is present. This 'unexpected' behaviour can be of value to the small-molecule crystallographer. We saw in the previous section that, in the absence of an anomalous scatterer, it was impossible to determine whether an inversion centre is present in a crystal solely from observation of a diffraction pattern. If, however, the crystal contains an anomalous scatterer and it is still observed that

$$I_{hkl} = I_{\overline{hkl}}$$

then we may correctly infer that an inversion centre is indeed present. Whilst this effect may not seem particularly relevant to protein crystals, it is of some importance for the centric zones of diffraction patterns, for in these zones, the anomalous-scattering effects balance out for the Friedel mates.

Consider the space group $P2_1$, which is very common for proteins. This has a 2_1 screw axis along $\mathbf{b}$, which gives rise to a centric zone in the $(h, 0, l)$ layer of the diffraction pattern. The general equivalent positions (defined in Chapter 3, Section 3.8) for this space group are (x, y, z) and $(-x, y + \frac{1}{2}, -z)$. The structure factor equation for the $(h, 0, l)$ layer is

$$F_{h0l} = \sum_{j=1}^{j=N/2} \left(f_j e^{2\pi i (hx_j + lz_j)} + f_j e^{-2\pi i (hx_j + lz_j)} \right)$$

where N is the total number of atoms in the unit cell and $N/2$ is therefore the number in the asymmetric unit. Since

$$\cos \theta = \frac{e^{i\theta} + e^{-i\theta}}{2} \tag{2.33}$$

it follows that

$$F_{h0l} = 2 \sum_{j=1}^{j=N/2} \cos 2\pi (hx_j + lz_j)$$

which is real, i.e. the imaginary terms containing i have disappeared.

On an Argand diagram, real numbers are drawn on the horizontal axis, which implies that their phases are either 0 or π; these values correspond to the positive and negative directions, respectively, of the x axis. These arguments indeed apply to diffraction from crystals possessing a centre of symmetry in real space and, for such crystals, all of the reflections will have phases of either 0 or π. If a reflection hkl has a phase of 0, then by use of equation (9.29) we see that the Friedel mate $\overline{hkl}$ will also have a phase of 0. Similarly, for a reflection hkl with a phase of π, its Friedel mate will have the same phase. This is important because it means that F_{hkl} and $F_{\overline{hkl}}$ will both lie on the horizontal axis of the Argand diagram and will both point in the same direction. In this situation, the breakdown of Friedel's law due to the presence of an anomalous scatterer, as depicted in Fig. 9.14, does not occur, and even for non-centrosymmetric crystals containing an anomalous scatterer, Friedel's law is still obeyed in the centric zones of the diffraction pattern.

In general, the breakdown of Friedel's law in a non-centrosymmetric crystal is of extreme importance in the solution of the phase problem, as we shall see later.

9.13 Friedel's law and electron density calculations

Computing the electron density $\rho(x, y, z)$, which is a real number corresponding to each xyz coordinate in the unit cell, requires that we eliminate the imaginary terms from equation (9.15). Friedel's law provides an elegant way to do this for crystals that do not contain anomalously scattering elements.

The electron density is given by

$$\rho(x, y, z) = \frac{1}{V} \sum_h \sum_k \sum_l F_{hkl} e^{-2\pi i(hx+ky+lz)} \tag{9.15}$$

By considering Friedel's law, we may write

$$\rho(x, y, z) = \frac{1}{2V} \sum_h \sum_k \sum_l \left[F_{hkl} e^{-2\pi i(hx+ky+lz)} + F_{\overline{hkl}} e^{2\pi i(hx+ky+lz)} \right]$$

$$= \frac{1}{2V} \sum_h \sum_k \sum_l \left[|F_{hkl}| e^{i\phi_{hkl}} e^{-2\pi i(hx+ky+lz)} + |F_{\overline{hkl}}| e^{i\phi_{\overline{hkl}}} e^{2\pi i(hx+ky+lz)} \right]$$

From Friedel's law, we know that

$$|F_{hkl}| = |F_{\overline{hkl}}| \tag{9.28}$$

and

$$\phi_{\overline{hkl}} = -\phi_{hkl} \tag{9.29}$$

Hence

$$\rho(x, y, z) = \frac{1}{2V} \sum_h \sum_k \sum_l \left[|F_{hkl}| e^{i\phi_{hkl}} e^{-2\pi i(hx+ky+lz)} + |F_{hkl}| e^{-i\phi_{hkl}} e^{2\pi i(hx+ky+lz)} \right]$$

$$= \frac{1}{2V} \sum_h \sum_k \sum_l |F_{hkl}| \left[e^{i[\phi_{hkl} - 2\pi(hx+ky+lz)]} + e^{-i[\phi_{hkl} - 2\pi(hx+ky+lz)]} \right]$$

$$= \frac{1}{V} \sum_h \sum_k \sum_l |F_{hkl}| \cos\left(\phi_{hkl} - 2\pi(hx + ky + lz) \right)$$

or

$$\rho(x, y, z) = \frac{1}{V} \sum_h \sum_k \sum_l |F_{hkl}| \cos\left(2\pi(hx + ky + lz) - \phi_{hkl} \right)$$

This equation contains no imaginary terms and therefore expresses the electron density equation (9.15) as a trigonometric series. This can be calculated directly from the diffraction pattern, assuming that we have determined each $|F_{hkl}|$ and ϕ_{hkl} and that the crystal does not contain atoms which give significant anomalous scattering, which is usually the case.

9.14   Systematic absences

The previous sections have discussed the relationship of the diffraction pattern to the presence within the crystal of those symmetry elements which do not involve a translation. We shall now discuss the effect of translational symmetry elements (e.g. screw axes), but before we do this we shall investigate the more fundamental translational symmetry, as determined by the real-lattice type.

Our first example is to calculate the expected structure factors for a body-centred lattice, which, by definition, contains identical lattice points at the corners and at the body centre. As we saw in Table 3.2, a body-centred lattice has two independent lattice points, indexed as $(0, 0, 0)$ and $(1/2, 1/2, 1/2)$. Any molecule at the general position (x, y, z) will have a symmetry-related counterpart at $(x + 1/2, y+1/2, z + 1/2)$. The expression for the structure factor F_{hkl} therefore has the following terms:

$$F_{hkl} = \sum_{j}^{j=N/2} \left(f_j e^{2\pi i [hx_j + ky_j + lz_j]} + f_j e^{2\pi i [h(x_j+1/2)+k(y_j+1/2)+l(z_j+1/2)]} \right)$$

$$= \sum_{j}^{j=N/2} f_j e^{2\pi i [hx_j + ky_j + lz_j]} \left(1 + e^{\pi i [h+k+l]} \right) \tag{9.30}$$

It is interesting to interpret equation (9.30) for various values of h, k and l. Since the indices h, k and l are necessarily integral, the sum $h + k + l$ is also an integer, which must be either even or odd. If this sum is even, and equal to, say, $2n$, where n is any integer, then

$$e^{i\pi(h+k+l)} = e^{2\pi i n} = 1$$

Therefore

$$F_{hkl} = 2 \sum_{j}^{j=N/2} f_j e^{2\pi i \left(hx_j + ky_j + lz_j \right)} \text{ if } h + k + l = 2n$$

On the other hand, if $h + k + l$ is odd, say $2n + 1$, then

$$e^{i\pi(h+k+l)} = e^{\pi i(2n+1)} = e^{2\pi i n} e^{i\pi} = -1$$

and so

$$F_{hkl} = 0 \text{ if } h + k + l = 2n + 1$$

The cases of the sum $h + k + l$ being even and being odd cover all diffraction maxima, and we see that F_{hkl} is finite or zero depending on whether $h + k + l$ is even or odd.

Since the diffraction pattern records intensities, and these are equal to $|F_{hkl}|^2$, the above discussion implies that half the diffraction maxima which we would expect to see are in fact 'missing', in that they have zero intensity as a result of the body-centred structure of the unit cell of the real lattice.

Diffraction maxima which turn out to have zero intensity as a result of the symmetry of the structure of the unit cell are called *systematic absences* or *extinctions*. Since we shall meet the word 'extinction' in a rather different context in later chapters, to avoid confusion, we shall use the term 'systematic absence' exclusively.

As a result of the above discussion, we may now interpret further aspects of a diffraction pattern. Suppose that a particular diffraction pattern possesses peaks indexed h, k and l such that $h + k + l$ is always an even number, for example points given by the indices $(0, 0, 0)$, $(1, 1, 0)$, $(2, 2, 2)$ and so on. If the remaining reciprocal-lattice points, where $h + k + l$ is an odd number, have zero intensity then we may infer directly that the unit cell is body-centred.

As another example, we shall predict the systematic absences for a face-centred lattice, with identical atoms at the corners and face centres. Reference to Table 3.2 shows that we must consider four independent lattice sites, indexed as $(0, 0, 0)$, $(\frac{1}{2}, \frac{1}{2}, 0)$, $(\frac{1}{2}, 0, \frac{1}{2})$ and $(0, \frac{1}{2}, \frac{1}{2})$.

By the same treatment as given above, we have

$$F_{hkl} = \sum_{j}^{j=N/4} f_j e^{2\pi i [hx_j + ky_j + lz_j]} \left(1 + e^{\pi i [h+k]} + e^{\pi i [h+l]} + e^{\pi i [k+l]}\right) \qquad (9.31)$$

We may distinguish the following cases:

(a) h, k, l all even. In this case the sums $h + k$ and so on are even, and

$$F_{hkl} = 4 \sum_{j}^{j=N/4} f_j e^{2\pi i [hx_j + ky_j + lz_j]}$$

(b) h, k, l all odd. It is still true that the sums $h + k$ are even, and so again

$$F_{hkl} = 4 \sum_{j}^{j=N/4} f_j e^{2\pi i [hx_j + ky_j + lz_j]}$$

(c) One of h, k, l is even. If h is even, $k + l$ is even, but $h + k$ and $h + l$ are odd, giving $F_{hkl} = 0$.

(d) one of h, k, l is odd. If k is odd, $h + l$ is even but $h + k$ and $k + l$ are odd, and therefore $F_{hkl} = 0$.

For a face-centred lattice, therefore, we have that

$$F_{hkl} = 4 \sum_{j}^{j=N/4} f_j e^{2\pi i [hx_j + ky_j + lz_j]} \text{ if } h, k, l \text{ all odd or all even}$$

$$F_{hkl} = 0 \text{ if } h, k, l \text{ mixed}$$

and the above four cases cover all combinations of h, k and l.

This pattern of systematic absences is different from that given earlier for a body-centred lattice. Hence, given a diffraction pattern in which systematic absences are characterised by triplets such as $(1, 1, 0)$ and $(1, 2, 3)$, we may therefore construe that the crystal has a face-centred lattice.

We may carry out similar analyses for each of the other 12 Bravais lattices, and derive the conditions for the systematic absences. Calculation shows that particular systematic absences correlate with each of the Bravais lattices, and so the lattice type of any crystal may be unambiguously identified. The conditions for the lattice types relevant to protein crystals are shown in Table 9.1.

Translational symmetry elements associated with certain symmetry operators such as screw axes in protein crystals also give rise to systematic absences. Whereas the lattice type may be identified using all three indices h, k and l, screw axes can be identified if we restrict our attention to certain groups of diffraction maxima. We shall derive the appropriate conditions for the common protein space group $P2_1$ again, for

Table 9.1 Conditions on h, k and l for systematic absences.

Class of reflection	Condition for presence	Interpretation	Symbol
hkl	None	Primitive	P
	$h + k + l = 2n$	Body-centred	I
	$h + k = 2n$	C-face-centred	C
	$k + l = 2n$	A-face-centred	A
	$h + l = 2n$	B-face-centred	B
	$\left.\begin{array}{l} h + k = 2n \\ k + l = 2n \\ h + l = 2n \end{array}\right\}$	Centred on all faces	F
	$-h + k + l = 3n$	Rhombohedral, obverse	R
	$h - k + l = 3n$	Rhombohedral, reverse	R
$h00$	$h = 2n$	Screw axis $\parallel$ **a**	$2_1, 4_2$
	$h = 4n$		$4_1, 4_3$
$0k0$	$k = 2n$	Screw axis $\parallel$ **b**	$2_1, 4_2$
	$k = 4n$		$4_1, 4_3$
$00l$	$l = 2n$	Screw axis $\parallel$ **c**	$2_1, 4_2, 6_3$
	$l = 3n$		$3_1, 3_2, 6_2, 6_4$
	$l = 4n$		$4_1, 4_3$
	$l = 6n$		$6_1, 6_5$
$hh0$	$h = 2n$	Screw axis $\parallel$ [1 1 0]	2_1

which the general equivalent positions are (x, y, z) and $(-x, y + 1/2, -z)$. The structure factor F_{hkl} is then given by

$$F_{hkl} = \sum_{j}^{j=N/2} \left(f_j e^{2\pi i \left(hx_j + ky_j + lz_j \right)} + f_j e^{2\pi i (-hx_j + k(y_j + 1/2) - lz_j)} \right)$$

Considering only the row of diffraction spots for which $h = 0$ and $l = 0$, i.e. the $0k0$ axis of the diffraction pattern,

$$F_{0k0} = \sum_{j}^{j=N/2} f_j e^{2\pi i k y_j} \left(1 + e^{\pi i k} \right)$$

For the $0k0$ axis, when k is even, F_{0k0} will be non-zero,

$$F_{0k0} = 2 \sum_{j}^{j=N/2} f_j e^{2\pi i k y_j} = \sum_{j}^{j=N} f_j e^{2\pi i k y_j}$$

but when k is odd, F_{0k0} will be zero. Thus for space group $P2_1$, diffraction spots on the $0k0$ axis are observed only when k is even.

Hence systematic absences are extremely useful in determining the lattice type and space group of a crystal. The full details are shown in Table 9.1. It should be noted that whereas systematic absences allow unambiguous determination of the lattice type, those corresponding to screw axes identify the direction of the axis, but there may still be some ambiguity as to the exact nature of the symmetry element; for example, $P4_1$ and $P4_3$ have the same systematic-absence conditions. However, this ambiguity can be resolved at a later stage of the structure analysis.

9.15 The determination of crystal symmetry

We have now shown that the symmetry of an X-ray diffraction pattern is characterised by the symmetry of the reflection intensities and by the systematic absences. M. J. Buerger has calculated that there are 122 different possible symmetry types for all X-ray diffraction photographs, and so all crystals must give a pattern corresponding to one of these.

The ultimate objective of a symmetry determination is to specify the space group of a crystal, but since there are 230 space groups but only 122 diffraction pattern types, the space group cannot always be unambiguously identified.

In this section, we shall see just what symmetry information can be derived from the symmetry of an X-ray diffraction photograph.

Firstly, the systematic absences corresponding to the lattice type unambiguously determine the Bravais lattice. As regards the point group, of the 32 possible point groups, only 11 are centrosymmetric. Since in the absence of an anomalous scatterer Friedel's law is upheld, the point group of the diffraction pattern is necessarily one of these 11, and these are referred to as the 11 *Laue groups*. We may therefore identify the Laue group of the real lattice from the diffraction pattern. Laue groups can be elegantly represented by stereograms (as shown in Chapter 3, Section 3.7), and these greatly aid the visual interpretation of diffraction data.

Information on screw axes (and glide planes) is obtained to a certain extent from the systematic absences, and Buerger has shown that of the 122 diffraction symmetries, 58 uniquely define a single space group. Should a particular diffraction pattern be one of these 58, the space group is then known. As regards the other 64 diffraction symbols, there is an ambiguity as to the space group, but those space groups corresponding to each diffraction symbol are tabulated, for instance in Volume A of the *International Tables for Crystallography*. Once the diffraction symmetry is known, the space group is then reduced to one of up to four or five possibilities, and the ambiguity must be removed by some other means; we shall mention some of these in Part III of this book.

It is interesting to note that the space groups which are more difficult to identify are those of relatively high symmetry—space groups of low symmetry are, in general, much more easily identified. In the case of a biological macromolecule, the complex structure of the molecule itself suggests that the space group of the crystal is likely to be of relatively low symmetry, partly because mirror planes and inversion centres are impossible. It is therefore often quite easy to identify the space groups of biologically interesting molecules.

The preceding discussion of crystal symmetry and the way in which we may identify space groups from an X-ray photograph has, of necessity, been brief. There are many points which have not been mentioned and, in general, details have been glossed over. For further information, the reader is referred to the texts mentioned in the general bibliography at the end of the book.

Summary

Diffraction of X-rays by the contents of the unit cell is determined by *Thomson scattering* of the incident X-ray beam by the electrons within the unit cell. Each electron scatters radiation according to the *electronic scattering factor* f_e, and the diffraction pattern of an arbitrary distribution of electrons as specified by the electron density function $\rho(\mathbf{r})$ is given by

$$F(\Delta\mathbf{k}) = \int\limits_{\text{all } \mathbf{r}} f_e\rho(\mathbf{r})e^{i\,\Delta\mathbf{k}\cdot\mathbf{r}}\,d\mathbf{r} \tag{9.7}$$

One way of describing the contents of the unit cell is by means of the *electron density function* $\rho(x, y, z)$, where x, y and z are the dimensionless fractional coordinates of a point within the unit cell. The diffraction pattern of the unit cell is given by

$$F_{\text{rel}}(\Delta\mathbf{k}) = V\int\limits_0^1\int\limits_0^1\int\limits_0^1 \rho(x, y, z)e^{i\,\Delta\mathbf{k}\cdot(x\mathbf{a}+y\mathbf{b}+z\mathbf{c})}\,dx\,dy\,dz \tag{9.10}$$

where $F_{\text{rel}}(\Delta\mathbf{k})$ is the relative scattering ability of the unit cell contents as compared with a single electron.

Since the unit cell is associated with the lattice, the diffraction pattern of the contents of the unit cell is sampled at the reciprocal-lattice points *hkl*. The diffraction pattern function associated with a particular reciprocal-lattice point *hkl* is given by the *structure factor* F_{hkl},

$$F_{hkl} = V \int_0^1 \int_0^1 \int_0^1 \rho(x, y, z) e^{2\pi i(hx+ky+lz)} \, dx \, dy \, dz \qquad (9.11)$$

and the corresponding intensity is $|F_{hkl}|^2$.

The set of structure factors F_{hkl} is the Fourier transform of the electron density function, and so $\rho(x, y, z)$ may be expressed in terms of the structure factors F_{hkl} as

$$\rho(x, y, z) = \frac{1}{V} \sum_h \sum_k \sum_l F_{hkl} e^{-2\pi i(hx+ky+lz)} \qquad (9.15)$$

The numerical computation of $\rho(x, y, z)$ using F_{hkl} data is called a *Fourier synthesis*, and this is the ultimate objective of a crystal structure analysis. The function $\rho(x, y, z)$ is usually plotted as a three-dimensional contour map, called an *electron density map*, from which atomic positions and molecular structure information may be derived.

An alternative description of the contents of the unit cell is in terms of the electron distribution as referred to the atomic arrangement within the unit cell.

The *atomic electron density function* $\rho_j(\mathbf{R})$ specifies the electron distribution about the centre of the *j*th atom. The overall electron density within the unit cell is the sum of the atomic electron density functions. If the *j*th atom has coordinates (x_j, y_j, z_j), we may write the structure factor F_{hkl} as

$$F_{hkl} = \sum_j f_j e^{2\pi i(hx_j+ky_j+lz_j)} \qquad (9.21)$$

where f_j is the atomic scattering factor,

$$f_j = \int_{\text{atom}} \rho_j(\mathbf{R}) e^{2\pi i \mathbf{S}_{hkl} \cdot \mathbf{R}} \, d\mathbf{R} \qquad (9.20)$$

The term f_j represents the scattering ability of an atom as compared with a single electron. As defined in equation (9.20), the value of f_j obtained on integration is that which is associated with the reciprocal-lattice point *hkl*. The integral may be calculated on the assumption of some model for $\rho_j(\mathbf{R})$, of which the most usual is that the function is spherically symmetrical. In this case, f_j is a function of $(\sin \theta)/\lambda$, where θ is the appropriate Bragg angle for a particular diffraction peak.

As well as ultimately allowing us to determine the structure of the unit cell, inspection of the diffraction intensities gives us information on the crystal symmetry. The analysis of the symmetry of the diffraction pattern is facilitated by the use of equation (9.21) for F_{hkl}.

The symmetry information we may obtain is the following:

The symmetry of the intensity maxima determines the *Laue group* of the crystal, which will be one of the 11 point groups which are centrosymmetric.
Systematic absences identify the Bravais lattice type unambiguously and give much information as to the presence and nature of screw axes.

An important feature of the symmetry of diffraction patterns is stated by *Friedel's law*:

In the absence of an anomalous scatterer, the intensities I_{hkl} and $I_{\overline{hkl}}$ are equal.

In the absence of an anomalous scatterer, diffraction patterns conform to 122 types, of which 58 specify the space group of the crystal unambiguously. Space group identification of the other types requires more information.

An atom which shows *anomalous scattering* has a complex atomic scattering factor f_j, and this causes a breakdown of Friedel's law.

Review II

The structure of a crystal may be described in terms of the real lattice, the crystal symmetry and the contents of the unit cell. In the last three chapters we have seen how information on all these aspects of crystal structure may be derived from an X-ray diffraction pattern. The aim of Chapter 7 was to show how information on the lattice and motif manifested itself in a diffraction pattern. In Chapter 8, we derived the formalism for the diffraction pattern of a real lattice, and showed that this could be interpreted in terms of the *reciprocal lattice*. By using the *Ewald sphere* construction, we may correlate diffraction maxima with reciprocal-lattice points, and so measurements of the spacing between diffraction maxima allow us to identify the parameters of the unit cell. In the last chapter, we saw how the intensities of the diffraction maxima were determined by the electron distribution within the unit cell. By carrying out a *Fourier synthesis* on the F_{hkl} data, we may generate the electron density function, and so the contents of the unit cell may be determined. The intensities of the diffraction maxima also specify the symmetry of the crystal, and since this may be seen from inspection of the diffraction pattern, this is usually the first analysis of a diffraction pattern which is undertaken.

It therefore seems as if the problem of crystal structure analysis is solved. But, in practice, there remain two major experimental difficulties:

(a) *The phase problem.* Our data are intensities $I_{hkl} = |F_{hkl}|^2$, and not the complex quantities $F_{hkl} = |F_{hkl}| e^{i\phi_{hkl}}$. For the Fourier synthesis, we need the structure factor F_{hkl} in both magnitude and phase.

(b) The intensity as measured on, for instance, an electronic detector is not $I_{hkl} = |F_{hkl}|^2$, but only proportional to this quantity. In order to derive I_{hkl} we must correct the observed intensities for various physical effects such as diffraction geometry, absorption and beam fluctuations.

Fortunately, the correction of intensity data is relatively straightforward and yields the amplitudes of the structure factors $|F_{hkl}|$. In contrast, however, the phase problem is indeed a problem, if not *the* problem of X-ray crystallography.

The scheme for the next part of this book is as follows.

Part III, 'Structure solution', comprises Chapters 10–16. The methods for expressing, purifying and crystallising proteins are reviewed in Chapter 10, followed by the methods by which diffraction intensity data are collected and corrected in Chapter 11. Chapters 12–14 examine methods for solution of the phase problem in protein crystal structure determination, and Chapter 15 looks at methods for refining the resulting crystal structure, i.e. methods to make the structure agree as much as possible with the experimental measurements. This chapter also covers the important applications that protein crystallography has in analysing ligand complexes and site-directed mutants. In the final chapter, we look at how protein X-ray crystallography is complemented by other methods of diffraction analysis.

Part III
Structure solution

10

Experimental techniques: sample preparation

The objective of this chapter is to provide a whistle-stop tour of the methods involved in the expression, purification and crystallisation of proteins. Consequently, the ensuing sections will describe commonly used laboratory techniques in an unashamedly anecdotal manner, but further details, including specific references to successful use of each method, may be found in excellent review articles referred to in the text and in the wider literature.

10.1 Protein expression

Whilst some proteins can be obtained in large quantities from their natural sources such as blood, liver and muscle, most need to be expressed from the genes encoding them in order to obtain sufficient material for structure analysis to proceed. The sequence of bases forming the gene will usually be available from prior biochemical and molecular genetic studies. It is then a matter of amplifying the gene so that we can clone it into an expression vector or plasmid—a circular piece of DNA that can be taken up by cells of the organism which will be used to express the gene (the expression host). The expression host of first choice is usually *Escherichia coli* owing to the relative ease of working with this well-studied bacterium, and strains have been engineered specifically for such work; for example, the genes for various proteinases have been knocked out to avoid degradation of any expressed proteins. A problem arises with eukaryotic genes, since these often contain non-coding regions that are not processed correctly by *E. coli*. However, the corresponding coding region for the gene of interest can be obtained from a cDNA library. The general scheme of cloning a gene and preparing an expression construct is indicated in Fig. 10.1, and the principal steps are outlined below. Further details of the essential microbiological techniques and the underlying molecular biology can be found in specialised textbooks.

First of all, the gene of interest has to be amplified from the DNA of its native organism by a process known the *polymerase chain reaction* (PCR), which involves a thermostable DNA polymerase that can perform multiple cycles of replication in an instrument known as a thermal cycler. The enzyme needs short pieces of DNA, which pair up with the DNA to be replicated and act as primers to which the new DNA bases are attached, one at a time. The primers have to be designed and then synthesised by a dedicated machine before the experiment, and this is often outsourced to a commercial service. Since only about half of the primer needs to pair up with the parent

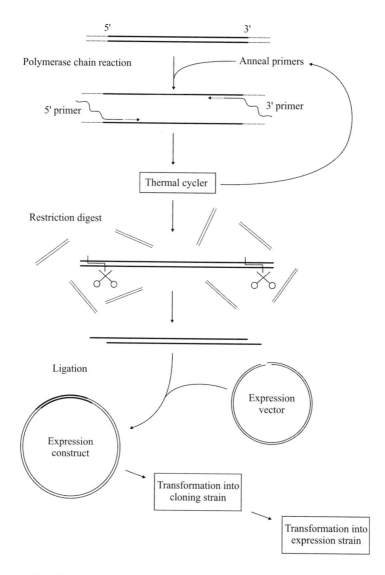

Fig. 10.1 *A flow diagram of gene cloning.* DNA containing the gene of interest (drawn as thick lines) is shown at the top. The gene can be amplified by raising the temperature so that the two strands separate, and specific primers can then be annealed to them by lowering the temperature. The primers are then extended by a thermostable DNA polymerase so that new copies of both DNA strands of the gene are made. The primers are present in huge excess so that on reheating the sample to separate the strands, fresh primers will anneal to the DNA, and these can then be elongated to generate a second round of copies. Repeating this process 20–30 times in a thermal cycler results in a large amplification of the gene. The DNA is then cut at each end with specific restriction enzymes so that it can be ligated with the expression vector, previously cut at its multiple cloning site by the same enzymes. Following ligation, the resulting DNA can be transformed into a general cloning strain of *E. coli* so that further quantities of the expression construct may be prepared for transformation into the expression host.

DNA, the remainder, or 'overhang', can have a sequence designed to include cleavage sites for enzymes known as restriction endonucleases. These cleavage sequences will then appear at the ends of the newly amplified DNA and can be cut with these restriction endonucleases following the PCR. If we also cleave the expression vector with the same enzymes, we can purify the resulting linear DNA and add some of the amplified gene to it. The two DNA molecules can then be joined together, or ligated, by use of an enzyme known as DNA ligase. The expression vector will be engineered so that the point at which the gene is inserted into it (known as the multiple cloning site) will have an appropriately placed promoter region and ribosome-binding site. The promoter allows the gene to be expressed at a high level following addition of a certain chemical known as an *inducer* to the bacterial culture; commonly, the lactose analogue isopropyl thiogalactoside (IPTG) is used for this purpose. Using two different restriction enzymes to cleave both the vector and the insert DNA allows the latter to be inserted into the plasmid in a unique orientation, which is very important for expression of the gene. The vector can also include another protein or specific sequence of amino acids (e.g. polyhistidine), which will be joined to the protein of interest, and these are known as 'tags' since they allow the expressed *fusion protein* to be purified relatively easily from the final cell extract.

Each of the steps involved requires a lot of care, patience and sometimes repetition if problems are encountered; ligation is notorious in this respect. The integrity of the gene of interest in the final construct should be checked by DNA sequencing prior to expression studies, since the PCR is somewhat error-prone, although DNA polymerases with high accuracy are now available. The latter enzymes also allow specific mutations of the gene to be made later on in the analysis by using primers that include a base mismatch.

Following ligation of the gene into the expression vector, it can be taken up by bacterial cells in a process known as transformation, and usually a general cloning strain will be used at this stage. The plasmid will possess a gene that confers resistance on the cells to a certain antibiotic, such as ampicillin. Growing the cells on agar containing this antibiotic will ensure that only those cells containing the expression vector will grow. The resulting colonies can be picked and grown in small cultures, also containing the antibiotic, to form cell stocks, and these can be used either for preparation of further quantities of the vector DNA or for expression studies. Note that some plasmids need to be transformed into a special expression strain in an additional step. Usually, trial expression studies involve growing small volumes of cells by shaking at 37°C to the point where the cultures appear to be thickening up appreciably and the cells are dividing at their highest rate (known as the mid-log phase, where the absorbance of the culture at 600 nm will be approximately 0.5). At this point the cultures can be split into two tubes and the chemical inducer added to one of the tubes. The cells are then grown for a longer period of 3–18 hours, and samples of the culture are taken for analysis by polyacrylamide gel electrophoresis, which will reveal any differences in protein expression between the induced and uninduced cultures. Successful induction should be indicated by the presence of a protein of the correct molecular weight in the induced cultures. Sometimes expression may not be apparent, which may be due to the codon usage of the parent organism being different from that of the *E. coli* expression host. Commercial *E. coli* strains which provide extra tRNA molecules for the rarely used codons are available (e.g. Rosetta^{TM}) for use in this situation. In some cases, the protein can be shorter than would be expected from its gene sequence, and this may stem from the fact that eukaryotic genes contain sequences which can terminate transcription prematurely in *E. coli*.

If induction is successful, the experiment can be repeated using larger-scale cultures (e.g. 1 litre), which are induced in the usual manner at mid-log phase and shaken at 37 °C for a further few hours or overnight (the latter can sometimes double the yield of expressed protein).

The resulting cells must be sedimented by centrifugation and then resuspended and lysed by either sonication (ultrasound treatment) in ice, passage through a narrow orifice under high pressure (French press), freeze-thawing, lysozyme treatment or some other method or combination of methods which ensures that the cell walls are adequately broken to release the intracellular proteins. Ultracentrifugation at typically 40 000*g* will then sediment the unwanted cell debris, and gel electrophoresis will verify whether the protein of interest is present in the supernatant or the pellet. A protein in the supernatant can be purified by the methods described in the next section. If, instead, the protein is present in the pellet, it is most likely incorrectly folded as *inclusion bodies* and one can then either try to refold it or, preferably, try to modify the expression conditions so as to increase the solubility of the expressed protein. Growing the cells to mid-log phase and then growing them at low temperature (16–18 °C) following induction can be successful, as can giving them a mild heat shock (e.g. at 42 °C) prior to adding the inducer, and subsequently growing the cells at low temperature. If a low temperature is used, the cells should, of course, be grown for significantly longer periods following induction. Many other methods, such as use of lower concentrations of the inducer, are available if expression of the protein in soluble form is problematic.

If expression in *E. coli* proves intractable, yeasts such as *S. cerevisiae* or *P. pastoris* can be tried and can achieve similar expression yields to *E. coli*, i.e. tens of milligrams of protein per litre of culture. Bacteria do not introduce post-translational modifications such as glycosylation and cleavage of the main chain, which might normally be introduced by cells of the organism from which the gene originates. This can be an advantage for crystallisation, particularly if the modifications are heterogeneous, as glycosylations usually are. Insect cells provide an excellent system for expression of eukaryotic proteins, and the associated baculovirus technology is routine in many laboratories. Whilst some post-translational modifications are carried out by yeasts, more extensive modifications occur in insect cells, which are also able to express genes containing introns (non-coding regions). Greater automation of these methods has been achieved through use of ligation-independent cloning—methods which allow the gene of interest to be inserted into the expression plasmid, usually by recombination, and a range of expression vectors with different affinity tags can be generated. More details of expression and purification for laboratory-scale and high-throughput applications are given by Skelly *et al.* (2007) and Owens *et al.* (2007).

10.2   Protein purification

The requirement for a protein to be highly purified in order to grow well-diffracting crystals cannot be overemphasised: the general aim is to obtain a sample in which the molecule of interest forms at least 95% of the total protein present and it should be present in milligram quantities.

The traditional prerequisite for purification is that an *assay*, which measures some specific function of the desired protein, such as catalytic activity or the ability to react with certain antibodies, is available. A catalytic assay could involve the use of a substrate which undergoes a change in spectroscopic absorption or fluorescence, and often such assays involve more than one enzyme (coupled assays), designed so that one of the components catalyses a reaction that is easily monitored. Inspection of electrophoretic gels is an excellent guide as to the level of purity, with the caveat that sometimes another protein of the same molecular weight can be purified in error, thus emphasising the need for an additional confirmatory test. However, developments in molecular biology and associated affinity purification methods have partially circumvented the requirement for a specific biochemical assay, since the molecule of interest can be given a 'tag' which allows its purification from whole cell extract in almost a single step, although it is still important to use a functional assay to be certain that the protein is in a correctly folded state. The nature of the final purified protein can be confirmed relatively inexpensively by proteomics— essentially a mass-spectrometric sequence analysis of the peptides yielded by a tryptic digest.

Since purification relies on separating the molecule of interest from others on the basis of their physicochemical properties, it is common for an initial crude step to be undertaken such as salt precipitation, a pH change or (for heat-stable proteins) a heat treatment step. The aim of this crude step is to give an initial purification without significant loss of the desired protein prior to more sophisticated chromatographic separation stages. At low ionic strength, raising the salt concentration generally increases a protein's solubility, since the salt ions help to solvate the protein molecules. However, as the salt concentration is raised further, the solubility of the protein tends to decrease, since the amount of water available to solvate the protein molecules decreases. This is known as *salting-out* and is exploited in the purification of many proteins, as well as in their eventual crystallisation (as we shall see later), with ammonium sulphate being the preferred salt. In purification, the ionic strength of the solution can be increased in stages, allowing those contaminants which precipitate to be removed by centrifugation, thus leaving the molecule of interest in solution. A point will, of course, be reached when the protein we are attempting to purify also precipitates, and at this stage the precipitate obtained from centrifugation of the solution will contain the molecule of interest, often with quite good purity. A protein is least soluble at its *isoelectric point*, or pI, this being the pH at which the protein possesses no net charge and has the minimum tendency to be solvated. This effect can be exploited in salting-out to further enhance the purification by selectively precipitating the molecule of interest. Certain impurities can be precipitated by use of specific additives; for example, DNA can be removed from the solution by addition of streptomycin sulphate. Sometimes the molecule of interest will exhibit significant thermostability, and the solution can therefore be heated for a period of time to precipitate impurities.

Virtually all purifications involve some form of *chromatography*, in which a mixture of substances in the liquid, or mobile, phase flows over a stationary phase that causes the components to migrate at different rates. In preparative protein purification, the stationary phase is usually enclosed within a column, through which the crude protein mixture is pumped; a typical laboratory set-up is shown in Fig. 10.2 The initial chromatographic step is usually *ion exchange*, in which the proteins are retained by the stationary phase by virtue of electrostatic interactions with it, and the strength with which an individual protein interacts with the column depends on its net charge. The columns are made of a polyionic material such as diethylaminoethylcellulose, or DEAE-cellulose, which is basic, i.e. positively charged at neutral pH, and therefore able to bind negatively charged proteins. Another such material is carboxymethylcellulose, or CM-cellulose, which is acidic, i.e. negatively charged at neutral pH, and able to bind positively charged proteins. If a mixture of proteins is applied to the column, molecules which possess a charge opposite to that of the column will be retained to a greater extent than those with the same charge or no charge at all. Molecules that do not bind will simply flow through the column. After the sample has been loaded, the column is washed with buffer to remove weakly bound components, i.e. those not retained by favourable electrostatic interactions with the solid phase. The proteins bound to the column can then be eluted in stages by gradually increasing the ionic strength of the elution buffer using a gradient maker. Increasing the salt strength will raise the dielectric constant of the solution and progressively weaken the interactions between the proteins and the column. Weakly bound proteins will elute first, and the strongly bound ones will be removed last. Thus purification is achieved on the basis of how strongly the individual components of the mixture interact with the column. Popular ion exchange columns include the GE Healthcare Mono Q column (so called because

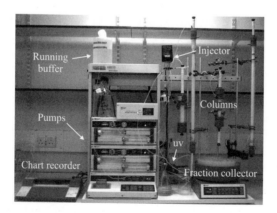

Fig. 10.2 *An FPLC chromatography workstation.* A classic Pharmacia fast protein liquid chromatography (FPLC) system, showing the pumps, injection system, columns, ultraviolet detector, chart recorder and fraction collector for the samples.

of its quaternary amine groups) and the Mono S column (which possesses sulphonic acid groups); Mono Q and Mono S are high-resolution anion and cation exchangers, respectively. Pellicular ion exchange columns, in which the retentive material forms a layer on an impermeable bead, are experiencing a renaissance, as are methods involving highly porous media.

It is important to remember that the charge of a protein varies with pH—the molecule will have a net negative charge above its pI and a net positive charge below the pI. To ensure that the protein of interest binds well to the ion exchange column, the pH of the running buffer should be at least one pH unit above or below the pI of the protein, depending on whether an anion or a cation exchanger is to be used. The pI can be predicted with reasonable accuracy from the amino acid sequence and used as a trial value for deciding on a suitable pH for the running buffer. The pH can be varied in successive runs, if required, to improve the separation of the proteins, and sometimes a pH gradient is used instead of a salt gradient for elution.

Differences in physicochemical properties of individual proteins are also exploited in *hydrophobic interaction chromatography* (HIC), where, instead of having an ionic stationary phase, the beads are coated with hydrophobic groups such as butyl, octyl or phenyl groups. The protein is loaded into the column in the presence of a high salt concentration. The 'salting-out' effect causes the exposed hydrophobic groups on the protein to bind to the immobilised hydrophobic groups on the column. The bound proteins can then be eluted differentially by running a gradient of decreasing ionic strength through the column—the most polar molecules will elute first and the most hydrophobic ones last. A typical HIC salt gradient might be from 1 M NaCl down to 0 M NaCl, and since the protein will be in high salt after ion exchange or ammonium sulphate precipitation, HIC is conveniently done at these stages. Whilst HIC is less commonly used than ion exchange in protein work, it has some parallels with reverse phase chromatography, which is the most commonly used purification method in chemistry laboratories owing to its high reproducibility and the robust nature of the columns.

Following purification by one or both of the above methods, it is common to further fractionate the proteins by use of *gel filtration chromatography*, also known as size-exclusion chromatography. This method involves the use of a column packed with hydrated gel beads made of materials such as dextran (Sephadex) and agarose (Sepharose). The beads possess numerous pores, the sizes of which can be controlled during synthesis. As the protein mixture moves through the column, the proteins small enough to fit into the pores are retarded relative to those that are too big to permeate the gel beads. The largest molecules therefore move fastest and are eluted first, followed by proteins of progressively smaller molecular weight. To get as good a separation as possible, it is sensible to choose a gel with the appropriate pore size for the protein of interest. For example, the GE Healthcare Superdex 75 and 200 columns fractionate proteins of molecular weights up to approximately 70 kDa and 600 kDa, respectively. The Superdex beads consist of dextran covalently bonded onto cross-linked agarose and offer the advantages of fast separation, high resolution, reproducibility and stability to high pressure.

Compared with other methods, the resolving power of gel filtration is relatively low, thereby necessitating its use after an initial purification, for example by ion exchange. Both methods may be combined using column materials such as DEAE-Sephadex and CM-Sephadex, which allow proteins that are retained at the top of the column by ionic forces to be separated by gel filtration during the elution step. Note that gel filtration columns have a residual ion exchange activity, and so to separate proteins purely on the basis of their size requires a certain minimum salt or buffer concentration.

Most chromatographic work for protein purification is done using *fast protein liquid chromatography* (FPLC) systems, which typically consist of an electronic controller, two pumps which push solutions through the column and can form a gradient, an injection system for loading the sample, an ultraviolet detector with a chart recorder for reporting on the elution of molecules from the column, and, importantly, a fraction collector for the eluted samples (see Fig. 10.2). Whilst FPLC systems can run columns at pressures of up to 2–4 MPa (20–40 atmospheres), pressures at least tenfold higher can be achieved with high-performance liquid chromatography (HPLC) systems, although the latter are not used to such an extent as FPLC for protein work. The use of higher pressures has the advantage of improved resolution owing to reduced diffusion in the mobile phase. The composition of the mobile phase can remain fixed, and this is referred to as isocratic elution. Alternatively, the mobile phase can be run with a gradient, which offers the advantage of mitigating against the progressive broadening of the peaks which are last to elute from the column. It is also common to run columns where the mobile-phase composition is changed in steps, as opposed to varying continuously, and this is important in affinity chromatography, as we shall see later.

Before carrying out ion exchange, it is sometimes necessary to desalt a protein, since a low initial salt concentration is needed for this type of chromatography to work. Gel filtration can be used to desalt a protein, since the salt ions will elute long after the protein owing to their smaller size. Alternatively, dialysing the protein against a low-salt buffer overnight will desalt it effectively. Use of a centrifugal or pressure concentrator will achieve the same result in less time, and these appliances, which are often used at the end of a purification, also allow the protein to be concentrated to the level required for crystallisation, which is usually in the range of 3–30 mg/ml.

Selective purification of a specific protein can be achieved in a single step by *affinity chromatography*. This involves chemically coupling a known ligand to an inert porous material such as agarose, usually by using cyanogen bromide. When a mixture of proteins is applied to the column, the protein molecules which bind the ligand are retained, whereas those that do not bind will simply flow through the column during the loading and subsequent washing steps. The protein of interest is then eluted by adding free ligand to displace the protein from the column. One caveat is that if the immobilised ligand and protein form a very tight complex, it may be impossible to elute the protein from the column without harsh treatment. Immunoaffinity chromatography involves having an antibody to the protein of interest coupled to the column, and elution of the molecule recognised by the antibody is achieved by changing the salt strength or pH.

By far the most commonly used forms of affinity chromatography involve nickel or glutathione affinity matrices. Nickel ions bound to a solid matrix (by nitrilotriacetic acid (NTA) or a related chelating group) are excellent at binding the polyhistidine tails (His-tags) that are frequently introduced into recombinant proteins by the expression vector. Likewise, the coupling of the protein of interest to the enzyme glutathione-S-transferase (GST) by the choice of cloning vector allows the expressed protein to be purified by passage through a column onto which glutathione has been immobilised. The GST tag has good solubility and refolding properties. Elution of the desired protein from the column is achieved by the application of free glutathione, typically at a concentration of 10 mM. In contrast, elution of a specifically bound protein from a nickel column requires the application of a high concentration of imidazole, typically 200–500 mM. Prior to elution of the protein of interest, the column is washed with a low concentration of imidazole to remove weakly bound species. A number of proteins have a high affinity for nickel NTA (e.g. *E. coli* histidine-rich protein and the membrane protein AcrB), which may therefore contaminate the final sample, but these are usually present at low levels and can be removed by gel filtration if required.

Removal of the tag on a fusion protein can usually be achieved by the addition of a highly specific proteinase (such as thrombin), which cleaves at a target sequence that is engineered into the expression vector between the tag and the protein of interest. Cleavage with the proteinase will yield a mixture consisting of the desired protein and the tag, as well as small amounts of the proteinase itself. However, the tag can be removed by passage of the mixture through the affinity column again (providing that the eluting agent has been removed from the mixture by prior dialysis or ultrafiltration), and the proteinase can be removed by passage through a separate affinity column specifically for the proteinase; for example, a benzamidine column can be used for serine proteinases such as thrombin. A slightly simpler process is to perform the cleavage reaction on the column during the main purification run, i.e. instead of eluting the fusion protein, we add the proteinase and incubate the column to allow cleavage to occur. The protein of interest can then be removed from the column simply by washing it with binding buffer, leaving the tag bound to the column. The proteinase will also wash out of the column at the same time, but this can be removed by passage through its own affinity column (in fact, the two columns can be connected in series for this purpose). Finally, the unwanted tag can be eluted from the main affinity column in the usual manner and discarded. Note that there are other proteinases in use which have greater specificity than thrombin and their own affinity tags to aid removal after cleavage.

As a substitute for a number of chromatographic methods, such as affinity chromatography and ion exchange, it is possible to perform *batch binding*, in which the stationary phase is simply stirred into the mixture of proteins and incubated with it to allow binding to occur. The mixture can then be filtered to remove proteins that have not bound to the stationary phase or it can be centrifuged to pellet the beads with the protein of interest bound to them. The desired protein can then be separated from the stationary phase by addition of whichever reagent would normally be required to elute it from the equivalent chromatography column.

Once a sample has been purified, a sensible precaution to prevent degradation of the protein is to add a proteinase inhibitor such as phenylmethylsulphonyl fluoride (PMSF) or aminoethylbenzenesulphonyl fluoride (AEBSF) or a cocktail of such inhibitors, along with 0.01% azide to prevent microbial growth in the sample. Depending on the vulnerability of the sample, the compounds may need to be added earlier in the purification, bearing in mind that proteinase inhibitors could interfere with removal of the tag from a fusion protein.

10.3   Crystallisation

Following purification and concentration of the protein, crystal growth is usually achieved by gradually lowering the solubility of the molecule to the point where crystal nuclei form and grow, eventually reaching dimensions of 0.1 mm or larger. The solubility of a protein is affected by its concentration, as well as by pH and ionic strength, for reasons that were covered earlier. Whilst salts such as ammonium sulphate and sodium chloride are commonly used as precipitants, polyethylene glycols have a similar ability to dehydrate protein molecules and induce precipitation or crystallisation under favourable conditions. Protein solubility is also affected by temperature, the nature of the buffer and the presence of additives such as certain ions, organic solvents or specific ligands. With integral membrane proteins, a significant proportion of the outer surface is hydrophobic and a detergent is therefore an essential additive; β-octyl glucoside or dodecyl maltoside is frequently used for this purpose.

Since it is currently impossible to predict the conditions in which complex macromolecules such as proteins will crystallise, suitable conditions have to be found by trial and error. Usually, the purified protein will be subjected to a commercial *sparse matrix crystallisation screen* (also known as an *incomplete factorial screen*), which is a collection of conditions that are based on those known to work for other proteins, with some permutation of the constituents to ensure that all variables are sampled efficiently with a relative small number of trial conditions (Carter and Carter, 1979; Jancarik and Kim, 1991). Depending on the level of automation or time available, upwards of 100 or 1000 conditions can be tested. If the experiment works, small crystals will appear for a number of conditions, and these can be tested to see if they are protein or salt, for example by placing them in an X-ray beam or by use of a dye which stains proteins. Promising 'hits' can then be optimised by systematically varying the concentration of the precipitant or protein and varying the pH until crystals suitable for X-ray diffraction studies are achieved (see e.g. Fig. 10.3). It can also be very helpful to systematically vary or omit each component of the mixture to establish which ones are critical for crystallisation and which are not. This is known as a factorial screen and can be conducted along the lines shown in Fig. 10.4; any components that are found to be non-essential can then be omitted from further crystallisation trials. The aim of screening is to find conditions which yield a small number of good-sized crystals from a few microlitres of solution. If a large number of crystals which are too small for data collection are obtained, it usually helps simply to reduce the protein concentration.

Fig. 10.3 *Crystals of a typical protein after optimisation of the growth conditions.* The largest crystals shown are approximately 0.5 mm in their longest dimension.

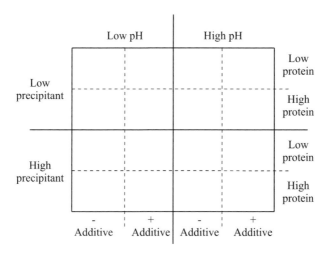

Fig. 10.4 *A factorial crystal screen.* Trial crystallisation conditions, such as might be obtained from a commercial screen, can be optimised by establishing which factors are critical for crystallisation. A full factorial screen allows all components to be tested systematically by altering their concentrations (indicated by 'low' or 'high') or omitting them entirely (e.g. the '+ additive' and '− additive' conditions), as indicated on the grid shown.

With well-behaved proteins, it may be possible to find crystallisation conditions by a classical 'hunch and crunch' approach. A typical start point might be to take 5 μL of protein and add saturated ammonium sulphate solution or a polyethylene glycol (PEG) solution in 1 μL steps until the point is reached where cloudiness or slight precipitation occurs. The corresponding concentration of ammonium sulphate or PEG would then serve as a trial condition that could be varied stepwise on a grid along with other variables such as protein concentration, pH, buffer and additive.

The most common method of crystal screening is the *hanging-drop method*, shown in Fig. 10.5(a), in which a small quantity of protein, typically 2–5 μL, is mixed with an equal volume of the screen solution on a siliconised glass coverslip. The coverslip is then inverted, placed on top of a 'well' of the screen solution and sealed with grease. Since the precipitant concentration in the droplet will be exactly half that of the well solution, water vapour will slowly diffuse from the droplet into the well solution,

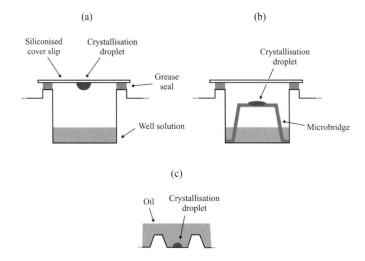

Fig. 10.5 *Different crystallisation methods.* (a) The hanging-drop method involves dispensing a droplet of protein and screen solution onto a siliconised glass coverslip and sealing the mixture above a well of the screen solution. Vapour diffusion from the droplet to the well causes a slow increase in protein and precipitant concentration and, in ideal conditions, this will lead to the formation of crystals. (b) The sitting-drop method relies on the same principle, but the droplet can be larger since it is supported on a bridge or ledge above the well solution. (c) The microbatch method involves pipetting a droplet of protein solution under a layer of oil and mixing it with the screen solution.

thereby reducing the volume of the drop and gradually increasing the protein and precipitant concentrations to the point where crystallisation occurs. The gradual increase in concentration of the protein and precipitating agent as the system approaches equilibrium are key features of the *vapour diffusion method*. One variant, which allows the use of larger volumes of protein, is the *sitting-drop method*, in which the droplet rests on a small plastic bridge or ledge above the well solution, as shown in Fig. 10.5(b).

With vapour diffusion it is sometimes necessary to slow the approach to equilibrium, and this can be achieved by layering water-permeable silicone oil on the surface of the well solution (Chayen, 1997). Temporarily opening the wells of a vapour diffusion tray and diluting the well solution some time after nucleation has occurred may allow the nuclei to grow at a slower rate than they otherwise would, with beneficial effects on crystal quality. Crystallisation in silica gels to inhibit convection in the mother liquor and sedimentation of the crystals is reported to improve crystal quality, as is crystallisation under microgravity conditions aboard space missions (Kundrot et al., 2001).

Another common crystallisation technique is known as the *batch method*. This involves mixing the protein with the precipitant such that their final concentrations are just below those required to cause turbidity. In the traditional batch method, millilitre quantities of protein and precipitant solutions are sealed in tubes and left for periods of weeks or months for crystals to appear. Sometimes batch crystallisations benefit from a volatile solvent additive, such as acetone—as the additive evaporates slowly, the solubility of the protein decreases to the point where crystallisation occurs. A more economical approach is provided by the *microbatch method* (Fig. 10.5(c)), in which microlitre quantities of protein and precipitant solutions are mixed together under a layer of oil on a plastic tray and left to crystallise.

A phase diagram indicating the effects of protein and precipitant concentration on crystallisation is given in Fig. 10.6. For a particular protein, a rough indication of the phase diagram can be determined experimentally by varying the protein and precipitant concentrations in the manner shown and observing the conditions which lead to turbidity of the solution. Conditions close to the region of turbidity are useful

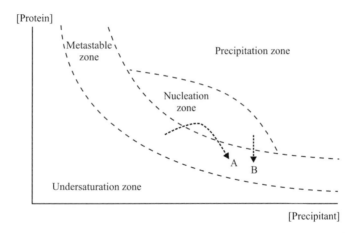

Fig. 10.6 *A crystallisation phase diagram.* A plot indicating the phase transition behaviour of a protein solute as a function of its concentration on the y axis and the concentration of precipitant on the x axis. Large-scale precipitation of the protein occurs spontaneously in the precipitation zone. In contrast, in the nucleation zone, a smaller number of crystal nuclei tend to form. Such precipitation or nucleation, as we have just described, does not occur in the metastable zone, but any nuclei which 'enter' this zone can grow to sizeable proportions, forming ordered crystals. The aim of any crystallisation experiment is for the system to enter the nucleation zone, such that nuclei form, and for it then to 'move' into the metastable zone so that the nuclei can grow. In a vapour diffusion experiment (shown by trajectory A), evaporation of water from the droplet means that the concentrations of both the protein and the precipitant increase until the system enters the nucleation zone. At this point, nuclei form and the concentration of protein in the liquid phase begins to decrease as the protein enters the solid phase. This shifts the system into the metastable zone, and the nuclei which formed in the nucleation zone then grow to form crystals, ideally of respectable proportions without further nucleation occurring. In a batch experiment (shown by trajectory B), the system starts in the nucleation zone such that as the nuclei form and the protein concentration begins to decrease, the system moves into the metastable zone, where crystals will grow from the nuclei.

starting points for batch or vapour diffusion experiments. The routes which the system will take through the phase diagram in the batch and vapour diffusion methods are indicated in Fig. 10.6, and awareness of these trajectories can help to plan subsequent crystallisation experiments.

Crystals can also be grown by dialysis, in which the protein and precipitant solutions are separated by a semi-permeable membrane that allows slow diffusion of the precipitant into the protein solution over a period of several hours. Small-scale dialysis can be done with capillaries or button cells. Free interface diffusion, in which the protein and precipitant are put in contact with each other, for example in a capillary tube, and allowed to mix by diffusion alone, is increasing in popularity. This method is particularly amenable to automation by the use of microfluidic chips that use nanolitre quantities of protein for each condition of a screen, thus enabling a large number of conditions to be sampled with just a few microlitres of protein.

Methods involving microlitre or nanolitre quantities of protein are excellent for screening a large number of conditions and subsequently optimising the 'hits'. However, preparative crystallisation for X-ray work usually involves working with somewhat larger volumes of solution once the conditions have been optimised, and even then all conditions should be varied slightly and replicated multiple times, given that reproducibility is a well-known concern in this field.

Small crystals which might be too small for a diffraction experiment may be used as 'seeds' that are transferred to a hanging or sitting drop or a batch vial, and left to grow. Small fibres or pieces of hair can also act as crystal nuclei. *Seeding* may be achieved simply by pipetting typically $1\,\mu L$ of a diluted seed solution into a $10\,\mu L$ droplet of protein and precipitant solution or by streaking a cat's whisker through the seed solution and then through the crystallisation droplet. Note that some workers etch their seeds first, by transfer to a solution which causes dissolution of the seed surface.

Problems with crystallisation often stem from poor purity or heterogeneity of the protein under study, and, sometimes, if the protein is glycosylated, removal of the sugar chains with glycosidases can be pivotal to crystal growth. Many more details can be found in excellent reviews such as McPherson (2004).

10.4 Crystal mounting

Collection of X-ray data from crystals requires that they are held with the minimum of supporting material to reduce X-ray absorption. Hence crystals are mounted either in very thin-walled glass capillaries or, more commonly, in minute loops made of thin plastic or hair. Both methods require the use of a microscope, and some practice with an unimportant protein that crystallises very easily is essential for beginners. For routine protein crystallographic studies, the use of loops in metal holders is by far the predominant practice. Consequently, capillary mounting is largely of historical interest only, although it still plays an important role in studies of proteins or larger assemblies that cannot be frozen to liquid nitrogen temperatures without loss of order in the crystal. Capillary, or wet, mounting involves picking a crystal out of a droplet of mother liquor (the solution from which the crystal has grown) by holding one end of the capillary near the crystal. The other end of the capillary can be attached to a syringe for drawing the crystal into the tube. Alternatively, licking one's index finger and using it to seal one end of the tube while the other end is dipped into the droplet will prevent the mother liquor running into the tube by capillary action! When the lower end of the tube is brought close to a crystal, removing one's index finger to release the seal will allow the capillary action to draw the crystal up into the tube. Once inside the glass tube, the crystal can be moved to a suitable position by very gentle tapping or by use of a syringe with an appropriate adaptor. The crystal can then be dried by using a Hamilton syringe to remove most of the surrounding mother liquor, followed by insertion of thin strips of filter paper into the tube; an indication of these steps is given in Fig. 10.7. Prior to the very final drying step, a droplet of mother liquor must be introduced at one end of the tube, either with a syringe or by dipping the tube in mother liquor again, to maintain the crystal in a hydrated state. It is essential that the mother liquor immediately surrounding the crystal is removed thoroughly to prevent any slippage of the crystal in the capillary during data collection. Final sealing of the tube with molten wax or resin will preserve the crystal for many months or even years.

Mounting crystals in loops is comparatively simple, since a small loop made of mohair or plastic held in a special holder (usually metallic) can be used to pick up the crystal under a microscope (see Fig. 10.8). Samples mounted in this way must be cooled rapidly, and this requires that the solution containing the crystal should contain an additive, or *cryoprotectant* (antifreeze), to prevent ice formation when the crystal is cooled. Rapid, or 'flash', cooling of the sample then ensures that the mother liquor forms a vitrified glass rather than crystalline ice. Typically, 30–40% v/v glycerol or a somewhat lower amount of isopropanol or ethylene glycol is suitable, as are low-molecular-weight precipitants such as 2-methyl-2,4-pentanediol (MPD), which may be present in the mother liquor anyway. It can help to experiment with growing

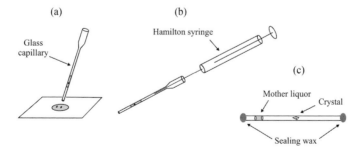

Fig. 10.7 *Wet mounting of a crystal in a capillary.* (a) shows a crystal being drawn, from the droplet in which it grew, into a glass tube by capillary action. (b) A Hamilton syringe can be used to remove excess mother liquor prior to drying the crystal with thinly cut strips of filter paper. (c) Finally, a droplet of mother liquor is introduced at one or both ends of the capillary tube to maintain the crystal in a hydrated state, and the tube is sealed with molten wax or hard-setting glue.

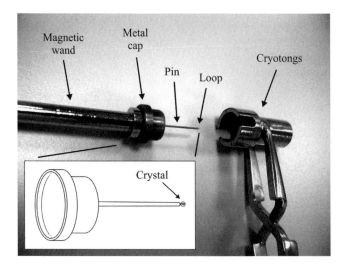

Fig. 10.8 *Loop mounting.* A crystal will normally be mounted in a loop at the tip of a pin held by a metal cap. During data collection, the cap will be held firmly in place by a magnet.

the crystals in a solution containing such compounds for ease of freezing the crystals subsequently. Alternatively, a loop can be used to transfer the crystal from the droplet in which it grew to another 10 µL droplet of well solution. Subsequently stirring in three or four 1 µL droplets of glycerol with a needle, avoiding direct contact of the glycerol with the crystal, works in the vast majority of cases. It is also common to make a series of droplets of well solution containing progressively higher concentrations of the cryoprotectant, for example increasing in steps of 5% v/v glycerol, and to transfer the crystal from one droplet to the next with a loop.

The now cryoprotected crystal can be picked up again with the loop and frozen under a cryostream of cooled nitrogen gas; initially, the cryostream is blocked with a small piece of card or plastic, which is removed sharply when the crystal is in place. Rapid cooling of the crystal to the temperature of the gas stream will then occur. Most X-ray instruments have a cryocooler incorporated, thereby allowing data to be collected from freshly frozen crystals. Alternatively, the flash-cooled crystal can be transferred to liquid nitrogen using special metal cryotongs, previously cooled under liquid nitrogen. Once under liquid nitrogen the crystal is placed in a plastic cryotube, using a specialised magnetic rod or 'wand', for storage and eventual transport to an X-ray facility. Transporting frozen crystals requires use of a dry-shipper—a metal dewar containing a lining of absorbent material which holds sufficient liquid nitrogen to maintain the samples at cryogenic temperatures for several weeks. For safety, the dewar must be drained of liquid nitrogen before shipping—sufficient cryogen will be retained by the absorbent lining to maintain a low temperature for considerably longer than the duration of the journey. The cryotongs and wand can also be used to transfer the crystal onto the instrument to be used for X-ray data collection and for saving the crystal at the end of the experiment. Further details are given in Garman and Doublie (2003).

Numerous variations of the above scheme are possible, including the use of 'economy' glass and plastic sample holders made of disposable laboratory items, as shown in Fig. 10.9. However, the use of standard sample mounts is important for data collection at multi-user facilities where robotic sample changers are employed. It is also possible to use the 'wand' to dip the loop in a small metal cup of liquid ethane or propane mounted in or above a bath of liquid nitrogen, in an effort to achieve a higher cooling rate and thus reduce ice formation in the mother liquor surrounding the crystal. The potential for greater cooling rates with liquid ethane or propane stems from their high heat capacities, which can be appreciated if one receives even a small splash of such liquids! It is common to use a guillotine device which drops the crystal in a controlled manner into a bath of liquid ethane or propane—the high velocity of the crystal in the

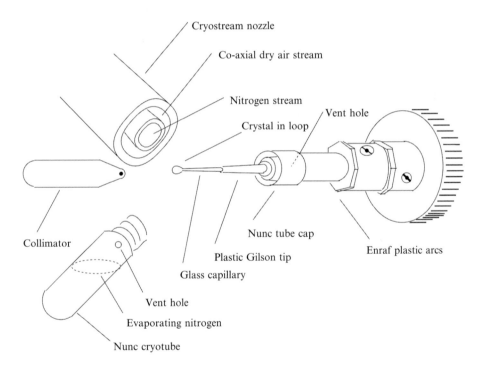

Fig. 10.9 *An 'economy' crystal-mounting system.* The loop is mounted in a strong glass capillary held in a plastic pipette tip that is glued into the lid of a cryotube, with an appropriate vent hole for liquid nitrogen escape. Tweezers can be used to hold the cap for freezing the crystal in liquid ethane or propane, and ordinary laboratory test-tube tongs can be used to hold a cryotube full of liquid nitrogen (with vent holes at the top) for transfer of the sample to the data collection instrument and, later, for retrieval and storage of the crystal. Note, with caution, that this system is not compatible with robot sample changers. The vent holes (indicated) are important to prevent the entrapment of liquid nitrogen and explosion due to the large increase in volume of around 700-fold upon boiling, which can cause injury or at least damage to the sample. Figure courtesy of Dr P. T. Erskine (UCL).

cryogen further increases its cooling rate. After flash cooling by one of these methods, the crystal is either transferred to the X-ray instrument for data collection or placed in a cryotube for storage under liquid nitrogen, in the usual manner. Note that liquid nitrogen itself is not used to freeze crystals directly, because it boils rapidly on contact with a crystal at ambient temperature and the resulting gas bubbles form an insulating layer around the crystal that slows its cooling rate significantly.

Of course, handling liquids at cryogenic temperatures requires protective clothing, gloves and glasses, and experimenters will be aware of the asphyxiating nature of nitrogen gas and the explosive nature of others mentioned above. Note also that cryotubes filled with liquid nitrogen can explode on warming unless they possess appropriate vent holes to allow the gas inside to escape.

A wide range of commercial loop designs cater for most shapes and sizes of crystal, and particularly fragile ones can be mounted on miniature plastic grids. It is also possible to freeze crystals under oil by using a loop to slide the crystal from the mother liquor into an adjacent droplet of oil, where further careful manoeuvring will encourage the remains of the mother liquor surrounding the crystal to depart. When the crystal is completely free of aqueous solution, it can simply be picked up from the oil with the loop and frozen by one of the methods described above.

Crystals mounted or frozen in these ways are ready for X-ray data collection using either an in-house facility or an external synchrotron source, more details of which will be covered in the next chapter.

Summary

Obtaining a protein in sufficient quantity for crystallisation commonly involves expressing the corresponding gene in bacteria or another suitable organism. For this purpose, the gene can be amplified by the *polymerase chain reaction* and ligated into an *expression vector*. The expression conditions or host organism may need to be varied in order to obtain the protein in soluble form rather than insoluble *inclusion bodies*.

Purification of the expressed protein commonly involves *ion exchange chromatography, hydrophobic interaction chromatography* and *gel filtration chromatography*. The expression vector may be chosen to provide the protein with an *affinity tag* such as polyhistidine or glutathione-S-transferase, which facilitates purification, often in a single step.

Crystals are usually obtained by screening a large number of crystallisation conditions based on those which work for other proteins, followed by extensive optimisation. *Vapour diffusion* methods involving *hanging drops* or *sitting drops* are commonly employed. The *microbatch* method and *free interface diffusion* are also widely used, as is crystal seeding. The crystals obtained are mixed with a *cryoprotectant* and mounted in loops before being frozen to liquid nitrogen temperatures for data collection and storage.

References

Carter, C. W., Jr and Carter C. W. (1979) Protein crystallization using incomplete factorial experiments. *J. Biol. Chem.* **254**, 12219–23.

Chayen, N. E. (1997) A novel technique to control the rate of vapour diffusion, giving larger protein crystals. *J. Appl. Crystallogr.* **30**, 198–202.

Garman, E. F. and Doublie, S. (2003) Cryocooling of macromolecular crystals: optimisation methods. *Methods Enzymol.* **368**, 188–216.

Jancarik, J. and Kim, S.-H. (1991) Sparse matrix sampling: a screening method for crystallization of proteins. *J. Appl. Crystallogr.* **24**, 409–11.

Kundrot, C. E., Judge, R. A., Pusey, M. L. and Snell, E. H. (2001) Microgravity and macromolecular crystallography. *Crystal Growth and Design* **1**, 87–99.

McPherson, A. (2004) Introduction to protein crystallization. *Methods* **34**, 254–65.

Owens, R. J., Nettleship, J. E., Berrow, N. S., Sainsbury, S., Aricescu, R., Stuart, D. I. and Stammers, D. K. (2007) High-throughput cloning, expression and purification. In *Macromolecular Crystallography*, ed. Sanderson, M. and Skelly, J. V. Oxford University Press, Oxford, pp. 23–41.

Skelly, J., Sohi, M. K. and Batuwangala, T. (2007) Classical cloning, expression and purification. In *Macromolecular Crystallography*, ed. Sanderson, M. and Skelly, J. V. Oxford University Press, Oxford, pp. 1–22.

11

Experimental techniques: data collection and analysis

In this chapter, some of the commonly used experimental techniques for X-ray data collection will be discussed, starting with a description of X-ray sources and some of the arrangements for diffraction intensity measurement and methods of data processing, along with assessment of the quality of the data. Finally, we will look at some of the caveats of crystal structure analysis, such as misindexing and twinning, which can be revealed by inspection of the processed data.

11.1   The origin of X-rays

X-rays are a form of electromagnetic radiation of wavelength characteristically in the 0–10 Å range, and may be derived principally from two effects. Firstly, if fast-moving charged particles suffer violent accelerations or decelerations, electromagnetic radiation is emitted. If the velocity changes are particularly rapid, the wavelength of this radiation is in the X-ray range. This effect occurs when, for example, fast-moving electrons collide with a solid material. On impact, the electrons are abruptly brought to a halt, causing the emission of X-rays.

Secondly, if an electron within an atom is ejected by some means, then it is possible for another electron in an orbital of higher energy within the same atom to 'fall' into the vacant orbital. During this transition, the electron which leaves the higher orbital to enter the lower one loses energy, and this energy is emitted as electromagnetic radiation. When the energy change accompanying the electronic transition is sufficiently great, the emitted radiation is in the X-ray range. In practice, X-rays are emitted only when a vacancy arises in the innermost energy levels of an atom, as in Fig. 11.1.

The electronic energy levels within an atom have well-defined energies, and so the energy of the radiation emitted as a result of an electronic transition is characterised fairly precisely. The wavelength range of the emitted radiation is therefore small, so that the X-rays resulting from a particular transition may be regarded as being essentially monochromatic. Since the energy levels of an atom are specific to that atom, the wavelengths of the X-radiation derived from a given atom are characteristic of that atom.

Laboratory sources are such that X-rays are emitted as a result of both of the above two effects. Another means of generating X-rays relies on deflecting an electron beam from its linear path by use of a powerful magnetic field. The electrons therefore

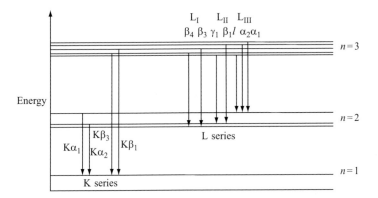

Fig. 11.1 *An atomic energy level diagram.* The horizontal lines represent the energies of the various electronic states of an atom. These are grouped into families of states, each with the same principal quantum number n. If an electron is removed from one of the lower-lying states then an electron in an upper state may drop into the vacancy, thereby losing energy, which is emitted as electromagnetic radiation. If the transitions are to the two lowest states, X-rays are produced, transitions to the lowest state ($n = 1$) giving rise to the K series, and those to the $n = 2$ group the L series.

experience a sudden change in direction, which is analogous to the deceleration we have described above and causes the emission of X-rays. This effect is exploited by synchrotron X-ray sources, which are notable for their much greater intensity than that which can be achieved with a laboratory source.

11.2   Laboratory X-ray sources

A schematic representation of a laboratory X-ray source is shown in Fig. 11.2. The operation of the instrument is as follows. The tungsten filament is connected to a power supply, so that an electrical current flows through it. This creates a heating effect, causing the tungsten to reach a high temperature. When any metal is heated, a process called *thermionic emission* takes place, which is the spontaneous emission of those electrons which have sufficient energy to break free from the structure of the metal. This is somewhat analogous to the evaporation of a liquid as its temperature is raised. The number of electrons emitted per second as a result of thermionic emission is strongly dependent on temperature, and so the filament is heated to as high a temperature as possible, close to the melting point—indeed, the metal tungsten is chosen as the filament material because of its high melting point. When the temperature of the filament is steady, an electric field is switched on so that the tungsten filament is negative with respect to a *target*, which is a block of a heavy metal such as copper or molybdenum. The electrons released from the tungsten filament are accelerated towards the target, into which they collide. If the potential difference between the filament and the target is V volts (this is typically some $40\,\text{kV}$), then the electrons hit the target with an energy eV joules, where e is the electronic charge.

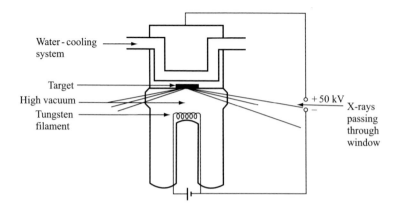

Fig. 11.2 *An X-ray source.* Electrons thermionically emitted from a heated tungsten filament are accelerated through a high potential towards the target. On impact with the target, X-rays are emitted.

The instrument is operated under high-vacuum conditions to ensure no loss in energy due to collisions of electrons with air molecules prior to hitting the target and to prevent oxidation of the anode and filament.

On impact with the target, the electrons are violently slowed down, and undergo repeated collisions with the target atoms. Some of these collisions are sufficiently energetic to eject electrons from the innermost levels of the target atoms, thereby making intra-atomic electronic transitions possible. Electromagnetic radiation is therefore emitted in two ways. Firstly, the electrons from the tungsten filament emit radiation as a result of the deceleration on impact with the target; and secondly, radiation is derived from the intra-atomic electronic transitions. The radiation is emitted in all directions from the target and passes out through the windows in the X-ray tube.

If the intensity of the emitted radiation from an X-ray tube of the type described above is investigated as a function of wavelength and as a function of the accelerating potential V, for any given target material, we obtain spectra such as those in Fig. 11.3.

Curve (a) corresponds to a particular accelerating potential V_1. The form of the spectrum is a continuum, on which are superposed some sharp spikes, which form well-defined groups. The continuous spectrum has a definite cut-off wavelength λ_{min}, below which no radiation is emitted; the spectrum rises to a maximum intensity at some wavelength λ_{max}, and then tails off slowly. If the potential is increased to some value V_2, we derive the spectrum of curve (b). This second curve has the same overall features as curve (a), in that there is a continuous spectrum on which spikes are superposed. On comparing the continuous spectra, we see that, in general, the intensity emitted is greater at all wavelengths for the higher accelerating potential V_2. The high-potential curve is also characterised by a low-wavelength cut-off, λ_{min}, which occurs at a shorter wavelength than that of curve (a), and the wavelength of maximum emitted intensity, λ_{max}, is also displaced towards the shorter wavelengths.

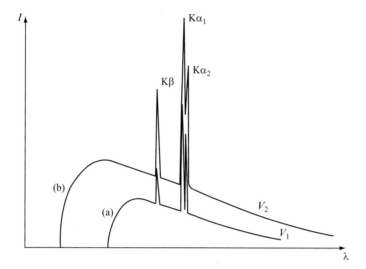

Fig. 11.3 *Typical X-ray spectra.* Shown are the spectra likely to be obtained from a typical X-ray source at two different accelerating potentials, V_1 (curve (a)) and V_2 (curve (b)), such that $V_2 > V_1$. Note that the characteristic spikes are at the same wavelengths for both curves. The wavelengths of the low-wavelength cut-off and of the maximum intensity of the continuous spectrum both decrease as the accelerating potential increases.

When we consider the spikes, however, we see that the same number of spikes occur at exactly identical wavelengths for both accelerating potentials, and only the intensity is changed. The wavelengths at which the spikes occur are therefore independent of the accelerating potential V, and this is highly suggestive that the spikes are a property of the target material, and not of the electrons emitted from the tungsten filament. This is indeed the case, for the spectra of Fig. 11.3 may be regarded as a superposition of two spectra, of the type shown in Fig. 11.4.

The continuous spectrum of Fig. 11.4(b), called the *white radiation* or *bremsstrahlung*, is the radiation emitted by the electrons thermionically released from the tungsten filament as a result of their slowing down as they pass through the target material. Since the electrons arrive at the target with an energy eV joules, the maximum amount of energy an electron may lose in a single event is also simply eV joules, and so we may write that the energy lost by a single electron in any single event satisfies the inequality

$$\text{energy loss per event} \leq eV \text{ J}$$

If the energy lost from the electron in any event is emitted as a quantum of radiation of wavelength λ and frequency ν, by Planck's law, we may write

$$\text{energy of radiation} = h\nu = \frac{hc}{\lambda}$$

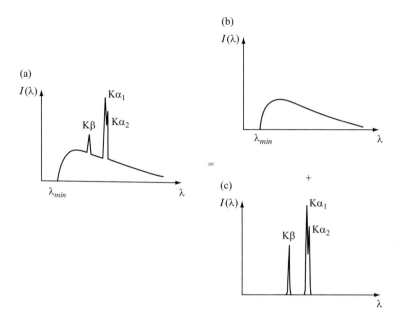

Fig. 11.4 *The structure of X-ray spectra.* The X-ray spectrum as derived from an X-ray tube (a) is a superposition of the bremsstrahlung spectrum (b) derived from the slowing-down of the electrons thermionically emitted from the tungsten filament, and the characteristic spectrum (c) of the target element, corresponding to allowed transitions between electronic energy levels.

in which c is the speed of electromagnetic radiation and h is Planck's constant. We now have

$$\frac{hc}{\lambda} \leq eV$$

$$\therefore \lambda \geq \frac{hc}{eV}$$

This equation predicts that the continuous radiation will be characterised by a low-wavelength cut-off, since the radiation of the smallest wavelength allowed is that for which $\lambda_{min} = hc/eV$. Emission at wavelengths lower than this is impossible, since the energy of the radiation would exceed that of the electrons hitting the anode. We can also see that operating the X-ray generator at a higher voltage will lower the value of λ_{min}, but a practical limit of 40–50 kV is chosen to maximise the lifetime of the filament.

The line spectrum of Fig. 11.4(c) is the spectrum of radiation emitted by electronic transitions between the electronic energy levels of the target atoms. Since the energies of allowed electronic levels within an atom are well defined, whenever a transition occurs, the emitted radiation is essentially monochromatic, and so we have a spectrum characterised by sharp lines. Each main group of lines corresponds to transitions to energy levels of common principal quantum number n. The fine structure is due to the variation of the energy level with parameters such as the angular-momentum quantum number. It is out of place to enter into a quantum mechanical discussion of electronic energy levels here, for all that is required for our purposes is that the line spectra are composed of sets of sharp lines. Each line is given a name, for instance the $K\alpha_1$, $K\alpha_2$ and $K\beta$ lines marked in Fig. 11.4(c). With increased resolution, it is possible to identify further separate lines within each group. Since the wavelengths of the lines are dependent on the energy differences between electronic levels within any atom, different X-ray line spectra may be obtained from different atoms. In general, the heavier the element, the lower the wavelengths of the emission lines, and also the more intense the emission.

In practice, the target material must satisfy certain operational requirements. Firstly, we wish to have an intense emission at a wavelength suitable for a particular experiment. Also, much of the energy of the electrons incident on the target from the tungsten filament is wasted as heat, and so the target becomes very hot. It is therefore desirable that the material of the target be of high thermal conductivity to dissipate the heat, and also to have a high melting point so that it does not melt. To facilitate the cooling, a water circulation system is included in the design of the instrument as in Fig. 11.2. Also, the target should be a good conductor of electricity so that the electric current brought by the incident electrons is readily carried away, and a low electrical resistance also minimises Joule heating. In practice, these requirements demand that the target should be a metal, preferably of high atomic number. Copper is the most common form of target material for macromolecular work, and has characteristic wavelengths given by

$$K\alpha_2 = 1.54433 \text{ Å}$$

$$K\alpha_1 = 1.54051 \text{ Å}$$

$$K\beta = 1.39217 \text{ Å}$$

Molybdenum is another useful choice:

$$K\alpha_2 = 0.71354 \text{ Å}$$

$$K\alpha_1 = 0.70926 \text{ Å}$$

$$K\beta = 0.63225 \text{ Å}$$

If another target material is to be used, a thin sheet of the required metal is usually fixed to a copper block, utilising the excellent thermal and electrical properties of copper in order to satisfy the operational requirements described above.

When an X-ray source is used for a long time, certain 'unexpected' wavelengths may be produced. This spurious radiation is usually derived from the deposition on the target of tungsten evaporated from the filament. The emitted radiation is therefore a mixture of the copper and tungsten emission spectra. During any experiment, it is desirable that the intensities of the emitted X-rays be constant over relatively long periods of time. This calls for considerable expertise in designing electronically stabilised power supplies.

For a long time, the preferred laboratory source for protein crystallography has been the rotating-anode generator, in which the anode is a cylinder that is rotated at high speed by an electric motor (for more details see Sanderson, 2007). The instrument is arranged so that the electrons collide orthogonally with the curved outer surface of the anode. The rotation of the anode ensures that the part of it which is in contact with the electron beam is constantly changing, and this allows more effective cooling of the target and permits higher currents to be used. The emitted X-rays leave the evacuated chamber through beryllium windows. Whilst a modern sealed-tube generator might have a practical limit of 40 kV, 40 mA (around 1.5 kW), a rotating-anode generator can be routinely operated at 50 kV, 100 mA (5 kW), and this has the benefit of giving a greater X-ray beam intensity. The disadvantage of the rotating-anode generator over the sealed-tube source is the greater practical complexity of spinning the anode at high speed and maintaining it under a very high vacuum.

Improvements in X-ray optics have led to recent popularity of sealed-tube systems owing to their greater reliability and lower cost overhead. The widely used microfocus system (Bloomer and Arndt, 1999) uses an ellipsoidal mirror to focus X-rays from the filament onto the sample.

11.3   Synchrotron sources

These are large and elaborate X-ray sources, which are usually established as national or international facilities at which researchers can apply for beam time that is usually awarded in days or 8 hour shifts. Such instruments rely on accelerating an electron or positron beam into a storage ring, where its energy is maintained at a high value (typically 2–6 gigaelectronvolts, or GeV) by radio frequency magnets (Fig. 11.5) (Helliwell, 1992). The electron beam is generated by a linear accelerator and is often fed into a smaller synchrotron or 'booster ring' to gain energy before being used to top up or refill the main storage ring of the facility. The storage ring is a large polyhedron, with devices known as bending magnets for bending the otherwise straight beam at each corner and quadrupole magnets for focusing it. The arrangement of bending and focusing magnets is referred to as the 'lattice' of the synchrotron. Typically, the main storage ring will be refilled once or twice a day to maintain the circulating electrons.

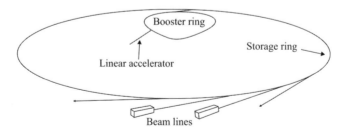

Fig. 11.5 *The general layout of the European Synchrotron Radiation Facility (ESRF).* Electrons originating from the linear accelerator are supplied to a booster ring, from where they are injected into the main storage ring, from which the experimental beam lines emanate.

When these charged particles change direction at the corners, the centripetal acceleration causes them to emit energy, some of which is in the form of X-rays. The emitted radiation emanates from the storage ring as a narrow cone or fan, which can be directed into different experimental beam lines. In the straight sections between the bending magnets, additional devices can be inserted to produce more intense radiation. These 'insertion devices' are known as *undulators* and *wigglers* and both consist of two banks of magnets, one above and one below the beam pipe. The magnets in each bank have their poles pointing in alternate directions as shown in Fig. 11.6, causing the direction of the electron beam to oscillate and emit a fan of radiation at each bend which is of higher energy (and lower wavelength) than that produced by the bending magnets. Wigglers produce a wide fan of energy which can supply several beam lines. In contrast, the magnets in undulators are arranged so that the deflections of the beam are less than those that occur in a wiggler. With an undulator, the fans of emitted radiation interfere constructively to generate intense radiation in a narrow wavelength band that is parallel to the direction of the main beam. The wavelength of the undulator radiation can be varied by changing the gap between the banks of magnets. Note that harmonics of the fundamental frequency are also generated, but these can be removed by the design of the optical components of each beam line.

Evacuation of the storage ring is essential to prevent loss of energy due to collisions of the electrons with air molecules and, likewise, individual beam lines also require a high vacuum to be maintained. To minimise the time needed to achieve a high vacuum after initial start-up of a new storage ring, its components are baked *in situ* using removable heating elements to evaporate any volatile material in the beam pipes!

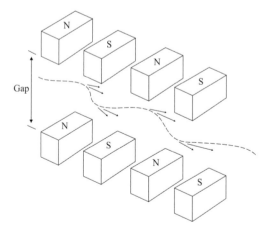

Fig. 11.6 *A schematic diagram of a synchrotron insertion device.* Insertion devices are generally classed as either undulators or wigglers because the regular arrangement of magnets above and below the electron beam (shown dashed) causes the beam to be deflected multiple times and emit 'cones' of intense radiation (indicated by small arrows) at each bend. The energy (and hence wavelength) of the emitted radiation can be controlled by the gap between the banks of magnets (shown).

The vacuum of the main ring is protected from any fluctuations that might occur in individual beam lines by beryllium windows or by valves that close rapidly if the beam line vacuum fails. The unwanted radiation from the ring is blocked by water-cooled absorbers and masks, some of which are movable to allow the beam into individual beam lines when required. In view of the harmful effects of ionising radiation, all synchrotrons (and other X-ray sources) have electronic safety interlocks that require the user to search and verify that no one is present in the experimental hutch before a beam is allowed to enter it. The hutch will be lined with lead to prevent the dose in surrounding areas from exceeding background levels, and many synchrotrons do not require visiting users to wear film badges. Not only do X-rays harm people, but they can also damage the X-ray detector if the direct beam is allowed onto its active surface. Hence a small piece of lead, known as a *backstop*, is always mounted somewhere between the crystal and the detector to prevent such damage from happening. An increasing number of synchrotrons allow users to run the experiment remotely via an Internet connection using crystals that are sent in advance by courier.

The electrons in the main storage ring move in bunches, the number and size of which can be controlled for different types of experiment. Multi-bunch mode is preferable for achieving the highest-intensity beam but single-bunch mode is of great value for time-resolved crystallographic studies, for example of ligands binding to proteins, where the reaction can be triggered with a laser flash that is synchronised with the beam pulse (e.g. Bourgeois *et al.*, 2007).

11.4 Optimising the X-ray beam

All the formal theory of X-ray diffraction which we have expounded in this book has been expressed in terms of a single wavelength parameter λ. In order to obtain easily interpreted data from an experiment, it is necessary to pass the radiation obtained from an X-ray source through some sort of filter in order that monochromatic or nearly monochromatic X-rays may be incident on the crystal. Two filtering devices are used, called *absorption filters* and *crystal monochromators*.

Any material absorbs X-rays in a manner characterised by a linear absorption coefficient μ. This varies in a complicated way with wavelength, as in Fig. 11.7(a). The sharp discontinuities in μ are called *absorption edges*, and the significance of these is that radiation on the short-wavelength side of the absorption edge is very much more strongly absorbed than that on the long-wavelength side. This sharp discontinuity in absorption properties may be utilised in the following manner. For any practical investigation, the radiation incident on the crystal should be intense and of a well-known, characterised wavelength. This criterion is most adequately satisfied by the most intense line of the characteristic spectrum of the target material, for instance the copper $K\alpha_1$ line. Our purpose is therefore to pass the radiation from an X-ray tube with a copper target through some filter which will transmit predominantly the copper $K\alpha_1$ line, or at least the $K\alpha_1 K\alpha_2$ doublet. Reference to Fig. 11.7(b), which shows the copper emission spectrum, indicates that we wish to eliminate the copper $K\beta$ line. Suppose we pass the radiation from the X-ray tube through a material which has an absorption edge at a wavelength between those of the Cu $K\beta$ and Cu $K\alpha$ lines. The

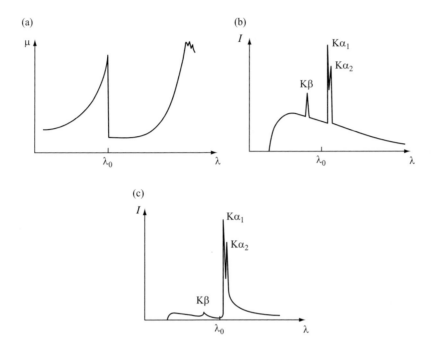

Fig. 11.7 *The absorption filter.* The absorption spectrum of some element is shown in (a), and is characterised by a sharp absorption edge at λ_0. The spectrum of an X-ray source is as in (b), and has predominant $K\alpha_1$, $K\alpha_2$ and $K\beta$ lines. If the absorption edge of the filter is at a wavelength between the $K\alpha$ doublet and $K\beta$ line of the source, then the $K\beta$ line is selectively reduced in intensity as compared with the $K\alpha$ doublet. The spectrum transmitted by the filter is as in (c).

Cu $K\beta$ will be selectively absorbed, and the form of the spectrum passing through the filter is shown in Fig. 11.7(c). The Cu $K\alpha_1 K\alpha_2$ doublet is transmitted with a small reduction in intensity, but the Cu $K\beta$ and much of the background continuum are suppressed. This is the principle of the *absorption filter*.

Since the Cu $K\alpha$ doublet is the predominant feature of the transmitted radiation, this may be allowed to fall directly onto the crystal. The most intense diffraction effects will be due to the Cu $K\alpha_1$ line, with some weak effects due to the Cu $K\alpha_2$ line, but this is acceptable for most purposes.

The absorption edges of the absorption spectrum of a material are related to those wavelengths which cause electronic transitions between the energy levels of the material, and so vary from element to element. A useful general rule is that if we wish to filter out the $K\beta$ line of an element of atomic number Z, then we use a filter of atomic number $Z - 1$. This usually allows for the selective absorption of the $K\beta$ line, while the $K\alpha$ doublet is transmitted. For a heavier target element of atomic number Z, materials of atomic number $Z - 1$ or $Z - 2$ are effective selective filters. Thus we would use a nickel filter for a copper target, and a zirconium filter for a molybdenum target.

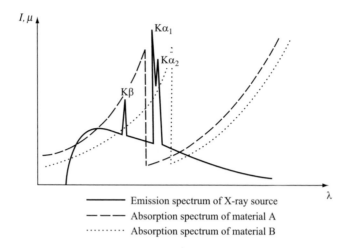

Fig. 11.8 *The balanced filter.* Two materials A and B are chosen such that their absorption edges are on each side of the Kα doublet of the source. The balanced filter therefore transmits selectively only the band between the two absorption edges.

A more sophisticated technique is the use of the balanced filter. This is an absorption filter made of two materials chosen such that their absorption edges are on either side of, and as close as possible to, the Kα doublet of the target material. The effect of this is illustrated in Fig. 11.8. If the thicknesses of the components of the filters are chosen so that the magnitude of the absorption is the same for both materials, then essentially only that radiation between the two absorption edges is transmitted, and since this band has been chosen to contain the required radiation, a balanced filter is an effective monochromator.

The second filtering technique makes use of one of the important results of X-ray diffraction theory, namely Bragg's law, equation (8.31):

$$\lambda = 2d_{hkl} \sin \theta \qquad (8.31)$$

One interpretation of Bragg's law was in terms of the 'reflection' of X-rays from crystal planes separated by the spacing d_{hkl}, as discussed in Section 8.15.

Suppose a polychromatic X-ray beam is incident onto a crystal at some angle θ, as in Fig. 11.9. There may be a set of planes within the crystal separated by a characteristic spacing d_{hkl} which satisfies Bragg's law for the angle θ. In this case, a diffracted beam will emerge from the crystal at the same angle θ as made by the incident beam, corresponding to Bragg reflection. But since the parameters d_{hkl} and θ are fixed, Bragg's law is satisfied by only a single wavelength λ. Hence the radiation which is Bragg-reflected is composed simply of that single-wavelength component of the incident radiation which satisfies Bragg's law. The Bragg-reflected waves are therefore monochromatic, and we have derived monochromatic radiation from a polychromatic source. This is the principle of the *crystal monochromator*.

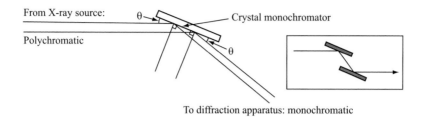

From X-ray source:

Polychromatic

Crystal monochromator

To diffraction apparatus: monochromatic

Fig. 11.9 *The crystal monochromator.* Polychromatic radiation from an X-ray source is incident onto a crystal at some angle θ. The waves which emerge from the crystal at the same angle θ are necessarily those suffering Bragg reflection within the crystal. This condition is satisfied only by certain wavelengths $\lambda, \lambda/2, \lambda/3$ and so on. The $\lambda/2, \lambda/3, \ldots$ components can be filtered out, allowing monochromatic waves to be incident onto the experimental sample. The inset figure shows how a double-crystal monochromator can be used to maintain a constant beam direction.

Reference to Section 8.15 will show that the general form of Bragg's law is

$$n\lambda = 2d_{hkl}\sin\theta \qquad (8.31)$$

in which n is an integer specifying the order of the Bragg reflection. If the radiation incident on the monochromator crystal has wavelength λ, with a range of impurities including radiation of wavelength $\lambda/2$, the second-order diffraction (i.e. $n = 2$) of the $\lambda/2$ component will have the same θ as the first-order diffraction ($n = 1$) of the λ component. A crystal monochromator therefore passes waves of wavelength $\lambda, \lambda/2,$ $\lambda/3$ and so on. In order to derive truly monochromatic radiation, the $\lambda/2, \lambda/3, \ldots$ components must be filtered out. Since the wavelengths $\lambda/2, \lambda/3, \ldots$ are well separated from the main component λ, it is usually possible to find an absorption filter which will selectively pass only the wavelength λ. Alternatively, we may consider the choice of the crystal in the following terms. As we saw in Chapter 8, the diffraction of the λ wavelength component corresponds to Bragg reflection by a set of planes of spacing d_{hkl}, where h, k and l refer to the indices of this set of lattice planes in the monochromator crystal. The intensity of this radiation is characterised by the structure factor F_{hkl}, and so the waves of wavelength λ have an intensity determined by the structure factor F_{hkl}, those of wavelength $\lambda/2$ by $F_{2h\,2k\,2l}$ and so on. If we select a monochromator crystal for which one of the F_{hkl} is large but all the structure factors of the type $F_{nh\,nk\,nl}$ are small, then the Bragg-reflected radiation may be regarded as purely monochromatic. It is possible to choose crystals and angles of incidence so that the $K\alpha_1$ line of a given target material may be selected as the incident wave for a diffraction experiment; full details are to be found in the *International Tables for X-ray Crystallography*. Graphite has been the material of choice for many 'in-house' protein crystallography X-ray sources.

An interesting side effect of the use of a crystal monochromator concerns the fact that the Bragg-reflected waves are partially polarised even though the incident waves are non-polarised. The reason for this is that the component of the incident wave polarised perpendicular to the effective planes within the crystal is absorbed to a greater extent than the component parallel to the planes, implying that the Bragg-reflected waves are enhanced in the parallel-polarised component. This is the same physical explanation as that given to account for the partial polarisation of light after reflection from, for instance, the surface of water (it is this effect which gives the well-known glare from such surfaces). When using a crystal monochromator, we must therefore pay due attention to the polarisation factor when correcting intensity data.

The main problem with monochromators is that they attenuate the beam significantly. An alternative is to use X-ray mirrors which are coated with nickel to filter out the Kβ radiation. By using two orthogonal curved mirrors and arranging them so that the incident beam strikes both of them at grazing incidence, a monochromatic beam can be produced and also focused in both the vertical and the horizontal direction (Phillips and Rayment, 1985). Ideally, these double-mirror systems should be bathed in helium gas to prevent oxidative damage caused by the beam. In synchrotron applications, the monochromator crystal can be bent so that its concave surface focuses the beam (Helliwell, 1992). Alternatively, focusing can be achieved by use of bent and/or curved mirrors. Such components of a synchrotron beam line require cooling, often by use of liquid nitrogen, to prevent distortions due to the significant thermal load of the beam.

More recent developments for in-house sources include multilayer optic devices which consist of multiple layers of material that reflect and focus the beam in one direction (Kusz and Bohm, 2002). Two multilayer blocks can be glued to each other at 90°, to provide focusing in both directions. These OSMIC^{TM} mirror devices focus the beam and thereby achieve high intensity with good Kβ and harmonic suppression.

By using one or more of the above devices, we can obtain an almost monochromatic beam. This is then passed through a collimator of the type shown in Fig. 11.10, so that plane monochromatic waves will be incident on the crystal. Here, two equal apertures A and B define a uniform beam, but since scattering will occur at B, a third aperture C is present. The diameter of C is slightly larger than that of A and B, and C is designed to prevent the propagation of waves scattered at B, but to pass the direct beam unhindered. The collimated monochromatic beam may then impinge on a crystal, which is placed in one of the instruments to be discussed in the next section.

In many instruments, the collimation is done by sets of horizontal and vertical slits forming a square or rectangular aperture, the dimensions of which may be controlled by the user, either by micrometer adjustment or by use of control motors, the latter

A B C

Fig. 11.10 *A collimator.* The two apertures A and B define a collimated beam. Since scattering at B will occur, a further aperture C is positioned as shown. C serves to absorb the waves scattered at B and to pass the straight-through beam without further scattering.

(a)

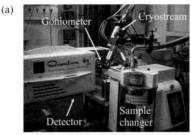

(b)

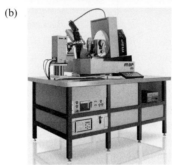

Fig. 11.11 *Data collection instrumentation.* (a) shows the main components of an experimental hutch at a synchrotron beam line (in this case ID14-EH2 at the ESRF). (b) A Marresearch image plate detector with a 'desk-top beam line' mounted on a sealed-tube generator, with microfocusing optics (reproduced with permission).

being preferable to avoid exposure of the user to X-rays—something which is essential at synchrotron sources. At such installations, electric stepping motors control all of the beam line optics, as well as the positioning of the camera in the experimental hutch. Between data collection runs, it is common to run a program which optimises the flux (photons per second) of the beam reaching the sample (as measured with an in-line ionisation chamber) by systematically adjusting the beam line optics (e.g. the mirror tilt) and the camera position. Bench-top instruments are now available in which the ability to make many of these adjustments is integrated into the design of the machine, as exemplified in Fig. 11.11.

11.5 The rotation method

Macromolecular data collection is conducted by slow rotation of the crystal through small angles about a fixed horizontal axis perpendicular to the beam (the ϕ axis), and this is therefore known as the rotation method with normal beam geometry (Fig. 11.12). Often each image is recorded by repeating the rotation several times to minimise the effects of any variations in the intensity of the incident beam and in the motor speed during the exposure and, for this reason, this mode of data collection is often referred to as the oscillation method (Arndt *et al.*, 1973; Wilson and Yeates, 1979). Typically the rotations are through 1–2°, but fine-slicing, in which

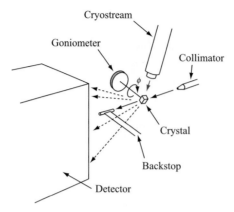

Fig. 11.12 *The rotation method in protein crystallography.* The crystal is rotated slowly about a fixed, usually horizontal axis perpendicular to the beam, and the diffraction data are recorded on a flat detector. The crystal-to-detector distance can be varied so that the diffraction pattern fills the active area of the detector. In some systems, the detector can be swung out on a 'theta-arm' or translated perpendicular to the beam to record high-resolution diffraction spots, although this is usually not necessary with the short X-ray wavelengths provided at synchrotron sources. The crystal is usually cooled by a liquid nitrogen cooler or cryostream. The arrangement shown roughly parallels that of Fig. 11.11(a).

data are collected in 0.1–0.2° steps, is also possible and can provide an improved signal-to-noise ratio. Diffraction images collected in this manner have the appearance of a series of curves, or 'lunes', which are formed by numerous regularly spaced diffraction spots. Each lune arises from the intersection of one layer of the reciprocal lattice with the Ewald sphere, as covered in Chapter 8. (e.g. see Fig. 8.18)

Whilst other more complicated arrangements have been developed and used very effectively, the simple arrangement of the protein rotation method involving a normal beam and a flat detector (Fig. 11.12), which can be moved closer to or further away from the crystal, allows high precision to be obtained. Care needs to be taken to ensure that the ϕ step size and the crystal-to-detector distance are such that the spots are well separated (overlapped reflections are usually rejected in data processing) and that they fill the active area of the detector. A smaller step size will reduce the number of overlaps, and a larger crystal-to-detector distance will enlarge the recorded image of the diffraction pattern. One disadvantage of rotating the crystal about one axis only is that the spots which are close to the rotation axis in the reciprocal lattice will never cross the Ewald sphere. These spots form a narrow cone in reciprocal space known as the 'blind region', which usually constitutes fewer than 10% of the reciprocal-lattice points. The blind region can be recorded by rotating the crystal through a small angle (independent of the rotation axis) and repeating the data collection again or by using another randomly oriented crystal. For crystals with moderate to high symmetry, the blind region is usually not a problem, since the spots in it will probably be related by symmetry to those in the recorded region of reciprocal space.

One common pitfall is to collect data with the detector too close to the crystal such that the diffraction spots are clustered around the centre of the image and much of the active area of the detector is wasted. The indexing and data processing tend to be more stable when full use is made of the whole active area of the detector to record diffraction, with the additional advantage that spot separation will be improved. The crystal will usually be cooled to 100 K by a liquid nitrogen cryostream, some components of which may lie between the crystal and the detector such that at very short crystal-to-detector distances, some obstruction of the diffraction pattern can occur. Indeed, the physical constraints imposed by the need to avoid collision of the detector with the cryostream and goniometer inevitably restrict the shortest possible crystal-to-detector distance and hence the highest possible resolution to which data could be collected. This is occasionally a problem with crystals that diffract to atomic resolution and, although such crystals tend to be rare, an increasing number of such studies are being undertaken. Collecting data is something that requires constant attention, for example to check for radiation damage, instrument failures and any accumulation of ice around the crystal, which causes the appearance of ice diffraction spots and rings in the images. Often decisions have to be made as to whether the data is worth collecting at all or whether effort is better spent on improving the crystal quality. Processing of images 'on the fly' during collection has become popular (Leslie *et al.*, 2002) and has the advantage that the user can monitor the quality of the data as it is collected and decisions can be made as to whether to collect further data on a new crystal, etc. Determination of the unit cell and symmetry prior to the main data collection can allow the user to determine a ϕ range to optimise the completeness of the dataset (Leslie, 2006; Leslie *et al.*, 2002), with the caveat that mistakes can sometimes be made at this stage and it may well be safer to collect 180° (or even 360° for anomalous data). As an example of the problems that can arise, monoclinic crystals often have a β angle very close to 90°, which can mislead the experimentalist into thinking that the space group is orthorhombic. In this case, collecting 90° of data (the minimum ϕ range required for orthorhombic crystals) instead of 180° (the usual requirement for monoclinic) could lead to a dataset in which only about half of the unique reflections have been measured.

Data collection usually requires cooling of the crystal using a device known as a cryostream, which blows a laminar stream of cooled nitrogen gas over the crystal to maintain its temperature at, typically, 100 K. A robot sample changer is usually available so that pre-frozen crystals can be transferred to the cryostream without warming to room temperature; otherwise, the equivalent step can be performed by the experimentalist using special pre-cooled metal 'cryotongs' or some other simple device or manual procedure to shield the crystal from the surrounding air (Garman, 1999). Cryocooling to 100 K significantly reduces radiation damage to the crystal, although at very high doses specific damage to disulphides and carboxylate groups can still be observed in electron density maps (Burmeister, 2000; Ravelli and McSweeney, 2000). The use of temperatures much above 100 K is not recommended, owing to phase transitions in water which occur at around 136 K and 150 K and could lead to ice formation in and around the crystal. At significantly higher temperatures, usually around 200–220 K, most proteins undergo a dynamic transition, i.e. an appreciable increase in thermal motion as reflected in something called the temperature factor, which we will cover later on (Wood *et al.*, 2008). Even lower temperatures, such as 20–40 K, can be achieved with helium cryocooling, which has the advantage of mitigating against the reduction of metalloprotein redox centres stemming from the production of photoelectrons in the irradiated crystal (Corbett *et al.*, 2007).

11.6   Electronic detectors

Any X-ray detector must have a large active area and good spatial resolving power to be able to measure a large number of closely spaced diffraction spots, which are typically of the order of 1 mm in diameter at the surface of the detector. The detector must have a high counting rate and must have a large dynamic range, i.e. it must be able to record intensities where the smallest and largest differ by several orders of magnitude and, ideally, give a linear response for most of this range. Since X-ray film has a dynamic range of only two orders of magnitude, protein crystallographers in the past used to stack three films together so that spots which were overloaded on the first film would be recorded accurately on the second or third film in the pack. Electronic detectors must also have a high detective quantum efficiency, or DQE—a term which refers to the signal-to-noise ratio of the response from the detector divided by the signal-to-noise ratio of the incident X-ray beam (strictly speaking, the DQE is the square of this ratio). The commonly used X-ray detectors generally fall into two classes, namely *image plates* and *charge-coupled devices* (CCDs). Both eliminate the extremely tiresome processes of developing, scanning and processing countless X-ray films, which, in days when computer resources were limited and expensive, could represent many weeks of work per dataset.

Image plates were initially developed to replace medical X-ray film and consist of a sheet of plastic which is coated with a europium-containing emulsion ($BaFBr:Eu^{2+}$) or phosphor (Amemiya and Miyahara, 1988). X-rays cause localised excitation within the phosphor layer, which persists for a substantial period of time (half-life ~ 8 days). In essence, the absorbed X-rays cause ionisation of Eu^{2+} to Eu^{3+}, and the liberated electrons are trapped in bromine vacancies that are introduced in the manufacturing process. Following exposure to X-rays, the plate is scanned with a He–Ne laser, which releases the trapped electrons. These convert the Eu^{3+} to excited Eu^{2+} ions, which emit light of a shorter wavelength than that of the stimulating laser. The emitted light is known as photostimulated luminescence, or PSL, the intensity of which is proportional to that of the incident X-ray beam. The PSL is recorded by a photomultiplier and stored on a computer, thus providing an electronic image of the diffraction pattern. The residual image within the phosphor can then be erased by illuminating it with a bright light, and the plate can be reused in this manner almost indefinitely. The image plate can be scanned with reasonably high spatial resolution—typically, the pixel size is 0.1 or 0.15 mm. The dynamic range of the image plate is of the order of 10^5, and it gives a linear response between 8 X-ray photons per pixel and 4×10^4 photons per pixel. Image plates have a high quantum efficiency of greater than 80% in the useful X-ray wavelength range. Part of the sensitivity of the detector stems from the extremely low background, which corresponds to around 1 X-ray photon per pixel—something to compare with the equivalent background, or 'chemical fog', of X-ray film, which is around 1000 X-ray photons per pixel. The response of the detector is uniform across its surface and, unlike other detectors that have been widely used in the past, the image does not suffer from distortion. Image plate devices for protein crystallography usually have a scanner integrated into the instrument. For example, the Marresearch detector (see Fig. 11.11) uses a circular plate. At the end of each exposure, the plate is spun at high speed about its centre, and the laser read-head scans the image by moving in a radial manner towards the image centre. The measured PSL data are essentially a spiral bitmap and are converted to a Cartesian image format. Since scanning and subsequent erasing of the plate take around 1–2 minutes in total, depending on the amount of the plate that is scanned, the instrument has a small amount of 'idle time', although this is generally not significant compared with the exposure time on in-house X-ray sources. Other manufacturers have developed systems to eliminate the delay time by having three image plates on a flexible metal 'belt', which can be rotated between exposures such that while one plate is being exposed, another is being scanned and the third is being erased. Nevertheless, these instruments are not intended for modern synchrotron sources, where exposure times are typically only a few seconds and the 'read' and 'erase' steps of even a multi-plate system would represent a significant time limitation.

In contrast, CCDs are solid-state detectors that provide a very rapid readout time. For X-ray diffraction work, the detector consists of a large phosphor screen, which is attached to a fibre-optic taper that transmits an image of the scintillation in the phosphor to the CCD chip itself—a component borrowed from digital photography (Westbrook and Naday, 1997). As shown in Fig. 11.13, the fibre-optic taper essentially shrinks the image of the diffraction pattern from the phosphor so that it fits onto the surface of the chip, although a 2×2 or 3×3 array of CCD chips is more often used, each with its own fibre-optic taper. The CCD converts the incident photons

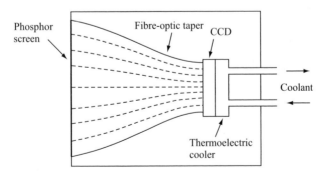

Fig. 11.13 *Components of a charge-coupled device (CCD) detector.* The phosphor screen converts the X-ray diffraction pattern into a visible light image, which is conveyed to the CCD chip (or an array of multiple CCD chips) by a fibre-optic taper. In some applications, CCDs can be used to record the X-ray data directly without the phosphor–fibre-optic coupling, but the arrangement shown is that most commonly used for protein crystallography.

into electrical charge, which is stored in miniature capacitors that form a 2D array on the surface of the chip. The charge stored in each capacitor is proportional to the intensity of incident light at that location. At the end of each exposure, the control circuits cause the charge in each of the capacitors to be 'shifted' sequentially from one capacitor to the next and finally 'dumped' into an amplifier, giving a voltage that is digitised and stored. With the ADSC 3×3 CCD array, an entire diffraction image of 6144×6144 pixels can be read in half a second. The pixel size is around $50 \times 50\,\mu\text{m}$, but the image is usually recorded with 2×2 'binning', in which the charges in adjacent pairs of pixels are combined by shifting them together. This enhances the dynamic range and the signal-to-noise ratio, as well as minimising data storage requirements. Whilst image plates work well at ambient temperature, CCD detectors need cooling to around $-40\,^{\circ}\text{C}$ to minimise the dark current and maximise the sensitivity of the instrument. The phosphor–fibre-optic coupling gives the CCD a good dynamic range that is comparable to that of the image plate, and the quantum efficiency of CCDs is generally superior, particularly for high intensities. The response of a CCD, or indeed of any detector, can vary somewhat across its surface or from pixel to pixel, and this is corrected for in calibration of the instrument by flooding the active surface of the detector with a uniform field of X-rays. The response of each pixel is recorded by the calibration software and used to correct subsequent measurements. Distortions introduced by the fibre optic are generally small, and are corrected for by separate calibration of the instrument using a drilled mask in front of the detector that is irradiated with a uniform X-ray field. The noise level of CCD detectors is low, and the response is linear except close to saturation. With good crystals, CCD detectors frequently allow entire datasets to be collected to high resolution within a few minutes at synchrotron beam lines. The CCD is an essential tool for fine-slicing work owing to its very low readout time, and its sensitivity makes it an essential detector for use with microfocus in-house sources, which operate at comparatively low power.

The future developments in the field most probably lie in instruments known as pixel detectors (Huelson *et al.*, 2006), which are semiconductor devices that, like CCDs, convert the energy of the X-ray beam into an electrical charge. However, pixel detectors do this by directly recording incident X-rays without recourse to a phosphor–fibre-optic coupling and have the additional advantage that each pixel behaves as a separate detector, each with its own amplifier and counter circuit, i.e. the charge can be read out directly without shifting it from one pixel to the next as happens with CCDs. The readout time of just a few milliseconds endows the pixel detector with the ability to record diffraction data almost in 'real time' as the crystal is rotated, and obviates the need for doing numerous separate X-ray exposures, each followed by scanning or readout. In contrast, the traditional method of data collection requires a fast X-ray shutter that is accurately synchronised with the ϕ axis motor. Since the discrimination level of each pixel can be set, the pixel detector has the ability to count incident photons above a certain minimum energy level and the instrument can thus be used to discern the wavelength of the diffraction peaks—something which is useful in Laue data collection (see later). Pixel detectors score very highly in terms of their dynamic range ($\sim 10^6$), their low background and the other criteria that we have discussed above for other detectors.

11.7 Other aspects of data collection

Crystal setting

We shall say a few words here about the mounting of the crystal which build on the practical aspects of wet mounting and flash cooling of crystals that we described in Chapter 10. When a crystal has been transferred to an instrument for data collection, it has to be accurately centred in the X-ray beam. This is a fairly intricate procedure, and hence the crystal is often mounted on a special instrument, called a *goniometer head*, which allows very fine movements of the crystal. The construction of a goniometer head is depicted in Fig. 11.14. A microscope or video microscope attached to the machine allows the user to translate the crystal in mutually orthogonal directions using the goniometer so that the crystal stays within the beam as it is rotated on the ϕ axis. With a removable goniometer head, such as the one shown in Fig. 11.14, centring of the crystal requires manual adjustment of the screw heads with a special key. In contrast, more modern data collection instruments usually have the goniometer built,

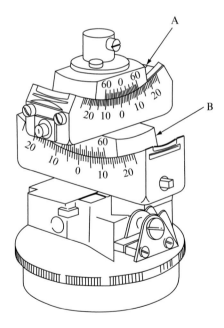

Fig. 11.14 *A goniometer head.* The crystal will be sealed in a glass capillary or, more commonly, frozen in a magnetic loop and mounted at the top of this device. The cradles A and B rotate about two mutually perpendicular directions for orienting the crystal, and translations or sledges, also operated by screw threads, allow for fine positioning of the crystal in the incident beam. (Courtesy of Stoe and Co.)

to a greater or lesser extent, into the instrument itself, and translations of the crystal are done by computer-controlled motors (Fig. 11.11). Occasionally, we may wish a particular crystallographic direction to be perpendicular or parallel to the incident beam. This is achieved by orienting the crystal in the incident beam prior to data collection. Since the position of the crystal is limited by the arc sizes of the cradles on the goniometer, it is necessary to place the crystal on the goniometer head in approximately the required orientation, and then alter its position by fine setting of the cradles. The general methods for the fine setting of the crystal usually invoke some aspect of the symmetry of a diffraction photograph as recorded for various crystal settings. If the crystal is misaligned, the spots making up the diffraction pattern will appear as groups of curves, or 'lunes', oriented in various directions, and so the crystal orientation is adjusted to obtain a perfectly symmetrical picture of concentric circles. A number of instruments incorporate a special type of goniometer, known as a κ-*goniometer*, which allows rotation of the sample about an axis (the kappa, or κ, axis) which is offset from the main ϕ rotation axis and allows the crystal to be partially reoriented in the beam, for example for collection of the blind region.

The Laue method

The very first X-ray diffraction experiment carried out by Friedrich and Knipping used a single stationary crystal, and unfiltered radiation obtained directly from an

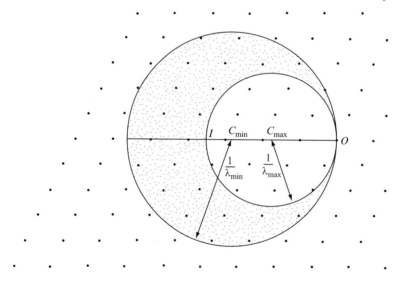

Fig. 11.15 *A section through the Ewald sphere construction for the Laue method.* Since unfiltered radiation is used, all wavelengths between λ_{min} and λ_{max} may give diffraction effects. Two Ewald spheres, one corresponding to each wavelength extreme, are drawn, and all reciprocal-lattice points in the shaded region between the two spheres satisfy the diffraction condition. Note that with protein crystals, the reciprocal-lattice points will be much more closely spaced than those shown here. Nevertheless, a significant proportion of the diffraction maxima are recorded at a single crystal setting.

X-ray tube. This experimental technique is now known as the Laue method, and has been applied in the protein crystallography field using intense synchrotron radiation for very rapid data collection to allow study of short-lived enzyme intermediates.

The diffracted waves are intercepted by a film or detector. Since the incident waves are essentially unfiltered, all wavelengths between some λ_{min} and a λ_{max} may be diffracted. The Ewald sphere construction takes the form of Fig. 11.15, in which we draw two spheres corresponding to each of the limiting wavelengths. Any reciprocal-lattice points in the region between the two spheres satisfy the diffraction condition and, in general, many diffracted beams are produced. In the rotation method, the diffraction pattern is recorded as the crystal rotates around a given axis. In contrast, in the Laue method, the diffraction pattern is recorded as a series of still images with the crystal at a series of angles separated by typically 15° on ϕ. Since a large number of reciprocal-lattice points lie between the limiting spheres corresponding to λ_{min} and λ_{max}, many spots are recorded at a single orientation of the crystal.

11.8 Data processing

The basic problem encountered with the quantitative interpretation of rotation images concerns the fact that they are two-dimensional projections of the three-dimensional reciprocal lattice. If we wish to obtain detailed data from a diffraction experiment, we must associate each diffraction spot unambiguously with three integers h, k and l, and this is obviously not easy on a two-dimensional surface with a randomly oriented crystal.

The first step of data processing involves determination of the crystal unit cell parameters and Bravais lattice, as well as the orientation of the unit cell axes with respect to the instrument. The second step is to refine the orientation and unit cell parameters along with the crystal mosaicity—a parameter that describes the broadness of the diffraction spots. This second step is referred to, somewhat confusingly, as 'post-refinement'. The final step involves integration of the detector pixels for every diffraction spot that occurs in each image so that the intensities I_{hkl} and their standard deviations $\sigma(I_{hkl})$ be calculated. Note that whilst the intensities and the structure factor amplitudes derived from them constitute the data required to solve the structure, the standard deviations reflect how precisely each intensity has been measured, and are therefore used to determine weights which are used in the phasing calculations and refinement. To determine the intensity of every spot in the dataset, we need to predict which spots occur on each image, and to do this we need the best possible estimates of the unit cell and orientation parameters as well as a good estimate of the spot size. Hence, it is usual to refine these parameters again using a narrow range of images during the actual data integration. This local fine-tuning of the crystal unit cell and orientation parameters during processing can provide very accurate spot predictions even in situations where the crystal slips during data collection.

One of the biggest advances in data processing has been the advent of automatic indexing, which allows the unit cell dimensions and crystal orientation to be determined from diffraction images obtained from a randomly oriented crystal

(Kabsch, 1988a). Prior to the availability of automatic indexing, it was necessary to align the crystal precisely on the goniometer, usually by taking alignment photographs—short X-ray exposures from which the crystal orientation can be deduced, and corrections can then be applied by adjusting the arcs on the goniometer (as described in Section 11.7). The ability to collect and process data from randomly oriented crystals has been one of the most important methodological developments in the field.

During each exposure of a data collection, some diffraction spots will arise from reciprocal-lattice points that pass completely through the Ewald sphere, and these are referred to as *fully recorded reflections*. Others arise from reciprocal-lattice points that cross the Ewald sphere at the start or end of the exposure and these spots will therefore appear on two successive images, or sometimes more. The latter set of spots are known as *partially recorded reflections*. The proportion of spots which are partially recorded (as opposed to fully recorded) will be greater for crystals with high mosaic spread (broad reciprocal-lattice points) and when smaller oscillation angles ($\Delta\phi$) are used—in the extreme case of data collection by fine-slicing (where $\Delta\phi$ is typically $0.1°$), all spots will be partially recorded. Calculating the net intensity of a partially recorded spot involves summing its components on adjacent images. This involves careful scaling of the images, which is something that will be covered later in this chapter. During data collection, some reciprocal-lattice points will happen to cross or merely graze the Ewald sphere almost tangentially and thus spend a large amount of time in their diffracting position. This set of spots are those that appear to be close the rotation axis in the diffraction images, and these spots can be present in many successive diffraction images, making the process of summing their components and correcting the resulting intensity very difficult and error-prone. For this reason, the small proportion of spots that happen to lie close to the rotation axis in the diffraction images are usually rejected in data processing.

Calculating the intensity of each reciprocal-lattice point involves subtracting the background counts from the pixels that make up the diffraction spots. The background should ideally be uniform, but often spurious effects such as scattering from the backstop, poorly aligned optics, spot overlap and diffraction rings from ice surrounding the crystal, as well as splitting of the crystal and other disorder, can lead to variations in the background intensity which require careful treatment in data processing. We will now look at more details of various steps of the data processing.

Autoindexing is done as follows. To determine the unit cell and crystal orientation from a diffraction image, we can make the initial assumption that all spots within the image have a ϕ angle corresponding to the midpoint of the ϕ range. Usually, a few hundred strong spots are chosen automatically from two images 90° apart on ϕ, although the method can work with just a single image. The method is based on the fact that we can calculate the reciprocal-lattice coordinates of spots from their coordinates on the detector (X_d, Y_d) and the crystal-to-detector distance D (Leslie, 2006). The spot coordinates are measured relative to the position of the direct beam, which is usually at the centre of the image. The detector coordinate system is shown in Fig. 11.16, from where it can be seen that the distance r of the recorded diffraction spot from the crystal is $(X_d^2 + Y_d^2 + D^2)^{1/2}$. The diffracted beam is, of

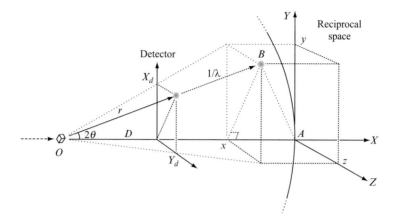

Fig. 11.16 *The detector and reciprocal-space coordinate systems.* The choices of axes for the detector and reciprocal space are shown on the left and right, respectively. The incident beam originates from the left-hand side, and a diffracted ray from the crystal at O will make an angle 2θ with the incident beam direction. The corresponding diffraction spot, shown as a grey patch on the detector, occurs at B in reciprocal space, which has A as its origin. For diffraction to occur, B must intersect the Ewald sphere (shown), and hence $OA = OB = 1/\lambda$.

$$x = -(OA - OB\cos 2\theta) = -\left(\tfrac{1}{\lambda} - \tfrac{1}{\lambda}\cos 2\theta\right) = -\tfrac{1}{\lambda}\left(1 - \tfrac{D}{r}\right)$$

$$\therefore x = \tfrac{1}{\lambda}\left(\tfrac{D}{r} - 1\right)$$

We may also state that $y = X_d/\lambda r$ and $z = Y_d/\lambda r$.

course, at an angle 2θ to the incident beam, where $\cos 2\theta = D/r$. The right-hand side of Fig. 11.16 shows the corresponding spot in reciprocal space crossing the Ewald sphere, and the axial system of the reciprocal lattice, which differs from that of the detector. Since in reciprocal space the spot is at a distance $1/\lambda$ from the crystal, we can scale the coordinates in the reference frame of the detector to those of the reciprocal lattice by multiplying by a factor of $1/r\lambda$. Hence we can see that two of the spot coordinates in reciprocal (x, y, z) space follow easily:

$$y = \frac{X_d}{r\lambda}$$

$$z = \frac{Y_d}{r\lambda}$$

Finally, consideration of the triangle OAB shown in Fig. 11.16 reveals that

$$x = \frac{1}{\lambda}\left(\frac{D}{r} - 1\right)$$

Hence, for any diffraction image, we can calculate the position (x, y, z) of each spot in reciprocal space, and we initially approximate the ϕ angle as being the midpoint of the ϕ range. This usually gives sufficient information to determine the crystal unit cell and its orientation. If the autoindexing does not work, we can try again with more images included, at say 45° or preferably 90° away on ϕ. From each of these images, we can calculate the reciprocal-lattice coordinates of the strong diffraction spots. By applying a rotation matrix corresponding to the ϕ angle separating these images, we can essentially reconstruct thin slices of reciprocal space where reciprocal-lattice points have crossed the Ewald sphere.

In the rules that we derived in Chapter 8 concerning the relationship between the real and reciprocal lattices, we showed that the principal zones in the reciprocal lattice are perpendicular to the real axes. For example, if the **a** axis of the crystal was parallel to the incident X-ray beam, then the reciprocal-lattice axes $\mathbf{b}^*$ and $\mathbf{c}^*$ would be perpendicular to the beam, i.e. the $0kl$ layer of the diffraction pattern would be perpendicular to the beam, as would the higher levels, for example $1kl$, $2kl$, $3kl$, etc. In this situation of a perfectly aligned crystal, the diffraction image will consist of concentric circles of spots due to each of the above layers intersecting the Ewald sphere. Hence, if we were to project the coordinates of all spots in reciprocal space onto the **a** axis, they would cluster at regular intervals along this axis and the clusters would be separated by a constant interval corresponding to the vector between the reciprocal-lattice planes. In crystals with orthogonal unit cell vectors, the spacing of the clusters would be equal to a^*. From Chapter 8, we know that the real and reciprocal lattices are related by a Fourier transform. Hence, if we calculate the Fourier transform of the projected vectors, we obtain a function which has peaks separated by the real cell dimension a.

Now, instead of having a major axis of the crystal parallel to the beam, let us imagine a situation where the zone axis is oriented away from the incident beam. Projecting the reciprocal-lattice points onto the direct beam axis will no longer yield a regularly repeating pattern, because the projected vectors will have different lengths. Likewise, the Fourier transform of the projected vectors will not have a clear set of maxima. Hence, only when the axis onto which we project the reciprocal-lattice points is oriented parallel to a major crystal axis will the Fourier transform contain large peaks. In the first example, we considered a situation where the spots were projected onto the direct-beam axis. However, we can project the spots onto any axis of our choosing, and this forms the basis of the autoindexing method proposed by Stellar *et al.* (1997), in which the direction of the projection axis is varied over a grid of small angular steps, typically 2°, to systematically cover all possible unique orientations (i.e. a hemisphere centred on the beam direction). At each orientation of the projection axis, the Fourier transform of the projected vectors is calculated and the peak heights are analysed. Three directions of the projection axis that give large peaks in the Fourier transform and are at reasonable angular separations are selected as potential unit cell axes. The orientations of these axes are then optimised by a finer angular search. The putative unit cell will be triclinic and most probably will not reflect the true symmetry of the crystal, but it should allow all of the diffraction spots to be assigned h, k and l indices. The axes of the initial cell are then transformed to all possible lattice types and a penalty is presented to the user, who can then decide on the likely crystal system. In general, the lattice type with the highest symmetry and lowest penalty is selected and the unit cell parameters are refined using the spot positions. Further details can be found in Powell (1999).

Whilst autoindexing allows the unit cell and its orientation on the diffractometer to be determined, it does not permit the full determination of the crystal symmetry. In most cases, this can be ascertained from the symmetry within the diffraction pattern, which is usually determined when the processed data from all images are scaled and merged together at a later stage. In general, the aim is to find the highest-symmetry space group which allows the data to be scaled together such that a parameter known as the R_{merge} has a low value, and this is something we will return to later in this chapter. Hence space group determination is often a simple practical matter of merging the data in a range of likely space groups and selecting the one with highest symmetry that gives good agreement statistics, i.e. a low R_{merge}. Any remaining ambiguity in the symmetry of the crystal is resolved during structure analysis.

Success in the automatic-indexing step will be indicated by good agreement between the observed and predicted spot positions, which must be checked both visually and by inspection of the rms deviation between the calculated and observed spot coordinates on the detector; this deviation should, ideally, be of the order of the pixel size or less. When the predictions are checked visually, it is often found that not all of the spots are predicted, particularly those at the edges of the lunes, and this is usually due to underestimation of the mosaic spread—initially, it may be assumed to be zero. The mosaicity is refined later on, but an initial estimate can be obtained in the program MOSFLM (Leslie, 2006) by calculating the total spot intensity in an image as the mosaicity is increased at regular intervals. The total intensity will reach a plateau when the mosaicity is such that all of the diffraction spots are predicted—something which should also be checked by eye. Success with the indexing and refinement of the mosaicity can be assessed using a graphical user interface, which all data-processing programs provide (see e.g. Fig. 11.17). This also allows the user to define regions of the detector within which the data should not be processed, for example the shadow of the backstop.

Automatic indexing relies heavily on accurate knowledge of the direct-beam position in each image, which is, of course, hidden behind the backstop to prevent damage to the detector. However, if there is any doubt about the main-beam coordinates, they can be measured by fitting a circle to diffraction rings from a sample of wax or even to the ice rings from a poorly frozen crystal. Errors in the direct-beam position can lead to misindexing, i.e. the unit cell is determined correctly but the diffraction spots have the wrong *hkl* values assigned to them, often by just one or two reciprocal-lattice units on one of the axes. The dataset can still be processed, but the resulting agreement statistics will be very poor.

During data processing, the unit cell parameters, orientation matrix, detector position, beam position, divergence and crystal mosaic spread are refined using the spot coordinates and the relative intensities of the components of partially recorded reflections, both of which give orientational information. From the trial unit cell and orientation of the crystal, determined by autoindexing, we can predict what percentage of the intensity of each partially recorded reflection will be recorded on the respective images. The orientation matrix can therefore be adjusted to improve the agreement between the observed and predicted partialities. This is referred to as post-refinement, since it requires that the data have already been through one round of integration.

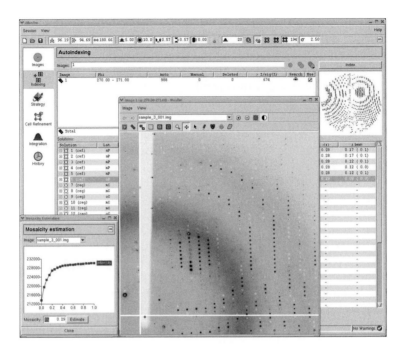

Fig. 11.17 *The data-processing program iMOSFLM.* The graphical user interface of iMOSLFM (Leslie *et al.*, 2006) at the stage of indexing a typical protein dataset.

In the case of fine-slicing, the ϕ angles of the centre of each diffraction spot can be determined experimentally and used as observations in the refinement of the crystal and instrumental parameters. In MOSFLM, the spot positions are used to refine the detector parameters (e.g. the crystal-to-detector distance and the tilt and twist of the detector), and the partial reflections are used to 'post-refine' the cell, orientation and mosaic spread. Following post-refinement using two segments of data widely separated on ϕ, the errors in the unit cell parameters should be of the order of 0.1% or less, for crystals that diffract to high resolution. Once the unit cell parameters have been refined to this level of precision, they can often be fixed for the remainder of the data processing, during which only the detector parameters are refined.

To determine the intensity of each diffraction spot, it is necessary to subtract the background counts from those due to the diffraction spot itself. The background under the spot itself cannot be measured, but we can estimate it from the pixels surrounding the diffraction peak by use of a mask such as that shown in Fig. 11.18. The mask can be either two-dimensional, such as those used by the programs MOSFLM (Leslie, 2006) and DENZO (Otwinowski and Minor, 1997), or three-dimensional for fine-sliced data, such as those used by XDS (Kabsch, 1988a,b) and d*TREK (Pflugrath, 1999). A plane is fitted to those pixels in the background region of the mask surrounding the spot, and this is used to estimate the background in the peak region. The dimensions of the mask can be specified by the user or, preferably, determined automatically. When

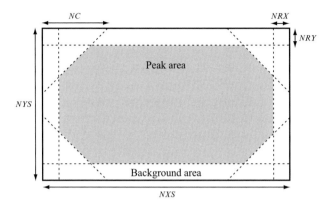

Fig. 11.18 *A typical spot integration mask.* The area assigned to the peak of the diffraction spot is shown shaded. In MOSFLM (Leslie, 2006), the overall size of the mask is defined by the numbers of horizontal and vertical pixels, shown as *NXS* and *NYS*, respectively, and the outer background region is delimited by the parameters *NRX*, *NRY* and *NC*, as shown.

spots are close together or the mask is too big, the background region may contain neighbouring spots, and the affected pixels have to be eliminated from the calculations, which is done automatically. Once the background plane has been determined, we can estimate the intensity of the diffraction spot by summing the pixels making up the peak and then subtracting the counts for the calculated background plane in the peak area of the mask. This is referred to as the *integrated intensity* of the spot. It has been found that for weak spots, more accurate estimates of their intensities can be obtained by a different method, known as *profile fitting* (Diamond, 1969, Rossmann, 1979). In this method, the background plane is calculated in the manner just described, but the intensity is determined from a 'profile' representing the average spot shape of reflections nearby in the same part of the detector. The profile is determined by dividing the detector up into different areas and taking a weighted sum of the equivalent pixels making up the spots in these areas on several adjacent images (Leslie, 2006). The intensity of each spot is proportional to the factor by which the standard profile must be multiplied to match the observed profile of the spot. This is determined by least-squares minimising the residual R in the below equation for each spot:

$$R = \sum_{\text{peak pixels}} w_i \left(X_i - KP_i \right)^2 \qquad (11.1)$$

where X_i is the background-subtracted intensity at pixel i, P_i is the value of the standard profile at the same pixel, w_i is a weighting factor and K is the desired factor, which is directly proportional to the intensity of the spot. Profile fitting benefits the weak reflections most but can occasionally be problematic when the strong spots are very sharp and coarsely sampled by the raster, i.e. the spots are only a few pixels across. MOSFLM and DENZO use two-dimensional spot profiles, whereas XDS and d*TREK use three-dimensional profiles, i.e. they analyse how the spot shape changes over successive images, and are therefore better for processing fine-sliced data. DENZO performs autoindexing using one image only and, during the integration step, locally averages the profiles of surrounding diffraction peaks for each spot that is being analysed (Otwinowski and Minor, 1997).

Profile fitting is also useful to estimate the intensities of overloaded reflections, i.e. reflections for which the intensity exceeds the dynamic range of the detector. The outer regions of the profile can be fitted to those pixels which are not overloaded. However, this is usually a desperate way to measure such reflections, since a far better practice is to remeasure the dataset with a shorter exposure time and/or a lower beam intensity specifically to record just the overloaded reflections, and attenuators can also be inserted into the beam for this purpose. This second dataset can then be scaled to the first one by methods that we will cover later in this chapter.

The data reduction program XDS (Kabsch, 1988a,b, 1993) operates in a rather more automated manner than the others, requiring the user to specify only the approximate unit cell dimensions and crystal space group in order to process the dataset, although the latter crystallographic parameters can be determined automatically if desired. The program works by locating strong diffraction spots in the first 5° of data and using these images to estimate the background. The strong spots are used for automatic indexing, in which a reduced cell is determined by searching for all possible vectors between reciprocal-lattice points. The shortest vectors usually correspond to the reciprocal unit cell axes, which can be defined accurately by searching for clusters of short vectors in reciprocal space. Transformations of the corresponding 'reduced cell' to all possible symmetries are considered, and a quality index for each allows the lattice of highest symmetry which is most consistent with the reduced cell to be selected. Following this, the program returns to the first image and begins assigning each pixel to the nearest diffraction spot. Pixels close to the predicted spot positions are used to estimate spot profiles, and the remainder are used to estimate the background. Partially recorded reflections are used to estimate three-dimensional profiles, which are constantly updated during the processing, as is the background. The profiles are transformed into reciprocal-space coordinates to minimise the distortions of the diffraction spots that are inherent in having a single rotation axis with a flat detector. The program then scales the data from all images to compensate for absorption and crystal decay, using methods that will be covered later in this chapter.

The program d*TREK (Pflugrath, 1999) has been developed principally for processing fine-sliced data where the oscillation angle is less than the mosaicity of the crystal. In addition to the improved spot separation, this approach has a number of elegant advantages, including a lower background in the images and a greater $I/\sigma(I)$ ratio of the intensities. It is more demanding in terms of data storage and its requirement for instrumental precision than is the conventional 'thick-slicing' approach, where the oscillation angle is generally greater than the mosaicity and a large proportion of the spots are fully recorded. In the fine-slicing approach, a ϕ step of exactly half the mosaicity is recommended (Pflugrath, 1999), i.e. for a crystal with a mosaicity of 1°, an oscillation angle of 0.5° should be used initially and decreased if necessary to reduce spot overlap.

All data-processing programs must estimate the standard deviations of the spot intensities as well as the intensities themselves. This is important, since the standard deviations $\sigma(I)$ are used as weights in scaling and merging the data from different images together, as well as in the subsequent structure analysis and refinement. On the surface of the detector, the arrivals of individual X-rays which form a particular diffraction spot are essentially random and independent events. For this reason, we may apply Poisson statistics to estimate the $\sigma(I)$ of each diffraction spot. For a quantity obeying the Poisson distribution, its standard deviation is simply the square root of that quantity, which in this case is the number of X-ray photons arriving at the detector and forming a particular diffraction spot. The proof of this is rather lengthy, but for the interested reader it is given by Bright Wilson (1990), where more details of the Poisson distribution can be found. For data from an area detector, it is important to convert the value of each pixel into the corresponding number of X-rays, and the conversion factor is referred to as the *gain*, which is defined as the pixel value per incident X-ray photon and can be estimated by the software from the variation of pixel values in the background region. We can demonstrate the role of Poisson statistics most easily for the simplified case of a spot with a constant background that has been determined from an area equal to that of the peak. The intensity I of the spot is given by the total of the peak counts P minus the sum of the background counts B under the peak as follows:

$$I = P - B \qquad (11.2)$$

In general, if a variable u is a function of independent variables x, y, z etc., we can apply the following formula for the propagation of errors:

$$\sigma^2(u) = \left(\frac{\partial u}{\partial x}\right)^2 \sigma^2(x) + \left(\frac{\partial u}{\partial y}\right)^2 \sigma^2(y) + \left(\frac{\partial u}{\partial z}\right)^2 \sigma^2(z) + \dots \qquad (11.3)$$

Applying this rule to the intensity given by equation (11.2), we obtain

$$\sigma^2(I) = \sigma^2(P) + \sigma^2(B)$$

But since P and B each obey the Poisson distribution,

$$\sigma(P) = \sqrt{P}$$

and

$$\sigma(B) = \sqrt{B}$$

Therefore we obtain

$$\sigma^2(I) = P + B$$
$$\therefore \sigma(I) = \sqrt{P + B} \qquad (11.4)$$

Hence the standard deviation of the intensity can be determined easily from the square root of the sum of the peak and estimated background counts in the region of the spot. Note that, in practice, the actual calculations are somewhat more complicated. Nevertheless, because the assumptions used in deriving the respective formulae (such as the independence of the variables) are never fully met, the $\sigma(I)$ values that we obtain tend to be underestimated and, consequently, they have to be deliberately inflated during the scaling and merging of the data so that they match the actual discrepancies between the observed intensities of the symmetry-related reflections. However, equation (11.4) indicates that if, say, we measure a diffraction spot with a peak plus background count of 100 photons, the estimated error will be 10 photons, or 10% of the total counts. As we will see later, in a diffraction experiment we aim to achieve better than 10% precision in the intensity measurements, and so 100 counts is around the minimum acceptable intensity. It is easy to verify that the greater the total intensity, the higher the signal-to-noise or $I/\sigma(I)$ ratio will be; for example, a spot of 10 000 counts would have an $I/\sigma(I)$ ratio of 100 and an estimated error of 1%.

There are a number of advantages to processing a dataset as it is being collected, since the crystallographer can monitor data quality more objectively, and this is an option at most data collection facilities (Leslie *et al.*, 2002).

11.9   The basis of intensity data corrections

In Chapter 9, we learnt that the relative amplitude of the diffraction spot associated with the reciprocal-lattice point *hkl* is given by the structure factor F_{hkl},

$$F_{hkl} = \sum_j f_j e^{2\pi i \left(hx_j + ky_j + lz_j \right)} \tag{9.21}$$

and the relative intensity I is

$$I_{hkl} = |F_{hkl}|^2 \tag{9.13}$$

I_{hkl} represents the intensity of the scattering by a unit cell as compared with that by a single free electron, and so I_{hkl} is a relative, not absolute, measurement of intensity. In this chapter, we shall discuss how the observed intensity is related to the true intensity I_{hkl} defined above.

In order to derive the correct value of I_{hkl} from the observed intensity data, we must apply certain correction factors that allow for various geometrical and physical factors which we have up to now ignored. In this section, we shall discuss each of these factors in turn, but in order to avoid confusion, we shall state now the relationship between the structure factor magnitude $|F_{hkl}|$ and a quantity E_{hkl} which, as we shall presently see, represents the total energy scattered by the crystal during a diffraction event, and is the quantity which may be recorded by a detector. Our expression is

$$E_{hkl} = \frac{I_0}{\omega} K L_{hkl} p_{hkl} A'_{hkl} e_{hkl} \left| (F_{hkl})_T \right|^2 V \tag{11.5}$$

in which I_0 is the intensity of the incident X-ray beam; ω is the angular velocity of rotation of the crystal; K is a factor concerning fundamental physical constants, defined below in equation (11.8); L_{hkl} is the Lorentz factor; p_{hkl} is the polarisation factor; A'_{hkl} is a correction for absorption of X-rays and an effect known as secondary extinction; e_{hkl} is a correction for an effect known as primary extinction; $|(F_{hkl})_T|^2$ is the relative intensity defined by equations (9.21) and (9.13), corrected for thermal and other disorder effects; and V is the volume of the crystal.

We shall discuss each of these terms in due course, but rather than giving a detailed mathematical analysis of each factor, we shall describe a simple physical model so that the principles behind each correction factor may be understood.

The integrated reflecting power

Our first discussion concerns the diffraction of X-rays by a very small volume element dV of the crystal, chosen to be so small that we may ignore any effects such as absorption of the incident beam, which we shall discuss later.

There are a number of factors that cause diffraction spots to appear slightly blurred rather than as infinitely sharp points, and these can be understood if we consider Bragg's law in the form

$$\lambda = 2d_{hkl} \sin \theta \tag{8.31}$$

On taking differentials, we obtain

$$\frac{d\lambda}{d\theta} = 2d_{hkl} \cos \theta$$

$$\therefore d\theta = \frac{d\lambda}{2d_{hkl} \cos \theta} \tag{11.6}$$

No incident X-ray beam is perfectly monochromatic, so that in practice there are always waves of a range of wavelengths propagating through the crystal, and this wavelength variation may be represented by $d\lambda$. According to equation (11.6), associated with a wavelength range $d\lambda$ there is an angular range $d\theta$ over which the Bragg diffraction condition is satisfied. Consequently, even an ideal mathematical reciprocal-lattice point will diffract over a range of angles $d\theta$, and this may be interpreted as being due to a blurred reciprocal-lattice point. Furthermore, the incident beam is never ideally collimated, and so the incident waves impinge on the crystal over a narrow range of incident angles depending on the degree of divergence of the incident beam. The factors of incident-beam divergence and lack of monochromaticity both contribute to the blurring of the reciprocal-lattice points and, as we shall see presently, there is an additional contribution due to disorder within the crystal.

Diffraction corresponding to a particular reciprocal-lattice point *hkl* occurs whenever that point intersects the Ewald sphere, but in view of the fact that each reciprocal-lattice point is smeared out, the intersection of any given point with the Ewald sphere is not a precise event occurring at the single instant when the Bragg condition is exactly satisfied. This is why all practical procedures for data collection involve rotation of the crystal about some axis with a constant angular velocity ω. As the crystal rotates, the reciprocal lattice rotates with the same constant angular velocity ω about its origin. Any reciprocal-lattice point therefore sweeps through the Ewald sphere, and diffraction occurs over a range of angles corresponding to those orientations of the crystal at which the reciprocal-lattice point intersects the Ewald sphere. This is represented in Fig. 11.19, in which (a) shows the intensity diffracted by the intersection of an ideal mathematical point with the Ewald sphere and (b) represents the distribution of the intensity caused by a reciprocal-lattice point of finite size as it passes completely

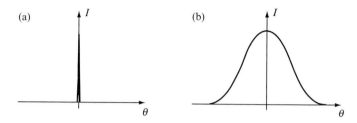

Fig. 11.19 *Diffracted intensities.* The passage of an ideal reciprocal-lattice point through the Ewald sphere gives a very sharp diffraction event as in (a). In reality, all reciprocal-lattice points are smeared out, and so diffraction occurs over an angular range, with the maximum intensity being recorded at the exact Bragg angle as in (b).

through the diffracting position. If the diffraction event were recorded by an X-ray detector, the situation corresponding to Fig. 11.19(a) would give rise to a very sharp diffraction peak, whereas that corresponding to Fig. 11.19(b) would appear as a rather diffuse spot.

If a particular reciprocal-lattice point has a very broad shape, the diffraction event will take place over a fairly wide angular range, so that a very diffuse spot is recorded by the detector as in Fig. 11.20(a). In contrast, a sharper spot will have the appearance of that shown in Fig. 11.20(b). Such a case often arises at large Bragg angles θ, for reference to equation (11.6) shows that when θ is large, $\cos\theta$ is small and the angular spread $d\theta$ due to the spread of wavelengths $d\lambda$ alone is large. On the other hand, very compact reciprocal-lattice points (such as those occurring at low θ) will give more compact spots, as in Fig. 11.20(b). Although the spot shown in Fig. 11.20(b) has a greater peak height than that in (a), a greater area is enclosed by curve (a) than by curve (b), implying that the total energy diffracted by the broad, diffuse spot is greater than that scattered by the sharper spot. The fact that all crystals necessarily give diffraction spots of a more or less diffuse nature demonstrates that measurement of the total energy scattered during the diffraction event by integrating under the curve, rather than merely measuring the peak height, is essential.

Our measurement of diffracted intensity is therefore the total energy diffracted during the complete diffraction event, which we represent as E_{hkl}. Before we quote the expression for E_{hkl}, we must point out that there is a continuous background scattering of X-rays due to scattering by the air, Compton scattering and so on. As described earlier, this is corrected for during data processing by measuring the intensity recorded by the detector in between diffraction maxima, usually close to the diffraction peak being integrated. All subsequent discussion assumes that we have corrected for background scattering.

For an elementary volume dV of a crystal which is small enough that absorption effects may be neglected, it may be shown that the total diffracted energy E_{hkl} associated with the reciprocal-lattice point hkl is given by an expression of the form

$$E_{hkl} = \frac{I_0}{\omega} K L_{hkl} \, p_{hkl} \, |(F_{hkl})_T|^2 \, dV \qquad (11.7)$$

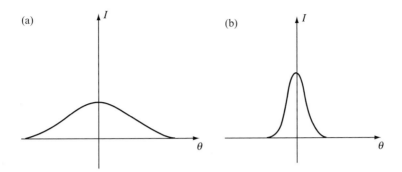

Fig. 11.20 *The integrated reflecting power.* Although the diffraction peak in (b) has a greater peak height than that in (a), the total area under curve (a) is greater than that under (b). This implies that a greater total energy is diffracted by the reciprocal-lattice point giving rise to (a) than that giving rise to (b). Our measure of the total energy is proportional to the area under the intensity curve, and is called the integrated reflecting power.

where

$$K = \left(\frac{e^2}{4\pi\varepsilon_0 mc^2}\right)^2 N^2\lambda^3 \tag{11.8}$$

in which e is the electronic charge; m is the electron mass; ε_0 is a fundamental electro-magnetic constant, of value 8.85416×10^{-12} F m^{-1}; c is the speed of electromagnetic radiation; N is the total number of unit cells within the volume dV; and λ is the wavelength of the incident radiation. Note that equation (11.7) is similar to, but not the same as, equation (11.5): the conceptual difference is that equation (11.5) defines the diffracted energy of a macroscopic crystal of volume V, whereas equation (11.7) refers to the diffracted energy of an infinitesimal crystal of volume dV. Mathematically, the two equations differ by virtue of the presence in equation (11.5) of two additional terms, A'_{hkl} and e_{hkl}, both of which relate to phenomena present only in a macroscopic structure.

The Lorentz factor, polarisation factor and temperature/disorder correction will be discussed in due course, but for the present we shall concentrate our attention on the factor K, and discuss the relevance of the terms contained within the definition of K in equation (11.8).

Equation (11.7) is more meaningfully written as

$$E_{hkl} = I_0 \left(\frac{e^2}{4\pi\varepsilon_0 \frac{1}{\lambda} mc^2}\right)^2 N^2 \left(\frac{\lambda L_{hkl}}{\omega}\right) p_{hkl} |(F_{hkl})_T|^2 \, dV \tag{11.9}$$

in which we associate λ/ω with the Lorentz factor, and $1/\lambda$ with the denominator of the term containing the fundamental constants e, m, ε_0 and c.

The rationale behind the form of equation (11.9) is as follows. Firstly, we expect the total energy diffracted to be proportional to the intensity of the incident beam, and so we have the term I_0. Reference to Chapter 9, equation (9.2), shows that we defined the electronic scattering factor f_e as

$$f_e = \frac{e^2}{4\pi\varepsilon_0 rmc^2} \tag{9.2}$$

where the symbols e, m, ε_0 and c are as defined above, and r represents the distance from the electron at which we sample the diffracted waves. f_e specifies the scattering ability of a single electron, and is defined as the ratio of the amplitude E_{scat} of the field scattered by an electron to that incident on the electron, E_{in}:

$$E_{scat} = f_e E_{in}$$

Since the structure factor F_{hkl} is the scattering ability of a unit cell as compared with that of a single electron, the amplitude of the wave scattered by a unit cell in absolute terms is simply $f_e |F_{hkl}| E_{in}$. To obtain the scattered intensity, we square the amplitude, giving the intensity as $f_e^2 |F_{hkl}|^2 I_0$. Thus the factor $f_e^2 I_0$ converts the relative quantity $|F_{hkl}|^2$ into an absolute quantity. Comparison of equation (9.2) with equation (11.9) shows that the spatial parameter r corresponds to the factor $1/\lambda$. The significance of this is that the diffraction event is defined by the intersection of a reciprocal-lattice point with the Ewald sphere. The Ewald sphere has a radius of $1/\lambda$, and so $1/\lambda$ is the appropriate spatial measure of the energy received at the surface of the Ewald sphere.

The association of the factor λ/ω with the Lorentz factor will become clear when we discuss the Lorentz factor itself. For the moment, it is significant to note that the quantity ω, which represents the angular velocity of rotation of the crystal, appears in the denominator of equation (11.9). Since ω has the dimensions of $[\text{time}]^{-1}$, the term $1/\omega$ therefore expresses the time during which a particular reciprocal-lattice point is in the diffracting position, and clearly the total energy diffracted will be proportional to the time taken for the diffraction event to occur.

The remaining factor which we have to justify is the N^2 term, where N is the total number of unit cells within the volume dV of the crystal. Simplistically, we can say that since each unit cell scatters radiation with a relative amplitude $|F_{hkl}|$ then N unit cells will scatter with a relative amplitude $N|F_{hkl}|$. On squaring to obtain intensities, we derive the requisite N^2. This argument is a convenient one, but is not quite the true origin of the N^2 term. The full analysis derives this term from an integration concerned with the shape factor of the crystal, the calculation of which we shall not go into here.

If we rewrite equation (11.7) as

$$\frac{E_{hkl}\omega}{I_0} = KL_{hkl}\, p_{hkl}\, |(F_{hkl})_T|^2\, dV \tag{11.10}$$

then we may define the quantity $E_{hkl}\omega/I_0$ in terms of an absolute factor K and the various other correction factors, and the significance of this is as follows.

At any instant during the diffraction event, the reciprocal-lattice point hkl intersects the Ewald sphere and diffracts an energy dE_{hkl}. The power diffracted in the time dt taken for an element of the reciprocal-lattice point to pass through the Ewald sphere is given by dE_{hkl}/dt. We now define the *reflecting power* $P(\theta)$ as the ratio of the power of the diffracted beam to the intensity of the incident beam:

$$P(\theta) = \frac{1}{I_0} \frac{dE_{hkl}}{dt}$$

The reflecting power $P(\theta)$ is a function of the Bragg angle θ only, and as the reciprocal-lattice point hkl sweeps through the Ewald sphere with constant angular velocity ω, then the total power diffracted by the crystal, compared with the intensity of the incident beam, is given by

$$\int P(\theta) \, d\theta = \int \frac{1}{I_0} \frac{dE_{hkl}}{dt} \, d\theta$$

where the integration is taken over the angular range corresponding to the entire diffraction event. Since the angular velocity ω of rotation of the crystal is equal to $d\theta/dt$, we have

$$\int P(\theta) \, d\theta = \frac{1}{I_0} \int \omega \, dE_{hkl}$$

But ω is a constant, and so

$$\int P(\theta) \, d\theta = \frac{E_{hkl}\omega}{I_0}$$

where E_{hkl} is the total energy diffracted. The quantity $E_{hkl}\omega/I_0$ is therefore a measure of the total diffracted power as compared with the incident beam intensity I_0, and is called the *integrated reflection*, or *integrated reflecting power*. The latter nomenclature is more self-explanatory, and is that which we shall use in this volume.

According to equation (11.10), the integrated reflecting power depends on the volume dV of the crystal element, but is independent of the crystal shape. Also, the dimensions of the integrated reflecting power are those of [angular velocity] [energy]/[intensity], namely, $[\text{time}]^{-1}[\text{energy}]/[\text{energy}]$ $[\text{area}]^{-1}$ $[\text{time}]^{-1} = [\text{area}]$. Specifically, the integrated reflecting power is independent of time, and hence of the angular velocity with which the crystal rotates. The explanation of this is that if the angular velocity ω changes as the crystal rotates, then the number of X-ray quanta which are given the opportunity to be diffracted is inversely proportional to the angular velocity.

11.10   The polarisation factor

Our first discussion of the correction factors concerns the *polarisation factor* p_{hkl}. As was mentioned in Section 9.2 during our discussion of Thomson scattering, the actual expression for the amplitude of the wave scattered by a single electron depends on a function of the scattering angle 2θ which was derived from the state of the polarisation of the incident beam. In the case of an unpolarised incident beam, the polarisation factor p_{hkl} is given by

$$p_{hkl} = \frac{1 + \cos^2 2\theta}{2} \tag{11.11}$$

If a crystal monochromator (see Section 11.4) is used to obtain monochromatic radiation, then the polarisation factor takes a form different from that of equation (11.11).

11.11   The Lorentz factor

As we have stated, the recorded image of the diffraction event is due to the intersection of a smeared-out reciprocal-lattice point with the Ewald sphere. As the point passes through the Ewald sphere, the area of intersection of the surface of the sphere with the reciprocal-lattice 'point' will vary according to the shape of the reciprocal-lattice point and the precise geometry of the intersection. Reference to Fig. 11.21 will demonstrate this.

In Fig. 11.21 are two reciprocal-lattice points, which we assume to have the same shape. The point P intersects the equatorial plane of the Ewald sphere, whereas the other point Q passes through the Ewald sphere at an upper level. The passage of P through the Ewald sphere as the crystal rotates is a fairly sharp event, but the point Q intersects the Ewald sphere at a glancing angle, and so, at any given time, the opportunities for the diffraction events from P and Q are different. In addition to this geometrical effect, there is a further factor which will distinguish between the

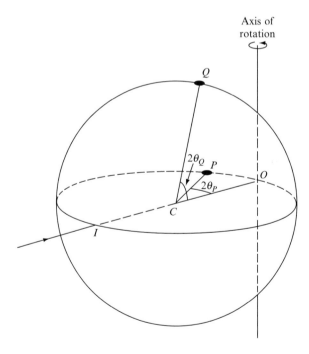

Fig. 11.21 *The intersection with the Ewald sphere.* The point P intersects the Ewald sphere in the equatorial plane, but Q grazes the top surface. As the reciprocal lattice rotates about O, the point P will cut the Ewald sphere quite quickly, whereas Q will intersect the Ewald sphere surface over a wide angular range.

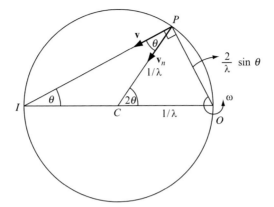

Fig. 11.22 *The Lorentz factor.* As the reciprocal lattice rotates with angular velocity ω about an axis through O, perpendicular to the plane of the paper, the reciprocal-lattice point P will sweep through the equatorial plane of the Ewald sphere with a linear velocity $\mathbf{v}$ in the direction $\overrightarrow{PI}$, such that $v = OP\omega$. The component of this velocity normal to the Ewald sphere surface is represented by $\mathbf{v}_n$, and is directed along the radius $\overrightarrow{PC}$.

opportunities for the points P and Q to diffract. The point P is closer to the rotation axis than is the point Q, and as the reciprocal lattice rotates, the linear velocity of Q will be greater than that of the point P. The point Q therefore crosses the Ewald sphere surface in a shorter time than does the point P, and so the time opportunities for the two points to diffract are different.

It is the *Lorentz factor*, in the form L_{hkl} (see equation (11.9)), which takes account of both the time opportunity and the geometrical opportunity for different reciprocal-lattice points to diffract. As an example of the calculation of the Lorentz factor L_{hkl}, we shall investigate the intersection of a reciprocal-lattice point P with the equatorial plane of the Ewald sphere as shown in Fig. 11.22.

As the crystal rotates with constant angular velocity ω, the reciprocal lattice rotates with constant angular velocity ω about its origin O. The reciprocal-lattice point P will move in the direction shown in Fig. 11.22, and if we represent the magnitude of the vector $\overrightarrow{OP}$ by OP, then the magnitude v of the linear velocity $\mathbf{v}$ of the point P as it crosses the Ewald sphere is given by

$$v = OP\,\omega$$

This velocity is directed along the vector $\overrightarrow{PI}$, and so the component of this velocity perpendicular to the surface of the Ewald sphere is the component of $\mathbf{v}$ as resolved along the radius PC. If we represent the velocity component perpendicular to the Ewald sphere surface as $\mathbf{v}_n$ and its amplitude as v_n, then we have

$$v_n = v\cos\theta = OP\omega\cos\theta$$

But the radius of the Ewald sphere is $1/\lambda$.

$$\therefore OP = \frac{2 \sin \theta}{\lambda}$$

Since

$$\frac{\omega}{v_n} = \frac{1}{OP \cos \theta}$$

$$\frac{\omega}{v_n} = \frac{\lambda}{2 \sin \theta \cos \theta} = \frac{\lambda}{\sin 2\theta}$$

We now define the Lorentz factor L_{hkl} to be

$$L_{hkl} = \frac{1}{\lambda} \frac{\omega}{v_n}$$

$$\therefore L_{hkl} = \frac{1}{\sin 2\theta} \tag{11.12}$$

For the geometry of Fig. 11.22, the Lorentz factor L_{hkl} is simply $1/\sin 2\theta$, where θ is the Bragg angle appropriate to the diffraction maximum indexed as hkl. We should note that this formula is for the special case of reciprocal-lattice points in the equatorial plane of the Ewald sphere, i.e. the plane perpendicular to the crystal rotation axis where it intersects the origin of the reciprocal lattice.

Since the Lorentz factor will vary according to the diffraction spot we are considering, we write the Lorentz factor as L_{hkl} so that we remember that its value depends on the indexing hkl.

The definition of the Lorentz factor, equation (11.12),

$$L_{hkl} = \frac{1}{\lambda} \frac{\omega}{v_n}$$

may be rewritten as

$$\frac{1}{v_n} = \frac{\lambda L_{hkl}}{\omega}$$

As seen in equation (11.9), the Lorentz factor L_{hkl} is associated with the λ/ω term, and so we conclude that the Lorentz factor is really due to a consideration of the normal velocity with which a reciprocal-lattice point passes through the Ewald sphere. To see why the total energy E_{hkl} of the diffraction event should be proportional to $1/v_n$, we may consider the following intuitive argument. Firstly, the total energy diffracted is proportional to the time during which the reciprocal-lattice point intersects the Ewald sphere, and so this justifies the $1/\omega$ term. Now v_n is given by $v \cos \theta$, where v is the magnitude of the linear velocity of the reciprocal-lattice point and θ is the Bragg angle. Reference to Fig. 11.22 will show that reciprocal-lattice points which have glancing intersections with the Ewald sphere are also associated in general with large Bragg angles θ, and when θ is large, $\cos \theta$ is small. These reciprocal-lattice points also have a relatively large linear speed v. We now see that reciprocal-lattice points which intersect the Ewald sphere at glancing angles have large values of v but small values of $\cos\theta$, whereas those which have more normal intersections have small values of v and large values of $\cos\theta$. By forming the product $v\cos\theta$, we may compare the opportunities for diffraction for different reciprocal-lattice points more validly. We now appreciate that the proportionality of E_{hkl} to $1/v_n$ serves two purposes. The inverse speed implies that E_{hkl} is proportional to the time during which a reciprocal-lattice point intersects the Ewald sphere, and the use of the amplitude of the normal component v_n of the velocity in particular allows for a more appropriate comparison of different diffraction peaks.

One advantage of the rotation method is that the Lorentz correction varies very slowly across the surface of the detector except in the vicinity of the rotation axis, where the spots will be in diffracting position for longest and can be present in many successive images. For the rotation method with normal beam geometry, the Lorentz factor has the following form:

$$L = \frac{1}{\sqrt{\sin^2 2\theta - \zeta^2}} \tag{11.13}$$

where ζ is the cylindrical polar coordinate of the reciprocal-lattice point measured along the rotation axis from the origin of reciprocal space in reciprocal-lattice units (i.e. $1/\lambda$) (Buerger, 1940; Jeffrey, 1971; Milch and Minor, 1974). At low θ, the denominator in equation (11.13) is proportional to the distance of the reciprocal-lattice point from the rotation axis. An important thing to note is that for all spots which appear close to the rotation axis in the diffraction images (regardless of resolution), the value of ζ^2 will be very close to $\sin^2 2\theta$ and the Lorentz factor will become very large. Our estimate of $|F_{hkl}|^2$ for these spots can be seriously in error since they occur on successive images, and, consequently, the affected diffraction peaks are usually rejected during data processing. Nevertheless, this is unlikely to be a major problem, since symmetry-equivalent spots will usually be recorded elsewhere on the detector.

The nature of the Lorentz and polarisation corrections is such that they can be applied by use of a precise mathematical formula, and this is usually done during the data integration step, as covered in Section 11.8. We will now look at a number of terms in equation (11.5) for which the corrections are very sample-dependent, i.e. they usually cannot be calculated by an exact formula and must therefore be determined empirically by analysis of the data itself.

11.12 Absorption

As an X-ray beam passes through any material, the electric field causes electronic excitations which increase the thermal energy of the electrons, and, theoretically, the temperature of the material will rise. The thermal energy of a crystal therefore increases at the expense of the energy of the incident X-ray beam, and so the specimen attenuates the incident beam; this effect is known as *absorption*. Experiment shows that the reduction $-dI$ in the intensity of the incident beam is proportional to the product of the distance dx through which the beam passes and the local intensity I:

$$-dI \propto I\,dx$$

The constant of proportionality is a property of the material of the specimen, and is called the *linear absorption coefficient* μ.

$$\therefore -dI = \mu I\,dx$$

By the same mathematics as is used to derive the familiar Beer–Lambert law in spectrophotometry, we obtain

$$I\,(x) = I_0 e^{-\mu x} \tag{11.14}$$

Equation (11.14) gives the intensity $I(x)$ of the X-ray beam at any distance x within the crystal, and I_0 is the incident intensity. The linear absorption coefficient is a property of the material considered, and varies with the wavelength of the X-radiation in a somewhat erratic manner, as emphasised earlier in Section 11.4 regarding the design of X-ray filters.

All our previous calculations have assumed that there was no absorption of the X-rays, for we have considered the case of an element of volume dV, in which all linear dimensions are too small to cause the exponential of equation (11.14) to differ significantly from unity. The correction for absorption A_{hkl} is a rather difficult matter. As shown in Fig. 11.23, X-rays passing through different parts of the crystal have different path lengths within the crystal, and reference to equation (11.14) shows that in order to calculate the absorption factor A_{hkl} we need the total path length x for each different diffraction spot; this requires extensive knowledge of the crystal's geometry, which is not possible to determine accurately, given the manner in which protein crystals are mounted. However, X-ray tomography may provide a solution to this problem in the long term (Brockhauser *et al.*, 2008). A number of elegant empirical methods have been devised for determining the absorption correction for protein crystals by measuring the intensity of reflections which have their scattering vectors parallel to the rotation axis, as the crystal is rotated (North *et al.*, 1968). Corrections can also be determined by measuring the transmission of the direct beam

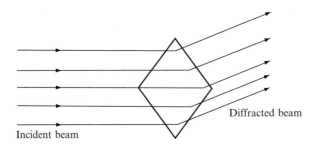

Incident beam

Diffracted beam

Fig. 11.23 *Absorption.* Different waves passing through the crystal have different path lengths within the crystal. Hence each wave is absorbed by a different amount, and this makes the calculation of the overall absorption difficult except in the very simplest geometrical cases.

as a function of the orientation of the crystal, and the absorption correction for each reflection is determined from the transmissions in the incident and diffracted beam directions (Schwager *et al.*, 1973).

Absorption is estimated to introduce errors of around 15% into the intensity data. Whilst mineralogists of the past used to grind and polish their crystals to a perfectly spherical shape to eliminate the problem of determining A_{hkl}, this is clearly not practical for protein crystals. Hence, the absorption correction is applied in an empirical manner during data scaling, which will be covered later in this chapter.

Mosaic theory

The intensity corrections so far discussed were appreciated at an early stage in the development of X-ray crystallography, and originally it was thought that the following equation, which takes account of all the corrections mentioned above, was a correct relationship between the observed intensities and those calculated from known structural parameters:

$$E_{hkl} = \frac{I_0}{\omega} K L_{hkl}\, p_{hkl}\, |(F_{hkl})_T|^2\, A_{hkl}\, dV \tag{11.15}$$

Experiment showed, however, that this was not the case, and, especially at the exact Bragg angle, the observed intensity was much less than that predicted by equation (11.15). The situation was clarified by C. G. Darwin (grandson of the famous biologist) in his dynamical theory of X-ray diffraction, about which we shall say a few words here.

An interesting effect arises when X-rays are incident onto a set of planes at exactly the Bragg angle, as shown in Fig. 11.24, which shows three parallel planes. The incident X-ray beam undergoes partial Bragg reflection at plane 2, so that only part of the incident energy is transmitted towards plane 3 and some is deflected back towards plane 1. But since the planes are accurately parallel, the wave 'reflected' at plane 2 strikes plane 1 exactly at the Bragg angle, and may undergo a second Bragg reflection as shown. This 'twice-reflected' wave is now parallel to that part of the original wave which was transmitted in the first Bragg reflection event.

Now, although this has not been stated explicitly until now, the full theory of diffraction shows that the Bragg-reflected wave is $\pi/2$ out of phase with the transmitted wave at any Bragg reflection event. Since this effect applies to all Bragg reflections alike, it does not affect any of our discussions in preceding chapters, and therefore has been ignored up till now. The twice-reflected wave (in Fig. 11.24) is therefore exactly π out of phase with the originally transmitted wave, and since these two waves are now propagating in the same direction, destructive interference occurs. Consequently, when the Bragg reflection condition is exactly satisfied, the direct and twice-reflected waves interfere destructively, and the intensity of the wave which is propagated through the crystal is diminished. Similarly, the 'waves' which are Bragg-reflected multiple times at plane 2 are exactly out of phase, so that the diffraction maxima are weakened by destructive interference—an effect known as *primary extinction*. Hence, for a truly perfect crystal (i.e. one with accurately parallel lattice planes throughout its volume), the effect of primary extinction on the diffraction

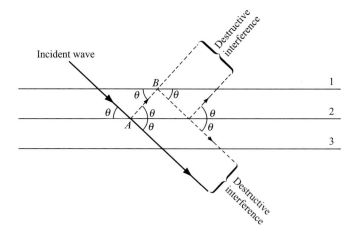

Fig. 11.24 *Primary extinction.* Waves are incident at the exact Bragg angle θ, and partial Bragg reflection occurs at A. The reflected wave strikes B at the exact Bragg angle and so another Bragg reflection event occurs. The 'twice-reflected' wave is now parallel to that part of the original wave which was transmitted at A. These two parallel waves are exactly out of phase, and destructive interference occurs. This effect reduces the intensity of the waves, and is called primary extinction.

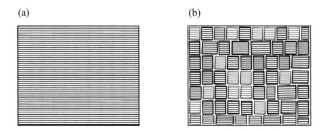

Fig. 11.25 *Ideally perfect and mosaic crystals.* In (a), the structure of an ideally perfect crystal is such that the lattice planes are accurately parallel throughout the crystal's macroscopic volume. Since real crystals are observed to diffract more strongly than would be predicted from such a model, Darwin proposed the mosaic model, in which ideally perfect crystallites or domains are packed to form a lattice with approximate regularity within the crystal, as in (b). The effect of primary extinction is much less in the mosaic crystal.

intensities would be large. However, real crystals are observed to diffract more strongly than would be predicted if they possessed this degree of internal order. Hence it is considered that the high level of order that is required to produce diffraction extends only over short regions, or domains, within the crystal which have a reasonable degree of positional and orientational consistency, such as that shown in Fig. 11.25, thus reducing the extent of primary extinction.

Whilst primary extinction and also another effect known as secondary extinction (the progressive loss of energy of the incident beam caused by diffraction events as it moves from one domain to the next) are interesting physical phenomena, their effects are likely to be small, and for practical purposes they are ignored during data processing—these effects are largely offset during scaling of the measured data, as will be seen later. Nevertheless, it is important to appreciate the mosaic nature of crystals, since the same effect contributes to broadening of the diffraction spots, something which we referred to earlier as mosaic spread (see Section 11.8), and it is very important in understanding the subject of crystal twinning, which will be dealt with later on.

11.13 The temperature factor

The derivation of the diffraction pattern of the contents of the unit cell as carried out in Chapter 9 included the assumption that all atoms were located at single points in space. This was explicitly stated in the derivation of the structure factor F_{hkl} in the form of equation (9.21),

$$F_{hkl} = \sum_j f_j e^{2\pi i \left(hx_j + ky_j + lz_j \right)} \tag{9.21}$$

for the coordinates (x_j, y_j, z_j) of the centre of the jth atom define a mathematical point. However, the assumption that the atoms are fixed at single points is not valid, and we must make the appropriate correction to the theory of Chapter 9.

The physical model which we use to derive the correction factor partly concerns the vibrational motion of an atom about its equilibrium position. At the absolute zero of temperature, we may assume that the potential energy of interaction with the neighbouring atoms is at a minimum. As the temperature increases, the ambient thermal energy causes the atoms to oscillate to a greater extent about their equilibrium positions, and so we cannot strictly assign point coordinates (x_j, y_j, z_j) to the centre of any atom. Rather, (x_j, y_j, z_j) should be taken as the *average* position of the jth atom. Since the thermal motions are very rapid, we may represent their effect by associating the atom centre with a position characterised by (x_j, y_j, z_j) but which in reality is somewhat blurred, and not an ideal mathematical point. As the atom oscillates about its average position, the electrons associated with the atom also move, and so the interaction of the electrons with the incident X-rays is changed. One way of looking at this would be to suggest that, over a period of time, for an oscillating atom, the total quantity of electric charge is spread over a greater volume than for a stationary atom, and so the probability of interaction of an electron with an X-ray is reduced. Alternatively, at any point in space, there is rather less effective electric charge present instantaneously, and so the scattering effect of the atom is reduced. Either model results in the same conclusion, namely, that the effect of thermal vibrations is to reduce the effective scattering power of an atom. A similar effect arises from static disorder in the crystal, such that small differences in the position of an atom (x_j, y_j, z_j) from one unit cell to the next, or indeed from one asymmetric unit to the next, will give a reduction in the scattering power of that atom, similar to that caused by thermal motion.

The fundamental physical event in diffraction is the scattering of X-rays by electrons, and for any given atom j, this is expressed mathematically in terms of the atomic scattering factor f_j. Reference to Section 9.7 will show that all the methods by which f_j may be calculated assume that the atom is stationary. If we recast equation (9.21) in terms of the scattering vector of the reflection ($\mathbf{S}_{hkl}$) and the position vector of each atom ($\mathbf{r}_j$), we obtain

$$F_{hkl} = \sum_j f_j e^{2\pi i \mathbf{S}_{hkl} \cdot \mathbf{r}_j}$$

If an atom is 'blurred' owing to uncorrelated thermal and static-disorder effects, then the expression for f_j must be modified to take account of this. If atom j is displaced by a vector $\mathbf{u}_j$, equation (9.21) becomes

$$F_{hkl} = \sum_j f_j e^{2\pi i \mathbf{S}_{hkl} \cdot (\mathbf{r}_j + \mathbf{u}_j)}$$

$$F_{hkl} = \sum_j f_j e^{2\pi i \mathbf{S}_{hkl} \cdot \mathbf{r}_j} e^{2\pi i \mathbf{S}_{hkl} \cdot \mathbf{u}_j} \tag{11.16}$$

One important simplification arises from the fact that X-ray diffraction is sensitive only to motions of atoms perpendicular to the reflecting plane, and this can be understood from Chapter 9, Section 9.6 where we saw that the component of any F_{hkl} due to atom j has a phase $\mathbf{S}_{hkl} \cdot \mathbf{r}_j$. Since the scattering vector $\mathbf{S}_{hkl}$ is perpendicular to the reflecting plane, any displacement of the atom parallel to the plane will make no contribution to the scalar product. Hence, if we redefine u_j as the displacement of atom j in the direction of $\mathbf{S}_{hkl}$, the scalar product $\mathbf{S}_{hkl} \cdot \mathbf{u}_j$ becomes simply $S_{hkl} u_j$ and we can simplify equation (11.16) to

$$F_{hkl} = \sum_j f_j e^{2\pi i \mathbf{S}_{hkl} \cdot \mathbf{r}_j} e^{2\pi i S_{hkl} u_j}$$

Since F_{hkl} is measured over a long period of time, it is sensitive to the average value of $S_{hkl} u_j$.

$$\therefore F_{hkl} = \sum_j f_j e^{2\pi i \mathbf{S}_{hkl} \cdot \mathbf{r}_j} \overline{e^{2\pi i S_{hkl} u_j}}$$

Assuming that u_j is small, an expansion given in Chapter 2, Section 2.15 (Maclaurin's theorem), gives

$$\overline{e^{2\pi i S_{hkl} u_j}} \approx 1 + 2\pi i \overline{S_{hkl} u_j} - 2\pi^2 \overline{(S_{hkl} u_j)^2}$$

Since S_{hkl} is constant, we may state

$$\overline{e^{2\pi i S_{hkl} u_j}} \approx 1 + 2\pi i S_{hkl} \overline{u_j} - 2\pi^2 S_{hkl}^2 \overline{u_j^2}$$

The atomic vibration is symmetric over time, and therefore $\overline{u}$ is zero. Hence

$$\overline{e^{2\pi i S_{hkl} u_j}} \approx 1 - 2\pi^2 S_{hkl}^2 \overline{u_j^2}$$

and using Maclaurin's theorem again gives

$$\overline{e^{2\pi i S_{hkl} u_j}} \approx e^{-2\pi^2 S_{hkl}^2 \overline{u_j^2}}$$

Expanding the exponential term gives

$$-2\pi^2 S_{hkl}^2 \overline{u_j^2} = -2\pi^2 \left(\frac{2\sin\theta}{\lambda}\right)^2 \overline{u_j^2}$$

$$= -8\pi^2 (\sin\theta/\lambda)^2 \overline{u_j^2}$$

If we now define a parameter B_j as the *temperature factor* (or *thermal parameter*), where

$$B_j = 8\pi^2 \overline{u_j^2} \tag{11.17}$$

we see that

$$\overline{e^{2\pi i S_{hkl} u_j}} \approx e^{-B_j\left(\sin^2\theta/\lambda^2\right)}$$

and we may therefore define the 'temperature-corrected' atomic scattering factor $(f_j)_T$:

$$(f_j)_T = f_j e^{-B_j\left(\sin^2\theta/\lambda^2\right)} \tag{11.18}$$

Substituting this into equation (9.21) gives the 'temperature-corrected' structure factor for all atoms in the model as

$$(F_{hkl})_T = \sum_j (f_j)_T \, e^{2\pi i\left(hx_j + ky_j + lz_j\right)} \tag{11.19}$$

$$(F_{hkl})_T = \sum_j f_j e^{-B_j\left(\sin^2\theta/\lambda^2\right)} e^{2\pi i\left(hx_j + ky_j + lz_j\right)} \tag{11.20}$$

Equations (11.19) and (11.20) are the results we have been seeking: expressions which modify the theoretical atomic scattering factors f_j and the structure factor F_{hkl} to take into account both thermal motion and static disorder in terms of the temperature factor B_j.

An implicit assumption in this treatment is that the thermal vibrations of the atom j are spherically symmetric. For an ionic solid such as NaCl, it is probably true that the ionic vibrations are spherically symmetric, but for bonded atoms within an organic or a protein molecule, there are directional constraints on the thermal motion, and so the assumption of spherical symmetry is more likely to be a first approximation. The anisotropy of thermal vibrations, however, is not of concern initially, and the form of equations (11.18) and (11.20) is quite adequate. We shall meet the techniques of investigating non-spherically symmetric oscillations during our study of refinement in Chapter 15.

Equations (11.18) and (11.20), which take account of thermal motion and other disorder effects, do so by introducing a negative exponential. Equation (11.17) shows that B_j is necessarily positive, as is the term $\sin^2\theta/\lambda^2$. Hence $(f_j)_T$ is always less than

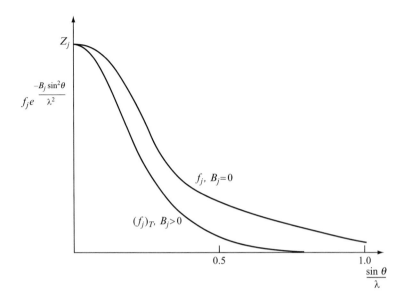

Fig. 11.26 *The temperature factor B_j. The temperature factor B_j is used to define the 'temperature-corrected' atomic scattering factor $(f_j)_T$, which falls off rather more steeply with $\sin\theta/\lambda$ than does the quantity f_j. Note that when $\theta = 0$, $(f_j)_T = f_j = Z_j$, the total number of electrons within the atom j.*

f_j, except for when θ is zero, when $(f_j)_T$ equals f_j. Also, the value of the exponential decreases quite rapidly as θ increases, causing $(f_j)_T$ to fall off sharply as compared with f_j. This behaviour is depicted in Fig. 11.26, and is seen to agree with the physical argument outlined above in which we predicted that an oscillating atom would be less effective a scattering centre than a stationary one.

For protein structures solved at high resolution $(d \sim 2\,\text{Å})$, the B-values for well-defined atoms will be in the region of 20–30 Å^2 and the corresponding mean-squared amplitudes in the region of 0.3–0.4 Å^2. Significantly lower B-values will be obtained with atomic-resolution analyses $(d \sim 1\,\text{Å})$.

The magnitude of the thermal disorder effect can be seen from a consideration of typical values of B_j. Since the quantity $B_j \sin^2\theta/\lambda^2$ must be dimensionless, we see that the dimensions of B_j are [length]^2, which is consistent with equation (11.17). Whilst B_j generally takes values around 20 Å^2, for atomic-resolution structures it can have values as low as 2 Å^2. In the latter case, for a value of $\sin\theta/\lambda$ of 0.2 Å^{-1}, corresponding to a Bragg angle of about 18° for Cu Kα radiation of wavelength 1.54 Å, $(f_j)_T$ is about $0.9f_j$, but for a Bragg angle approaching 90°, the value of $(f_j)_T$ drops to below a half of f_j. When B_j is 20 Å^2, the corresponding values of $(f_j)_T$ are $0.45f_j$ and $0.0002f_j$, respectively, demonstrating the drastic effect of thermal motion and disorder on the diffraction pattern at high θ angles.

Since B_j deals with a fundamental physical event, it should be possible to express it in terms of fundamental physical parameters. This is a problem for the physicists, which may be tackled by considering the nature of atomic vibrations within a solid. Of the various theoretical models which have been proposed for this calculation, the most successful is that of Debye and Waller (Kittel, 2004). Having taken into account the quantum nature of vibrations within solids, Debye and Waller expressed B_j as a function of the absolute temperature T and various other fundamental constants:

$$B_j = \frac{6h^2}{mk\Theta_j} \left(\frac{\phi\left(\Theta_j / T\right)}{\left(\Theta_j / T\right)} + \frac{1}{4} \right)$$

where m is the atomic mass, h is Planck's constant, k is Boltzmann's constant and

$$\phi\left(\frac{\Theta_j}{T}\right) = \frac{T}{\Theta_j} \int\limits_{0}^{\Theta_j/T} \frac{x}{e^x - 1} \, dx$$

The integral defined as $\phi(\Theta_j/T)$ is called the *Debye integral*, and the parameter Θ_j, known as the *Debye temperature*, is a measure of the quantum nature of atomic vibrations, and varies according to the solid under study. In fact, physicists refer to the quantity $\exp(-B_j \sin^2 \theta / \lambda^2)$ as the *Debye–Waller factor*. As is evident from the expression for B_j, its calculation is a considerable problem in theoretical physics, which need not concern us here. However, it should be noted that B_j increases with temperature, implying that the amplitude of atomic vibrations also increases, as is intuitively obvious. Concomitantly, $(f_j)_T$ falls off more sharply.

What is of direct importance to the crystallographer is the magnitude of the temperature factor and its use. So far, we have discussed only the temperature factor correction applied to the atomic scattering factors of single atoms; we shall now discuss how multi-atom arrays are handled. Earlier we defined the 'temperature-corrected' structure factor $(F_{hkl})_T$ as

$$(F_{hkl})_T = \sum_j f_j e^{-B_j \left(\sin^2 \theta / \lambda^2 \right)} e^{2\pi i \left(hx_j + ky_j + lz_j \right)} \tag{11.20}$$

If we assume that the B_j's for all the atoms present are approximately equal, then, writing the average value of B_j simply as B, we may factor a term $\exp(-B \sin^2 \theta / \lambda^2)$ out of equation (11.20), obtaining the following approximation:

$$(F_{hkl})_T \approx e^{-B \left(\sin^2 \theta / \lambda^2 \right)} \sum_j f_j e^{2\pi i \left(hx_j + ky_j + lz_j \right)}$$

$$\approx e^{-B \left(\sin^2 \theta / \lambda^2 \right)} F_{hkl}$$

Since the intensity I_{hkl} is proportional to $|(F_{hkl})_T|^2$, we may write

$$I_{hkl} = c \left| (F_{hkl})_T \right|^2 \tag{11.21}$$

where c is a constant containing all the relevant correction factors of equation (11.5). Hence

$$I_{hkl} \approx c \, |F_{hkl}|^2 \, e^{-2B\left(\sin^2 \theta / \lambda^2\right)}$$

and

$$\log_e \left[\frac{I_{hkl}}{|F_{hkl}|^2} \right] \approx \log_e c - 2B \frac{\sin^2 \theta}{\lambda^2} \tag{11.22}$$

In this equation, I_{hkl} represents a measured intensity, and $|F_{hkl}|^2$ the corresponding intensity as would be observed if the atoms were at rest. Since we do not know the structure at this stage of the analysis, we have to estimate $|F_{hkl}|^2$ from the atomic composition of the unit cell, which is usually known, and assume that the atoms are located at random positions. Since

$$F_{hkl} = \sum_j f_j e^{2\pi i \left(hx_j + ky_j + lz_j \right)}$$

then

$$|F_{hkl}|^2 = F_{hkl} F_{hkl}^*$$

$$= \left[\sum_i f_i e^{2\pi i (hx_i + ky_i + lz_i)} \right] \left[\sum_j f_j e^{-2\pi i \left(hx_j + ky_j + lz_j \right)} \right]$$

$$= \sum_i \sum_j f_i f_j e^{2\pi i (h[x_i - x_j] + k[y_i - y_j] + l[z_i - z_j])}$$

We can separate out the terms where $i = j$, since the complex exponential becomes unity for these terms:

$$|F_{hkl}|^2 = \sum_j f_j^2 + \sum_i \sum_{j \neq i} f_i f_j e^{2\pi i (h[x_i - x_j] + k[y_i - y_j] + l[z_i - z_j])}$$

$$= \sum_j f_j^2 + \sum_i \sum_{j > i} f_i f_j \left[e^{2\pi i (h[x_i - x_j] + k[y_i - y_j] + l[z_i - z_j])} + e^{2\pi i (h[x_j - x_i] + k[y_j - y_i] + l[z_j - z_i])} \right]$$

$$= \sum_j f_j^2 + \sum_i \sum_{j > i} f_i f_j \left[e^{2\pi i (h[x_i - x_j] + k[y_i - y_j] + l[z_i - z_j])} + e^{-2\pi i (h[x_i - x_j] + k[y_i - y_j] + l[z_i - z_j])} \right]$$

$$\therefore |F_{hkl}|^2 = \sum_j f_j^2 + 2 \sum_i \sum_{j > i} f_i f_j \cos 2\pi \left(h [x_i - x_j] + k [y_i - y_j] + l [z_i - z_j] \right)$$

If the atoms are randomly arranged, then the terms $(x_i - x_j)$, $(y_i - y_j)$ and $(z_i - z_j)$ are random. If we were to calculate the average $|F_{hkl}|^2$ over many hkl values of similar resolution, then the cosine terms, which can be either positive or negative owing to the nature of the cosine function, would tend to sum to zero. Thus we can say

$$\overline{|F_{hkl}|^2} \approx \overline{\sum_j f_j{}^2}$$

where the overline indicates that we are taking an average.

Since $|F_{hkl}|^2$ can be estimated from the scattering factors of the atoms known to be present in the structure, even though we do not know their positions, a graph of $\log_e \left[\overline{I_{hkl}} \big/ \sum_j f_j^2 \right]$ against $\sin^2 \theta / \lambda^2$ can now be drawn. Inspection of equation (11.22) shows that this should be a straight line of gradient $-2B$, thus allowing the parameter B to be estimated, and giving us an indication of the effect of thermal motion and other disorder. Thus a graph of $\log_e \left[\overline{I_{hkl}} \big/ \sum_j f_j^2 \right]$ against $\sin^2 \theta / \lambda^2$, which is known as a *Wilson plot* (Wilson, 1942), can be drawn, and a value for B can be estimated from the slope of the resulting line, as shown in Fig. 11.27. Also, the Wilson plot is useful in estimating the value of the constant of proportionality c which occurs in equation (11.21) from the intercept on the y axis, thereby allowing the experimental data to be placed on an absolute scale, and it is for this reason that the Wilson plot is of great importance in data processing. In principle, when data from more than one crystal are available, we can use a Wilson plot to obtain a scale factor and a

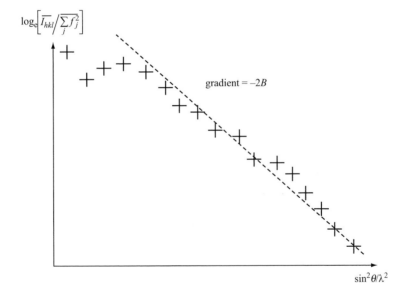

Fig. 11.27 *A Wilson plot.* A semi-logarithmic graph of $\log_e \left[\overline{I_{hkl}} \big/ \sum_j f_j^2 \right]$ against $\sin^2 \theta / \lambda^2$ allows the mean temperature factor for the crystal to be determined from the gradient $(-2B)$. Commonly, with macromolecules, the plot is non-linear at low $\sin^2 \theta / \lambda^2$ owing to solvent scatter, and so these points should be omitted. Sharp spikes in the linear region of the Wilson plot indicate the presence of ice diffraction rings in the dataset, suggesting a poorly frozen crystal.

temperature factor for each crystal, which is important for merging the data from each crystal into a combined dataset. When the data are collected as a series of diffraction images, it is common for the mean diffraction intensity to vary from one image to another owing to various factors, including radiation damage, absorption effects and beam decay, and we are then faced with the problem of scaling the data in each image so that a final merged set of intensities can be obtained. In principle, one can perform a Wilson plot analysis for the data from each image to determine a scale factor and temperature factor for it, and these parameters would then allow the data from all images to be combined. However, there are more sophisticated ways of scaling the data from multiple images and crystals which empirically correct for experimental factors such as fading of the diffraction pattern or incident beam, absorption, and crystal anisotropy. These methods work by using the intensities of symmetry-related reflections (which we know must be equal), and the relevant details will be covered in the next section. In crystal structure analysis, we do not usually apply the temperature factor determined from the Wilson plot as a correction to the measured intensity data. Instead, the scale factor and B-value are used when it is necessary to put structure factors calculated from a model of the structure (F_c's) on the same scale as the observed structure factor data (F_o's). It is also useful to have an idea of the scale of disorder in the crystal at an early stage in the analysis, which can be revealed by the overall B-factor—values less than $\sim$30 Å^2 indicate a high-quality crystal. A Wilson analysis is performed by the CCP4 program TRUNCATE (CCP4, 1994).

The fact that the temperature factor determined by the Wilson method is not applied to the intensity data may slightly dismay some readers. But as we shall soon discover, during the solution of the phase problem, many other errors arise which are more significant than those associated with the approximation of B. The fact that many errors are possible early in a crystal structure analysis implies that the initially derived electron density map may not be a very accurate representation of the true structure. For this reason, refinement techniques (see Chapter 15) are applied, during which successive electron density maps are derived in a systematic fashion so that each is more reliable than the previous one. To anticipate a little, refinement techniques usually involve the calculation of an electron density map from the observed set of structure factors; the identification of atomic sites from the electron density map; and then the recalculation of the structure factors using equation (11.20), in which the coordinates (x_j, y_j, z_j) used are those identified from the map. Comparison of the calculated structure factors with the originally observed ones indicates how accurately the current model represents the true structure. Since this refinement process involves the calculation of structure factors from a set of known atomic coordinates, it is here that the temperature factor for each atom within the structure may be refined. The overall B-factor of the measured dataset can also be determined more accurately during refinement of the structure. Thus temperature and disorder effects on the diffraction intensities are dealt with at two stages of a structure analysis. Firstly, a correction is applied to the measured dataset to approximately scale the F_o's so that they are consistent with the atomic composition of the crystal, in what is known as a Wilson analysis. Secondly, once the structure has been determined, temperature factors for each atom can be determined, providing that the diffraction data are of sufficient resolution. As we shall see in Chapter 15, it is possible, in small-molecule structures or in atomic-resolution macromolecular structures, to map out the volume over which thermal vibrations occur.

11.14   Scaling and merging intensity measurements

When data are collected from a crystal, there are a number of practical factors affecting the measured intensity of a given spot, such as the primary beam intensity, the exposure time and the extent of radiation damage. During a data collection, the intensity of the primary beam can change appreciably, and the crystal can experience significant radiation damage. In certain orientations of the crystal, absorption can also reduce the quality of the diffraction pattern. To compensate, we may need to stop the data collection and increase the exposure time by a factor of two or three, realign the instrument to improve the beam intensity, or move the crystal such that an undamaged part is exposed to the beam. The effect of all these changes is that the measured intensity of a particular spot recorded at the start of the data collection can be very different from a measurement of the same spot made later on in the experiment. This is important, since in the data processing we need to average the intensities of those spots which should be of equal intensity by the symmetry of the diffraction pattern (or Laue group) so as to obtain a unique set of data. However, for all of the above reasons, the intensities of the symmetry-equivalent spots in the diffraction pattern can differ appreciably.

During data processing, it is therefore important to correct for these factors before the intensities of symmetry-related spots are finally averaged. This is done by assuming that the factors affecting the intensity are correlated with the point in time (or ϕ angle) at which each symmetry-equivalent spot is recorded. Since the spots within any given image will have been recorded more or less simultaneously, with the same beam intensity, exposure time, crystal orientation etc., it is common to determine a scale factor and a temperature factor for each diffraction image. There are more sophisticated scaling methods, which we will cover later on, but for now let us consider that data scaling is a process of determining a scale factor and a temperature factor correction for each image. When data are collected from several crystals of the same protein, we need to scale the intensities recorded from the second or third crystal to those recorded from the first, and this can be done by determining a scale factor and a temperature factor for each image, obtained from all of the crystals. Sometimes the data collection on one crystal is deliberately repeated with a longer or shorter exposure time and, again, the data from these different 'runs' can be scaled together by determining a scale and temperature factor for each image or 'batch' of data. The scale factors and temperature factors in this context are all relative—usually, the first image is chosen as the reference batch, and this is arbitrarily given a scale of 1.0. Thus the scale factors determined for the rest of the images reflect changes in beam intensity, absorption, radiation damage etc. relative to the start of the experiment. Likewise, the temperature factors indicate how the resolution-dependent weakening of the diffraction pattern varies as the crystal is rotated, and these can be calculated relative to the first image or, preferably, relative to whichever image exhibits the least intensity fall-off as a function of resolution.

In scaling, we determine relative scale factors and temperature factors which make the separate measurements of each unique reflection as consistent as possible, and these are determined by a least-squares method. For simplicity, we will initially consider the process of determining just the relative scale factors of the images and assume that there are no temperature factor differences. The details of least-squares minimisation will be given in Chapter 15, but for now we will merely describe the minimisation function ψ, which is as follows:

$$\psi = \sum_h \sum_j w_{hj} \left(I_{hj} - g_j \langle I_h \rangle\right)^2 \tag{11.23}$$

where I_{hj} is the observation of reflection h on image j, g_j is the scale factor for image j and w_{hj} is the weight $1/\sigma^2(I_{hj})$, which reflects the precision of the observation. The mean intensity $\langle I_h \rangle$ is given by

$$\langle I_h \rangle = \frac{\sum_j w_{hj} g_j I_{hj}}{\sum_j w_{hj} g_j^2} \tag{11.24}$$

Thus scaling involves determining the values of g_j which minimise the sum of the squares of the differences between the intensities of the symmetry mates of each reflection and the average intensity of that reflection. Least-squares refinement is such that we need to start with trial values for all of the unknown parameters, in this case the scale factors. Hence, initially all values of g_j are set to unity and the corresponding value of ψ is calculated. It is then possible to calculate shifts in the scale factors that will lower ψ, and these shifts are applied to the scales, except for the first one, which is fixed at unity as the reference image. The revised scale factors are used to calculate a new ψ value and new shifts, which are applied to the scales. The process is repeated until it converges, i.e. the shifts in the scale factors become smaller than their estimated errors.

In the above example, we have discussed only the refinement of the relative scale factors, in the full knowledge that this was a simplification and that it is more common to refine both a scale factor and a B-factor for each image. This has a number of advantages over scale factor refinement in that it compensates for the fact that absorption and radiation damage can cause the B-factors of the images to vary significantly. Whilst 'batch scaling' by this method is widely used and is probably the best way to scale a run of data if there are any discontinuities in it, for example changes in exposure time or beam intensity, there have been a number of refinements to it. For example, we can determine smoothed scale factors and/or temperature factors, providing that there are not any sharp discontinuities from one image to the next. Instead of refining a scale factor and temperature factor for each image (which can create the problem of having the components of a partially recorded reflection on different scales), equation (11.23) is reformulated so that the scale factor and/or temperature factor are smoothly varying functions of the ϕ angle

(Evans, 2006; Kabsch, 1988b). Likewise, the effects of absorption can be modelled empirically by mathematical functions known as spherical harmonics (Blessing, 1995). Whichever scaling model is used, the symmetry-related intensities are used to refine the parameters of the model in essentially the same manner that we described above for the simplest scale-factor-only model.

To assess the success of scaling, we need to inspect the values of a parameter known as the R_{merge} (sometimes called R_{sym}), which is given by

$$R_{merge} = \frac{\sum_h \sum_j |I_{hj} - \langle I_h \rangle|}{\sum_h \sum_j I_{hj}} \tag{11.25}$$

Note that in this equation, the observed I_{hj} values are assumed to have been correctly scaled (by $1/g_j$ in the case of the scale-factor-only model). Ideally, the R_{merge} should be less than 10%, but values that are slightly higher are acceptable. Values higher than 20% indicate that the resolution of the dataset has been overestimated or that the space group is wrong—in fact, reprocessing the data in different possible space groups and inspecting the resulting R_{merge} values is a valid way of determining the crystal symmetry. A high R_{merge} can also indicate that the origin of the reflection indexing is shifted from its correct position so that symmetry-equivalent spots are not being merged, and this requires that the data be reprocessed.

An improved version of the R_{merge} is the multiplicity-weighted R-factor, or R_{meas}, where

$$R_{meas} = \frac{\sum_h \left(\frac{N_h}{N_h - 1}\right)^{1/2} \sum_j |I_{hj} - \langle I_h \rangle|}{\sum_h \sum_j I_{hj}} \tag{11.26}$$

and N_h is the number of measurements of reflection h. The R_{meas} has the advantage that it better reflects the greater accuracy of data when there is high redundancy (or multiplicity), i.e. each unique reflection has been measured many times. In contrast, the R_{merge} tends to increase in these situations. The R_{merge} and/or R_{meas} should also be inspected as a function of image number, since very sharp discontinuities can indicate poor images due to loss of the beam or other factors. Both parameters should also be monitored as a function of resolution to determine the high-resolution cut-off of the data—values above 50% indicate that the intensities are no longer significant, which will probably be matched by a decrease in the $I/\sigma(I)$ ratio below 2.

When the data have been scaled, the separate measurements for each reflection have to be averaged. Since it is possible for individual measurements to contain significant errors due to spurious diffraction spots arising from ice or contaminating salt crystals and, possibly, sudden fluctuations in the beam intensity, it is usual to check how much each individual intensity measurement differs from the mean intensity of each spot. If outliers exist, the measurement which is the largest number of standard deviations from the mean is rejected, and the mean recalculated and rejection tests reapplied. This is repeated until no further rejections occur or the number of remaining measurements is too small to test for outliers.

In Section 11.8, we described how the standard deviations for the measured intensities can be calculated by use of equation (11.4). This equation provides an experimental standard deviation for each individual intensity measurement. As explained in the previous paragraph, the intensities of symmetry-related reflections are averaged following various rejection tests which use the experimental standard deviations. A minor problem arises because the experimental standard deviations are calculated by assuming that the photon counts obey Poisson statistics. Since this assumption is not fully valid, the experimental standard deviations tend to be underestimated and, as a result, we may find that valid measurements are failing the rejection tests simply because their estimated standard deviations are too low. Hence, in scaling, it is common to inflate the experimental standard deviations slightly such that they match the actual standard deviations calculated from the individual intensity measurements by the standard formula for statistical variance. Following scaling, the formula commonly used for obtaining modified experimental standard deviations (σ_{mod}) is as follows:

$$\sigma_{mod}(I_{hj}) = SD_{fac}\left[\sigma^2(I_{hj}) + (SD_{add}I_{hj})^2\right]^{1/2} \tag{11.27}$$

The constants SD_{fac} and SD_{add} are determined empirically such that the estimated standard deviations are on a par with the actual standard deviations in all resolution shells. The resulting σ_{mod} values will then be more in accord with the observed scatter of individual intensity measurements for each reflection than are the unmodified σ values.

11.15   Conversion of intensities to structure factor amplitudes

Following scaling and merging of the diffraction data, we need to convert the resulting intensities I_{hkl} and their standard deviations $\sigma(I_{hkl})$ into structure factor amplitudes $|F_{hkl}|$ and associated standard deviations $\sigma(|F_{hkl}|)$. We showed in Chapter 9 that the structure factor amplitudes are simply the square roots of the intensities, as follows:

$$|F_{hkl}| = \sqrt{I_{hkl}} \tag{11.28}$$

There are a number of caveats with this simple formula, and we will look at refinements to it a bit later on, but for now let us assume that it is fully valid. To obtain an expression for the standard deviation of the structure factor amplitude, we can use the formula for the propagation of errors which was introduced earlier in Section 11.8:

$$\sigma^2(u) = \left(\frac{\partial u}{\partial x}\right)^2\sigma^2(x) + \left(\frac{\partial u}{\partial y}\right)^2\sigma^2(y) + \left(\frac{\partial u}{\partial z}\right)^2\sigma^2(z) + \dots \tag{11.3}$$

If we apply this formula to equation (11.28), we can derive the following:

$$\sigma^2\left(|F_{hkl}|\right) = \left(\frac{1}{2\sqrt{I_{hkl}}}\right)^2 \sigma^2\left(I_{hkl}\right)$$

$$\therefore \sigma\left(|F_{hkl}|\right) = \frac{\sigma\left(I_{hkl}\right)}{2\sqrt{I_{hkl}}}$$

or

$$\sigma\left(|F_{hkl}|\right) = \frac{\sigma\left(I_{hkl}\right)}{2\,|F_{hkl}|} \tag{11.29}$$

Whilst this formula allows the standard deviation of the structure factor amplitude to be estimated for strong diffraction spots, it suffers from the fact that as $|F_{hkl}|$ tends to zero, the corresponding $\sigma(|F_{hkl}|)$ tends to infinity. This problem arises from the fact that the formula for the propagation of errors is an approximation which ignores higher-order terms, and these terms become significant when $|F_{hkl}|$ is small. Whilst an improved formula which does not suffer from this effect is given by

$$\sigma\left(|F_{hkl}|\right) = \sqrt{I_{hkl} + \sigma\left(I_{hkl}\right)} - \sqrt{I_{hkl}} \tag{11.30}$$

a more common solution is for the standard deviations to be estimated by a method of statistical inference (known as Bayesian statistics), which we will touch on briefly in this section because it relates to how the weakest intensities are treated in data processing. Bayesian statistics will be covered in more detail later on in the chapters on phasing and refinement of crystal structures (Chapters 14 and 15).

One problem with the process of calculating the $|F|$'s by square-rooting the corresponding intensities is that some of the weak intensities will have negative values. This is, of course, physically impossible and also adds the complication that the square root of a negative number is imaginary, whereas we know that a structure factor amplitude is a real number. Negative intensities arise from the use of equation (11.2), which we described in Section 11.8:

$$I = P - B \tag{11.2}$$

For very weak diffraction spots, if the sum of estimated background counts under the peak (B) happens to be higher than the sum of the peak pixels (P), I will be negative. In a good dataset, the majority of the spots should be strong, i.e. $I \geq 3\sigma(I)$, so there should be very few intensities which have negative values. Since we know that they must simply be very weak spots, a purely practical solution is to reset the negative intensities to zero or, preferably, to a small number, for example a fraction of the smallest positive intensity, although this clearly introduces bias. An alternative approach, which is based on Bayesian statistics and was developed by French and Wilson (1978), is to analyse the distribution of intensity values in the dataset and, essentially, to determine small increments to each intensity which narrow the distribution slightly and thereby eliminate the negative intensities. Of course, narrowing the distribution involves increasing the weak and negative intensities at the same time as decreasing the strong ones such that the mean is unaffected, although for the strong data the corrections are very small in comparison with the reflection intensities. This Bayesian correction is applied in the CCP4 program TRUNCATE (CCP4, 1994), which yields corresponding values of $|F_{hkl}|$ and also estimates the $\sigma(|F_{hkl}|)$ values by the same method.

11.16   Normalised structure factors

The tendency of the diffraction pattern to fade at high θ angles (or high resolution) is reflected in the B-factor of the dataset. It is possible to negate this resolution-dependent fall-off by normalising the data such that the average (or root mean square) intensity is uniform as a function of resolution. Normalised structure factors or E-values can be calculated by dividing each measured structure factor amplitude $(I_{hkl})^{1/2}$ by the value calculated from the expected atomic composition of the crystal and the Wilson plot B-factor, as shown below:

$$|E_{hkl}| = \frac{(I_{hkl})^{1/2}}{\left(\sum_j f_j^2\right)^{1/2} e^{-B\sin^2\theta/\lambda^2}} \tag{11.31}$$

where the summation j is over all atoms in the unit cell. Note that the normalised structure factor E_{hkl} is distinct from the total diffracted energy, for which we used the same abbreviation in Section 11.9. The form of the denominator of equation (11.31) introduces inaccuracies with protein data and, in practice, E-values are calculated from the ratio of each structure factor amplitude to the mean amplitude of all diffraction spots of the same resolution as follows:

$$|E_{hkl}| = \sqrt{\frac{|F_{hkl}|^2/\varepsilon}{\left\langle|F_{hkl}|^2/\varepsilon\right\rangle}} \tag{11.32}$$

where the term ε is a symmetry factor to compensate for the fact that some classes of reflection, for example those along symmetry axes, generally have higher intensity than others. To calculate the mean $|F_{hkl}|$ values as a function of resolution (d_{hkl}), the diffraction spots are divided into thin shells in reciprocal space. The advantage of E_{hkl} is that the average value of $|E_{hkl}|^2$ is unity within any resolution shell, i.e. there is no longer any θ-dependent fall-off in the mean intensity. It has been shown that the use of E_{hkl} is tantamount to regarding the atoms within a structure as points which do not suffer from thermal motion.

One example of the use of E-values is in the determination of the Laue symmetry of the diffraction pattern from the processed images (Evans, 2006). If we calculate the E-values for each image of the dataset, the effect is to normalise out both the resolution-dependent and the ϕ-angle-dependent variations in the mean diffraction intensity. The mean intensity in each resolution shell, which forms the denominator of equation (11.32), is calculated from each consecutive 'run' of images that are assumed to be on approximately the same scale, with a correction for ϕ-dependent variations. The corresponding resolution-dependent mean intensities are used to calculate

E-values for all reflections measured in that run. It is then possible to calculate a correlation coefficient for the normalised intensities that are related by one of the possible Laue group symmetries, and this will indicate how likely it is that the diffraction data belong to that Laue group. We can recalculate the correlation coefficient for all possible Laue groups and thus assess likely symmetries for the crystal at an early stage of data analysis. This is the basis of the program POINTLESS (Evans, 2006), which must be applied to data obtained directly from the image-processing software (e.g. MOSFLM) prior to any scaling and merging, which will, by definition, impose a symmetry chosen by the operator on the data set. Of course, when a decision as to the crystal space group has been made, the data must be scaled and merged accordingly.

11.17 Completeness of the data

Once a dataset has been subjected to scaling and merging, it is important to assess what proportion of the total number of unique reflections to any given resolution has actually been measured during the experiment. Factors that militate against completeness include radiation damage, which can lead to a significant proportion of the spots being unmeasurably weak towards the end of the collection. The crystal may be anisotropic, for example diffracting to 2 Å resolution in one direction and 3 Å in another, and there may be a significant blind region, especially if the crystal is rotated about one axis only and the detector is moved to collect high-θ diffraction spots in one direction only. Ideally, the dataset that we use to solve the crystal structure should be in excess of 90–95% complete, and when highly redundant data are collected from a good crystal, 100% completeness to the desired resolution limit may be achieved.

Whilst completeness of the dataset is always reported by the scaling and merging software, a useful rule of thumb to estimate it quickly can be derived simply. In Chapter 8, Section 8.7, we derived formulae for the volume V^* of the reciprocal unit cell and showed that

$$V^* = \frac{1}{V}$$

where V is the volume of the real unit cell. In reciprocal space, the diffraction spots within a certain resolution limit d_{min} will fill a sphere of radius $1/d_{min}$, and the volume of this sphere is therefore

$$\frac{4}{3}\pi \left(\frac{1}{d_{min}}\right)^3$$

We can estimate the number of reciprocal unit cells filling this sphere as follows:

$$N_{spots} = \frac{4}{3}\pi \left(\frac{1}{d_{min}}\right)^3 \frac{1}{V^*}$$

$$\therefore N_{spots} = \frac{4}{3}\pi \left(\frac{1}{d_{min}}\right)^3 V$$

For orthogonal unit cell vectors, the formula can be simplified to

$$N_{spots} = \frac{4}{3}\pi \left(\frac{1}{d_{min}}\right)^3 abc$$

We saw in Chapter 9, Sections 9.10 and 9.11, that the diffraction pattern is symmetric such that each diffraction spot has at least one symmetry-equivalent spot owing to Friedel's law, i.e. diffraction spots hkl and $\overline{hkl}$ are usually of equal intensity, and hence there is redundancy due to symmetry in the dataset of at least a factor of 2. For crystals of higher symmetry than $P1$, there will of course be greater symmetry in the diffraction pattern; for example, with monoclinic crystals, any spot hkl will have equal intensity to $\overline{h}kl$, $h\overline{k}l$ and $\overline{hk}l$, and therefore the redundancy due to symmetry in the common space group $P2_1$ is fourfold. For primitive orthorhombic symmetry, the redundancy is eightfold, and for tetragonal symmetry, the redundancy can likewise be eightfold or 16-fold depending on the point group. The general formula for the number of unique spots is therefore

$$N_{unique} = \frac{4}{3}\pi \left(\frac{1}{d_{min}}\right)^3 \frac{V}{n} \tag{11.33}$$

where n is the redundancy factor due to diffraction pattern symmetry. The value of n will also have to be increased by an additional factor if a large proportion of the diffraction spots are systematically absent; for example, if half of them are absent owing to lattice centring (as in the example given in Section 9.14), we need to use $2n$ in the denominator of equation (11.33). The completeness can be calculated from the number of reflections in the merged dataset (N_{merge}) from the ratio

$$\frac{N_{merge}}{N_{unique}}$$

and expressed as a fraction or, more usually, a percentage.

11.18   Estimating the solvent content

Prior to analysing the structure of a molecule, it is essential to know how many copies of it are present in the asymmetric unit of the crystal. For proteins, this can be estimated from the finding that the volume of the crystal occupied by solvent is usually in the range of approximately 40–70%, with the majority of crystals having a solvent content in the region of 45–60% (Matthews, 1968; Kantardjieff and Rupp, 2003). The estimation of the number of molecules in the asymmetric unit also relies on the fact that protein molecules generally have a constant density, which is reflected in a quantity known as the partial specific volume $\overline{V}_p$, i.e. the increase in volume of the solution per unit mass of solute (in this case protein) that is added to it. The inverse relationship between the density and partial specific volume of the solute should be apparent. Proteins typically have partial specific volumes of around $0.74\,\mathrm{cm}^3/\mathrm{g}$, and this value is much higher than that of nucleic acids, which typically have $\overline{V}_p$ values of around $0.5\,\mathrm{cm}^3/\mathrm{g}$ owing to the presence of the heavier element phosphorus, which confers a greater density on them.

If we consider the unit cell of a crystal containing n copies of a protein of molecular weight M (in daltons), the mass of protein in grams in a single unit cell is nM/N, where N is Avogadro's number, or 6×10^{23}. The volume in cm^3 occupied by the protein molecules in the unit cell is therefore

$$\frac{nM\overline{V}_p}{N}$$

The volume of the unit cell that is occupied by solvent, V_s, is therefore given by

$$V_s = V - \frac{nM\overline{V}_p}{N}$$

where V is the volume of the unit cell in units of cm^3. The fraction of the unit cell that is occupied by solvent is therefore

$$\frac{V_s}{V} = 1 - \frac{nM\overline{V}_p}{NV} \tag{11.34}$$

To simplify the expression, we should note that NV in cm^3 is simply the unit cell volume in $\mathring{A}^3$ multiplied by 0.6. The partial specific volume $\overline{V}_p$ can be assumed to be 0.74 cm^3/g. Thus equation (11.34) can be simplified to

$$\frac{V_s}{V} = 1 - 1.23\frac{nM}{V} \tag{11.35}$$

The ratio nM/V can be readily calculated from the unit cell parameters (in $\mathring{A}$) and molecular weight (in daltons) of the protein, assuming a trial value for n. If there is one molecule in the asymmetric unit, then n will be equal to the number of symmetry-related molecules in the unit cell and will therefore be dictated by number of general equivalent positions (GEPs) for that space group, which can be found in the *International Tables for Crystallography*. If there are, for example, two molecules in the asymmetric unit, n will be twice the number of GEPs. Equation (11.35) then allows us to calculate the solvent content of the crystal as a fraction (or percentage) using the trial value of n. If, for example, the value of n corresponding to one molecule in the asymmetric unit gives an estimated solvent content that is in the typical range for protein crystals (40–70%), we may indeed conclude that there is indeed only one molecule per asymmetric unit. If instead we obtain an estimate of the solvent content that is unrealistically high, say 80% or above, this indicates that we most probably have more than one molecule in the asymmetric unit, and the calculation must therefore be repeated with $2n$, $3n$ etc. until a reasonable solvent content value is obtained. Sometimes there is some ambiguity as to which value of n is correct, and this has to be resolved at a later stage of the structure analysis. Generally, crystals that diffract very strongly tend to have low solvent contents, and this can help to ascertain the likely number of molecules in the asymmetric unit if there is any ambiguity. The solvent content and number of molecules per asymmetric unit can be assessed using a number of software tools, including one which calculates the relevant probabilities based on the resolution of the diffraction (Kantardjieff and Rupp, 2003). Note that in the literature it is common to refer to the ratio V/nM as V_M, which is defined as the crystal volume in $\mathring{A}^3$ per unit molecular weight. V_M, like the solvent content of protein crystals, tends to have a restricted range of values, usually between 2.0 and 3.0 $\mathring{A}^3$/Da (Matthews, 1968).

11.19   Misindexing and twinning

Problems can arise with data from crystals of certain symmetries owing to the fact that there are different ways in which the h, k and l axes can be assigned to the diffraction pattern. Consider a crystal belonging to the space group $P4$ in which the constituent molecules are arranged around the fourfold axis in the manner shown for the triangular flags in Fig.11.28(a). If data were collected from a crystal mounted in this manner, we could assign the h, k and l axes in the diffraction pattern to correspond with the right-handed x, y, and z axis of the unit cell, as shown. If, just by chance, the crystal had been mounted in the opposite orientation such that the arrowheads pointed the opposite way (see Fig. 11.28(b)), the unit cell and symmetry that we would determine from the diffraction pattern would be exactly the same as in the first situation. During data processing, we would assign the h, k and l axes corresponding to x, y and z as shown in Fig.11.28(b). However, since the crystals are oriented differently in the two situations shown, the diffraction intensities assigned to any particular h, k, l will generally be completely different in these two cases.

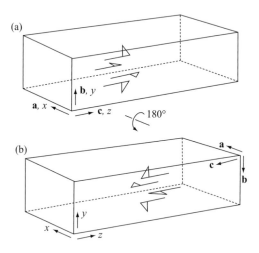

Fig. 11.28 *Misindexing.* (a) shows a tetragonal unit cell containing triangular flag motifs representing the asymmetric units of the crystal that are related by the fourfold axis parallel to z. During data collection, we might slowly rotate the crystal around a fixed axis, such as the crystal z axis, and the diffraction data would allow us to determine the unit cell vectors **a**, **b** and **c** and diffraction intensities I_{hkl}. (b) shows an alternative orientation of the crystal from which we would determine exactly the same unit cell and orientation parameters, but a completely different set of diffraction intensities. This effect is known as misindexing and can also give rise to crystal twinning.

This gives rise to two problems in crystallography, one of which can be solved relatively easily by reindexing the data, whereas the other one is much more difficult to solve, and we will leave the harder problem till a bit later on. The first problem, usually tractable, arises if we happen to have a native crystal belonging to one of the affected space groups and we then make a heavy-atom derivative or a ligand complex. The resulting derivative dataset may then be indexed differently from the native dataset even though both have exactly the same unit cell and symmetry. In these situations, we need to measure small intensity differences between the derivative and the native crystal in order to determine the binding site(s) of the ligand. If, however, the two datasets being compared are indexed differently, the difference in intensity for each *hkl* in the two datasets will be large and meaningless. To assess whether misindexing has occurred, one needs to calculate an *R*-factor for the two datasets (e.g. by using equation (14.1) in Chapter 14), and a high value for this parameter will indicate that the data from the two crystals have been indexed differently. It is then a practical matter of reindexing the data by applying a transformation of the axes ($h\bar{k}\bar{l}$ in the case shown in Fig. 11.28) and recalculating the *R*-factor to check that the data have been reindexed successfully. This problem commonly arises with space groups belonging to the *P*3 and *P*4 point groups and, in some space groups, several transformations may need to be tested until the correct one is found. For many of the common space groups, misindexing is not a concern, but it can arise in unexpected situations, for example with monoclinic space groups, where certain combinations of the unit cell parameters *a*, *c* and *β* allow an alternative diagonal unit cell to be drawn with almost identical cell parameters. More details are shown in Fig. 11.29.

Whilst misindexing is one manifestation of certain symmetries or unit cell parameters, a more serious problem can arise for the same space groups when the crystal

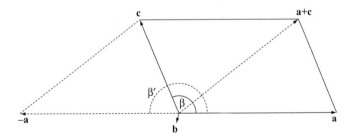

Fig. 11.29 *The diagonal unit cell of a monoclinic crystal system.* The solid lines shows the plane formed by the **a** and **c** vectors of a monoclinic unit cell with the unique axis **b** perpendicular to the plane of the page. An alternative unit cell, shown by dashed lines, can be drawn using the diagonal. The alternative cell has the same **b** vector as the previous one but the other two unit cell vectors are now given by **a** + **c** and -**a**. In this case, the two alternative cells are easy to distinguish; for example, the second choice has a larger *β* angle (shown as *β′*). However, certain combinations of *a*, *c* and *β* cause the two choices of unit cell to be virtually indistinguishable, which can cause problems of misindexing and even crystal twinning.

contains regions, or domains, in which the molecules are packed in different orientations. This relates to the mosaic model of crystals that we described in Section 11.12. In the best crystals, the domains making up the crystal will be packed together in almost exactly the same orientation. However, in certain space groups, it is possible for the domains to have significantly different orientations and still pack together in the manner of an extensive lattice. This arises from 'mistakes' during crystal growth, and the vulnerable space groups are the same ones that are also susceptible to misindexing problems. Consider Fig. 11.28 again, where two possible orientations for a tetragonal unit cell are shown. If the crystal were to consist of domains with their unit cells in either of the two possible orientations shown, we would have great difficulty in interpreting the diffraction data from the crystal, since the molecules would not be arranged regularly within it. This is known as *twinning*, and crystals which suffer from it are said to be *twinned*—the subject is covered very well by (Parsons 2004).

The principal effect of twinning is that the diffraction spots from the differently oriented domains of the crystal overlap, i.e. the intensity of each diffraction spot is the sum of the diffraction intensities of each domain, weighted by its relative abundance in the crystal, known as the *twinning ratio*. One effect of twinning on the diffraction pattern is that it can give the impression that the crystal has a higher symmetry than it really has; for example, if two domains related in the manner shown in Fig. 11.28 were present in equal proportions in the crystal, the diffraction pattern would correspond to that of a crystal with a perfect twofold axis perpendicular to the fourfold axis, i.e. it would have $4/mmm$ symmetry. This effect will be less for crystals where the twinning ratio deviates from 50 : 50. The symmetry operator relating the two domains of the twin (in this case a twofold axis) is referred to as the *twin operation*, and it is very important that this is determined in order to refine the structure. In the example shown in Fig. 11.28, we can see from the directions of the axes that the twin operation in reciprocal space is $(h, -k, -l)$, although this is more often quoted as $(k, h, -l)$, which is equivalent by tetragonal symmetry. This type of twinning, in which the different domains are related in such a way as to confer higher symmetry on the diffraction pattern than is actually present within each domain, is referred to as *merohedral twinning*. In cases where there are only two components to the twinning (such as in the above example), it is referred to as *hemihedral twinning*, and this is the most common type. Twinning can also occur when the unit cell parameters appear to be of higher symmetry than the space group; for example, an orthorhombic unit cell with $a \approx b$ might allow the domains to pack with pseudo-tetragonal symmetry. Likewise, monoclinic cells with β close to 90° can pack with pseudo-orthorhombic symmetry. This type of twinning is referred to as *pseudo-merohedral.*

So far, we have discussed how twinning can give rise to an exact superposition of two diffraction patterns arising from each component of the twin. However, it is possible for twinning to occur in such a way that only a subset of the diffraction spots overlap, and this is referred to as *non-merohedral twinning* (Parsons, 2004; Yeates and Fam, 1999). An example of an orthorhombic cell where one of the cell dimensions happens to be approximately half the other is shown in Fig. 11.30. This allows the cells to pack in separate domains which are related by a 90° rotation. Since the horizontal cell dimension of the right-hand domain is twice that of the left-hand domain, the

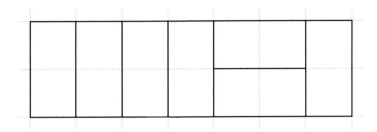

Fig. 11.30 *Non-merohedral twinning.* This shows a twinned crystal lattice consisting of rectangular unit cells (adapted from Parsons, 2004). In the dominant component of the twin, shown on the left-hand side, these cells are arranged vertically. However, the fact that one unit cell dimension is approximately half the other allows the cells to pack in an alternative manner, shown as the two horizontal cells towards the right. The diffraction patterns arising from the two components of this non-merohedrally twinned crystal are partially overlapped, i.e. some spots will overlap exactly but others will not. The data may 'index' on the 'tetragonal supercell' (formed from two adjacent rectangular cells) but the structure would probably be impossible to solve.

reciprocal unit cell dimension of the right-hand domain is half that of the left-hand domain. Consequently, only diffraction spots from the right-hand domain of even order on the horizontal axis ($h = 2n$) will overlap with those from the left-hand domain. The converse situation arises in the vertical direction, and the net effect is to give a pseudo-tetragonal diffraction pattern with inexplicable systematic absences. There are several types of non-merohedral twinning, including *epitaxial twinning,* in which the domain lattices overlap in fewer than three dimensions, such that no spots may overlap at all. Non-merohedral twinning can sometimes be detected in the physical appearance of a crystal under a microscope when the domains segregate appreciably, for example when boundaries, hollows or slightly disjointed regions occur within the crystal. A purely practical solution to the problem is to find crystallisation conditions that reduce the extent of the problem, for example the use of lower temperatures or an additive. Sometimes carefully splitting a twinned crystal along a visible boundary within it and using the resulting fragments as separate crystals for data collection can work perfectly well.

The presence of merohedral twinning, where the diffraction patterns of the twin components overlap exactly, can be revealed by a branch of crystallography known as intensity statistics. The superposition of the diffraction patterns from the constituent domains of the crystal means that it is a relatively common situation for a strong spot from one domain to be superimposed on a weak spot from the other domain. For this reason, the diffraction patterns of twinned crystals contain relatively few reflections at both the very weak and the very strong extremities of the intensity range, thus affecting the overall intensity distribution. A very useful graph for detecting twinning is the cumulative intensity distribution, which shows the fraction $N(z)$ of all intensity values that are lower than a certain value z as a plot of $N(z)$ versus z. For the reasons outlined above, the cumulative intensity distribution of a twinned

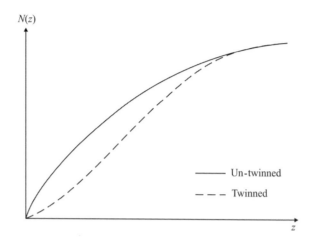

Fig. 11.31 *The cumulative intensity distribution.* In this distribution, we plot the proportion of reflections, $N(z)$, with an intensity lower than a certain cut-off value z as a function of z. The form of the relationship for non-twinned acentric data is shown as a solid line and that for merohedrally twinned data is indicated as a broken line. Twinning effectively narrows the distribution of intensities, and this causes the cumulative intensity distribution to have a sigmoid form.

dataset tends to be sigmoid rather than hyperbolic, as shown in Fig. 11.31, and the mean $||E|^2 - 1|$ tends to be appreciably lower than the value expected for a non-centrosymmetric crystal, which is 0.736. These analyses are performed by the CCP4 program TRUNCATE (CCP4, 1994). Twinning may also be indicated if the data can be processed in a high-symmetry space group and yet the calculated solvent content and V_M (see Section 11.18) are unrealistically low.

Providing the twin operation is known, there are a number of statistical ways that allow the twin ratio to be estimated from the intensities, and this allows the data to be 'de-twinned' (Dauter, 2003). However, these methods work badly when the twin ratio approaches 50:50; and, conversely, if the extent of twinning is low, de-twinning may not be necessary to solve the structure. Providing that the twinning operator is known, twinned structures can be handled in refinement (which is covered in Chapter 15), and the twin ratio can be refined as a free variable.

Summary

The X-radiation derived from a conventional source is polychromatic, consisting of a background on which are superposed several sharp spikes. These correspond to the emission of radiation when an allowed electronic transition takes place within an atom of the target. The Kα doublet is the most intense feature of the spectrum, and for most purposes it is this radiation which we wish to be incident onto the crystal to be studied. The unwanted radiation may be filtered out by using an absorption filter, a balanced filter, a crystal monochromator or a combination of these.

Synchrotron facilities are intense X-ray sources that operate by circulating electrons at high energy in an evacuated storage ring. The emission of X-rays is caused by the deflection of the electron beam by the use of bending magnets, as well as by the use of insertion devices such as wigglers and undulators, which yield more intense X-ray beams. Each synchrotron beam line has a number of optical components to monochromate, focus and collimate the X-ray beam.

Routine data collection is performed by slowly rotating the crystal about an axis perpendicular to the beam (normal beam geometry), and the diffraction spots are recorded by an electronic area detector. Laboratory X-ray sources often utilise an image plate detector, which stores an image of the diffraction pattern in an erasable phosphor which is scanned by a laser at the end of each exposure. In contrast, synchrotron sources generally utilise detectors with direct readout such as the charge-coupled device (CCD) and the pixel detector, which allow faster data collection. Radiation damage to the crystal is minimised by cooling it to liquid nitrogen temperatures, usually 100 K.

Analysing the data collected by these instruments requires that the unit cell and orientation matrix of the crystal are determined, usually in a process known as autoindexing, which normally benefits from the use of images separated by 90° on the rotation axis (ϕ axis). Following estimation of the mosaic spread (essentially the spot width), the diffraction intensities in selected groups of images can be measured, and the partially recorded reflections are used to refine the orientation matrix accurately, in a process known as post-refinement. The final orientation matrix is used to predict the positions of all spots in all images such that the final set of diffraction intensities I_{hkl} can be obtained by direct integration or by profile fitting.

The intensities obtained depend on a number of instrumental parameters, such as the polarisation of the incident beam and the geometry with which each reciprocal-lattice point intersects the Ewald sphere (represented by the Lorentz factor). Whilst the measured intensities can be corrected for these effects by precise formulae, other physical factors such as absorption cannot be calculated directly, and such corrections are usually made empirically during scaling of the data.

Analysis of the mean diffraction intensity as a function of resolution allows us to determine the extent of disorder and thermal motion within the crystal in the form of a temperature factor B, which is related to the mean square displacement $\overline{u^2}$ by the relationship $B = 8\pi^2 \overline{u^2}$. B can be determined from the gradient $(-2B)$ of a Wilson plot, a graph of $\log_e \left[\overline{I_{hkl}} / \sum_j f_j^2 \right]$ versus $\sin^2 \theta / \lambda^2$.

Following initial data processing, the measurements of the intensities from different diffraction images must be placed on a common scale, which usually involves determining a scale factor and temperature factor either for each individual image or as smoothly varying functions of the ϕ angle (when there are no discontinuities in the beam intensity or exposure time). The quality of the scaled dataset can be assessed by the R_{merge}, which should generally be less than 10%. The point group symmetry of the crystal can often be determined empirically by scaling the dataset in different possible Laue groups—that with highest symmetry which gives an acceptable R_{merge} is likely to be correct. The dataset quality should also be assessed by its completeness for the chosen space group.

The end product of data processing will be a set of unique structure factor amplitudes and their associated standard deviations, $|F_{hkl}|$ and $\sigma(|F_{hkl}|)$. Normalised structure factors E_{hkl}, in which the resolution-dependent fall-off in amplitude is eliminated, can also be determined for various aspects of the subsequent structure analysis. Some indication of the complexity of this task will be indicated by the solvent content of the crystal and the number of molecules in the asymmetric unit, which can be estimated once the unit cell parameters and symmetry are known.

In some space groups, a dataset may require reindexing so that the assignment of the *hkl* indices is consistent with other previously collected datasets. Such space groups also tend to be prone to problems of merohedral twinning, in which the constituent domains of the crystal pack in different orientations that are not consistent with the space group symmetry. In this type of twinning, the twin components give diffraction patterns that superimpose exactly, although in favourable cases the intensity data can be 'de-twinned' for structure analysis. In non-merohedral twinning, the twin components give diffraction patterns that only partially superimpose, and the structure may be solvable if one of the twin components can be selected in the initial data processing.

References

Amemiya, Y. and Miyahara, J. (1988) Imaging plate illuminates many fields. *Nature* **336**, 89–90.

Arndt, U. W., Champness, J. N., Phizackerley, R. P. and Wonacott, A. J. (1973) A single-crystal oscillation camera for large unit cells. *J. Appl. Crystallogr.* **6**, 457–463.

Blessing, R. H. (1995) An empirical correction for absorption anisotropy. *Acta Crystallogr. A* **51**, 33–8.

Bloomer, A. C. and Arndt, U. W. (1999) Experiences and expectations of a novel X-ray microsource with focusing mirror. *Acta Crystallogr.* **55**, 1672–80.

Bourgeois, D., Schotte, F., Brunori, M. and Vallone, B. (2007) Time-resolved Laue crystallography as a tool to investigate photo-activated protein dynamics. *Photochem. Photobiol. Sci.* **6**, 1047–56.

Bright Wilson, E. (1990) *An Introduction to Scientific Research.* Dover, New York, pp. 191–3.

Brockhauser, S., Di Michiel, M., McGeehan, J. E., McCarthy, A. A. and Ravelli, R. B. G. (2008) X-ray tomographic reconstruction of macromolecular samples. *J. Appl. Crystallogr.* **41**, 1057–66.

Buerger, M. J. (1940) The correction of X-ray diffraction intensities for Lorentz and polarization factors. *Proc. Natl. Acad. Sci. USA* **26**, 637–42.

Burmeister, W. P. (2000) Structural changes in a cryo-cooled protein crystal owing to radiation damage. *Acta Crystallogr. D* **56**, 328–41.

CCP4 (Collaborative Computational Project Number 4) (1994) The CCP4 suite: programs for protein crystallography. *Acta Crystallogr. D* **50**, 760–3.

Corbett, M. C., Latimer, M. J., Poulos, T. L., Sevrioukova, I. F., Hodgson, K. O. and Hedman, B. (2007) Photoreduction of the active site of the metalloprotein putidaredoxin by synchrotron radiation. *Acta Crystallogr. D* **63**, 951–60.

Dauter, Z. (2003) Twinned crystals and anomalous phasing. *Acta Crystallogr. D* **59**, 2004–16.

Diamond, R. (1969) Profile analysis in single-crystal diffractometry. *Acta Crystallogr. A* **25**, 43–54.

Evans, P. (2006) Scaling and assessment of data quality. *Acta Crystallogr. D* **62**, 72–82.

French, S. and Wilson, K. (1978) On the treatment of negative intensity observations. *Acta Crystallogr. A* **34**, 517–25.

Garman, E. (1999) Cool data: quantity and quality. *Acta Crystallogr. D* **55**, 1641–53.

Helliwell, J. R. (1992) *Macromolecular Crystallography with Synchrotron Radiation.* Cambridge University Press, Cambridge.

Huelson, G., Broennimann, C., Eikenberry, E. F. and Wagner, A. (2006) Protein crystallography with a novel large-area pixel detector. *J. Appl. Crystallogr.* **39**, 550–7.

Jeffrey, J. W. (1971) *Methods in X-ray Crystallography.* Academic Press, London and New York.

Kabsch, W. (1988a) Automatic indexing of rotation diffraction patterns. *J. Appl. Crystallogr.* **21**, 67–71.

Kabsch, W. (1988b) Evaluation of single-crystal X-ray diffraction data from a position-sensitive detector. *J. Appl. Crystallogr.* **21**, 916–24.

Kabsch, W. (1993) Automatic processing of rotation diffraction data from crystals of initially unknown symmetry and cell constants. *J. Appl. Crystallogr.* **26**, 795–800.

Kantardjieff, K. A. and Rupp, B. (2003) Matthews coefficient probabilities: improved estimates for unit cell contents of proteins, DNA and protein–nucleic-acid complex crystals. *Protein Sci.* **12**, 1865–71.

Kittel, C. (2004) An Introduction to Solid State Physics, 8th edn. Wiley, New York.

Kusz, J. and Bohm, H. (2002) Performance of a confocal multilayer X-ray optic. *J. Appl. Crystallogr.* **35**, 8–12.

Leslie, A. G. W. (2006) The integration of macromolecular diffraction data. *Acta Crystallogr. D* **62**, 48–57.

Leslie, A. G. W., Powell, H. R., Winter, G., Svensson, O., Spruce, D., McSweeney, S., Love, D., Kinder, S., Duke, E. and Nave, C. (2002) Automation of the collection and processing of X-ray diffraction data – a generic approach. *Acta Crystallogr. D* **58**, 1924–8.

Matthews, B. W. (1968) Solvent content of protein crystals. *J. Molec. Biol.* **33**, 491–7.

Milch, J. R. and Minor, T. C. (1974) The indexing of single-crystal X-ray rotation photographs. *J. Appl. Crystallogr.* **7**, 502–505.

North, A. C. T., Phillips, D. C. and Matthews, F. S. (1968) A semi-empirical method of absorption correction. *Acta. Crystallogr. A* **24**, 351–9.

Otwinowski, Z. and Minor, W. (1997) Processing of X-ray diffraction data. *Methods Enzymol.* **276**, 307–26.

Parsons, S. (2004) Introduction to twinning. *Acta Crystallogr. D* **59**, 1995–2003.

Pflugrath, J. W. (1999) The finer things in X-ray diffraction. *Acta Crystallogr. D* **55**, 1718–25.

Phillips, W. C. and Rayment, I. (1985) A systematic method for aligning double-focussing mirrors. *Methods Enzymol.* **114**, 316–29.

Powell, H. R. (1999) The Rossmann Fourier autoindexing algorithm in MOSFLM. *Acta Crystallogr. D* **55**, 1690–5.

Ravelli, R. B. G. and McSweeney, S. M. (2000) The fingerprint that X-rays leave on structures. *Structure* **8**, 315–28.

Rossmann, M. G. (1979) Processing oscillation diffraction data for very large unit cells with an automatic convolution technique and profile fitting. *J. Appl. Crystallogr.* **12**, 225–38.

Sanderson, M. R. (2007) In-house macromolecular data collection. In *Macromolecular Crystallography*, ed. Sanderson, M. and Skelly, J. V. Oxford University Press, Oxford, pp. 77–86.

Schwager, P., Bartels, K. and Huber, R. (1973) A simple empirical absorption-correction method for X-ray intensity data films. *Acta Crystallogr. A* **29**, 291–5.

Stellar, I., Bolotovsky, R. and Harrison, S. C. (1997) An algorithm for automatic indexing of oscillation images using Fourier analysis. *J. Appl. Crystallogr.* **30**, 1036–40.

Westbrook, E. M. and Naday, I. (1997) Charge-coupled device-based area detectors. *Methods Enzymol.* **276**, 244–68.

Wilson, A. J. C. (1942) Determination of absolute from relative X-ray intensity data. *Nature* **150**, 152.

Wilson, K. and Yeates, D. (1979) On the treatment of data measured on the oscillation camera. *Acta Crystallogr. A* **35**, 146–57.

Wood, K., Frolich, A., Paciaroni, A., Moulin, M., Hartlein, M., Zaccai, G., Tobias, D. J. and Weik, M. (2008) Coincidence of dynamical transitions in a soluble protein and its hydration water: direct measurements by neutron scattering and MD simulations. *J. Am. Chem. Soc.* **130**, 4585–7.

Yeates, T. O. and Fam, B. C. (1999) Protein crystals and their evil twins. *Structure* **7**, 25–9.

12

The phase problem and the Patterson function

12.1 The nature of the problem

The ultimate aim of an X-ray diffraction experiment is the computation of the electron density function $\rho(x, y, z)$ according to the Fourier synthesis of equation (9.15),

$$\rho(x, y, z) = \frac{1}{V} \sum_h \sum_k \sum_l F_{hkl} e^{-2\pi i (hx + ky + lz)} \tag{9.15}$$

Every structure factor F_{hkl} is a complex number, which may be represented as

$$F_{hkl} = |F_{hkl}| e^{i\phi_{hkl}} \tag{12.1}$$

implying that

$$\rho(x, y, z) = \frac{1}{V} \sum_h \sum_k \sum_l |F_{hkl}| e^{i\phi_{hkl}} e^{-2\pi i (hx + ky + lz)}$$

This summation consequently requires us to know both the magnitude $|F_{hkl}|$ and the corresponding phase ϕ_{hkl} for all the structure factors F_{hkl}. As we saw in Chapter 11, however, the quantity directly observed in a diffraction experiment is the intensity, from which, after applying the appropriate correction and scaling factors, we may determine relative values of $|F_{hkl}|^2$. Our experimental measurements therefore enable us to derive only the magnitudes $|F_{hkl}|$, and not the corresponding phases ϕ_{hkl}. This implies that we cannot perform the Fourier synthesis of equation (9.15) directly from our experimental data, and that the goal of computing the electron density function $\rho(x, y, z)$ cannot be achieved until we have devised a way of determining the appropriate phase angle ϕ_{hkl} for each of the measured magnitudes $|F_{hkl}|$. The derivation of the correct values of the phases ϕ_{hkl} constitutes the *phase problem*.

The nature of the problem is vividly illustrated in the Argand diagram of Fig. 12.1, in which the complex number F_{hkl} is represented by a vector in the complex plane of magnitude $|F_{hkl}|$, making an angle ϕ_{hkl} with the real axis. For different values of the phase angle ϕ_{hkl}, the vector in the complex plane sweeps out a circle of radius $|F_{hkl}|$, and since $|F_{hkl}|$ is our directly observed quantity, all that our experimental data can give us is the radius of the circle in the complex plane on which the structure factor F_{hkl} must lie. Knowledge of $|F_{hkl}|$ alone therefore allows all phase angles from 0 to 2π, yet for each reciprocal-lattice point hkl, there is one phase angle ϕ_{hkl}.

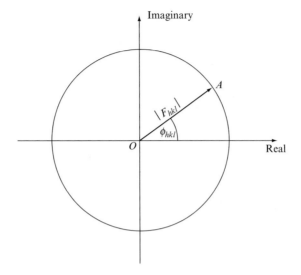

Fig. 12.1 *The phase problem.* The structure factor F_{hkl} may be represented on an Argand diagram by a vector in the complex plane of length $|F_{hkl}|$, making an angle ϕ_{hkl} with the real axis as in the case of *OA*. If only the magnitude $|F_{hkl}|$ is known, then the structure factor may be any vector from the origin *O* to the circle of radius $|F_{hkl}|$, and the phase angle ϕ_{hkl} is unknown.

12.2   Why is phase not detectable?

Before we investigate the methods which help us to solve the phase problem, it is useful to discuss the reasons why the phase problem exists, and so we now answer the question 'Why is phase not detectable?'

The answer to this question is threefold. Firstly, X-rays have a very high frequency, of the order of 10^{18} Hz. In order to determine the phase of a wave, we must be able to monitor the wave completely and accurately. This implies that an instrument capable of measuring the phase of an X-ray beam must be able to detect changes in an electric field occurring in times of the order of 10^{-18} s. This is an extremely fast response time and, at present, no recording device is known which can respond to any influence as fast as 10^{18} times a second.

Knowledge of the phase angle ϕ_{hkl} for any diffracted beam also implies that we must be able to monitor the phase of any given wave with respect to that of a reference wave, say that of the straight-through beam, to which we assign zero phase. Consequently, phase determination requires that we compare the phases of the diffracted wave and the reference wave, and in order to do this we must know the distance between the site at which we sample the diffracted wave and that at which we measure the reference wave. If our phase measurement is to be meaningful, this distance must be known accurately to very much less than a single X-ray wavelength, which means that we must have a ruler capable of measuring accurately down to 10^{-11} m. Once again, this is stretching the limits of present technology.

The third reason why we cannot detect the phases of the diffracted waves concerns the actual nature of the scattering process. Accurate phase measurement implies that the phases of the diffracted waves are correlated over relatively long periods of time. Now, the scattering of X-rays by a crystal is a passive process, and the time course of the scattering depends on the instantaneous behaviour of the incident X-ray beam. Any time correlations in X-ray scattering therefore imply the presence of time correlations in the incident X-rays. It so happens that all presently available X-ray sources are such that the emission of an X-ray is a random event, and no two X-rays are propagated with a well-defined phase relationship. Such a source is called an *incoherent source*, as opposed to a *coherent source*, for example a laser, which emits radiation with very highly correlated phase relationships. Since the incident X-rays come from an incoherent source, there are no well-defined phase relationships between the incident waves, and this prevents us from measuring phase relationships between the diffracted waves.

For the three reasons stated above, it is not possible to detect the phases of the diffracted waves, and until such phase measurements can be made, the phase problem will remain. One point worth noticing, however, is that each of the three reasons concerns technology as opposed to pure physics. We are unable to measure phase directly because we lack three instruments—a detector with a very fast time response, a highly accurate ruler and a coherent X-ray source. Should these three instruments ever be invented, then the phase problem will pass into history. The fact that our experimental data are limited by technological difficulties suggests that there is some hope for the future. The detection of phase does not contravene any physical laws, and so it may be merely a matter of time before the requisite technology is available.

12.3 The Fourier transform of the intensities

As we saw above, the quantity directly observed in a diffraction experiment is the integrated reflecting power, from which we may derive the relative intensity $|F_{hkl}|^2$ by applying the appropriate correcting and scaling factors. Our ultimate aim is to derive the electron density function $\rho(x, y, z)$, which is given by

$$\rho(x, y, z) = \frac{1}{V} \sum_h \sum_k \sum_l F_{hkl} e^{-2\pi i(hx+ky+lz)} \tag{9.15}$$

The Fourier synthesis of equation (9.15) requires the exact values of the complex quantities F_{hkl} in both magnitude and phase. Our data, however, give us only the real quantity $|F_{hkl}|^2$, and the task of obtaining the correct F_{hkl} from $|F_{hkl}|^2$ constitutes the phase problem. The purpose of the present chapter is to investigate the nature of the information which we may derive using only the relative-intensity ($|F_{hkl}|^2$) data. As we shall see, this information is of very great importance in solving the phase problem.

By analogy with equation (9.15), we need to consider a Fourier synthesis of the form

$$\frac{1}{V} \sum_h \sum_k \sum_l |F_{hkl}|^2 e^{-2\pi i(hx+ky+lz)}$$

The summation of this Fourier synthesis goes over the indices h, k and l, implying that the result of the synthesis will be a function of the spatial coordinates x, y and z. For reasons which will soon become apparent, we shall choose to express the functional dependence of the result of the summation of the expression in equation (12.2) below in terms of three spatial parameters u, v and w which have the same dimensions as x, y and z.

We now define the *Patterson function $P(u, v, w)$* as

$$P(u, v, w) = \frac{1}{V} \sum_h \sum_k \sum_l |F_{hkl}|^2 \, e^{-2\pi i (hu + kv + lw)} \tag{12.2}$$

The Patterson function is simply the Fourier synthesis of the relative-intensity data. This may be obtained directly from an experiment, and so the Patterson function $P(u, v, w)$ may be computed immediately, with no knowledge of the phases.

12.4   The Patterson function and the crystal structure

We know that the Fourier synthesis using the complex values of the structure factors F_{hkl} gives the electron density function $\rho(x, y, z)$, and we have just defined the result of the Fourier synthesis using the values of the relative intensities $|F_{hkl}|^2$ as the Patterson function $P(u, v, w)$. In this section, we shall derive the relationship between the Patterson function $P(u, v, w)$ and the electron density function $\rho(x, y, z)$.

In order to show the relevance of the Patterson function to a crystal structure analysis, it is useful to cast equation (12.2) into symbolic form using the notation T, which we defined in Chapter 5, Section 5.3, as meaning 'is the Fourier transform of'. Since the Patterson function $P(u, v, w)$ is the result of the Fourier synthesis of the relative intensities $|F_{hkl}|^2$, we may write

$$P(u, v, w) = \frac{1}{V} T\left(|F_{hkl}|^2\right) \tag{12.3}$$

Strictly speaking, the Patterson function is the inverse transform of the relative intensities, but since the direct and inverse Fourier transforms are quite symmetrically related, the notation of equation (12.3) is quite valid.

From equation (2.22), we may express $|F_{hkl}|^2$ as

$$|F_{hkl}|^2 = F_{hkl} F_{hkl}^*$$

where F_{hkl} is the true structure factor and F_{hkl}^* is the corresponding complex conjugate. We now have

$$P(u, v, w) = \frac{1}{V} T\left(F_{hkl} F_{hkl}^*\right) \tag{12.4}$$

At this stage, we invoke the convolution theorem (see Section 5.11) in the form
The Fourier transform of a product is the convolution of the individual Fourier transforms.
Equation (12.4) therefore becomes

$$P(u, v, w) = \frac{1}{V} \left[T\left(F_{hkl}\right) * T\left(F_{hkl}^*\right) \right] \tag{12.5}$$

Our next task is to evaluate the Fourier transforms of F_{hkl} and F^*_{hkl}. For the structure factor F_{hkl}, we know the result of the Fourier transformation, for equation (9.15) may be written symbolically as

$$\rho(x, y, z) = \frac{1}{V} T\left(F_{hkl}\right)$$

The analogous result for the complex conjugate structure factor F^*_{hkl} requires a little mathematics. Our starting point is the definition of the structure factor F_{hkl} in terms of the electron density function $\rho(x, y, z)$ given by equation (9.11),

$$F_{hkl} = V \int_0^1 \int_0^1 \int_0^1 \rho(x, y, z) e^{2\pi i(hx+ky+lz)} \, dx \, dy \, dz \tag{9.11}$$

F^*_{hkl} is the complex conjugate of F_{hkl}, and in order to derive F^*_{hkl}, we simply change the signs of all the imaginary terms in the definition of F_{hkl}. Since the electron density function $\rho(x, y, z)$ is a measurable quantity, $\rho(x, y, z)$ must be real, so that the only imaginary terms in the definition of F_{hkl} as given in equation (9.11) are in the complex exponential. F^*_{hkl} is therefore of the form

$$F^*_{hkl} = V \int_0^1 \int_0^1 \int_0^1 \rho(x, y, z) e^{-2\pi i(hx+ky+lz)} \, dx \, dy \, dz$$

$$= V \int_0^1 \int_0^1 \int_0^1 \rho(x, y, z) e^{2\pi i(h(-x)+k(-y)+l(-z))} \, dx \, dy \, dz \tag{12.6}$$

Equation (12.6) takes on a more meaningful form if we introduce a change of variables such that

$$X = -x, Y = -y, Z = -z$$

giving

$$F^*_{hkl} = -V \int_0^{-1} \int_0^{-1} \int_0^{-1} \rho(-X, -Y, -Z) e^{2\pi i(hX+kY+lZ)} \, dX \, dY \, dZ \tag{12.7}$$

On reversing the limits of the integral, the value of the integral changes sign and we obtain

$$F^*_{hkl} = V \int_{-1}^0 \int_{-1}^0 \int_{-1}^0 \rho(-X, -Y, -Z) e^{2\pi i(hX+kY+lZ)} \, dX \, dY \, dZ \tag{12.8}$$

We now consider the limits of the integral in equation (12.8). The integration over the variable X is from values of X from -1 to 0. These values of X define the limits of the unit cell, which are equally as well defined by values of X between 0 and 1, between 2 and 3 or between any ascending pair of integers. Since the unit cell is repeated exactly throughout the crystal, the value of the integral in equation (12.8) must take the same value no matter which pair of integers we use for the limits of X, as long as they are taken in the correct order. Exactly similar remarks may be made for the variables Y and Z, and so the integral in equation (12.8) may be written as

$$F^*_{hkl} = V \int_0^1 \int_0^1 \int_0^1 \rho(-X, -Y, -Z) e^{2\pi i (hX + kY + lZ)} \, dX \, dY \, dZ \tag{12.8}$$

But it does not matter what names we give to the variables of this integration, and so we may write the quantity F^*_{hkl} as

$$F^*_{hkl} = V \int_0^1 \int_0^1 \int_0^1 \rho(-x, -y, -z) e^{2\pi i (hx + ky + lz)} \, dx \, dy \, dz \tag{12.9}$$

We now contrast equation (12.9) with equation (9.11),

$$F_{hkl} = V \int_0^1 \int_0^1 \int_0^1 \rho(x, y, z) e^{2\pi i (hx + ky + lz)} \, dx \, dy \, dz \tag{9.11}$$

Whereas F_{hkl} is the Fourier transform of the electron density function $\rho(x, y, z)$, F^*_{hkl} is seen from equation (12.9) to be the Fourier transform of a function written as $\rho(-x, -y, -z)$. This new function $\rho(-x, -y, -z)$ is related to the electron density function $\rho(x, y, z)$ by a change of sign in all coordinates and so represents the *centrosymmetric image* of the electron density function $\rho(x, y, z)$. This brings us to the important result that

> F_{hkl} *is the Fourier transform of the electron density function* $\rho(x, y, z)$.
> F^*_{hkl} *is the Fourier transform of* $\rho(-x, -y, -z)$, *the centrosymmetric image of the electron density function* $\rho(x, y, z)$.

The state which we had reached in our discussion of the relationship of the Patterson function $P(u, v, w)$ to the crystal structure was

$$P(u, v, w) = \frac{1}{V} \left[T\left(F_{hkl}\right) * T\left(F^*_{hkl}\right) \right] \tag{12.5}$$

We now know that

$$\rho(x, y, z) = \frac{1}{V} T\left(F_{hkl}\right) \tag{9.11}$$

and that

$$\rho(-x, -y, -z) = \frac{1}{V} T\left(F^*_{hkl}\right)$$

and so we have

$$P(u, v, w) = V \left[\rho(x, y, z) * \rho(-x, -y, -z) \right] \tag{12.10}$$

Equation (12.10) tells us that

The Patterson function $P(u, v, w)$ is the convolution of the electron density function $\rho(x, y, z)$ and its centrosymmetric image $\rho(-x, -y, -z)$, scaled by V, the volume of the unit cell.

12.5 The form of the Patterson function

It is now convenient to write equation (12.10) in vector notation. The (x, y, z) coordinates are written in terms of a vector $\mathbf{r}$, and the (u, v, w) coordinates in terms of a vector $\mathbf{u}$. Equation (12.10) becomes

$$P(\mathbf{u}) = V \left[\rho(\mathbf{r}) * \rho(-\mathbf{r}) \right] \tag{12.11}$$

According to Section 5.10, the definition of the convolution of two functions $f(\mathbf{r})$ and $g(\mathbf{r})$ is given by equation (5.27),

$$f(\mathbf{r}) * g(\mathbf{r}) = \int_{\text{all } \mathbf{r}} f(\mathbf{r}) g(\mathbf{u} - \mathbf{r}) \, d\mathbf{r}$$

Identifying

$$f(\mathbf{r}) = \rho(\mathbf{r})$$
$$g(\mathbf{r}) = \rho(-\mathbf{r})$$

we have

$$P(\mathbf{u}) = V \int_{\text{unit cell}} \rho(\mathbf{r}) \rho(\mathbf{u} + \mathbf{r}) \, d\mathbf{r} \tag{12.12}$$

in which the integral goes over a single unit cell of the real lattice, for this is the domain of definition of the electron density function $\rho(\mathbf{r})$.

Inspection of equation (12.12) shows that the integrand is a function of the variables $\mathbf{u}$ and $\mathbf{r}$, and that the integration is over $\mathbf{r}$. The result of the integration is therefore a function of the variable $\mathbf{u}$, and it was for this reason that we introduced the variable $\mathbf{u}$ earlier.

Our full definition of the Patterson function $P(\mathbf{u})$, or $P(u, v, w)$, is contained in the following equations:

$$P(u, v, w) = \frac{1}{V} \sum_h \sum_k \sum_l |F_{hkl}|^2 \, e^{-2\pi i (hu + kv + lw)} \tag{12.2}$$

$$P(\mathbf{u}) = V \int_{\text{unit cell}} \rho(\mathbf{r}) \rho(\mathbf{u} + \mathbf{r}) \, d\mathbf{r} \tag{12.12}$$

Equation (12.2) is the formula for deriving the Patterson function $P(\mathbf{u})$ from the relative-intensity data $|F_{hkl}|^2$, but it is by means of equation (12.12) that we relate the Patterson function $P(\mathbf{u})$ to the structure of the unit cell, and it is with the properties of equation (12.12) that most of the remainder of this chapter will concern itself.

12.6   The meaning of the Patterson function

In this section, we shall discuss the meaning of the mathematics of equation (12.12),

$$P(\mathbf{u}) = V \int_{\text{unit cell}} \rho(\mathbf{r})\rho(\mathbf{u} + \mathbf{r}) \, d\mathbf{r} \tag{12.12}$$

The form of the integral in equation (12.12) is very similar to that of the convolution integral,

$$c(\mathbf{u}) = f(\mathbf{r}) * g(\mathbf{r}) = \int_{\text{all } \mathbf{r}} f(\mathbf{r})g(\mathbf{u} - \mathbf{r}) \, d\mathbf{r}$$

and for this reason we discussed the Patterson integral in Chapter 5, Section 5.12, and it is suggested that the reader review that section at this point.

The structure of equation (12.12) implies that given the function $\rho(\mathbf{r})$, for some value of $\mathbf{u}$ we form the function $\rho(\mathbf{u} + \mathbf{r})$, which corresponds to shifting the shape of $\rho(\mathbf{r})$ by the distance $\mathbf{u}$. The variable $\mathbf{u}$ therefore specifies the relative displacement of two functions of identical shape. As the parameter $\mathbf{u}$ varies, the degree of overlap between the functions $\rho(\mathbf{r})$ and $\rho(\mathbf{u} + \mathbf{r})$ varies, and the value of the integral in equation (12.12) changes, thereby generating the function $P(\mathbf{u})$.

The general properties of the Patterson function $P(\mathbf{u})$ can be seen from consideration of the one-dimensional example which was given in Section 5.12, and which is reproduced in Fig. 12.2.

Figure 12.2(a) shows an arbitrary one-dimensional function $f(x)$ which consists of three δ functions of different weights asymmetrically positioned about the origin. The one-dimensional Patterson function $P(u)$ derived from this particular form of $f(x)$ is shown in Fig. 12.2(b), and the steps by which this was derived are explained in Section 5.12. Comparison of Figs 12.2(a) and (b) shows a one-dimensional example of a number of general relationships between any function $f(\mathbf{r})$ and its corresponding Patterson function $P(\mathbf{u})$:

(a) Even if $f(\mathbf{r})$ is non-centrosymmetric, the Patterson function $P(\mathbf{u})$ is always centrosymmetric.

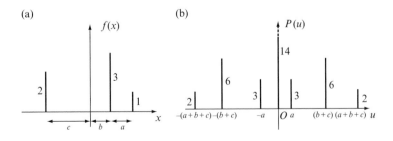

Fig. 12.2 *A one-dimensional function (a) and its Patterson function (b).*

(b) If $f(\mathbf{r})$ has a total of N peaks, then the Patterson function $P(\mathbf{u})$ has $N(N-1)$ peaks at points other than the origin, but in certain cases, some of these may overlap and not all peaks will be separately resolved.

(c) The peaks in the Patterson function $P(\mathbf{u})$ occur at values of $\mathbf{u}$ corresponding to all vectors between all pairs of peaks in the original function $f(\mathbf{r})$.

(d) The strength of each Patterson peak is the product of the strengths of the two peaks in the original function $f(\mathbf{r})$ which are separated by the corresponding value of $\mathbf{u}$.

(e) When the magnitude of $\mathbf{u}$ is zero, the Patterson function is derived from the square of the electron density function, and the value of the Patterson function is the sum of the squares of the total numbers of electrons associated with each atom.

The first point, concerning the fact that the Patterson function $P(\mathbf{u})$ is necessarily centrosymmetric, is interesting from the point of view of the theory of Fourier analysis. The definition of the Patterson function in equation (12.2) shows that the Patterson function is the Fourier synthesis of the relative intensities $|F_{hkl}|^2$, and so, conversely, the set of relative intensities $|F_{hkl}|^2$ is the Fourier transform of the Patterson function. Since intensity is an observable quantity, the relative intensities $|F_{hkl}|^2$ are real, implying that the Fourier transform of the Patterson function is always a real quantity. Reference to Section 5.9 will show that we proved that the Fourier transform of any centrosymmetric function is necessarily real, and the converse is also the case. Fourier theory therefore predicts that since the Fourier transform of the Patterson function is always real, the Patterson function itself is centrosymmetric. This is reasonable, since one way of interpreting the Patterson function is as the convolution of the electron density function with its centrosymmetric image, which convolution is itself centrosymmetric.

12.7 Patterson maps

We now extend the concepts behind the example depicted in Fig. 12.2 to a more realistic case. We know that electrons are not located at points in space, but are associated with atoms according to the appropriate atomic electron density functions.

For a one-dimensional unit cell containing three atoms, a possible form of the electron density function $\rho(x)$ is that shown in Fig. 12.3(a). Figure 12.3(a) is very similar to Fig. 12.2(a), the only difference being that we have smeared out the electron density about the atomic positions, corresponding to nuclei of atomic numbers Z_1, Z_2 and Z_3. By analogy with Fig. 12.2(b), the Patterson function of Fig. 12.3(a) is given in Fig. 12.3(b). The peaks of the Patterson function are in the same positions as previously, but instead of the peaks being mathematical δ functions, they are more spread out and somewhat lower. A significant point to note that the width of any peak in the Patterson function is equal to the sum of the widths of the two peaks in the electron density function which overlap to give the appropriate Patterson peak. The reason for this is evident from Fig. 12.4.

As can be seen, the Patterson function is first non-zero when the two electron density function peaks just overlap. This event will occur when the relative separation

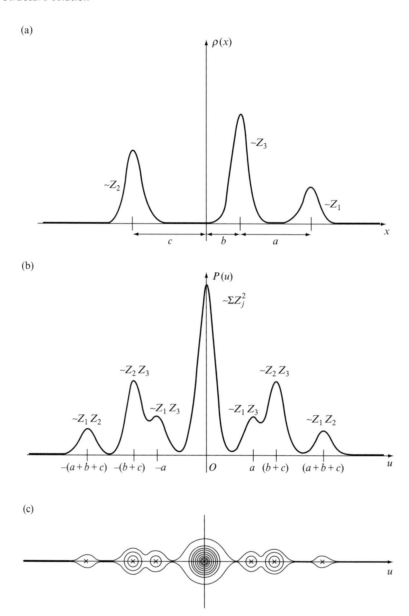

(a)

(b)

(c)

Fig. 12.3 *The Patterson function of a one-dimensional unit cell.* In (a) is the electron density function of a one-dimensional unit cell containing three atoms, whose positions are $x = -c$, b and $(a + b)$. The corresponding Patterson function is shown in (b), and the equivalent Patterson map in (c). Note that the strength of each peak is approximately equal to the product of the atomic numbers of the two atoms related by the appropriate spacing u. Only if the 'atoms' are δ functions is the peak exactly equal to the product of the appropriate atomic numbers.

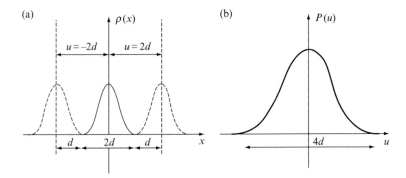

Fig. 12.4 *The width of Patterson peaks.* Shown in (a) is the overlap with itself of an electron density peak of width $2d$ at the origin. The Patterson function is formed by the overlap as the peak slides past itself, and overlap occurs over values of u from $-2d$ to $+2d$. The width of the Patterson peak is therefore $4d$, the sum of the widths of the two electron density peaks which were involved in the overlap.

of the centres of the two electron density functions is equal to the sum of their individual half-widths.

Inspection of Fig. 12.3(b) shows that the characteristics of the Patterson function are present, namely that there are $3 \times 2 = 6$ peaks not at the origin, and each peak corresponds to the interatomic distance between a pair of atoms. The height of each Patterson peak is once again determined by the product of the electron densities associated with the two atoms separated by the appropriate value of u, and the peak at the origin has a strength equal to the sum of the squares of the atomic numbers of all the atoms in the unit cell.

The Patterson function in Fig. 12.3(b) may be represented in terms of a form of contour map, as shown in Fig. 12.3(c). Whereas the 'view' of Fig. 12.3(b) is from the 'side of the hills', that of Fig. 12.3(c) is from 'above' as we 'look down on the contours'. Study of Figs 12.3(b) and (c) will readily show the relationship between these two different representations of the same function. The diagram in Fig. 12.3(c) is analogous to an electron density map, except that it is derived from a Patterson function, and such a diagram is called a *Patterson map* or *vector map*.

As a further example, we may consider the Patterson map of the two-dimensional unit cell shown in Fig. 12.5(a), which, for simplicity, contains only five atoms that are related by a mirror plane. To derive the Patterson map, we first draw all interatomic vectors as shown. We then plot these vectors with origins at the origin of the Patterson map, giving the distribution of Fig. 12.5(b). Associated with, and at the head of, these vectors is a peak with a strength determined by the product of the two electron density peaks separated by the appropriate vector. Peaks of the Patterson function occur centrosymmetrically about the origin, and the width of each peak is equal to the sum of the widths of the two relevant electron density peaks. There is a very strong peak at the origin, which corresponds to the value of zero for the parameter u, and so

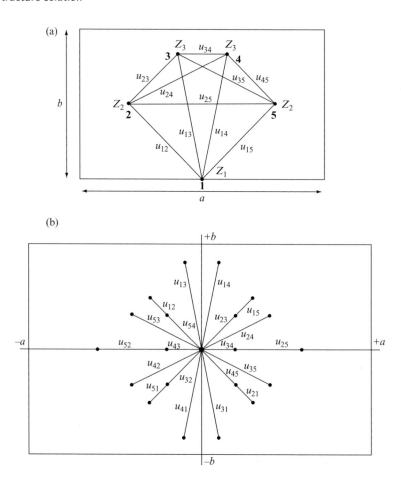

Fig. 12.5 *The Patterson map of a two-dimensional unit cell.* In (a) is drawn a two-dimensional unit cell in which the positions of five atoms are marked; (b) shows the positions of the peaks of the corresponding Patterson map. Patterson peaks occur at positions relative to the origin corresponding to all interatomic vectors in the real unit cell.

the five zero vectors from each atom to itself combine to give a peak of weight $\sum_j Z_j^2$, where Z_j is the atomic number of the jth atom.

Returning to equation (12.2),

$$P(u, v, w) = \frac{1}{V} \sum_h \sum_k \sum_l |F_{hkl}|^2 e^{-2\pi i(hu + kv + lw)} \tag{12.2}$$

we may use relative-intensity $|F_{hkl}|^2$ data to perform this Fourier synthesis in a manner exactly identical to that in which we calculated the electron density function $\rho(x, y, z)$ using the values of the structure factors F_{hkl} (see Section 9.5). The resulting Fourier synthesis may be plotted out as a three-dimensional Patterson map, which we are now in a position to interpret.

Firstly, all Patterson maps are centrosymmetric. Secondly, there will be $N(N-1)$ peaks at points other than the origin for any real unit cell containing N atoms. Since each Patterson peak is wider than the electron density peaks, it is unlikely that these $N(N-1)$ Patterson peaks will be separately resolved, for often many peaks coincide or overlap. This is one of the difficulties to be borne in mind when attempting to interpret a Patterson map. Thirdly, we know that each peak occurs at a value of **u** corresponding to the vector between an atomic pair. And, lastly, the strength of each Patterson peak is determined by the product of the atomic numbers of the two atoms separated by the appropriate value of **u**.

12.8   Patterson map symmetry

Before we discuss the primary use of the Patterson map, we shall review its symmetry properties. The structure of the Patterson map is derived from a set of vectors, corresponding to all interatomic vectors in the real unit cell. All these vectors are drawn radially from the Patterson function origin, and this implies that the symmetry of the Patterson map is rather restricted, since it cannot contain, for instance, glide planes or screw axes. It has been shown that of the 230 possible space groups, all Patterson maps conform to only 24 space groups. Despite this, it was pointed out by Harker that the symmetry of the Patterson map often arises in a highly systematic manner, which of itself can give useful information. Specifically, he showed that certain symmetry elements within the real crystal are associated with planes or lines within the Patterson map where there will be local concentrations of peaks. The reason for this may be appreciated from Fig. 12.5, in which we found the Patterson function of a two-dimensional unit cell which contained a mirror plane. Since the Patterson map contains peaks corresponding to all interatomic vectors, those atoms which are in symmetry-related positions will have symmetrically related interatomic vectors. With reference to Fig. 12.4, the presence of the mirror plane implies that there are several parallel interatomic vectors, each of which gives rise to a Patterson peak along the horizontal axis of the Patterson function. This axis therefore contains a cluster of peaks which implies the existence of a mirror plane in the real unit cell. Lines or planes within the Patterson map which contain local concentrations of peaks are called *Harker lines* or *Harker sections*, and identification of these is often useful in giving us further symmetry information in addition to that which we derive from the symmetry of the diffraction pattern and the recognition of systematic absences.

If we consider any space group, then we can determine what are its Harker lines or sections by consideration of the general equivalent positions (GEPs), which were discussed in Chapter 3 (Section 3.8). If we start with the very common space group $P2_1$, then we will recall that the 2_1 symbol signifies a screw axis in the y direction. This space group has GEPs (x, y, z) and $(-x, 1/2 + y, -z)$. Hence, for any atom at position (x, y, z), there will be another identical atom at position $(-x, 1/2 + y, -z)$. Hence there will be a vector running from position (x, y, z) to position $(-x, 1/2 + y, -z)$, and this vector $\mathbf{u}$ can be defined as

$$\begin{aligned}
\mathbf{u} &= (-x, 1/2 + y, -z) - (x, y, z) \\
&= (-x - (x), 1/2 + y - (y), -z - (z)) \\
&= (-2x, 1/2, -2z)
\end{aligned}$$

We know that the Patterson function has inversion symmetry, and so this vector is equivalent to $(2x, -1/2, 2z)$ and we can apply a unit cell translation on y to give us $(2x, 1/2, 2z)$. Therefore the $v = 1/2$ layer of the Patterson map is a Harker section, since it contains all of the vectors between atoms related by the 2_1 screw axis of the crystal. Let us now consider the higher-symmetry space group $P2_12_12_1$, which again is very common with biological macromolecules. This has GEPs (x, y, z), $(1/2 - x, -y, 1/2 + z)$, $(1/2 + x, 1/2 - y, -z)$ and $(-x, 1/2 + y, 1/2 - z)$. The vector $\mathbf{u}$ running from the first to the second GEP is

$$\begin{aligned}
\mathbf{u} &= (1/2 - x, -y, 1/2 + z) - (x, y, z) \\
&= (1/2 - x - (x), -y - (y), 1/2 + z - (z)) \\
&= (1/2 - 2x, -2y, 1/2)
\end{aligned}$$

Therefore the $w = 1/2$ layer of the Patterson function for this space group is a Harker section, since it contains peaks due to vectors between symmetry-related atoms. Likewise, by consideration of the vectors between the other GEPs, we can prove that the following are also Harker vectors for this space group:

$$(1/2, 1/2 - 2y, -2z) \quad \text{and} \quad (-2x, 1/2, 1/2 - 2z)$$

Thus we have established that the following are Harker vectors for space group $P2_12_12_1$:

$$(1/2 - 2x, -2y, 1/2), \quad (1/2, 1/2 - 2y, -2z) \quad \text{and} \quad (-2x, 1/2, 1/2 - 2z)$$

If we return to our consideration of the vector between the first and second GEP and think of the vector $-\mathbf{u}$ extending in the opposite direction, i.e. from the second to the first GEP, we have

$$\begin{aligned}
-\mathbf{u} &= (x, y, z) - (1/2 - x, -y, 1/2 + z) \\
&= (2x - 1/2, 2y, -1/2)
\end{aligned}$$

By applying unit cell translations along $\mathbf{a}$ and $\mathbf{c}$, we obtain $(1/2 + 2x, 2y, 1/2)$.

The same considerations for the other GEPs lead us to conclude that the following are also Harker vectors for this space group:

$$(1/2, 1/2 + 2y, 2z) \quad \text{and} \quad (2x, 1/2, 1/2 + 2z)$$

Likewise, by consideration of all possible vectors between the GEPs, it can be shown that a more generic description of the Harker vectors for this space group is

$$(1/2 \pm 2x, \pm 2y, 1/2), \quad (1/2, 1/2 \pm 2y, \pm 2z) \quad \text{and} \quad (\pm 2x, 1/2, 1/2 \pm 2z)$$

The fact that we can change the signs of the terms freely arises from the fact that the Patterson function for this space group contains three orthogonal mirror planes which intersect at the origin. This arises from the fact that the Patterson function is, by definition, centrosymmetric and, as we showed in Chapter 3 (Section 3.7), addition of an inversion centre to a system with three orthogonal twofold axes generates *mmm* symmetry.

12.9   The use of Patterson maps

Since the Patterson map does not represent the electron density function $\rho(x, y, z)$, but rather the convolution of this function with its centrosymmetric image, Patterson maps are often not easily interpreted directly. Firstly, whereas an electron density map of a unit cell containing N atoms will have N peaks, the Patterson map has $N(N-1)$ peaks other than that at the origin. Furthermore, each Patterson peak is roughly twice as wide as an electron density peak. When we consider that macromolecules of biological importance have upwards of 1000 non-hydrogen atoms, an electron density map would have about 10^3 major peaks, whereas the Patterson map would contain some 10^6 peaks, many of which would overlap. Thus the degree to which a Patterson map may be interpreted to give direct information about molecular structure depends very critically on the size of the molecule under study. For a small organic molecule containing perhaps ten or fifteen atoms, the Patterson map may be useful in determining the structure, especially if the atomic constitution of the compound is known. If the investigator has some chemical knowledge in advance, then this may guide the interpretation of the Patterson map, and one may eventually be able to identify a sufficient number of interatomic vectors to enable a reasonable model to be built. But this requires that many of the peaks of the Patterson map be resolved, and also calls for considerable insight on the part of the investigator. As the size of the molecule under study becomes larger, the Patterson map becomes progressively less distinct and harder to interpret directly. Thus, for biopolymers, it is quite impossible to derive detailed structural information directly from a Patterson map.

But this is not to say that the Patterson function is irrelevant to the study of biological macromolecules. On the contrary, its use forms an integral part of the overall scheme of a structure investigation, for one particular property of the Patterson map is of extreme importance. Earlier, we saw that the strength of each Patterson peak is determined by the product of the atomic numbers of the two atoms separated by the appropriate vector **u**. In general, the vast majority of biologically significant molecules contain mainly the atoms H, C, N, O, P and S, of atomic numbers 1, 6, 7, 8, 15 and 16, respectively. Most of the Patterson peaks will therefore have strengths of approximately the products of these numbers, from 1 for H–H atomic pairs up to 256 for S–S pairs. But suppose we add some heavy element, such as iodine ($Z = 53$) or mercury ($Z = 80$), and this binds to the macromolecule. With biologically important compounds, it is very unlikely that there will be more than a small number of atoms of the heavy element bound to each molecule. If, for example, we know that there are two mercury atoms within the molecule, then we should expect one Patterson peak of strength about $80 \times 80 = 6400$ occurring at a position corresponding to the inter-mercury vector distance within the molecule. There will be other Patterson peaks, of strengths of order $80 \times 10 = 800$ corresponding to Hg–O and Hg–N interatomic vector distances, and so on, but there is no question that by far the strongest Patterson peak will be due to the pair of mercury atoms themselves.

Examination of the Patterson map as calculated by Fourier synthesis using $|F_{hkl}|^2$ data will show a bewildering array of peaks, but some peaks will be manifestly more strong than all the others. These will be the peaks associated with the vectors between the pairs of heavy atoms, and identification of these strong peaks will allow us to establish the vectors between the heavy atoms and, ultimately, the heavy-atom substructure. As we shall see later in this book, the solution of the phase problem requires that the position of at least one atom be known, and it is by using the Patterson map (or, as we shall see, a closely related function called a *difference Patterson map*) that such a site may be identified.

Even if the asymmetric unit of the crystal contains only a single heavy atom, which we shall call R, there will be relatively strong Patterson peaks corresponding to the R–R vectors between adjacent protein molecules in the unit cell. These vectors therefore arise from heavy atoms which are related to each other by the symmetry of the crystal. Above, we considered the Patterson function for the space group $P2_12_12_1$ and derived equations for the vectors between symmetry-related atoms: the Harker vectors. It is by consideration of these that we may determine the heavy-atom coordinates from peaks in the Harker sections of the Patterson function. Normally, to interpret a Patterson map one would calculate just the unique part of it, which, for this space group, corresponds to $0 \leq u \leq 1/2$, $0 \leq v \leq 1/2$, $0 \leq w \leq 1/2$. To interpret this region of the Patterson map, we may consider the following Harker vectors: $(1/2 - 2x, 2y, 1/2)$, $(1/2, 1/2 - 2y, 2z)$ and $(2x, 1/2, 1/2 - 2z)$. If we see a strong peak in the $v = 1/2$ section, such as that shown in Fig. 12.6, we may measure its u and w coordinates. By using the equation for the corresponding Harker vector,

$$u = 2x$$

$$w = 1/2 - 2z$$

we can determine the x and z coordinates of the heavy atom. If inspection of another Harker section reveals another strong peak, use of the corresponding Harker vector equation will allow us to determine the y coordinate of the heavy atom. The heavy site will then be defined completely in three-dimensional space.

It is in the identification of the positions of certain heavy atoms that the Patterson map is most useful. For very simple structures, it is possible to use analytic methods to derive an electron density function $\rho(\mathbf{r})$ compatible with a given Patterson map, but with most organic and macromolecular structures, the Patterson map is too complicated to allow this. However, we may select from the array of Patterson peaks those main peaks which allow us to locate certain atomic positions correctly, and with

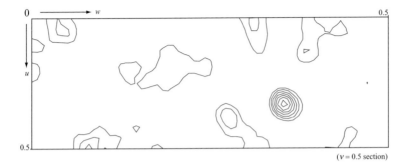

Fig. 12.6 *An experimental Patterson map.* Shown is a Harker section of the Patterson map of a protein with the heavy metal platinum bound to it. The positions of the main peaks in each of the Harker sections allow the heavy-atom coordinates to be determined, although in this case it is clear that only one heavy atom has bound.

this knowledge, we now have a weapon with which we may attack the phase problem. In addition, for a molecule which has a similar structure to another molecule that has already been solved by X-ray diffraction, we may use the known structure as a model with which to interpret the Patterson function of the unknown structure. This method will be covered in detail in the next chapter.

Summary

Every structure factor F_{hkl} has a magnitude $|F_{hkl}|$ and a phase ϕ_{hkl}, and so can be represented by

$$F_{hkl} = |F_{hkl}|\, e^{i\phi_{hkl}} \qquad (12.1)$$

Hence, to calculate the electron density distribution of a molecule by the equation below,

$$\rho(x, y, z) = \frac{1}{V} \sum_h \sum_k \sum_l F_{hkl}\, e^{-2\pi i(hx+ky+lz)} \qquad (9.15)$$

requires that we know both the amplitude and the phase of each structure factor. In contrast, the *Patterson function* $P(u, v, w)$ is defined by the following equation,

$$P(u, v, w) = \frac{1}{V} \sum_h \sum_k \sum_l |F_{hkl}|^2\, e^{-2\pi i(hu+kv+lw)} \qquad (12.12)$$

and may be calculated as the Fourier synthesis of the relative intensities $|F_{hkl}|^2$ alone, requiring no knowledge of the phases.

An alternative equation, below, expresses the Patterson function in terms of the convolution of the electron density function $\rho(\mathbf{r})$ and its centrosymmetric image $\rho(-\mathbf{r})$:

$$P(\mathbf{u}) = V \int_{\text{unit cell}} \rho(\mathbf{r})\rho(\mathbf{u}+\mathbf{r})\, d\mathbf{r} \qquad (12.12)$$

By performing the Fourier synthesis of equation (12.2) directly from experimental data, a three-dimensional Patterson map may be drawn. Theoretical investigation of equation (12.12) predicts that the Patterson map will have the following features.

(a) The Patterson map is centrosymmetric.
(b) The Patterson map contains $N(N-1)$ peaks at points other than the origin, where N is the number of peaks in the electron density function. In general, many of the Patterson peaks coincide or overlap.
(c) Patterson peaks occur at values of $\mathbf{u}$ corresponding to all interatomic vectors within the real unit cell.
(d) The strength of each Patterson peak is determined by the product of the atomic numbers of the two atoms separated by the appropriate value of $\mathbf{u}$.

The symmetry of the Patterson map corresponds to one of 24 of the 230 space groups. Symmetry information is therefore limited, but recognition of Harker lines or planes (these are local concentrations of Patterson peaks) leads to useful information.

The use of the Patterson map, in so far as deriving direct structural information is concerned, depends critically on the size of the structure under study. In principle, small organic structures may be solved by careful study of the Patterson map, especially if some information concerning the chemistry of the molecule is known. With experience, it is possible to identify the many peaks in the Patterson map as a set of interatomic vectors, and so a model structure may be proposed. But as the size of the molecule under study increases, the Patterson map becomes very much more complicated and indistinct, and certainly as regards biopolymers, it is quite impossible to deduce detailed structural information from a Patterson map alone, unless we already know the structure of a similar molecule.

The great use of the Patterson map to the crystallographer investigating large molecules is that it provides a means of unambiguously locating the positions of certain heavy atoms within the unit cell. This may be done by finding the strongest Patterson peaks, which are necessarily due to the overlap of the electron density functions of the atoms of highest atomic number. From the Patterson map, the values of $\mathbf{u}$ corresponding to the vectors between the heavy atoms may be estimated, allowing identification of the heavy-atom sites within the unit cell. This is vital for the success of several methods of solving the phase problem.

Bibliography

Buerger, M. J. *Vector Space*. Wiley, New York, 1959. Full discussion of the information to be obtained from Patterson techniques.

13
Molecular replacement

13.1 Solving the phase problem when the structure of a related protein is known

In this chapter, we will look at methods which can be used to solve the phase problem when the molecule being analysed is similar to another molecule which has already had its structure solved. These methods are referred to as 'molecular replacement'. Structure analysis by this method involves determining the orientation and position of the molecule in the unit cell using the previously solved structure as a 'search model'. The unknown structure is referred to as the 'target molecule'. The search and target molecules must have reasonable amino acid sequence identity (i.e. > 25%) for there to be a good chance of success (Dodson, 2008).

The original concept underlying the molecular replacement method was that it would provide a way of exploiting non-crystallographic symmetry in protein structures for *ab initio* solution of the phase problem. However, this has only been feasible with crystals that possess extraordinarily high non-crystallographic symmetry, such as viruses. The majority of structures possessing new folds have been determined by isomorphous replacement or MAD, in which the phases are determined experimentally. Once the tertiary structure of a member of any unique protein family has been determined, the other members can usually be analysed by molecular replacement. The success of the method is demonstrated by the fact that the vast majority of protein structures have been solved by molecular replacement.

Generally, there are two steps in molecular replacement, and these are known as the rotation function and the translation function. If the molecular replacement method is successful, a preliminary model of the target structure will be obtained by correctly orienting and positioning the search molecule in the unit cell of the target molecule. Phases can then be calculated from the model and used with the structure factor amplitudes measured from the target crystal to compute an electron density map. Finally, the search model has to be rebuilt, usually by computer graphics, so that it fits the electron density map as well as possible and its amino acid sequence can be converted to that of the target molecule.

Molecular replacement will not work as a means of solving the phase problem if the search model and target structure are not sufficiently similar. However, even in the absence of a suitable search molecule, molecular replacement can be used to determine the direction and nature of any non-crystallographic symmetry elements that might be present in the crystal of the target molecule.

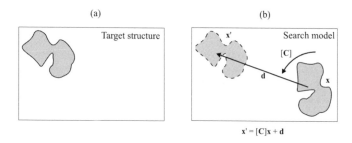

Fig. 13.1 *The fundamental concept of molecular replacement.* The unsolved, or target, structure is shown in (a) and we can make use of a similar structure as a search model, shown with a solid line in (b). The problem is to determine the rotation matrix [C] and translation vector **d** which superimpose the search model onto the target structure. The coordinate set of the search model is represented by **x**, and application of the rotation matrix and shift vector gives a new coordinate set **x′** which is superimposed on that of the target structure.

We will now consider the basic calculations involved in molecular replacement. The coordinates of the atoms in any molecule can be represented relative to orthogonal axes by a matrix **x**, where

$$\mathbf{x} = \begin{pmatrix} x_1 & x_2 & x_3 & \cdots & x_j & \cdots \\ y_1 & y_2 & y_3 & \cdots & y_j & \cdots \\ z_1 & z_2 & z_3 & \cdots & z_j & \cdots \end{pmatrix}$$

If we represent the coordinates of the search model by the matrix **x**, then solving the target structure (which we will refer to as the matrix **x′**) involves determining how much we have to rotate and translate the search molecule so that it is superimposed on the target molecule. This involves determining a rotation matrix [C] and a translation matrix **d**, which are applied to the coordinates of the search model **x** (Fig. 13.1) to give us a preliminary model of the target structure **x′**. Hence,

$$\mathbf{x}' = [\mathbf{C}]\mathbf{x} + \mathbf{d} \tag{13.1}$$

where

$$\mathbf{x}' = \begin{pmatrix} x_1' & x_2' & x_3' & \cdots & x_j' & \cdots \\ y_1' & y_2' & y_3' & \cdots & y_j' & \cdots \\ z_1' & z_2' & z_3' & \cdots & z_j' & \cdots \end{pmatrix}$$

$$[\mathbf{C}] = \begin{pmatrix} c_{11} & c_{12} & c_{13} \\ c_{21} & c_{22} & c_{23} \\ c_{31} & c_{32} & c_{33} \end{pmatrix}$$

and

$$\mathbf{d} = \begin{pmatrix} d_x \\ d_y \\ d_z \end{pmatrix}$$

where d_x, d_y and d_z represent the components of the translation vector along the x, y and z axes, and the c_{ij} are the individual terms of the rotation matrix. If we consider a typical atom with coordinates (x_j, y_j, z_j) in the search model, then equation (13.1) operates on its coordinates such that it is rotated about the origin and translated to a new position (x_j', y_j', z_j') given by the equations

$$x_j' = c_{11}x_j + c_{12}y_j + c_{13}z_j + d_x$$

$$y_j' = c_{21}x_j + c_{22}y_j + c_{23}z_j + d_y$$

$$z_j' = c_{31}x_j + c_{32}y_j + c_{33}z_j + d_z$$

Hence, in molecular replacement, the problem of determining $[\mathbf{C}]$ and $\mathbf{d}$ is six-dimensional because we are dealing with three-dimensional space, i.e. it involves rotation of the search model about three axes and translational components along each of these axes. A brute force approach to this problem would be to place the model in every possible orientation at each possible position in the cell, calculate structure factors at each step and assess their agreement with the diffraction pattern. However, the calculations involved, although they are feasible with modern computers, would be very time-consuming. Fortunately, the process can be simplified and speeded up greatly by separating the search into two stages, namely a rotation search and a translation search—a computational divide-and-conquer strategy.

13.2   The rotation function

The aim of the rotation function is to allow the orientation of the search molecule in the unit cell of the target molecule to be determined. We must remember that at this stage we have no knowledge of the phases of the target structure, and so we cannot calculate its electron density function. What we need is a method of determining which orientation of the search model is most consistent with the target data. In the previous chapter, we described the Patterson function and how this can be calculated from diffraction measurements in the absence of phases. Thus we can calculate the Patterson function $P(u, v, w)$ of the target structure from the measured structure factor amplitudes by use of the equation

$$P(u, v, w) = \frac{1}{V} \sum_h \sum_k \sum_l |F_{hkl}|^2 e^{-2\pi i(hu+kv+lw)} \tag{12.2}$$

Since the Patterson function will display all interatomic vectors that are present in the target structure, some of these vectors will arise from pairs of atoms within the same molecule, and these are known as intramolecular vectors. Other vectors will arise from pairs of atoms in different molecules within the crystal, and these are known as intermolecular vectors. We saw in the previous chapter that a subset of these intermolecular vectors, known as Harker vectors, are concentrated in lines or planes.

One feature of protein crystals is that they have large solvent channels between the constituent protein molecules and that the proteins themselves tend to be globular and compact. These features mean that, in general, the intermolecular vectors will be

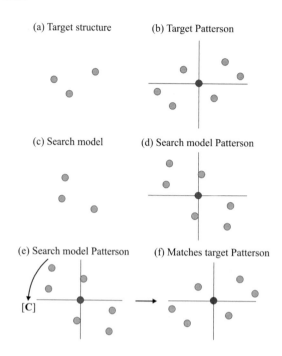

Fig. 13.2 *The rotation function for a simple three-atom structure.* (a) shows the target structure and (b) shows its Patterson function, which is formed by placing each atom in turn at the origin and depositing copies of the other atoms at each stage. The strong peak on the origin stems from the overlap of atomic self-vectors. (c) shows the search model and (d) shows the corresponding Patterson function. The rotation function involves applying a rotation matrix [**C**] to the search model Patterson function, as shown in (e), to put it in an orientation that matches the target Patterson function, as shown in (f).

longer than the intramolecular vectors. This is true for molecules of all sizes, but the high solvent content of macromolecular crystals means that this effect is particularly prominent. Hence, if we choose peaks in the target Patterson function which are reasonably close to the origin, we can be confident that these are predominantly due to intramolecular vectors. For intramolecular vectors, the translation vector **d** in equation (13.1) is irrelevant, since all intramolecular vectors are shifted to the origin in the Patterson function (Fig. 13.2). This has the effect that the peaks of the Patterson function which are close to the origin are more sensitive to the orientation of the molecule than to its packing arrangement in the crystal.

The above arguments are the crux of the molecular replacement method and allow the orientation of the search model to be defined prior to determination of its position in the cell. To determine the orientation of the search model requires that we calculate a *rotation function*. In this calculation, we start off with the search model in a given orientation and calculate the intramolecular vectors of the Patterson function. These are compared with the Patterson function of the unknown or target molecule calculated from its diffraction pattern. The Patterson function of the search model is then rotated by a small amount (e.g. 5 or 10° about one axis) and compared again with the Patterson function of the target molecule. This process is repeated by rotating the search model Patterson function in three dimensions and assessing its agreement with the target Patterson function at all possible unique orientations. It is easy to visualise that rotating a molecule will rotate its Patterson function by the same amount.

In cases where there is no search model, we may still use the rotation function to rotate a copy of the target Patterson function and compare it with another fixed copy of the same Patterson map. This is known as the self-rotation function and can be used to determine the direction and nature of non-crystallographic symmetry elements within the asymmetric unit. Proteins often crystallise with more than one molecule in the asymmetric unit. The multiple copies of the molecule in the asymmetric unit are related to each other by what is referred to as non-crystallographic symmetry (NCS). Before a molecular replacement analysis is attempted, it is essential to assess how many copies of the target molecule are present in the asymmetric unit and thus how many copies of the search model need to be correctly oriented and positioned. The commonly used method for estimating the number of molecules per asymmetric unit is described in Chapter 11, and the associated calculations are performed automatically by many of the contemporary molecular replacement programs.

The agreement between the two Patterson functions at any orientation of the search model can be assessed by the following product function, as proposed by Rossmann and Blow (1962):

$$R = \int P_T(\mathbf{u}) \Re P_S(\mathbf{u}) \, d\mathbf{u} \tag{13.2}$$

In this equation, $P_T(\mathbf{u})$ is the Patterson function of the target structure as a function of position, given by the matrix $\mathbf{u}$, and $\Re P_S(\mathbf{u})$ is the Patterson function of the search molecule which has been rotated by the matrix $[\mathbf{C}]$ which we discussed above.

For every point $\mathbf{u}$ in the target Patterson function, we need to calculate the value of the rotated search model Patterson function $\Re P_S(\mathbf{u})$, multiply the values of the two Patterson functions together and sum over all points. In any orientation where the two Patterson functions match well, the value of this sum will be high and, conversely, in orientations where the Patterson functions are not correlated, R will be low. For any orientation of the search model, we can calculate the value of $\Re P_S(\mathbf{u})$ from the original unrotated model Patterson function $P_S(\mathbf{u}')$, where the point $\mathbf{u}'$ in the unrotated model Patterson function corresponds to the point $\mathbf{u}$ in the rotated one. The rotation function thus becomes

$$R = \int P_T(\mathbf{u}) P_S(\mathbf{u}') \, d\mathbf{u} \tag{13.3}$$

The value of $\mathbf{u}'$ in the unrotated Patterson function corresponding to any value of $\mathbf{u}$ in the rotated one is given by

$$\mathbf{u} = [\mathbf{C}] \, \mathbf{u}'$$

$$\therefore \mathbf{u}' = \left[\tilde{\mathbf{C}}\right] \mathbf{u} \tag{13.4}$$

where $\left[\tilde{\mathbf{C}}\right]$ is the transpose of $[\mathbf{C}]$, which corresponds to applying the rotation in the opposite direction. We will look at the properties of rotation matrices later in this section. The rotation function can now be expressed as

$$R = \int P_T(\mathbf{u})P_S\!\left(\left[\tilde{\mathbf{C}}\right]\mathbf{u}\right)d\mathbf{u} \tag{13.5}$$

We must remember that the Patterson function contains a large peak at the origin due to the zero-length vectors that run from each atom to itself. Another important feature of any Patterson function is that it will contain intermolecular vectors between neighbouring molecules in the crystal lattice. These features do not provide useful information for determining the orientation of the molecule, and therefore contribute to noise in the rotation function. Since it is desirable that the rotation function be dominated by intramolecular vectors, the integral is usually calculated for a shell in Patterson space with inner and outer radii which allow exclusion of the Patterson origin peak and the majority of the intermolecular vectors arising from neighbouring molecules. The outer radius is usually referred to as the *radius of integration* and typically has a value of around 20 to 30 Å.

The rotation function can be calculated numerically in real space (i.e. using the Patterson densities) but this is slow, although it can be speeded up greatly by selecting just the strongest peaks from one of the Patterson functions (Huber, 1965), and this approach is employed in the program CNS (Grosse-Kunstleve and Adams, 2001). Rossmann and Blow (1962) proposed a reciprocal-space method for calculating R which is faster than a full real-space search, and related methods have been developed (e.g. Lattman and Love, 1970). However, what is perhaps the most widely used method, known as the *fast rotation function*, was developed by Crowther (1972), who showed that the Patterson functions being compared can be approximated by spherical harmonics and Bessel functions, neither of which we will elaborate on here. The rotation function can then be calculated from these functions by a computationally efficient form of Fourier transformation known as the fast Fourier transform, or FFT, which confers great speed on the calculations even for very modest computers. This method has been developed further and is used in several highly successful automated molecular replacement programs, for example AMoRe (Navaza, 1994, 2001) and MOLREP (Vagin and Teplyakov, 1997).

More recent mathematical developments involving the application of *maximum-likelihood methods* allow a rotation function to be calculated without recourse to Patterson space (Read, 2001; Storoni *et al.*, 2004; McCoy *et al.*, 2005). In this method, the search model is rotated through all possible unique orientations, and the probability that the set of observed structure factors $|F_o|$ would have been obtained for each orientation of the model is calculated. At each trial orientation of the model, the contribution made by each symmetry-related molecule to each structure factor can be calculated. To calculate the resultant structure factor F_c, we have to determine the vector sum of the contributions from each of the symmetry-related molecules. However, this requires that we know the positions of the model and its symmetry mates in the target cell, which, of course, we do not know at this stage. In spite of this, we can, with a number of assumptions, calculate a likelihood function, and this is presented to the user as a 'log-likelihood gain', or LLG. The LLG is a measure of the probability of obtaining the observed structure factor amplitudes from the trial structure relative to the probability of obtaining the observed amplitudes from a set of randomly positioned atoms. Interestingly, the underlying assumptions involved in calculating the LLG are most accurate for high-symmetry space groups and in situations where the model is poor or incomplete, which are generally the most difficult cases for traditional molecular replacement methods. A first approximation of the likelihood parameter can be calculated by means of a fast Fourier transform, which is several orders of magnitude faster than the full calculation. The most promising orientations of the model are then rescored by calculating the likelihood parameter more precisely, and the most promising solutions are then input to the translation function.

13.3   Choice of variables in the rotation function

When calculating a rotation function, it is worth bearing in mind that success may depend on a number of variables, which the crystallographer often has to choose with some care. These variables are considered below.

Radius of integration

When one is calculating the rotation function as given in equation 13.2, the integral is calculated for a spherical shell of Patterson space. Whilst the inner radius can be chosen to exclude the origin peak of both of the Patterson functions, a more critical parameter is the outer radius, which is referred to as the radius of integration. Any peaks in the two Patterson functions which are further than the radius of integration from the origin are excluded from the integral. Hence, when one is calculating a rotation function by Patterson methods, it is necessary to choose the radius of integration carefully so as to minimise the number of intermolecular vectors between neighbouring molecules within the sphere of integration.

In order to calculate structure factors for the search model, it is common for programs to place it in a hypothetical $P1$ cell with orthogonal unit cell axes of dimensions which separate the molecules by at least the chosen radius of integration. The choice of the space group as $P1$ is to keep the search model Patterson function as simple as possible. The resulting Patterson function calculated from the F_c values will have only intramolecular peaks inside the sphere of integration. For spherical molecules, radii of integration of between 40% and 80% of the molecular diameter are commonly used.

With elongated molecules, a problem arises because the Patterson density for the search molecule may overlap with an adjacent origin peak of the target Patterson function at certain values of [**C**], giving false peaks. This problem is one faced by nucleic acid crystallographers owing to the elongated and repetitive structure of DNA. The problem may be partly solved by using a smaller radius of integration or by treating the target molecule data so as to remove the origin peak (P_{000}). This can be done by subtracting the average, or mean, value of the intensity (i.e. $\langle I_{hkl} \rangle$ or $\langle |F_{hkl}|^2 \rangle$) from the intensity of each reflection, giving I_{hkl}':

$$I_{hkl}' = I_{hkl} - \left\langle |F_{hkl}|^2 \right\rangle \tag{13.6}$$

At the origin of the Patterson function (i.e. where $u = 0$, $v = 0$ and $w = 0$), the exponential term in equation (12.2) becomes unity, meaning that P_{000} is simply the sum of the I_{hkl}, i.e.

$$P(0,0,0) = \frac{1}{V} \sum_h \sum_k \sum_l |F_{hkl}|^2 = \frac{1}{V} \sum_h \sum_k \sum_l I_{hkl}$$

If, instead, the Patterson function is calculated with the modified terms I_{hkl}', then P_{000}', which is the sum of the I_{hkl}', must be zero at the origin. This follows from simple consideration of the properties of any average. However, in practice, it is better to perform the calculation that we have just described for groups of spots that fall within given resolution shells. The need for this arises from the fact that $\langle I_{hkl} \rangle$ varies appreciably from low-resolution diffraction spots to high-resolution ones. Therefore, performing the calculation given in equation (13.6) for separate resolution shells rather than the whole dataset ensures more effective origin removal in the Patterson function of the target molecule.

Resolution limits

As well as the radius of integration, the resolution limits used in the calculation affect the rotation function significantly. Exclusion of the very low-resolution terms ($d > 20$ Å) is sensible, as these are sensitive only to very gross features of the structure. For average-size proteins, the exclusion of high-resolution terms (i.e. $d < 2.5$ Å) is recommended, since these reflections are more sensitive to the finer details of the structure, which are likely to be different in the search and target molecules owing to non-conserved amino acids and small conformational differences. However, for very small proteins or peptides, inclusion of high resolution data may be essential in both the rotation and the translation functions. For the self-rotation function, there is no need to restrict the high-resolution limit, since NCS-related molecules tend to have a high degree of similarity.

In Chapter 9, Section 9.7, we showed that atoms scatter more weakly as the Bragg angle θ increases. Hence the strongest $|F|$ terms occur at low resolution and would be expected to dominate the rotation function, which is proportional to $|F|^4$. This is a major problem since, as we have discussed above, the low-resolution terms are sensitive only to the gross shape of the molecule rather than to its tertiary structure. There are a number of ways of minimising this problem, which include the use of normalised structure factors, or E-values (see Chapter 11), instead of structure factor amplitudes ($|F|$). E-values, by definition, provide a more even contribution from all resolution ranges. An alternative way to down-weight the very low-resolution terms is to modify the structure factor amplitudes by applying an artificial negative temperature factor. In essence, this involves multiplying each $|F_{hkl}|$ by an exponential factor of the form

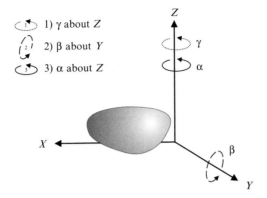

1) γ about Z

2) β about Y

3) α about Z

Fig. 13.3 *The Eulerian angular conventions for the rotation function.* The effect of the Eulerian rotations on a molecule (represented by the grey object) is shown with respect to fixed axes. The order in which the rotations are applied is also indicated.

$e^{-B(\sin^2 \theta / \lambda^2)}$ in which a negative number is assigned to B so as to compensate for the natural decrease in mean $|F_{hkl}|$ with increasing θ angle. Contemporary molecular replacement programs use various weighting schemes which can also correct for anisotropy in the target molecule's diffraction data (i.e. if the diffraction pattern is weak in certain directions) and can also take into account the similarity of the search and target sequences. All methods essentially aim to achieve the same effect, namely the down-weighting of the very low-resolution terms and the very high-resolution terms, which are likely to increase the noise level. The anisotropy correction to the E-values is particularly important in maximum-likelihood methods (McCoy, 2007).

Angular variables

The rotation function is usually calculated as a function of three angles, referred to as Eulerian angles, α, β and γ (Fig. 13.3). Note that these are distinct from the unit cell angles referred to in Chapter 3, Section 3.3. In order to calculate the rotation function, many programs essentially put the search model in an orthogonal $P1$ unit cell, i.e. a triclinic cell where the unit cell angles are 90°. For Crowther's fast rotation function program (Crowther, 1972), the Eulerian rotations can be defined as anticlockwise rotations of the molecule, as viewed looking towards the origin from the positive direction of each axis. These rotations are applied about fixed axes in the following order:

(1) γ about the Z axis;
(2) β about the Y axis;
(3) α about the Z axis.

In two dimensions, anticlockwise rotation of a point with rectangular coordinates (x, y) by an angle θ about an axis passing through the origin perpendicular to this plane moves the point to a new position (x', y') which is given by applying the rotation matrix below to the initial coordinates:

$$\begin{pmatrix} x' \\ y' \end{pmatrix} = \begin{pmatrix} \cos\theta & -\sin\theta \\ \sin\theta & \cos\theta \end{pmatrix} \begin{pmatrix} x \\ y \end{pmatrix} \tag{13.7}$$

The reader is invited to confirm that applying the transpose of the rotation matrix (i.e. swapping the rows and columns) is equivalent to rotating the point in the opposite direction (i.e. applying a rotation angle of $-\theta$), and this is also equivalent to applying the inverse of the rotation matrix.

The Eulerian angles as defined above, i.e. rotation of the molecule initially by γ about Z, then β about Y and finally α about Z, with respect to fixed axes, correspond to the following matrix operation:

$$\mathbf{R} = \mathbf{R}_z\left(\alpha\right)\mathbf{R}_y\left(\beta\right)\mathbf{R}_z\left(\gamma\right)$$

$$= \begin{pmatrix} \cos\alpha & -\sin\alpha & 0 \\ \sin\alpha & \cos\alpha & 0 \\ 0 & 0 & 1 \end{pmatrix} \begin{pmatrix} \cos\beta & 0 & \sin\beta \\ 0 & 1 & 0 \\ -\sin\beta & 0 & \cos\beta \end{pmatrix} \begin{pmatrix} \cos\gamma & -\sin\gamma & 0 \\ \sin\gamma & \cos\gamma & 0 \\ 0 & 0 & 1 \end{pmatrix} \tag{13.8}$$

In these matrices, we see the essential operation of equation (13.7) applied to the model as defined by the Eulerian convention. In each matrix, one diagonal term is unity, which represents the coordinate that is unaltered by that particular rotation. For completeness, the product of these three matrices is given below (Evans, 2001):

$$\mathbf{R} = \begin{pmatrix} \cos\alpha\cos\beta\cos\gamma - \sin\alpha\sin\gamma & -\cos\alpha\cos\beta\sin\gamma - \sin\alpha\cos\gamma & \cos\alpha\sin\beta \\ \sin\alpha\cos\beta\cos\gamma + \cos\alpha\sin\gamma & -\sin\alpha\cos\beta\sin\gamma + \cos\alpha\cos\gamma & \sin\alpha\sin\beta \\ -\sin\beta\cos\gamma & \sin\beta\sin\gamma & \cos\beta \end{pmatrix} \tag{13.9}$$

The main advantages of the Eulerian system are that it allows the rotation function to be calculated rapidly (by fast Fourier transform) and that the combined symmetry of the two Patterson functions appears directly in the rotation angles. The main disadvantage is that when β is close to 0 or 180°, the rotation function peaks become stretched along a diagonal. This arises because all rotations with the same $(\alpha + \gamma)$ are identical when β is zero, and all rotations with the same $(\alpha - \gamma)$ are equivalent when β is 180°. If this causes a problem, the search molecule may be rotated to a better β-angle or the distortion removed by calculating the map on a grid of $(\alpha + \gamma, \beta, \alpha - \gamma)$, which reduces the computation by $2/\pi$, although the symmetry is not so simple (Lattman, 1972).

The spherical polar coordinate system is also used (see Fig. 13.4, where the relevant angles ϕ, ψ and χ are defined). The spherical polar system is especially useful for detecting non-crystallographic symmetry; for example, the occurrence of a peak in the $\chi = 180°$ section would indicate the presence of a non-crystallographic twofold axis, and an example is shown in Fig. 13.5. A problem arises when non-crystallographic symmetry axes lie parallel to crystallographic axes, as the corresponding maxima can overlap. However, we will see later on that inspection of the Patterson map calculated from the target dataset can indicate if such NCS is present.

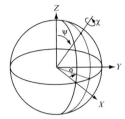

Fig. 13.4 *The spherical polar angular system.*

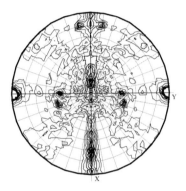

Fig. 13.5 *The spherical polar representation of a typical self-rotation function.* This shows the $\chi = 180°$ section, which has peaks in it due to the presence of twofold axes in the crystal. To interpret this diagram, one has to imagine looking down on a globe from above the North Pole. The concentric circles indicate lines of constant ψ, much like the lines of constant latitude on a globe, and the ϕ angle is shown in essentially the same way as longitude. The peaks indicate the directions of the twofold axes emanating from the centre of the sphere. In this case, the two strongest peaks, where the Y axis intersects the equator, are due to a crystallographic diad (the space group is $P2_1$). The other strong peaks, which lie away from the crystallographic axes, indicate the presence of non-crystallographic twofold axes. In fact, this structure possesses non-crystallographic 222 point group symmetry. This figure was output by MOLREP (Vagin and Teplyakov, 1997).

Whilst most molecular replacement programs sample Eulerian space on a regular grid so that each unique possible orientation of the model is sampled at least once, some employ stochastic methods which sample random orientations (and positions) of the model and assess their agreement with the experimental data (e.g. Kissinger *et al.*, 1999, 2001). The most promising solutions are selected and optimised in mathematical ways analogous to genetic evolution.

Search model

There is a tremendous range of bioinformatics tools available for searching sequence and structure databases for molecules homologous to the one being analysed. In general, it is sensible to use whatever search model has the greatest similarity to the target molecule in terms of both sequence identity and completeness. Search models having less than about 30% sequence identity are likely to cause problems, and so it may help to use more than one; and, in some cases, an ensemble of molecules superimposed upon one another can be beneficial. It can also help to align the search model sequence with that of the target molecule and truncate side chains of amino acids in the search model which are different in the target molecule. At least one program, MOLREP (Lebedev *et al.*, 2008), performs these steps automatically and also inflates the *B*-factors of exposed atoms, since surface amino acids are likely to adopt different conformations in the search and target molecules. When one is attempting to analyse molecules that possess several domains (compact folding units) that may adopt different orientations, the search model can be divided up into smaller units, and separate rotation and translation functions can be calculated for each fragment. A number of programs search the known database of structures for proteins or domains that are homologous to the target protein, and automatically prepare these for input to molecular replacement calculations. These programs also prepare multimeric forms of the search model if the known structures of homologues appear to consist of stable oligomeric forms. Such high-throughput approaches are clearly going to be increasingly important in the future with the ever-increasing pace of automation in the field.

13.4   Testing the rotation function

To assess the significance of a rotation function peak, the number of standard deviations above both the mean and the highest noise peak should be calculated. If the solution is noisy, the situation may be improved by varying the radius of integration or the resolution range, and the correct peak should consistently reappear at essentially the same angles. Finer angular sampling may help and should be used to refine a potential solution. Trying a different search model can also help in difficult cases, and each search model that is tested should be superimposed with the original one to ensure that the rotation functions are directly comparable.

The ultimate test of a rotation function solution is whether it gives a good peak in the translation function, and most molecular replacement programs will automatically test the highest rotation function peaks in the translation function. Since, occasionally, the correct rotation function solution will not be the highest peak, automatic testing of the top 10 to 20 rotation function solutions in the translation function is a sensible strategy.

13.5   Refining the rotation function solution

A powerful procedure that optimises potential rotation function solutions prior to the translation function, known as Patterson correlation (or PC) refinement, has been developed and is available in the program CNS (Brünger, 1990; Grosse-Kunstleve and Adams, 2001). It has often proved to be pivotal to the success of the translation function. In this method, the orientation of the search model that is obtained from the rotation function is altered successively in small steps, which are calculated to maximise the correlation coefficient of the target and search Patterson functions. The direction and size of each angular step are calculated by methods that are described later, in Chapter 15, on crystallographic refinement. The steps or increments are applied until convergence is achieved, i.e. there is no further improvement in the correlation coefficient. The search model can either be treated as one rigid body or be split into separate domains or secondary structure elements, and this approach can correct for misorientations of up to 15°. In practice, the highest rotation function peaks are selected and then automatically subjected to PC refinement with the aim of enhancing the correct solution, which should be clear from comparison of the Patterson correlation coefficients. The main advantage of PC refinement is with multi-subunit or multi-domain proteins, where the subunits or domains may have different relative orientations in different members of the family.

13.6 Symmetry of the rotation function

The symmetry of the rotation function arises from the symmetry of each Patterson function and from the inherent symmetry of the Eulerian system, i.e. it can be shown that the rotations (α, β, γ) and $(\alpha + \pi, -\beta, \gamma + \pi)$ are equivalent. The orders, or parities, of the axes parallel to z in the rotated and fixed Patterson functions (p_1 and p_0, respectively) determine the nature of this symmetry (Moss, 1985). Consideration of Section 13.3 confirms that an axis of order p_1 parallel to z in the search function gives a rotation function unit cell dimension of $c = 2\pi/p_1$, i.e. the unique range of γ is from 0 to $2\pi/p_1$. Similarly, an axis of order p_0 parallel to z in the target function gives a rotation function cell dimension of $a = 2\pi/p_0$. The dimensions of the rotation function unit cell are therefore $(2\pi/p_0, 2\pi, 2\pi/p_1)$, i.e. the higher the symmetry of both Patterson functions, the more compact the rotation function becomes, which may increase the noise level. Additional symmetry arises from axes of even order perpendicular to z (Moss, 1985). A special case arises with the self-rotation function, since the two Patterson functions are identical and the rotations (α, β, γ) and $(-\gamma, \beta, -\alpha)$ are equivalent. In the cross-rotation function, it is convenient to orient any axes of pseudo-symmetry in the search model parallel to z, since peaks arising from this extra symmetry will then appear in the same β-section.

13.7 The translation function

The rotation search, as described in the above few sections, defines the orientation of the search model in terms of the three Eulerian angles α, β and γ, from which a rotation matrix $[\mathbf{C}]$ can be calculated and applied to the coordinates of the search molecule. The shift vector $\mathbf{d}$, which is required to position the search molecule correctly relative to the symmetry elements of the target crystal, can be determined by one of a number of translation searches.

In some space groups, the translation problem is simplified or, in one case, eliminated altogether. In the space group $P1$, the position of the molecule is arbitrary, and moving the molecule from one position to another will change only the phases of the calculated structure factors and not their amplitudes. We can show this by reference to equation (9.21),

$$F_{hkl} = \sum_j f_j e^{2\pi i \left(hx_j + ky_j + lz_j \right)} \tag{9.21}$$

Let us imagine that we shift the molecule from its current position by increments Δx, Δy and Δz along the crystallographic axes. Substitution into the above equation gives

$$F_{hkl}' = \sum_j f_j e^{2\pi i \left(h(x_j + \Delta x) + k(y_j + \Delta y) + l(z_j + \Delta z) \right)}$$

and factoring out the exponential terms gives

$$F_{hkl}' = e^{2\pi i (h\Delta x + k\Delta y + l\Delta z)} \sum_j f_j e^{2\pi i \left(hx_j + ky_j + lz_j \right)}$$

This equation involves the multiplication of complex exponential terms, something which we covered in Chapter 2, Section 2.15. Since the first complex number in the above equation has unit amplitude and its exponential term is a constant for any given hkl, its effect, in an Argand diagram, is to simply rotate the structure factor through the angle $2\pi(h\,\Delta x + k\,\Delta y + l\,\Delta z)$. Hence changing the position of the molecule has altered the phases, but not the amplitudes of the structure factors. Thus, for triclinic crystals, the translation search is irrelevant; and, similarly, for other polar space groups the problem reduces to a lower-dimensional search. For example, in monoclinic cells only a two-dimensional search on x and z is needed, because the position of the molecule along the y axis is arbitrary. For space groups possessing more than one axis of symmetry, a full three-dimensional translation function is required.

As we mentioned in Chapter 9, Section 9.14, there are some space groups which are indistinguishable based on their systematic absences, for example $P4_12_12$ and $P4_32_12$. Such ambiguities can be resolved by calculating the translation function for each possible space group, and whichever one is correct will give the highest peak. This stems from the fact that the intermolecular vectors between the symmetry-related molecules will be different in the possible space groups and hence they are distinguishable by comparison of the search model Patterson function with the target Patterson function, which occurs when one is calculating a translation function.

13.8 Patterson-based translation methods

The traditional translation search measures the overlap of the intermolecular vectors in the target Patterson function with those calculated for the oriented search molecule as it ranges through the target cell. The method is based on the assumption that only one position of the search model will have intermolecular vectors compatible with the peaks in the target Patterson map. This concept is demonstrated for a very simple two-dimensional molecule in Fig. 13.6.

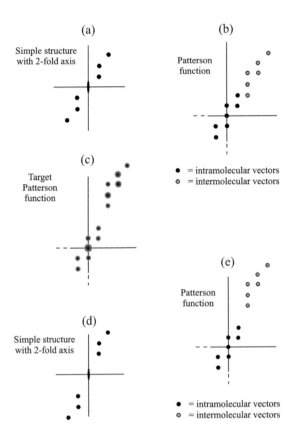

Fig. 13.6 *The translation function for a simple structure.* The translation function involves moving a correctly oriented search model on a grid and comparing the calculated intermolecular vectors with the observed Patterson function at each step. (a) A trial position for a small two-dimensional search model with a twofold axis perpendicular to the page and passing through the origin. The Patterson function for this structure is shown in (b), where the intramolecular and intermolecular vectors are indicated by solid and grey circles, respectively. This can be compared with the observed Patterson function shown in (c), and the reader is encouraged to do this by overlaying a tracing of (c) onto (b). It can be seen that whilst the intramolecular vectors match the observed Patterson function well (owing to the search model being in the correct orientation), the intermolecular vectors do not superimpose with the observed Patterson function very well. (d) shows the next trial position of the search model, and its corresponding Patterson vectors are shown in (e). In this position of the molecule, the calculated intermolecular vectors match the observed Patterson function shown in (c) very well.

Mathematically, the translation function can be defined as

$$T(\mathbf{t}) = \int P_c(\mathbf{u}, \mathbf{t}) P_o(\mathbf{u}) \, d\mathbf{u}$$

where $P_c(\mathbf{u}, \mathbf{t})$ is the search molecule Patterson function and $P_o(\mathbf{u})$ is the observed Patterson function of the target structure. Crowther and Blow (1967) expanded this in reciprocal space, giving

$$T(\mathbf{t}) = \sum_{\mathbf{h}} I_o(\mathbf{h}) \left[\sum_i^n \sum_j^n F_m(\mathbf{h}A_i) F_m^*(\mathbf{h}A_j) e^{-2\pi i \mathbf{h} \cdot \mathbf{t}_{ij}} \right] \qquad (13.10)$$

where $I_o(\mathbf{h})$ are the observed intensities, and the summation over i and j ($i \neq j$) is the calculation of the intermolecular vectors between each pair of search molecules (m) related by the n symmetry operators A_i and A_j. The vector $\mathbf{t}_{ij}$ is the intermolecular vector between the copies of the molecule generated by A_i and A_j. To allow calculation of the translation function as a single Fourier series, Crowther and Blow (1967) simplified equation (13.10) in terms of a single translation vector $\mathbf{t}$ defining the position of the search model. This function is referred to as the T_2 function, in which the subscript '2' signifies that all sets of intermolecular vectors are compared simultaneously with the target Patterson function. This differs from a simpler alternative, the T_1 function, which only considers one pair of symmetry operators at a time. Others have recast equation (13.10) in a form that allows its calculation by FFT with fewer approximations (Harada *et al.*, 1981; Tickle and Driessen, 1995).

To ensure that the T_2 function is dominated by intermolecular (or cross-) vectors, the intramolecular vectors calculated for all copies of the search molecule generated by the symmetry operators can be subtracted from the target Patterson function by calculation of I_{cross}:

$$I_{cross} = I_o(\mathbf{h}) - k \sum_i^n |F_m(\mathbf{h}A_i)|^2 \qquad (13.11)$$

where k is a scale factor given by the ratio of the sum of the observed intensities to the sum of the calculated intensities ($\sum I_o / \sum I_c$). In the T_2 function modified by Ian Tickle (Driessen *et al.*, 1991), the intramolecular vectors can also be subtracted from the search model Patterson function to improve the signal-to-noise ratio. Normalised structure factors (E-values) or use of a negative temperature factor can improve the signal-to-noise ratio of the translation function, and such treatments are applied in current highly automated Patterson-based molecular replacement programs such as AMoRe (Navaza, 1994, 2001), MOLREP (Vagin and Teplyakov, 1997) and CNS (Grosse-Kunstleve and Adams, 2001).

13.9   Reciprocal-space translation searches

Historically, a very common means of solving the translation problem has been to perform what is called an 'R-factor search'. This method was a very popular way of solving the translation problem, not least because of its conceptual simplicity.

It involves the use of a parameter known as the *R*-factor, which is shown in the following equation:

$$R = \frac{\sum_{h} ||F_{h}(\text{obs})| - k|F_{h}(\text{calc})||}{\sum_{h} |F(\text{obs})|} \tag{13.12}$$

The *R*-factor is calculated as the search molecule and its symmetry mates are moved through the unit cell of the target crystal. The correct position should give the lowest *R*-factor, since each $F_{h}(\text{obs})$ will agree closely with the $F_{h}(\text{calc})$ and the numerator in equation (13.12) will have a low value. Conversely, for positions that are unlikely, the *R*-factor will be high.

In the numerator of this equation, the difference between the observed structure factor amplitude $|F_{h}(\text{obs})|$ and the calculated structure factor amplitude $|F_{h}(\text{calc})|$ is obtained for each reflection, and the amplitudes of these differences are summed for the whole dataset. This is then divided by the sum of the observed structure factor amplitudes, and the result can be expressed either as a fraction or as a percentage. Often the $F_{h}(\text{obs})$ values will not be on the same scale as the $F_{h}(\text{calc})$ values, and so an overall scale factor k must be determined prior to the *R*-factor search so that the differences between $F_{h}(\text{obs})$ and $F_{h}(\text{calc})$ are meaningful. We looked at ways that datasets may be scaled together in Chapter 11, but in this case an approximate scale factor can be determined from the starting $F_{h}(\text{calc})$ values as $\sum |F_{h}(\text{obs})| / \sum |F_{h}(\text{calc})|$.

Other parameters, such as the correlation coefficient, can be used to measure the agreement between the $F_{h}(\text{obs})$ and $F_{h}(\text{calc})$ values or the corresponding intensities as the search model is moved from one grid point to the next. The correlation coefficient has the distinct advantage of being independent of the scale factor. The form of the correlation coefficient based on the intensities is shown in the equation below:

$$CC = \frac{\sum_{h} (I_{h}(\text{obs}) - \langle I_{h}(\text{obs})\rangle)(I_{h}(\text{calc}) - \langle I_{h}(\text{calc})\rangle)}{\sqrt{\sum_{h} (I_{h}(\text{obs}) - \langle I_{h}(\text{obs})\rangle)^{2}} \sqrt{\sum_{h} (I_{h}(\text{calc}) - \langle I_{h}(\text{calc})\rangle)^{2}}} \tag{13.13}$$

These methods require that the complete set of structure factors is calculated for the model and its symmetry mates at each grid position in the target unit cell, which is a very time-consuming process for the computer. However, there is an elegant way of speeding up these calculations. The contribution each symmetry-related search molecule makes to any calculated structure factor F_{c} can be obtained by calculating a hemisphere of structure factors for a single correctly oriented search model in the target cell assuming $P1$ symmetry. The so-called partial F_{c}'s for the symmetry-related search molecules are then obtained from the different asymmetric units of the diffraction pattern. This can be shown by consideration of equation (9.21), which is used for calculating the structure factors from the atomic positions and scattering factors:

$$F_{hkl} = \sum_{j} f_{j} e^{2\pi i (hx_{j} + ky_{j} + lz_{j})} \tag{9.21}$$

If a protein crystallises in the monoclinic space group $P2$, there will be identical copies of the molecule at the general equivalent positions (GEPs) (x, y, z) and $(-x, y, -z)$. If we calculate all structure factors for the single molecule in $P1$, then we generate a unique set of structure factors F_{hkl} with $h \geq 0$, but k and l can have positive and negative signs. These occupy a hemisphere in reciprocal space, the other hemisphere being identical owing to Friedel's law, which we examined in Chapter 9. Normally, when one performs a structure factor calculation in $P2$, we generate only a quadrant of reciprocal space, where $h \geq 0$, $k \geq 0$ and only l assumes positive and negative signs. In this space group, the contribution that the molecule at (x, y, z) makes to the overall structure factor is given by

$$F_{c1} = \sum_j f_j e^{2\pi i \left(hx_j + ky_j + lz_j\right)}$$

and the contribution that the molecule at $(-x, y, -z)$ makes is given by

$$F_{c2} = \sum_j f_j e^{2\pi i \left(-hx_j + ky_j - lz_j\right)}$$

The overall structure factor is the sum of the above two equations, i.e.

$$F_{hkl} = \sum_j f_j e^{2\pi i \left(hx_j + ky_j + lz_j\right)} + \sum_j f_j e^{2\pi i \left(-hx_j + ky_j - lz_j\right)}$$

Now, if we consider the structure factors that have been calculated in $P1$, it is easy to see that the reflection (hkl) corresponds exactly to the contribution of the molecule at (x, y, z) in $P2$ and the reflection $(\bar{h}k\bar{l})$ in $P1$ corresponds to the contribution of the molecule at $(-x, y, -z)$ in $P2$. Whilst reflection $(\bar{h}k\bar{l})$ appears to be outside the range of reciprocal space which we have calculated in $P1$, we can obtain $F\left(\bar{h}k\bar{l}\right)$ from its complex conjugate $F^*(h\bar{k}l)$, which is within the range that we have calculated. Thus the hemisphere of diffraction spots which we calculated assuming the space group to be $P1$ provides the contribution made by each asymmetric unit to the structure factor for the true space group $P2$. This approach can be applied to all space groups.

The speed enhancement stems from the fact that the effect on the structure factor of moving the molecules can be calculated by phase-shifting the partial F_c's:

$$F_{hkl} = F_{c1} \exp(2\pi i \mathbf{h} \cdot \mathbf{t}_1) + F_{c2} \exp(2\pi i \mathbf{h} \cdot \mathbf{t}_2) + \ldots \tag{13.14}$$

where $\mathbf{t}_1$ and $\mathbf{t}_2$ are the shift vectors applied to each symmetry mate of the search model. Calculating the effect on F_{hkl} of shifting the molecule on a fine grid is much faster than applying a shift to the coordinates and recalculating structure factors from scratch. This phase-shifting method is employed by many of the contemporary molecular replacement programs.

In this way, an R-factor, correlation coefficient or other translation function can be calculated for each position of the molecule on a grid scanning all possible positions in the asymmetric unit. It is sensible to exclude weak reflections from the search, owing to measurement errors and the small contribution that they make to the R-factor. The main disadvantage of the R-factor search is that it is slower than the T_2 search, which can be calculated by FFT. However, most molecular replacement programs will display the R-factor and correlation coefficient of potential solutions to help the user decide on their validity.

As with the rotation function, there has been much interest in the application of *maximum-likelihood methods* to the translation function (Read, 2001; McCoy *et al.*, 2005). In fact, its application to this problem requires fewer assumptions than with the rotation function, stemming from the fact that the orientation of the molecule has been determined by this stage. The maximum-likelihood translation function is a measure of the probability that the set of observed structure factors would have been obtained if the trial position of the search model was the correct position. It is calculated from the F_o and F_c data without consideration of Patterson space and is, with some approximations, amenable to rapid calculation by FFT methods. The promising solutions can then be evaluated by calculating a more precise likelihood function. The program PHASER (McCoy, 2007) calculates maximum-likelihood rotation and translation functions and can automatically solve heterogeneous complexes of proteins owing to a powerful method of screening solutions for each component and the fact that correctly orienting and positioning one component will enhance the signal of another.

13.10   Asymmetric unit of the translation function

The volume of the unit cell over which the translation function has to be calculated depends on the space group. The symmetry of the translation function for each space group is referred to as the Cheshire group—a term based on Lewis Carroll's Cheshire Cat (Hirshfeld, 1968). In *Alice's Adventures in Wonderland*, the Cheshire Cat has

Table 13.1 The asymmetric units of the crystallographic translation function for the different crystal classes and point groups.

Crystal system	Point group	Asymmetric unit of translation function
Monoclinic	2	$y = 0$ section; 1/2 of x and z (for **b** unique)
Orthorhombic	222	1/2 of x, y and z
Trigonal	3	$z = 0$ section; all x and y (3 asymmetric units)
	312	All x and y; 1/2 of z (3 asymmetric units)
	321	All x and y; 1/2 of z
Tetragonal	4	$z = 0$ section; 1/2 of x, all y (or all x, 1/2 of y)
	422	x and y as previous, 1/2 of z
Hexagonal	6	$z = 0$ section; all x and y
	622	All x and y, 1/2 of z
Cubic	23	1/2 of one of x, y or z; all of the other two
	432	Same as previous

Exceptions: space groups $F23$ and $F432$, 1/2 of x, y and z.

a habit of disappearing and at one point leaves only its grin visible, which has the same symmetry as the cat itself. Likewise, the Cheshire group is the symmetry of a space group diagram which remains when the molecule or motif has disappeared! The asymmetric units of the Cheshire groups relevant to proteins have been tabulated by Ian Tickle, and the relevant information is listed in Table 13.1.

13.11 Non-crystallographic symmetry

As we have mentioned several times previously, the asymmetric unit of real space may contain more than one copy of the molecule. In such a situation, the positions of molecules related by this non-crystallographic symmetry can be determined using the non-crystallographic translation function. The highly automated molecular replacement programs will perform the necessary calculations automatically, but there are also specialised translation function programs such as TFFC (Driessen *et al.*, 1991) which have the same capability and may be useful in difficult cases. In such a process, one of the oriented search models is used initially to calculate a crystallographic translation function which determines its position in the asymmetric unit of the target cell. Next, the positions of the NCS-related molecules have to be determined relative to the same origin. Just running a crystallographic translation function on each correctly oriented search model would probably give the wrong results. This problem arises because of the symmetry of the Patterson function, i.e. if a molecule is placed at position (x, y, z) in the unit cell, there are other places within the same asymmetric unit of the unit cell which will give exactly the same peaks in the Patterson function. This can be confirmed by reference to Chapter 12, Section 12.8, where we showed that for the space group $P2_12_12_1$, there are four general equivalent positions

$$(x, y, z), \quad (1/2 - x, -y, 1/2 + z), \quad (1/2 + x, 1/2 - y, -z) \text{ and } (-x, 1/2 + y, 1/2 - z).$$

The Harker vectors for the Patterson function are

$$(1/2 \pm 2x, \pm 2y, 1/2), \quad (1/2, 1/2 \pm 2y, \pm 2z) \quad \text{and} \quad (\pm 2x, 1/2, 1/2 \pm 2z)$$

Let us consider a molecule at the GEP (x, y, z), and if we imagine applying a half-unit-cell translation parallel to the z axis, the Harker vectors become

$$(1/2 \pm 2x, \pm 2y, 1/2), \quad (1/2, 1/2 \pm 2y, \pm 2(z + 1/2)) \text{ and } (\pm 2x, 1/2, 1/2 \pm 2(z + 1/2))$$

which simplifies to

$$(1/2 \pm 2x, \pm 2y, 1/2), \quad (1/2, 1/2 \pm 2y, \pm(2z + 1)) \quad \text{and} \quad (\pm 2x, 1/2, 1/2 \pm (2z + 1))$$

Because unit cell translations are redundant, the Harker vectors become

$$(1/2 \pm 2x, \pm 2y, 1/2), \quad (1/2, 1/2 \pm 2y, \pm 2z) \quad \text{and} \quad (\pm 2x, 1/2, 1/2 \pm 2z)$$

which are the same as the Harker vectors that were generated by having the molecule at (x, y, z). Exactly the same applies if we apply a half-unit-cell translation to any or all of the x, y and z coordinates—we will still get the same set of Harker vectors.

We can now return to the statement we made a few lines back where we said that there are several places where the molecule can be in the same asymmetric unit of the real unit cell that will give exactly the same peaks in the Patterson function. Since the asymmetric unit of the translation function for $P2_12_12_1$ is an eighth of the unit cell given by

$$0 \leq x \leq 1/2, \quad 0 \leq y \leq 1/2, \quad 0 \leq z \leq 1/2$$

the translation function will indicate a set of shifts to be applied to the oriented search model that are within the above range of x, y, and z. However, the asymmetric unit of space group $P2_12_12_1$ is a quarter of the unit cell, for example

$$0 \leq x \leq 1/2, \quad 0 \leq y \leq 1/2, \quad 0 \leq z \leq 1$$

Hence, within this volume, we could equally well place the molecule at (x, y, z) or $(x, y, z + 1/2)$ and we would get exactly the same Harker vectors. If we are searching for one molecule in the asymmetric unit, then the ambiguity does not matter and we can equally well place the molecule at either of these two positions. When we arbitrarily choose one of these positions, this is said to 'fix the origin' of the solution.

If there is more than one molecule in the asymmetric unit, then when we try to locate the second molecule, it is important that we determine its position correctly relative to the first molecule, whose origin we arbitrarily fixed. The problem arises because if we calculate a crystallographic translation function for the second molecule, there will again be more than one position where we can place it within the asymmetric unit of the real unit cell that will agree with the Harker vectors in the Patterson function. However, only one position will agree with the intermolecular vectors running between the molecules in the asymmetric unit of the crystal, and this represents the correct position of the second molecule relative to the first molecule.

Hence, in cases where there is more than one molecule in the asymmetric unit, the position of one molecule in the asymmetric unit is determined by a crystallographic translation function first. Following that, the relative positions of the other molecules have to be determined by calculation of non-crystallographic or partial translation functions. Only the non-crystallographic symmetry is used to calculate these partial translation functions, i.e. they search for vectors between NCS-related subunits only. For example, consider a tetramer of subunits A, B, C and D. Given that the orientations of each subunit have been determined by the rotation function, one would calculate a crystallographic translation function to determine the position of the A subunit. The position for the first molecule is then used to calculate a non-crystallographic translation function to determine the relative position of B. The position of molecule B can then be input and used in another partial translation function to determine the position of C, etc. When one is calculating any non-crystallographic translation function, it is important that it should cover all possible positions of the search model. Ian Tickle has tabulated the asymmetric units for each possible lattice type and, for completeness, these are shown in Table 13.2. The above complications may be somewhat off-putting, but the calculations are performed automatically by most of the contemporary molecular replacement programs and are completely transparent to the crystallographer in all but the most difficult cases, which require manual intervention.

Table 13.2 The non-crystallographic translation function has the following asymmetric units, depending on the lattice type.

Lattice type	Asymmetric unit of non-crystallographic translation function
P	All x, y and z.
R	All x, y and z (3 asymmetric units) (H lattice).
C	1/2 of x, all y and z (or 1/2 of y, all x and z).
I	1/2 of one of x, y or z; all of the other two.
F	1 in one of x, y, or z; 1/2 in the other two.

One common observation is that crystals possess non-crystallographic rotation axes that are parallel to the crystallographic axes. We mentioned in Section 13.1 that non-crystallographic symmetry should manifest itself in the self-rotation function. However, this is not the case when non-crystallographic rotation axes are parallel to crystallographic axes of the same order, since the peaks from both axes will overlap in the self-rotation function. However, such a situation can be resolved in the translation function by the methods that we have just described above. It can also be apparent from inspection of the Patterson function, which should possess very strong non-origin peaks. This arises because having a non-crystallographic rotation axis parallel to a crystallographic one of the same order generates non-crystallographic translational symmetry, i.e. pairs of molecules in the same or very similar orientations. The vectors running between the equivalent atoms of these molecules will coincide in the Patterson map and give a very strong 'pseudo-origin' peak, the position of which can indicate the location of the non-crystallographic rotation axis and the magnitude of any screw component.

13.12   The packing function

A packing function involves moving the correctly oriented search model on a fine grid throughout the target unit cell and, at each position, assessing the feasibility of the molecular packing between the symmetry-related molecules. Calculation of a packing function in this way can be useful in that it will indicate likely positions for the search molecule in the target unit cell. However, it will provide no discrimination between likely positions and so the X-ray data must, of course, be used for that purpose. Most molecular replacement programs have built-in algorithms that downweight translation function solutions which cause the molecule to interpenetrate with its symmetry-related neighbours in a sterically impossible manner. Potential solutions which cause minor collisions between symmetry-related molecules can be tolerated because there will almost certainly be differences between the search model and the target molecule in the surface loop regions, which usually are the most variable parts of any protein. In some cases, it may be sensible to remove protruding loop regions from the search model to prevent correct solutions from being rejected on packing criteria.

There are various ways to assess how well a molecule at any position packs with its symmetry-related partners, and the simplest is to count the number of C_α atoms clashing with each other. An alternative is to count the number of unfavourable van der Waals contacts between neighbouring protein molecules. The union of molecular volume can also be considered (Brünger, 1990). This can be estimated by summing the number of grid points in the unit cell that are within a short distance of any atom of the molecule or its symmetry mates. Since feasible molecular replacement solutions must involve a minimum of molecular overlap, potential translation function solutions that maximise the union of the molecular volumes are the most likely to be correct. Interestingly, Tickle and Driessen (1995) pointed out that the T_2 translation function described above in Section 13.8 will discriminate against solutions that cause molecular overlap, owing to the removal of intramolecular vectors from the target Patterson function. Positions of the search model which cause overlap will generate intermolecular vector peaks close to the origin of the search model Patterson function owing to the fact that many of the intermolecular vectors between the symmetry-related molecules are short in such a situation. Thus the model Patterson function will have strong peaks close to the origin which will have no corresponding peaks in the intramolecular-vector-subtracted target Patterson function.

13.13 Verifying the results

Very often, automated molecular replacement programs will produce the correct solution with no manual intervention. Provided that it has a good signal-to-noise ratio in the translation function (and, ideally, the rotation function too), then the resulting structure may be refined, initially as a rigid group and later with more degrees of freedom. We will look at methods of refining protein structures in Chapter 15. However, with difficult molecular replacement problems and in cases where the resulting structure has poor electron density or cannot be refined satisfactorily, there may be a number of alternative solutions (i.e. possible orientations and positions of the search model) that need to be tested. Increasing the resolution of the data used in the calculations and finer grid-sampling of the rotation and translation functions can make a dramatic improvement. Trial molecular replacement solutions should always be inspected for unfavourable crystal packing by use of a molecular graphics program. If the solutions are still ambiguous, it is necessary to refine each possible solution, and the correct one should produce an interpretable electron density map. The majority of molecular replacement programs automatically perform a simplified form of refinement in which the model is treated as a rigid group, known as rigid-body refinement.

One feature of the translation function is that it can be streaky, i.e. the peaks appear to be stretched and they tend to line up parallel to the major axes, making it hard to distinguish the solution peak from pseudo-solution peaks in difficult cases. Picking the wrong peak could give a structure in the correct orientation but with only two of its three translational parameters correct. Cases like this can refine convincingly and give deceptively good electron density maps. However, having a structure trapped in a false minimum should eventually be revealed by a lack of convergence in the refinement, as well as unexplained features of the electron density in the map and a lack of density in places where it should be present.

Another way of checking a solution is, of course, to repeat the process with a different molecular replacement program. However, some caution is necessary when comparing the results of different programs. It should be remembered that different programs will determine the position of the search model with respect to different origins and can apply different symmetry operations to it, giving solutions which appear to be different although they are in fact identical.

13.14   Wider applications of molecular replacement

Molecular replacement clearly has an important role to play in analysing the higher-order structures that some proteins form (see e.g. Fig. 13.7), as well as functional complexes with other molecules, be they small ligands or macromolecules such as other proteins or nucleic acids. These large multi-component complexes are also amenable to electron microscopy (EM), which, whilst generally providing electron density maps of lower resolution than X-ray crystallography (typically 10–30 Å), has the advantage that the complex does not have to be crystallised, which may be difficult or impossible. In such a situation, the X-ray (or NMR) structures of the individual components can be fitted to an EM map of the complex, thereby providing much more biologically relevant information than would have been available by either technique on its own. The importance of placing molecular structures such as those solved by X-ray crystallography in the context of subcellular components cannot be overemphasised.

Alternatively, an EM map of a protein can be used as a search model in molecular replacement to analyse the protein's X-ray structure. Once the low-resolution EM map has been correctly oriented and positioned, phases for the X-ray data can be

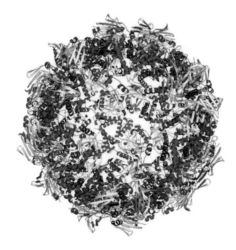

Fig. 13.7 *The structure of the protein MhpD solved by the automated molecular replacement program MOLREP (Vagin and Teplyakov, 1997) using a monomer as the search model.* This picture was kindly provided by Dr M. Montgomery (UCL Department of Medicine).

refined and extended using methods that we will cover in Chapter 14. In general, to use an EM map as either a search model or a target function in molecular replacement, structure factors have to be calculated from it by Fourier transformation. A number of molecular replacement programs have been adapted for using EM maps (Xiong, 2008; Navaza, 2008; Trapani and Navaza, 2008). These methods work particularly well when the structure possesses a high degree of non-crystallographic symmetry, such as that found in spherical viruses, where mathematical methods for averaging tens of thousands of images can provide EM maps of extraordinary quality.

Summary

Molecular replacement is the collective term for a group of methods of solving the phase problem for an unknown *target molecule* by using phase information derived from a *search model*, whose structure is already known. For molecular replacement to be successful, the target molecule and the search model must be reasonably similar in terms of amino acid sequence and tertiary structure. The atomic locations of the search model are usually different from the atomic locations required to form the maximum possible 'overlap' with the target molecule. The key computational task in molecular replacement is therefore to determine by how much the search model has to be rotated and then translated to form the optimum 'overlap'. This in turn requires the application of the *rotation function* and the *translation function*.

These two functions are generally calculated by comparing the Patterson function of the target molecule with the Patterson function of the search model so as to identify which specific orientation and location for the search molecule will most closely match the observed Patterson function of the target molecule. The resulting structure can then be refined, which will indicate whether the trial solution is correct or not.

References

Brünger, A. T. (1990) Extension of molecular replacement: a new search strategy based on Patterson correlation refinement. *Acta Crystallogr. A* **46**, 46–57.

Crowther, R. A. (1972) The fast rotation function. In *The Molecular Replacement Method*, ed. Rossmann, M. G. Gordon & Breach, New York, pp. 173–8.

Crowther, R. A. and Blow, D. M. (1967) A method of positioning a known molecule in an unknown crystal structure. *Acta Crystallogr.* **23**, 544–8.

Dodson, E. (2008) The befores and afters of molecular replacement. *Acta Crystallogr. D* **64**, 17–24.

Driessen, H. P. C., Bax, B., Slingsby, C., Lindley, P. F., Mahadevan, D., Moss, D. S. and Tickle, I. J. (1991) Structure of oligomeric β-B2-crystallin – an application of the T_2 translation function to an asymmetric unit containing 2 dimers. *Acta Crystallogr. B* **47**, 987–97.

Evans, P. R. (2001) Rotations and rotation matrices. *Acta Crystallogr. D* **57**, 1355–9.

Grosse-Kunstleve, R. W. and Adams, P. D. (2001) Patterson correlation methods: a review of molecular replacement with CNS. *Acta Crystallogr. D* **57**, 1390–6.

Harada, Y., Lifchitz, A., Berthou, J. and Jolles, P. (1981) A translation function combining packing and diffraction information: an application to lysozyme (high-temperature form). *Acta Crystallogr. A* **37**, 398–406.

Hirshfeld, F. L. (1968) Symmetry in the generation of trial structures. *Acta Crystallogr. A* **24**, 301–11.

Huber, R. (1965) Die automatisierte faltemolekulemethode. *Acta Crystallogr.* **19**, 353–6.

Kissinger, C. R., Gehlhaar, D. K. and Fogel, D. B. (1999) Rapid automated molecular replacement by evolutionary search. *Acta Crystallogr. D* **55**, 484–91.

Kissinger, C. R., Gelhaar, D. K., Smith, B. A. and Bouzida, D. (2001) Molecular replacement by evolutionary search. *Acta Crystallogr. D* **57**, 1474–9.

Lattman, E. E. (1972) Optimal sampling of the rotation function. *Acta Crystallogr. B* **28**, 1065–8.

Lattman, E. E. and Love W. E. (1970) A rotational search procedure for detecting a known molecule in a crystal. *Acta Crystallogr. B* **26**, 1854–7. Errata can be found in Lattman, E. E. and Love, W. E. (1971) *Acta Crystallogr. B* **27**, 1479.

Lebedev, A. A., Vagin, A. A. and Murshudov, G. N. (2008) Model preparation in MOLREP and examples of model improvement using X-ray data. *Acta Crystallogr. D* **64**, 33–9.

McCoy, A. J. (2007) Solving structures of protein complexes by molecular replacement with Phaser. *Acta Crystallogr. D* **63**, 32–41.

McCoy, A. J., Grosse-Kunstleve, R. W., Storoni, L. C. and Read, R. J. (2005) Likelihood-enhanced fast translation functions. *Acta Crystallogr. D* **61**, 458–64.

Moss, D. S. (1985) The symmetry of the rotation function. *Acta Crystallogr. A* **41**, 470–5.

Navaza, J. (1994) AMoRe: an automated package for molecular replacement. *Acta Crystallogr. A* **50**, 157–63.

Navaza, J. (2001) Implementation of molecular replacement in AMoRe. *Acta Crystallogr. D* **57**, 1367–72.

Navaza, J. (2008) Combining X-ray and electron-microscopy data to solve crystal structures. *Acta Crystallogr. D* **64**, 70–5.

Read, R. J. (2001) Pushing the boundaries of molecular replacement with maximum likelihood. *Acta Crystallogr. D* **57**, 1373–82.

Rossmann, M. G. and Blow, D. M. (1962) The detection of subunits within the crystallographic asymmetric unit. *Acta Crystallogr.* **15**, 24–31. Errata can be found in Tollin, P. and Rossmann, M. G. (1966) A description of various rotation function programs. *Acta Crystallogr.* **21**, 872–6.

Storoni, L. C., McCoy, A. J. and Read, R. J. (2004) Likelihood-enhanced fast rotation functions. *Acta Crystallogr. D* **60**, 432–8.

Tickle, I. J. and Driessen, H. P. C. (1995) Molecular replacement using known structural information. In *Crystallographic Methods and Protocols*, ed. Jones, C., Mulloy, B. and Sanderson, M. R., Methods in Molecular Biology, Volume 56, pp. 173–203.

Trapani, S. and Navaza, J. (2008) AMoRe: classical and modern. *Acta Crystallogr. D* **64**, 11–6.

Vagin, A. and Teplyakov, A. (1997) MOLREP: an automated program for molecular replacement. *J. Appl. Crystallogr.* **30**, 1022–5.

Xiong, Y. (2008) From electron microscopy to X-ray crystallography: molecular-replacement case studies. *Acta Crystallogr. D* **64**, 76–82.

14

Solving the phase problem experimentally

14.1   The techniques of solution

In Chapter 13, we saw how the technique of molecular replacement can be used to solve the phase problem with prior knowledge of a structure similar to the one that we are trying to solve. In this chapter, we will describe the principal methods which we may use in order to solve the phase problem by purely experimental means without reference to other known structures. These methods usually involve the chemical modification of the macromolecules within the crystal by the addition of elements, often electron-rich heavy atoms, that perturb the scattering of the X-rays. The effect that these additional elements have on the coherent scattering of the macromolecule within the crystal allows us to estimate the phases of diffraction spots by a method known as *isomorphous replacement*. Usually we need to make several such derivatives with different reagents to determine the phases unambiguously, and the method is then referred to as *multiple isomorphous replacement* (MIR). If we choose the element that we add to the crystal carefully or if we carefully select the wavelength of the incident X-rays, we may be able to exploit the anomalous scattering of the added element, or indeed of any element that happens to be present naturally within the molecule. In Chapter 9, Sections 9.8 and 9.12, we saw how anomalous scattering occurs when the incident beam causes electronic transitions in an atom, and this results in a breakdown of Friedel's law. The effect on the diffraction pattern provides information on the phases, and by changing the wavelength of the incident beam we can maximise the anomalous scattering effects. If we measure the diffraction pattern at a number of specially chosen wavelengths, this is referred to as *multi-wavelength anomalous diffraction* (MAD) phasing. Sometimes, the phases can be determined from anomalous scattering effects measured at a single optimised wavelength, and this is referred to *single-wavelength anomalous diffraction* (SAD) phasing. On balance, the multiple isomorphous replacement and multi-wavelength anomalous diffraction phasing methods are the most commonly used, with SAD gaining much ground in recent years. All involve either preparing heavy-atom derivatives of the crystal or incorporating elements into the protein that scatter anomalously at measurable X-ray wavelengths. With many heavy atoms, we can use both isomorphous and anomalous effects in solving the phase problem.

As emphasised above, it is necessary that we know exactly how we have modified the structure of the molecule. This implies that we must be able to determine the coordinates of any heavy atoms which bind to the molecule, in advance of the solution of the phase problem. As we saw in the Chapter 12, Patterson map methods allow us to determine the positions of heavy elements within the unit cell directly from relative intensity data without having to determine the phases first. For this reason, all chemical modification methods involve the preparation of a crystal containing a heavy element (often a metal), whose coordinates we determine using Patterson or other techniques. Identification of the site(s) of the heavy atom is a springboard from which the rest of the structure may be determined.

14.2 Isomorphism and the preparation of heavy-atom derivatives

In macromolecular crystallography, two crystals are said to be *isomorphous* if they have closely similar crystallographic structures, with essentially the same lattice parameters and symmetry properties. An obvious prerequisite for exploiting isomorphism in phasing is that the molecules within the isomorphous crystals should be very similar chemically and structurally. The relevance of isomorphism to the X-ray crystallographer is that, very often, a crystal of a large molecule and a crystal of a chemically modified form are very nearly isomorphous. The important consequence of this is that if the lattice structures are identical, then the reciprocal-lattice structures are also identical, implying that the positions of the spots in the X-ray diffraction patterns are the same for both crystals. In view of this, the same indices h, k and l are attributed to equivalent spots.

This is often valid for protein crystals, since we can adjust the conditions of the experiment such that the binding of the heavy atom does not perturb the crystal structure significantly. However, the presence of a heavy atom is liable to have an effect on its immediate environment and the conformation of the protein molecule can change, thus reducing the isomorphism.

Heavy atoms can be introduced into protein crystals either by soaking or by co-crystallisation. The former method is preferable, since it is generally much faster and sometimes it is found that inclusion of a heavy-atom reagent in the crystallisation conditions of a protein changes the unit cell and/or space group, giving rise to a serious lack of isomorphism. Soaking can also cause non-isomorphism, but it is usually possible to adjust the soaking conditions to minimise the problem, for example by reducing the concentration of the reagent and/or the duration of the soak to decrease the number of sites at which the reagent binds. In practice, a soak time of 12–18 hours (e.g. overnight) with a heavy-atom concentration of 1 mM in the mother liquor is good starting point. Very often, crystals grown by the hanging-drop method can be soaked in the well solution supplemented with a heavy-atom reagent for long periods of time. Soaks can be as short as 30 minutes, and the heavy-atom reagent can be as dilute as 50 µM, although with concentrations as low as this it is important to consider the stoichiometry, particularly if the derivative is being made by co-crystallisation. The protein concentration will typically be in the 0.1 mM range in the mother liquor, and so a heavy atom at a final concentration much lower than this would not be able to form a stoichiometric complex with the protein. Soaking in the dark is also recommended to avoid spurious photochemistry of some reagents (Petsko, 1985).

If the crystal is visibly damaged by soaking or the diffraction data indicate that it is non-isomorphous, then the conditions can be modified accordingly. Likewise, if the heavy atom does not bind, higher concentrations of the reagent and longer soak times can be used. Often the nature of the buffer is critical in the soaking experiment, since some buffers will bind the heavy atom tightly and precipitate. This commonly occurs with phosphate buffers, and other buffers which remain soluble in the presence of the heavy atom can still prevent the reagent from binding to the protein by ligand competition; for example, citrate can act in this way, as does the ammonia which is released from ammonium sulphate (a commonly used precipitant) at high pH values. The problem can usually be solved easily, since a crystal grown in one buffer can often be transferred to another of similar concentration and pH for soaking experiments. Sometimes crystals can be stabilised by chemical cross-linking with glutaraldehyde prior to soaking or transfer to a different buffer.

The side chains of cysteine residues very commonly react with mercury reagents (e.g. $HgCl_2$, K_2HgI_4 and *para*-chloromercuribenzenesulphonic acid, or PCMBS), and the use of high pH, which causes cysteine to deprotonate and form a negatively charge thiolate group, will assist binding of the mercury cation and other heavy-metal groups. Different mercury reagents allow some discrimination between the binding sites, sometimes allowing several useful derivatives to be made (Leslie, 1991). The other sulphur-containing amino acid, methionine, does not react with mercurials but it does react with some very useful platinum reagents, such as K_2PtCl_4, which are widely used and have the advantage of possessing different net charges depending on the groups ligating the metal (e.g. $PtCl_4^{2-}$ and $Pt(NH_3)_4^{2+}$), thereby allowing different platinum compounds to occupy different sites in the protein molecule by either dative or ionic bonding (Leslie, 1991). The tyrosine residues in a protein can react with iodine reagents (e.g. I/KI), and at high pH the side chain amino group of lysine becomes neutral, allowing it to bind various heavy-metal ions datively. The imidazole group of histidine is also reactive with heavy metals (such as gold, mercury and platinum) in its deprotonated state, which occurs above pH 6. It should be noted that many heavy-atom reagents are labile at high pH and form insoluble hydroxides (Petsko, 1985; Abdel-Meguid, 1995). At neutral pH, the carboxyl groups of aspartate and glutamate residues are negatively charged and can bind heavy-atom reagents by ionic bonding, for example the commonly used uranyl (UO_2^{2+}) group possessed by uranyl acetate $(UO_2)Ac_2$ and uranyl fluoride ($K_3UO_2F_5$). All of the compounds mentioned in this paragraph, together with potassium dicyanoaurate ($KAu(CN)_2$) and trimethyl lead acetate (Me_3PbAc), are amongst the most commonly used reagents and are worth trying in an initial screen for derivatives (Boggon and Shapiro, 2000). With large macromolecular assemblies, heavy-atom cluster compounds such as tetrakis(acetoxymercury)methane (TAMM) have been used very effectively. Anyone working with heavy-atom reagents will be well aware of their extreme toxicity and, in some cases, radioactivity and will take all appropriate precautions.

It is also possible to incorporate the 'heavy atom' at the synthesis stage; for example, with synthetic oligonucleotides it is common to incorporate brominated nucleotides, and with recombinant proteins it is possible to supply the bacteria with selenomethionine instead of methionine. In selenomethionine, the side chain sulphur of methionine is replaced by selenium, which, like bromine, has strong anomalous scattering at a convenient wavelength and allows the protein phases to be determined by the MAD or SAD method. A commonly used procedure for selenomethionine incorporation is to grow the bacteria to mid-log phase in Luria–Bertani medium and then to pellet and resuspend the cells in minimal medium, to which specific nutrients are added along with all of the standard amino acids, except for methionine, which is replaced by selenomethionine. Shortly afterwards, expression of the protein of interest can be induced in the usual way. It is possible to use a methionine auxotroph (i.e. an expression strain which cannot synthesise its own methionine) to ensure the highest possible level of selenomethionine incorporation, but this is usually unnecessary. For proteins that lack methionine or do not possess a sufficient number (the minimum number for MAD phasing is estimated to be 1 methionine per 100 amino acids), site-directed mutagenesis can be used to introduce additional methionine residues. Incorporation of selenium into a protein can be confirmed by mass spectrometry or some form of elemental analysis, and these approaches can also be useful for confirming the binding of heavy-atom reagents. Native gel electrophoresis can be used to screen for suitable heavy-atom reagents, since binding of the reagent can change the electrophoretic mobility of the protein appreciably under non-denaturing conditions (Boggon and Shapiro, 2000). In some cases it has been possible to make useful derivatives by soaking protein crystals for short periods of time ($< 1\,\mathrm{min}$) in high concentrations of bromide or iodide salts or by pressuring them with xenon gas (Dauter, 2002; Sauer *et al.*, 1997). The halide ions generally bind to the protein surface, and xenon tends to occupy pockets and channels inside the protein molecule.

To assess whether the protein has been modified by a heavy-atom reagent requires that X-ray data are collected from the crystal and that the intensities of its diffraction spots are compared with those of the native (unmodified) protein. Historically, this used to be done by visual inspection of X-ray precession photographs, but with advances in area detector technology it is much faster to collect and process the data and compare the native and derivative diffraction patterns computationally. Since we are interested only in derivatives that do not drastically alter the unit cell of the crystal, those that suffer from non-isomorphism (e.g. changes in unit cell parameters of more than a few per cent) should be rejected at this stage.

In the next sections we will look at methods for analysing heavy-atom-derivative data, firstly to assess whether or not the heavy-atom reagent has modified the protein and secondly to determine where the heavy atom has bound in the unit cell of the crystal. The usefulness of a heavy atom in solving the phase problem relies on it binding to the protein at the same place in a large proportion of the molecules within the crystal. Whilst there is some scope for disordered binding, reagents that modify the protein in a non-specific manner will not generate a useful derivative. Hence, the ultimate test of the usefulness of a derivative is not whether the diffraction intensities have changed upon binding of the reagent but whether we can determine where the heavy atom has bound in the asymmetric unit of the crystal.

14.3   Scaling and analysing derivative data

For reasons that we will cover later on, it is sensible to collect data on more than one heavy-atom derivative; in fact, the more derivatives that we analyse, the better the final electron density map should be, provided that the derivatives are isomorphous and have well-defined heavy-atom binding sites. Hence we need to screen putative derivatives to determine whether they are isomorphous and whether there are differences in the diffraction spot intensities from the native (unmodified) protein. To do this, we must collect X-ray data on each derivative by methods that we described in Chapter 11 and the data must be reduced to a set of structure factor amplitudes $|F_{PH}|$, where PH indicates the protein–heavy-atom derivative. These data are then scaled to the native amplitudes $|F_P|$. Scaling is essential, since data are very often collected using different instruments, different-sized crystals etc., so whilst the relative intensity of each diffraction spot with respect to the others in any one dataset is accurately determined, the numerical value of its intensity is on an arbitrary scale and could vary greatly from one measured dataset to another depending on the conditions of the experiment. The problem is partly circumvented by Wilson scaling (covered in Chapter 11), which applies a scale factor to the structure factor amplitudes such that they match the scattering to be expected from the atomic composition of the crystal. This effectively multiplies our measurements of the structure factor amplitudes ($|F_P|$, $|F_{PH}|$ etc.) such that they are put on an approximately absolute scale; in other words, they can be given units of 'electrons', i.e. each structure factor amplitude is defined relative to the scattering amplitude of a single electron. However, the scale determined from the Wilson plot is only approximate, and it is very common for a heavy atom to alter the temperature factor of the crystal; for example, diffraction from derivative crystals often falls off faster with increasing Bragg angle θ, and usually they diffract to lower resolution than the native crystal. It is important to correct for these systematic differences, since phasing requires that we measure small differences in structure factor amplitudes between the native and the derivative crystal as accurately as possible. Furthermore, any errors in the derivative-to-native scale factor can lead to holes or peaks in the electron density for the protein at the sites of heavy-atom binding, thus confusing our interpretation of the map (Bloomer, 1983).

In Chapter 11, we saw how the Wilson plot was constructed by calculating the ratio of the mean $|F|^2$ to the mean of the summed, squared atomic scattering factors in thin resolution shells and plotting the logarithm of the ratio as a function of $\sin^2\theta/\lambda^2$. From the intercept on y, we may determine the scale factor c, and the gradient $(-2B)$ gives us the B-factor. In scaling a derivative to a native dataset, we do not consider the atomic scattering factors; instead, we can calculate the ratio of the mean $|F_{PH}|^2$ to the mean $|F_P|^2$ and plot the logarithm of this as a function of $\sin^2\theta/\lambda^2$. Likewise, this gives us a scale factor K and a B-factor that can be used to correct for at least some of the systematic differences between the derivative and the native dataset. Whilst the Wilson plot has the advantage of conceptual simplicity, it is usual to determine these correction factors by least-squares minimisation, the theory of which we will cover in more detail in Chapter 15. The method of least squares requires that we have reasonably accurate trial values for the parameters to be refined and, in scaling, these can be determined from a Wilson analysis. The temperature factor may be refined as an isotropic B-factor (one parameter) or as an anisotropic B-factor (six parameters), the latter being useful if there are directional systematic differences between the two datasets, for example if the diffraction from the heavy-atom-soaked crystal fades off appreciably more in one direction than in others, as can happen. Such derivative-to-native scaling is applied in many programs, including specialised scaling software such as SCALEIT (CCP4, 1994) and automated structure solution packages such as SOLVE (Terwilliger and Berendzen, 1999a). In scaling two datasets together, it is generally sensible to exclude the very low-resolution terms (e.g. $d > 10$ Å), since these reflections are most affected by heavy atoms occupying the solvent channels in the crystal. To minimise this effect, it is possible to back-soak a derivative crystal with mother liquor, thereby removing the unbound or weakly bound heavy atoms prior to data collection.

Following successful scaling of two datasets, the mean of the $|F_{PH}|$ values should be approximately equal to the mean of $|F_P|$, and this should be reflected in a constant $\sum |F_{PH}| / \sum |F_P|$ ratio of approximately 1.0 across all resolution ranges. Scaling can be judged from calculation of a residual index, or R-factor, based on either the amplitudes of the structure factors or the intensities of the diffraction spots in the two datasets:

$$R = \frac{\sum ||F_{PH}| - |F_P||}{\sum |F_P|} \tag{14.1}$$

$$R_{int} = \frac{\sum ||F_{PH}|^2 - |F_P|^2|}{\sum |F_P|^2} \tag{14.2}$$

In these equations, the summations are done over all unique reflections that have been measured in the native and derivative datasets. The R-value given by equation (14.1) is sometimes referred to as the *fractional isomorphous difference,* and the lower its value, the better the two datasets have been scaled together. R-factors in excess of approximately 60% indicate that scaling has not converged or that there are major errors in at least one of the datasets. When data from a putative derivative in which no heavy atoms have actually bound to the protein are scaled to the native data, the R-factor based on the amplitudes, as defined in equation (14.1) will be in the region of 10% or slightly lower. R-factors as low as this are to be expected when the only differences between two datasets are due to measurement errors, for example independently measured datasets from different crystals of the exactly the same molecule. In contrast, fractional isomorphous differences in the vicinity of 20–30% indicate that there are small but significant differences between the two datasets that could be due to binding of the heavy atom to the protein. Alternatively, this could be due to non-isomorphism, which can often be diagnosed by inspecting the values of the R-factor as a function of resolution. With a good derivative, the R-factor is generally uniform across the resolution range, whereas non-isomorphism frequently results in an increase in the R-factor with increasing θ angle, since small differences in the sampling of the two molecular transforms will affect the high-resolution terms most. R-factors much higher than 30% generally indicate that there is significant non-isomorphism and that the derivative will probably not be useful in the phasing exercise. The reader should note that this discussion relates to the R-factor calculated from the amplitudes (equation (14.1)) and that similar arguments apply to that calculated from the intensities (equation (14.2)), although the approximate cut-off values we have described are slightly higher. Other statistical measures, such as normal probability analysis (Howell and Smith, 1992), can also be used to detect the presence of significant isomorphous differences ($|F_{PH}| - |F_P|$) between pairs of datasets.

The methods that we have described so far aim to determine a single, overall scale factor, along with an isotropic or anisotropic temperature factor, that will place the derivative dataset on the same scale as that of the native crystal. Sometimes there are systematic differences between the two datasets due to dissimilarities in crystal size, shape, strength of diffraction, data collection method etc., and these differences are not modelled well by a single scale factor and temperature factor. To perform more accurate scaling, the derivative dataset can be divided up into thin spherical shells in reciprocal space, and the spots within each shell are compared with their counterparts in the native dataset to determine a scale factor for that shell. In this way, resolution-dependent scale factors can be determined. We can divide the dataset up into as many spherical shells as we like, but for the scale factors to be accurate we should ensure that the shells contain a minimum of several hundred diffraction spots. It is also sensible for the shells to be of equal volume in reciprocal space to ensure that each contains roughly the same number of diffraction spots. Scaling of the derivative dataset to that of the native crystal in this manner requires the use of least-squares minimisation, which will be covered in more detail in Chapter 15, and the resulting shell scale factors can be smoothed to achieve a continuous resolution-dependent scaling of the data. Whilst scaling two datasets together in this way will cancel out systematic resolution-dependent differences, it will not cancel out directional effects, such as those which an anisotropic temperature factor might be able to correct for quite adequately.

The most intricate way of scaling two datasets together is to perform *local scaling*, in which a scale factor is determined for each diffraction spot based on the intensities of surrounding reflections (Matthews and Czerwinski, 1975; Tickle, 1983, 1992; Blessing, 1997). Typically, 25–30 of the surrounding reciprocal-lattice points are used and their intensities are weighted such that the further they are from the central spot, the less they contribute to the local scale factor. Since each local scale factor is determined from a much smaller number of spots than in the case of an overall scale factor, there is a danger that whilst we reduce systematic errors in the data by local scaling, we may introduce random errors due to the small number of measurements being used. It is also important that the central spot to which the scale factor is eventually applied does not contribute excessively to our estimate of that scale factor, or we risk diminishing any meaningful isomorphous differences ($|F_{PH}| - |F_P|$) that may exist between the two datasets.

Local scaling can be used to improve estimates of anomalous differences ($|F_{PH}(+)| - |F_{PH}(-)|$) between Friedel mates by treating the $F(+)$ and $F(-)$ data as separate datasets and scaling the spots in small volumes of reciprocal space around the Friedel partners together, in much the same way as one would locally scale a native and a derivative dataset in isomorphous replacement. One further complication worth pondering is that, in principle, local scaling will work best when used with unmerged data, but, in practice, it is rarely used in this way. It is generally considered that with good experimental measurements, local scaling is unnecessary; however, it is used by default to treat isomorphous-difference and anomalous-difference data in the automated structure solution package SOLVE (Terwilliger and Berendzen, 1999a).

The scaling methods that we have described so far assume that any real differences between the two datasets (e.g. due to the binding of a ligand in the derivative crystal) are negligible, and these methods prove to be adequate for most structure analyses. However, there are reasons why this assumption might be wrong, most notably if our derivatised protein contains a number of electron-dense heavy atoms. We can consider the effect of heavy atoms on the scale factor by taking the example of a small protein of, say, 1000 atoms. We can assume that all the atoms in the protein are nitrogen, i.e. they have an atomic scattering factor of 7, which is typical of the C, N and O atoms in proteins; these elements have 6, 7 and 8 electrons, respectively. The mean scattering intensity of the protein will therefore be 1000×7^2, or 49000. If a derivative is made containing a single mercury atom (80 electrons), this will contribute an additional 80^2, or 6400, to the mean scattering intensity, which represents an increase of approximately 13%, and it is estimated that these effects lead to errors of the order of 5–10% in scale factors determined by conventional scaling methods.

A more accurate way to determine scale factors that considers the heavy-atom contribution was devised by Kraut *et al.* (1962) and has been implemented in the program FHSCAL by Ian Tickle (CCP4, 1994). The derivation of the associated formulae is based on considering the height of the origin peak of the Patterson function for the heavy-atom derivative. Following correct scaling of the derivative data, the height of its Patterson origin peak must equal the sum of the origin peak in the native Patterson function and the origin peak in the Patterson function of the heavy-atom substructure. Hence

$$K^2 \sum |F_{PH}|^2 = \sum |F_P|^2 + \sum |F_H|^2 \tag{14.3}$$

At this stage in the analysis the K and F_H terms are, of course, unknown. However, it can be shown that, on average, $|F_H|^2$ is approximately twice the square of the difference between the derivative and native structure factor amplitudes. Hence

$$\sum |F_H|^2 \approx 2 \sum (K|F_{PH}| - |F_P|)^2 \tag{14.4}$$

This effect is one that we will examine in more detail in the next section. Note that the factor of 2 does not apply for centric reflections, because F_{PH} and F_P will have the same phase in this situation and so the factor of 2 may simply be dropped. For our current purposes, equation (14.4) gives us an independent equation for $\Sigma|F_H|^2$ that can be substituted into equation (14.3):

$$K^2 \sum |F_{PH}|^2 = \sum |F_P|^2 + 2 \sum (K|F_{PH}| - |F_P|)^2$$

$$= \sum |F_P|^2 + 2 \sum \left(K^2|F_{PH}|^2 + |F_P|^2 - 2K|F_{PH}||F_P|\right)$$

$$\therefore K^2 \sum |F_{PH}|^2 - 4K \sum |F_{PH}||F_P| + 3 \sum |F_P|^2 = 0 \tag{14.5}$$

Equation (14.5) is a quadratic equation for the unknown scale factor K, and by using the formula for solving quadratics (see Chapter 2, Section 2.10) we can prove that the solution of this equation is

$$K = \frac{2 \sum |F_P||F_{PH}| \pm \sqrt{4 \left(\sum |F_P||F_{PH}|\right)^2 - 3 \sum |F_P|^2 \sum |F_{PH}|^2}}{\sum |F_{PH}|^2} \tag{14.6}$$

If we consider the special case when F_{PH} and F_P are the same, the two solutions for K are 1 and 3, which originate from the minus and plus signs, respectively, in equation (14.6). Since the latter solution is obviously incorrect, we know that we will only get realistic solutions for K by using the formula below:

$$K = \frac{2 \sum |F_P||F_{PH}| - \sqrt{4 \left(\sum |F_P||F_{PH}|\right)^2 - 3 \sum |F_P|^2 \sum |F_{PH}|^2}}{\sum |F_{PH}|^2} \tag{14.7}$$

Again it should be noted that this formula applies to acentric reflections, but an equivalent formula for centric reflections can be derived in the same manner.

Finally, we must consider some general aspects of data scaling which apply to all of the methods described above. The scale factors that are determined from the structure factor amplitudes of any dataset also have to be applied to the associated standard deviations and anomalous differences too. When data have been measured from more than one derivative, it is essential to produce a file in which all of the derivatives have been scaled to the native dataset for subsequent phasing analysis. Some datasets will contain reflections with large errors in the amplitudes, and these can often be spotted from examining the distribution of isomorphous or anomalous differences after scaling, allowing the obvious outliers to be rejected from subsequent calculations. In a MAD experiment, there will be data collected from the same crystal at different wavelengths and we subsequently have to solve the phase problem by treating one of these wavelengths as the 'native' dataset and the others as 'derivatives'. The 'native' dataset is usually one collected at a wavelength remote from the absorption edge and will have the smallest anomalous scattering effects.

14.4   The difference Patterson function

The relevance of the Patterson map in experimental phasing is that identification of the strongest peaks can give information on the positions of the heavy atoms within the unit cell, which is pivotal in determining the protein phases. However, owing to the general complexity of Patterson maps, their use, as described in Chapter 12, happens to be a rather insensitive procedure. A far more effective method of identifying the positions of heavy elements involves calculating a difference Patterson function. The Patterson function $P(u, v, w)$ was defined in Chapter 12, and for the native protein data it would have the following form:

$$P_P(u, v, w) = \frac{1}{V} \sum_h \sum_k \sum_l |F_P(hkl)|^2 e^{-2\pi i(hu+kv+lw)} \tag{14.8}$$

The Patterson map drawn from the Fourier synthesis of equation (14.8) will have peaks corresponding to all interatomic vectors. If we calculate a Patterson function from the structure factor amplitudes for an isomorphous heavy-atom derivative as follows,

$$P_{PH}(u, v, w) = \frac{1}{V} \sum_h \sum_k \sum_l |F_{PH}(hkl)|^2 e^{-2\pi i(hu+kv+lw)} \tag{14.9}$$

we may categorise the vectors it will contain into three types. If we refer to all atoms which are not the heavy atom as 'non-H', then the Patterson peaks will be due to

'non-H'–'non-H' atom pairs
'non-H'–H atom pairs
H–H atom pairs

With the assumptions of isomorphism and the identity of the molecular structures of the protein components of the native and derivative crystals, the vector between any pair of 'non-H' atoms in the derivative (PH) is the same in both magnitude and direction as that between the equivalent atoms in the native crystal (P). Under these assumptions, a significant simplification occurs when we perform the subtraction

$$\Delta P = P_{PH} - P_P$$

This can be expressed as the following summation:

$$\Delta P(u, v, w) = \frac{1}{V} \sum_h \sum_k \sum_l \left(|F_{PH}(hkl)|^2 - |F_P(hkl)|^2 \right) e^{-2\pi i(hu+kv+lw)} \tag{14.10}$$

The form of equation (14.10) is valid because the values of the exponentials are the same for both PH and P, since the indexing hkl of the diffraction patterns is identical. The Fourier synthesis of equation (14.10) is called a *difference Patterson synthesis*, and the resulting map a *difference Patterson map*. The significance of this map is that all peaks common to the PH and P Patterson maps cancel out, so that the difference Patterson map indicates only the differences between the PH and P structures. Specifically, the peaks of the difference Patterson map correspond to

'non-H'–H atom pairs

H–H atom pairs

and the strengths of these peaks depends on the atomic numbers of the atom pairs concerned. Importantly, the large number of non-H–non-H peaks are not present, which reduces the cluttering of the map.

The relevance of the difference Patterson map concerns the fact that in a protein derivative molecule PH with more than 10^3 non-hydrogen atoms (hydrogen is too light to be shown on a Patterson map), only a few of these atoms will be of the heavy element H. The great majority of the Patterson peaks are therefore of the 'non-H'–'non-H' type, and since we are not particularly interested in these, they obscure much of the significant part of the Patterson map. When we take a difference Patterson map, however, those parts of the structures of PH and P which are in common—which account for most of the structure—cancel out, leaving a map with many fewer peaks, all of which contain information on the heavy atoms. A difference Patterson map is therefore much more easily interpreted than an ordinary Patterson map, and so the construction of a difference Patterson map is a much more sensitive method of locating heavy-atom positions.

Whilst this emphasises the paramount importance of the difference Patterson map, it should be noted that in practice it is more common to calculate this function from the isomorphous differences ($|F_{PH}| - |F_P|$, or Δ_{iso}), which provide an estimate of F_H:

$$\Delta_{iso} = |F_{PH}| - |F_P| \tag{14.11}$$

This is a reasonable assumption, since whilst the scattering amplitude of a heavy atom is very large compared with individual light atoms in proteins, the lighter elements are present in far greater abundance and, as a result, F_H is generally small in comparison with F_P (and F_{PH}). Consequently, F_{PH} and F_P have approximately the same phase, i.e. they tend to be almost parallel, which makes equation (14.11) a reasonable approximation, particularly for the largest isomorphous differences, which will tend to dominate the Patterson function anyhow. This assumption is fully valid for centric reflections where F_{PH} and F_P have exactly the same phase, although it breaks down for weak F_{PH} and F_P, where the binding of the heavy atom can cause the phase of F_{PH} to differ by 180° from that of F_P. The reader is invited to verify that in such situations, which are referred to as crossovers, the minus sign in equation (14.11) should be swapped for a plus. However, equation (14.11) should be valid for most of the data, and this allows us to calculate an isomorphous-difference Patterson function by use of the equation below:

$$\Delta P(u, v, w) = \frac{1}{V} \sum_h \sum_k \sum_l [|F_{PH}(hkl)| - |F_P(hkl)|]^2 \, e^{-2\pi i(hu+kv+lw)} \tag{14.12}$$

A Patterson function calculated in this way will contain peaks for the heavy-atom sites, with the caveat that the peak heights will tend to be at half their expected values. This is something that will be discussed very shortly.

The advantage of using the isomorphous-difference Patterson function as encapsulated in equation (14.12) over using that given by equation (14.10) is that by using isomorphous differences we mitigate against the 'non-H'–'H' vector peaks, which might confuse our interpretation of the Patterson map, whilst leaving the 'H'–'H'

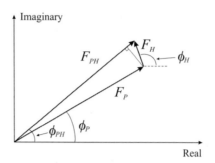

Fig. 14.1 *The isomorphous-replacement phasing triangle.* In this Argand diagram, the structure factors of the protein (F_P), the heavy-atom derivative (F_{PH}) and the heavy-atom substructure (F_H) are shown along with their corresponding phase angles.

atom peaks substantially unaffected (Dodson and Vijayan, 1971). This can be verified by considering Fig. 14.1, where F_{PH} is shown as the sum of F_P and F_H on an Argand diagram. Such figures are referred to as *Harker constructions* or *phasing diagrams*, and by simple geometric construction we can see that

$$|F_{PH}| = |F_P| \cos(\phi_{PH} - \phi_P) + |F_H| \cos(\phi_H - \phi_{PH})$$

Hence

$$|F_{PH}| - |F_P| = |F_P| \cos(\phi_{PH} - \phi_P) - |F_P| + |F_H| \cos(\phi_H - \phi_{PH})$$

$$= |F_P| (\cos(\phi_{PH} - \phi_P) - 1) + |F_H| \cos(\phi_H - \phi_{PH})$$

By using the trigonometric identity

$$\cos 2\theta = 1 - 2\sin^2 \theta$$

we may show that

$$|F_{PH}| - |F_P| = -2|F_P| \sin^2 \left(\frac{\phi_{PH} - \phi_P}{2} \right) + |F_H| \cos(\phi_H - \phi_{PH})$$

$$\therefore (|F_{PH}| - |F_P|)^2 = 4|F_P|^2 \sin^4 \left(\frac{\phi_{PH} - \phi_P}{2} \right) + |F_H|^2 \cos^2(\phi_H - \phi_{PH})$$

$$-4|F_P||F_H| \sin^2 \left(\frac{\phi_{PH} - \phi_P}{2} \right) \cos(\phi_H - \phi_{PH}) \qquad (14.13)$$

We need to remember that when this equation is used for calculating a Patterson function, each point in the Patterson function is calculated from a summation of a very large number of terms of the above form, typically several thousand. When F_H is small compared with F_P and F_{PH}, as we would expect with a useful isomorphous derivative, then $(\phi_{PH} - \phi_P)$ will also be small and the first and third terms of this equation will have low values owing to their sine terms. In the central term, the angle

$(\phi_H - \phi_{PH})$ will tend to take random values across the whole dataset. We can use another trigonometric identity,

$$\cos^2\theta = \frac{1}{2} + \frac{\cos 2\theta}{2}$$

to show that if θ adopts random values then, on average, the positive and negative values for $\cos 2\theta$ will tend to cancel out such that the mean value of $\cos^2\theta$ is 0.5. Therefore the middle term of equation (14.13) will, on average, have a value of $|F_H|^2/2$, and this tends to be by far the most dominant term in the equation. This discussion emphasises that the isomorphous-difference Patterson function will contain peaks due to the heavy atoms at approximately half their expected heights but with little contaminating noise from the protein or protein–heavy-atom vectors (Dodson and Vijayan, 1971). Since the difference Patterson function is calculated from differences between two sets of structure factors and these differences are generally small, the importance of having F_{PH} and F_P on a common scale, which was dealt with in the previous section, can be clearly seen. We have also justified the form of equation (14.4).

In Chapter 9, we saw how the presence of an anomalous scatterer causes a breakdown of Friedel's law, and in the case of an anomalously scattering heavy-atom derivative this can be expressed as

$$|F_{PH}(hkl)| \neq |F_{PH}(\overline{hkl})| \quad \text{or} \quad |F_{PH}(+)| \neq |F_{PH}(-)|$$

From the amplitudes of the Friedel-related structure factors $F_{PH}(+)$ and $F_{PH}(-)$, we can calculate a term referred to as the anomalous difference, or Δ_{ano}, which is given by the equation

$$\Delta_{ano} = |F_{PH}(+)| - |F_{PH}(-)| \tag{14.14}$$

A Patterson function calculated with anomalous differences will give information on the heavy-atom vectors. To prove this point, we need to consider the effects of anomalous scattering in an Argand diagram such as that shown in Fig. 14.2. We saw in Chapter 9, Section 9.8, that the anomalous scattering effects of an atom can be represented by the equation

$$f_j = f_{0j} + f'_j + if''_j \tag{14.15}$$

in which the quantity f_{0j} describes the classical scattering effects of the atom and is corrected by a real term f'_j and an imaginary part f''_j, which together represent the anomalous scattering effects.

In a heavy-atom derivative, the contribution of j anomalously scattering heavy atoms can be described by

$$\sum_j \left(f_{0j} + f'_j + if''_j\right)e^{2\pi i(hx_j + ky_j + lz_j)}$$

$$= \sum_j \left(f_{0j} + f'_j\right)e^{2\pi i(hx_j + ky_j + lz_j)} + i\sum_j f''_j e^{2\pi i(hx_j + ky_j + lz_j)}$$

$$= F_H + iF''_H$$

where

$$F_H = \sum_j \left(f_{0j} + f'_j\right) e^{2\pi i(hx_j + ky_j + lz_j)}$$

and

$$F''_H = \sum_j f''_j e^{2\pi i(hx_j + ky_j + lz_j)}$$

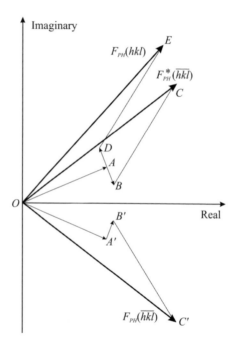

Fig. 14.2 *The effect of anomalous scattering in isomorphous replacement.* In this Argand diagram,

$$\overrightarrow{OA} = F_H\,(hkl) = \sum_j \left(f_{0j} + f'_j\right) e^{2\pi i(hx_j + ky_j + lz_j)}$$

$$\overrightarrow{OA'} = F_H\,(\overline{hkl}) = \sum_j \left(f_{0j} + f'_j\right) e^{-2\pi i(hx_j + ky_j + lz_j)}$$

$$\overrightarrow{AD} = iF''_H\,(hkl) = i\sum_j f''_j e^{2\pi i(hx_j + ky_j + lz_j)}$$

$$\overrightarrow{A'B'} = iF''_H\,(\overline{hkl}) = i\sum_j f''_j e^{-2\pi i(hx_j + ky_j + lz_j)}$$

$$\overrightarrow{AB} = -iF''^*_H\,(\overline{hkl}) = -i\sum_j f''_j e^{2\pi i(hx_j + ky_j + lz_j)}$$

$$\overrightarrow{BC} = \overrightarrow{DE} = F_P\,(hkl) = \sum_n f_n e^{2\pi i(hx_n + ky_n + lz_n)}$$

$$\overrightarrow{B'C'} = F_P\,(\overline{hkl}) = \sum_n f_n e^{-2\pi i(hx_n + ky_n + lz_n)}$$

The summations on j are for the anomalously scattering atoms and the summations on n are for the non-anomalous scatterers.

Hence F_H and F_H'' correspond to the real and imaginary components of the heavy-atom scattering, as shown in Fig. 14.2.

Since the heavy atoms are assumed to be of the same type, we may state

$$F_H + iF_H'' = |F_H|e^{i\phi_H} + i|F_H''|e^{i\phi_H}$$

We can now invoke equation (2.26), which states that

$$e^{i\theta} = \cos\theta + i\sin\theta$$

and if we substitute $\theta = \pi/2$ then we obtain

$$e^{i\pi/2} = i$$

Hence

$$F_H + iF_H'' = |F_H|e^{i\phi_H} + e^{i\pi/2}|F_H''|e^{i\phi_H}$$

$$= |F_H|e^{i\phi_H} + |F_H''|e^{i(\phi_H + \pi/2)} \tag{14.16}$$

This has a very important consequence, since it means that iF_H'' has a phase that is 90° greater than that of the real component F_H. Hence, on an Argand diagram, the F_H'' component of $F_{PH}(hkl)$ is drawn with a phase that is 90° larger than that of F_H, as is shown in Fig. 14.2.

The structure factor of the protein–heavy-atom complex can be written as

$$F_{PH}(hkl) = \sum_n f_n e^{2\pi i(hx_n + ky_n + lz_n)} + \sum_j (f_{0j} + f_j')e^{2\pi i(hx_j + ky_j + lz_j)}$$

$$+ i\sum_j f_j'' e^{2\pi i(hx_j + ky_j + lz_j)}$$

where the first summation is over the non-anomalously scattering atoms. The structure factor of the Friedel mate is given by

$$F_{PH}(\bar{h}\bar{k}\bar{l}) = \sum_n f_n e^{-2\pi i(hx_n + ky_n + lz_n)} + \sum_j (f_{0j} + f_j')e^{-2\pi i(hx_j + ky_j + lz_j)}$$

$$+ i\sum_j f_j'' e^{-2\pi i(hx_j + ky_j + lz_j)}$$

We can then express the complex conjugate of the Friedel mate as

$$F_{PH}^*(\bar{h}\bar{k}\bar{l}) = \sum_n f_n e^{2\pi i(hx_n + ky_n + lz_n)} + \sum_j (f_{0j} + f_j')e^{2\pi i(hx_j + ky_j + lz_j)}$$

$$- i\sum_j f_j'' e^{2\pi i(hx_j + ky_j + lz_j)}$$

and we see that the last term now contains a factor $-i$, which, as we can see by invoking equation (2.26), corresponds to a rotation of the anomalous scattering component F_H'' by $-90°$ for $F_{PH}^*(\bar{h}\bar{k}\bar{l})$. These operations are shown in Fig. 14.2, where we see that

$$\left|F_{PH}(\bar{h}\bar{k}\bar{l})\right| = \left|F_{PH}^*(\bar{h}\bar{k}\bar{l})\right|$$

as must be true for any pair of complex conjugates, but we also see that

$$\left|F_{PH}(hkl)\right| \neq \left|F_{PH}(\bar{h}\bar{k}\bar{l})\right|$$

since Friedel's law is broken for non-centrosymmetric crystals containing an anomalous scatterer. The contribution F_P of the native protein to $F_{PH}(hkl)$ and $F_{PH}^*(\bar{h}\bar{k}\bar{l})$ is the same in both magnitude and phase. Hence the sole difference between $F_{PH}(hkl)$ and $F_{PH}^*(\bar{h}\bar{k}\bar{l})$ is the change of sign in the imaginary part of the atomic scattering factor of the anomalous scatterer. This gives rise to the small but significant difference between the amplitudes of the Friedel mates, referred to as the anomalous difference.

When data are measured from any crystal, we will have a multitude of measurements for any hkl, some of which will be related by Friedel symmetry and some of which will be related by the symmetry elements of the crystal. For instance, in space group $P2$, in the absence of an anomalous scatterer, the reflections (hkl), $(\bar{h}k\bar{l})$, $(h\bar{k}l)$ and $(\bar{h}\bar{k}\bar{l})$ have identical amplitudes by virtue of the Laue group symmetry, which in this case is $2/m$. However, when an anomalous scatterer is present, this relationship breaks down and diffraction spots that are related by inversion or reflection operations will have different amplitudes. In contrast, those related by rotations alone will be identical. This can be summarised for $P2$ as

$$|F(hkl)| = |F(\bar{h}k\bar{l})| \neq |F(h\bar{k}l)| = |F(\bar{h}\bar{k}\bar{l})|$$

We can prove this relationship for $P2$ by considering the general equivalent positions (x, y, z) and $(-x, y, -z)$, which contribute to the calculated structure factor. Summing over all atoms in one asymmetric unit gives

$$F_{hkl} = \sum_j \left(f_j e^{2\pi i[hx_j + ky_j + lz_j]} + f_j e^{2\pi i[-hx_j + ky_j - lz_j]} \right)$$

$$= \sum_j \left(f_j e^{2\pi i k y_j} \left[e^{2\pi i(hx_j + lz_j)} + e^{-2\pi i(hx_j + lz_j)} \right] \right)$$

$$\therefore F_{hkl} = 2 \sum_j f_j e^{2\pi i k y_j} \cos 2\pi(hx_j + lz_j)$$

If the asymmetric unit contains a single anomalously scattering heavy atom, then the component of the structure factor contributed by this atom is given by

$$2 \left(f_{0H} + f_H' + i f_H'' \right) e^{2\pi i k y_H} \cos 2\pi(hx_H + lz_H)$$

From this we can see that the heavy atom's contribution will be the same for (hkl) and $(\bar{h}k\bar{l})$, since the cosine term will be unchanged. In contrast, it will be different for the other two equivalents when k changes sign, since the complex exponential term is affected and when any two complex numbers (e.g. a_1 and a_2) are multiplied, the product $a_1 a_2$ does not equal $a_1 a_2^*$.

At this stage, we do not know the phases of any of the structure factors shown in Fig. 14.2, and our reason for investigating the effects of anomalous scattering was the statement that a Patterson function calculated with anomalous differences will contain vectors due to the anomalously scattering atoms in the structure. This can be shown by consideration of Fig. 14.3, from which the following relationships can be derived by use of the cosine rule:

$$|F_{PH}(+)|^2 = |F_{PH}|^2 + |F_H''|^2 - 2|F_{PH}||F_H''|\cos(\phi_{PH} - \phi_H + 90°)$$
$$= |F_{PH}|^2 + |F_H''|^2 + 2|F_{PH}||F_H''|\cos(\phi_{PH} - \phi_H - 90°)$$

and

$$|F_{PH}(-)|^2 = |F_{PH}|^2 + |F_H''|^2 - 2|F_{PH}||F_H''|\cos(\phi_{PH} - \phi_H - 90°)$$
$$\therefore |F_{PH}(+)|^2 - |F_{PH}(-)|^2 = 4|F_{PH}||F_H''|\cos(\phi_{PH} - \phi_H - 90°) \tag{14.17}$$

From inspecting these formulae, the reader may be curious as to how the values of $|F_{PH}|$ are obtained when there are anomalous scatterers present, since our experimental measurements only give us $|F_{PH}(+)|$ and $|F_{PH}(-)|$. In general, $|F_{PH}|$ is approximated by use of the formula below,

$$|F_{PH}| \approx \frac{|F_{PH}(+)| + |F_{PH}(-)|}{2} \tag{14.18}$$

which is very reasonable if the anomalous differences are small. The amplitudes of the Friedel pairs $F_{PH}(+/-)$ can be regenerated from the anomalous difference easily by

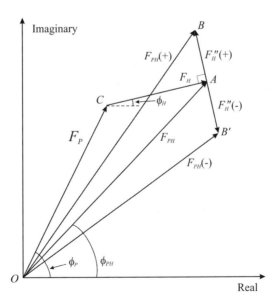

Fig. 14.3 *The phasing triangles for isomorphous replacement with anomalous scattering. Since angle OAC is $(\phi_{PH} - \phi_H)$, it follows that angle OAB is given by $(\phi_{PH} - \phi_H + 90°)$ and angle OAB' by $(\phi_H - \phi_{PH} + 90°)$. The amplitudes of F_H and F_H'' are exaggerated somewhat for clarity.*

use of the equations

$$|F_{PH}(+)| = |F_{PH}| + \Delta_{ano}/2$$
$$|F_{PH}(-)| = |F_{PH}| - \Delta_{ano}/2$$

By the difference of two squares, equation (14.17) becomes

$$|F_{PH}(+)|^2 - |F_{PH}(-)|^2 = (|F_{PH}(+)| + |F_{PH}(-)|)\,(|F_{PH}(+)| - |F_{PH}(-)|)$$
$$\approx 2|F_{PH}|\,(|F_{PH}(+)| - |F_{PH}(-)|)$$
$$\therefore 2|F_{PH}|\,(|F_{PH}(+)| - |F_{PH}(-)|) \approx 4|F_{PH}||F_H''|\cos(\phi_{PH} - \phi_H - 90°)$$

Hence,

$$|F_{PH}(+)| - |F_{PH}(-)| \approx 2|F_H''|\cos(\phi_{PH} - \phi_H - 90°)$$
$$\therefore \Delta_{ano} \approx 2|F_H''|\sin(\phi_{PH} - \phi_H)$$

and

$$\Delta_{ano}{}^2 \approx 4|F_H''|^2 \sin^2(\phi_{PH} - \phi_H) \tag{14.19}$$

Hence a Patterson map calculated with the anomalous differences Δ_{ano} as its coefficients would therefore contain vector peaks for the anomalously scattering heavy atoms at approximately twice the expected or theoretical height, owing to the fact that $\langle \sin^2 \theta \rangle = 0.5$. Note that the anomalous difference as given by equation (14.14) is sometimes referred to as the Bijvoet difference, although the same phrase can also refer to the difference in intensities of the Friedel mates as given by equation (14.17).

The size of the anomalous difference Δ_{ano} depends on the relative orientation of F_P and F_H. When they are perpendicular, the anomalous difference will be maximal $(2|F_H''|)$, and when they are collinear, the anomalous difference will be zero. The latter situation arises in centric zones of protein crystals, where F_P and F_H are collinear. Interestingly, the isomorphous difference shows exactly the opposite trends, i.e. when F_P and F_H are collinear, the value of Δ_{iso} is maximal, and when they are perpendicular, Δ_{iso} is at a minimum. This discussion suggests that we may be able to derive more information by combining the isomorphous and anomalous differences. However, to do this we need to compensate for the fact that they are on different scales owing to f'' generally being much smaller than the real component $(f_0 + f')$. The ratio of the real to the anomalous scattering is given by

$$k = \frac{f_0 + f'}{f''}$$

where k can be determined empirically in resolution shells from $\langle \Delta_{iso} \rangle / \langle \Delta_{ano}/2 \rangle$ and is therefore referred to as k_{emp}. We can then approximate $|F_H|^2$ as follows:

$$|F_H|^2 \approx \Delta_{iso}{}^2 + k_{emp}{}^2 \left(\frac{\Delta_{ano}}{2}\right)^2 \tag{14.20}$$

These terms can be used as coefficients in a Patterson synthesis and can yield improved density for the heavy-atom vectors. More rigorous formulations for combining isomorphous and anomalous differences are possible (Singh and Ramaseshan, 1966; Blundell and Johnson, 1976). However, in practice, anomalous differences tend to have high relative errors due to the low anomalous signal, unless the wavelength has been optimised, which often means that the isomorphous-difference Patterson function is cleaner and easier to interpret.

14.5   The methods of Patterson solution

By using an isomorphous-difference or an anomalous-difference Patterson map, we may identify the positions of the heavy atoms within the unit cell, and in this section we will deal with the automated methods for solution of a Patterson function. In Chapter 12, Section 12.9, we saw how the coordinates of a heavy atom may be determined by inspection of the peaks in the Patterson function, in particular the Harker sections, and these methods are now highly automated, with manifold benefits of speed and accuracy. Nevertheless, it is useful to inspect the Harker sections of the difference Patterson function to assess the quality of the derivative, for example whether or not it has discrete, preferably strong peaks. Sometimes errors in the data can manifest themselves as regularly spaced streaks or other artefacts in the Patterson function and if these are observed, the situation may be improved by rejecting spurious reflections from the calculation, normally by setting cut-offs on resolution or on the amplitude of Δ_{iso} or Δ_{ano}.

One method for automated interpretation of a Patterson function is the use of the symmetry function, which works by considering a grid of finely spaced points filling the asymmetric unit of the crystal. Each of these grid points is treated as a potential heavy-atom site, and the likelihood of it being a site is determined by calculating the Harker vectors for that site (Tickle, 1983). In Chapter 12, Section 12.8, we showed how the vectors running between heavy atoms related by the space group symmetry give rise to peaks in Harker sections of the Patterson function, known as Harker vectors. We also showed how the coordinates of these Harker vectors could be calculated from knowledge of a heavy atom's coordinates, and it is these formulae which are used by the program to calculate where the Harker vectors will appear in the Patterson function. If the program finds strong Patterson density at the corresponding positions in the Patterson function, then it is very likely that the grid point being tested actually does represent the binding site of a heavy atom or is very close to such a site. In contrast, if the program finds only noise at the corresponding positions in the Patterson function, it is very unlikely that a heavy atom resides at that particular grid point. There are various ways to assess the likelihood of each trial site actually being a heavy-atom site, and one of the most discriminating ways is to calculate a *minimum function*. If one imagines a space group with three Harker vectors, for each trial heavy-atom site the program will have to examine three Patterson density values P_1, P_2 and P_3, one for each Harker vector. The minimum function involves taking the minimum of the three Patterson densities P_1, P_2 and P_3, i.e. if all three are large, the minimum function will also be large, although it will equal the smallest of the three values. If one or two of the Patterson densities are small, then the minimum function will also be small. This makes the minimum function a sensible way of combining the information in this context, since it will often be the case that two of the trial heavy-atom coordinates (e.g. x and y) are correct while the third (e.g. z) is incorrect. In this situation, P_1 might be very high and the others (P_2 and P_3) would be low, and the minimum function, which returns the lowest of the three Patterson densities, correctly tells us that this is an unlikely solution. However, other statistical criteria such as the arithmetic mean (or average) would be far less discriminating, since they can be seriously offset by outliers and could give a falsely high result for such a solution.

The program will search for peaks in the symmetry function, and the coordinates of the highest peak can be chosen as the most promising candidate for a heavy-atom site. At this point we need to consider one important feature of the symmetry function, which is that it has a much higher symmetry than that of the space group of the crystal. In fact, if we have chosen our heavy-atom site with coordinates of, say, (x, y, z), there will be other positions in the asymmetric unit of the crystal where this heavy atom could be that will give exactly the same peaks in the Harker sections of the Patterson function. This is the same problem which we referred to in Chapter 13, Section 13.11, where we showed that when we are using molecular replacement to solve a structure that possesses non-crystallographic symmetry, there are several positions for the correctly oriented search model within the asymmetric unit that will give exactly the same peaks in the Patterson Harker sections. We showed, for the example of space group $P2_12_12_1$, that we can translate the correctly oriented search molecule exactly half a unit cell along any one of the three crystallographic cell axes and the resulting Harker vectors will be identical. If we are searching for only one molecule or heavy atom, then this ambiguity does not matter, but when we have more than one in the asymmetric unit, we need to ensure that all molecules that we locate are in positions where the cross-vectors between them correspond with peaks in the Patterson function. This problem is quite pronounced with heavy atoms because, more often than not, they will bind to the protein at several positions and, because a single atom is symmetric, unlike a protein search model in molecular replacement, there are yet more positions in the asymmetric unit where a heavy atom can be that will give exactly the same peaks in the Patterson function. In fact, for the example of space group $P2_12_12_1$ given in Section 13.11, if a heavy-atom site has been located at, say, position (x, y, z), the heavy atom can also be in any one of the following positions that will give exactly the same peaks in the Patterson function:

$$(\pm x \text{ or } \tfrac{1}{2} \pm x, \pm y \text{ or } \tfrac{1}{2} \pm y, \pm z \text{ or } \tfrac{1}{2} \pm z)$$

Hence there can be many positions within the asymmetric unit of the crystal where a heavy atom can be that will correspond exactly with a given set of Harker peaks. This means that we need to position each heavy atom at one of its possible sites such that all of the inter-heavy-atom cross-vectors agree with peaks in the Patterson function. The latter cross-peaks will not necessarily occur in Harker sections. The general procedure is to locate one of the heavy atoms by choosing the strongest or one of the strongest peaks in the symmetry function (avoiding peaks on the axes, as these tend to be artefacts) and to use this as a 'seed' site in subsequent analysis of the Patterson function. Hence the coordinates of the 'seed' site are input to the program, which will then search for further possible sites using a fine grid extending over the asymmetric unit of the crystal. As in the case of the symmetry function, each grid point is assessed as a putative heavy-atom site by calculating the corresponding Harker vectors, but this time the cross-vectors to the 'seed' site are also calculated. The Patterson densities of the Harker vectors and cross-vectors are then combined, for example by taking a minimum function. All peaks in the resulting map, which is referred to as a *superposition function*, are likely to be heavy-atom sites, and their coordinates will all be defined with reference to the same hand and origin as for the 'seed' site. Usually, the highest peak in the superposition function is selected and, along with the 'seed' site, is input to another superposition function, in which all grid points in the asymmetric unit are again screened for both Harker vectors and cross-vectors to the input sites. Additional sites can then be found and reinput to the program iteratively until no further sites are located. Note that the asymmetric unit of the superposition function is the same as the asymmetric unit of the space group, but since the symmetry function possesses much higher symmetry, its asymmetric unit is smaller.

Once the 'seed' site has been chosen, there is no ambiguity about the coordinates of the sites subsequently located in the superposition function. This can be shown by considering a 'seed' site at (x_1, y_1, z_1) and another site at (x_2, y_2, z_2). The cross-vector between these two sites is $(x_1 - x_2, y_1 - y_2, z_1 - z_2)$, and there will be a peak at the corresponding position in the Patterson map. However, the second site would also give exactly the same peaks in the Harker section if it were at an alternative position given by $(-x_2, -y_2, -z_2)$. In this case the cross-vector would be $(x_1 + x_2, y_1 + y_2, z_1 + z_2)$, which is not related by symmetry to the real cross-vector and so the possibility of the second site being at the alternative position is excluded by the superposition function, i.e. it will give a low score for that position. A more sophisticated Patterson superposition function has been developed (Sheldrick *et al.*, 1993) and is available in the SHELX program suite.

Derivatives with a small number of sites can often be solved manually with the assistance of a Patterson-searching program, but where the number of sites is large (e.g. > 6) it is preferable to use an automated algorithm. With an anomalous-difference Patterson function for a selenomethionine-containing protein, we have a very good idea of how many sites to expect from its amino acid sequence, and likewise for a mercury derivative of a cysteine-containing protein. However, with many heavy-atom derivatives, there is no *a priori* way to predict the number of binding sites, but an idea can be gained from assessing the complexity of the Harker sections. When the Patterson function has been solved, it is essential to check that the calculated Patterson function agrees well with the observed one.

An algorithm for solution of Patterson functions which essentially uses the above procedure but with a more rigorous method of assessing the sites based on the likelihood of their minimum function values having arisen by chance has been devised (Terwilliger and Eisenberg, 1987), and is available in the highly automated structure determination package SOLVE (Terwilliger and Berendzen, 1999a). This program is capable of automatically solving heavy-atom or anomalous-scattering substructures with dozens of sites present. The heavy sites are located by analysis of the difference Patterson function by the methods described above. The resulting heavy-atom substructures are generated and used as 'seeds' from which to calculate phases for the protein, by methods that we will describe later in this chapter. Once these phases are available, we can calculate a difference Fourier map to show up further sites that may have been missed in the initial Patterson search. All of the sites can be refined such that their coordinates, occupancies and temperature factors are adjusted to optimise the agreement between the calculated and the observed Patterson functions. The occupancy of a heavy-atom site refers to the percentage of protein molecules within the crystal which possess the heavy atom bound at that position. Occupancies can be significantly less than 100% with weakly bound heavy-atom reagents, although any site whose occupancy refines to zero will be deleted. It is also useful to test putative heavy-atom sites by deleting them one at a time from the substructure, calculating phases for the protein and checking that the omitted site returns in a difference map. Since a phased Fourier map will always be strongly biased towards the atoms from which the phases were calculated, systematically omitting each atom removes any bias in the phases arising from that site and gives an objective test as to its validity. The scoring functions in SOLVE allow it effectively to perform this test automatically on each putative site in the heavy-atom substructure.

14.6   Direct methods for locating sites

Direct methods are those techniques of phase determination which involve solely the comparison of structure factor magnitudes. In chemical crystallography, the structure factor amplitudes measured from a single crystal usually allow the phases to be determined without any reference to prior knowledge of likely structures or data from isomorphous derivatives. Thus direct methods represent an objective sequence of operations which will give phase information. However, on their own they cannot usually be applied to protein diffraction data for determination of the phases, owing to the large number of atoms involved and the relative paucity of the diffraction data compared with 'small molecules'. However, they can be applied to determination of the positions of heavy atoms and anomalous scatterers from isomorphous or anomalous differences using Δ_{iso} or Δ_{ano} data, respectively, which are treated in exactly the same manner as the diffraction amplitudes of a 'small molecule'. The recent application of stochastic methods has increased the size of the structures that can be determined by direct methods by an order of magnitude. Consequently, heavy-atom or anomalous-scattering substructures of the order of 100–150 or more atoms in size can be determined with a medium-resolution dataset provided that the distances between these atoms exceed the resolution of the data, which is usually the case. For proteins that diffract to atomic resolution ($\sim$1 Å), and many do at cryogenic temperatures with intense synchrotron radiation, direct methods on their own have been used to determine the whole structure. There appears to be an upper size limit of the order of

1000 atoms, but ongoing developments are likely to enhance the applicability of direct methods to ever larger problems.

Classically, direct methods relied on the assumptions that electron density is positive, that there is a discrete peak for each atom in the structure and that the atoms have the same or similar scattering factors, which is generally upheld for organic compounds. However, these constraints are not very strict and, providing that we are looking for atoms that are well resolved by the data, the methods work well.

Direct methods involve a comparison of the structure factor magnitudes, and in order to generalise the mathematics, we now define the *unitary structure factor* U_{hkl} as

$$U_{hkl} = \frac{F_{hkl}}{\sum_j f_j} \tag{14.21}$$

Since

$$F_{hkl} = \sum_j f_j e^{2\pi i (hx_j + ky_j + lz_j)} \tag{9.21}$$

we can see that F_{hkl} will be maximal if the component wavelets (such as those shown in Fig. 9.7 in Chapter 9) are collinear. This corresponds to the physically impossible situation of having all of the atoms superimposed on top of each other or to having every atom lying within the *hkl* Bragg plane, such that they all scatter in phase. Hence, for real crystals, the following applies:

$$|F_{hkl}| \le \sum_j f_j$$

and so

$$|U_{hkl}| \le 1$$

The expression for U_{hkl} can be made somewhat neater by defining the *unitary atomic scattering factor* n_j,

$$n_j = \frac{f_j}{\sum_j f_j} \tag{14.22}$$

allowing us to write

$$U_{hkl} = \sum_j n_j e^{2\pi i (hx_j + ky_j + lz_j)} \tag{14.23}$$

which is analogous to the expression for F_{hkl} in equation (9.21).

It should be noted that since the maximum value of $|F_{hkl}|$ is simply $\sum_j f_j$, the amplitude of U_{hkl} represents the fractional value of a given structure factor amplitude as compared with its maximum possible value.

Another useful quantity is the *normalised structure factor* E_{hkl},

$$|E_{hkl}| = \sqrt{\frac{|F_{hkl}|^2/\varepsilon}{\langle |F_{hkl}|^2/\varepsilon \rangle}} \qquad (14.24)$$

where ε is a symmetry enhancement factor to compensate for the fact that some classes of reflection (e.g. axial reflections) have a different average intensity from others. The advantage of E_{hkl} is that the average value of $|E_{hkl}|^2$ is unity, and it has been shown that the use of E_{hkl} is tantamount to regarding the atoms within a structure as points which do not suffer from thermal motion. We saw in Chapter 11 that, in practice, E-values are calculated in resolution shells.

Direct methods may be classified according to the means by which structure factor magnitudes are compared. Three categories may be assigned: (i) inequalities, (ii) equalities and (iii) probabilities.

Inequalities. The first direct methods to be introduced were the *Harker–Kasper inequalities*. These may be derived by the following method.

(a) Write equation (14.23) for U_{hkl}.
(b) Introduce into the expression for U_{hkl} information concerning the symmetry of the crystal. This will, in general, allow the expression for U_{hkl} to be simplified in certain ways.
(c) By making use of a mathematical relationship known as the *Cauchy inequality*, an inequality may be derived.
(d) After simplification of the algebra of the inequality, we obtain a relationship between the unitary structure factors.

As an example, let us take the simple case of a crystal containing only a centre of symmetry. Firstly, we write equation (14.23):

$$U_{hkl} = \sum_j n_j e^{2\pi i(hx_j + ky_j + lz_j)} \qquad (14.23)$$

Since the crystal is assumed to be centrosymmetric, all structure factors F_{hkl} (and hence all the U_{hkl}'s) are real and we obtain

$$U_{hkl} = \sum_j n_j \cos 2\pi \left(hx_j + ky_j + lz_j\right)$$

which we choose to write as

$$U_{hkl} = \sum_j \sqrt{n_j} \cdot \sqrt{n_j} \cos 2\pi \left(hx_j + ky_j + lz_j\right)$$

Cauchy's inequality states that

$$\left| \sum_j a_j b_j \right|^2 \leq \left(\sum_j |a_j|^2 \right) \left(\sum_j |b_j|^2 \right)$$

Identifying

$$a_j = \sqrt{n_j}$$

$$b_j = \sqrt{n_j} \cos 2\pi \left(hx_j + ky_j + lz_j \right)$$

and

$$\sum_j a_j b_j = U_{hkl}$$

we have

$$|U_{hkl}|^2 \le \sum_j n_j \sum_j n_j \cos^2 2\pi \left(hx_j + ky_j + lz_j \right)$$

But from equation (14.22) we can see that

$$\sum_j n_j = 1$$

$$\therefore U_{hkl}^2 \le \sum_j n_j \cos^2 2\pi \left(hx_j + ky_j + lz_j \right)$$

where, since U_{hkl} is real, we have dropped the modulus notation. Now,

$$\cos^2 \theta = \frac{1}{2}(1 + \cos 2\theta)$$

$$\therefore U_{hkl}^2 \le \sum_j \frac{n_j}{2} \left(1 + \cos 2\pi . 2 \left(hx_j + ky_j + lz_j \right) \right)$$

$$\le \frac{1}{2} \left(\sum_j n_j + \sum_j n_j \cos 2\pi . 2 \left(hx_j + ky_j + lz_j \right) \right)$$

$$\therefore U_{hkl}^2 \le \frac{1}{2} \left(1 + U_{2h\, 2k\, 2l} \right)$$

Rearranging, we obtain the result

$$U_{2h\, 2k\, 2l} \ge 2U_{hkl}^2 - 1 \tag{14.25}$$

This is the desired inequality, from which we may obtain certain phase information.

In that the crystal is centrosymmetric, the phase problem reduces to the allocation of the appropriate sign to each structure factor. A sign of $+1$ corresponds to a structure factor with a phase of 0, i.e. in an Argand diagram, the structure factor would lie on the real axis pointing in the positive direction. Hence,

$$F_{hkl} = |F_{hkl}|e^{i0} = +|F_{hkl}|$$

In contrast, a sign of -1 corresponds to a structure factor with a phase of 180°, i.e. it would point in the negative direction of the x axis. Hence

$$F_{hkl} = |F_{hkl}|e^{i\pi} = |F_{hkl}| \left(\cos \pi + i \sin \pi \right) = -|F_{hkl}|$$

The relation (14.25) relates information concerning the reciprocal-lattice point hkl to that for the reflection $2h\,2k\,2l$, and if the intensity at hkl is strong enough, then the sign of $U_{2h\,2k\,2l}$ may be unambiguously deduced. As an example, let us suppose that measurement of a diffraction pattern gives the following data for a given pair of intensity maxima:

$$|U_{hkl}| = 0.8$$

$$|U_{2h\,2k\,2l}| = 0.4$$

Relation (14.25) states that

$$U_{2h\,2k\,2l} \geq 2 \cdot (0.8)^2 - 1$$

$$\geq 0.28$$

In view of the fact that

$$U_{2h\,2k\,2l} = +0.4 \text{ or } -0.4$$

we conclude that only the positive sign is consistent with the inequality, and we identify

$$U_{2h\,2k\,2l} = +0.4$$

unambiguously.

As a second example, consider the case in which

$$|U_{hkl}| = 0.2$$

$$|U_{2h\,2k\,2l}| = 0.4$$

According to relation (14.25),

$$U_{2h\,2k\,2l} \geq 2 \cdot (0.2)^2 - 1$$

$$\geq -0.92$$

Since both $+0.4$ and -0.4 satisfy the inequality, the ambiguity still remains.

This example brings out one of the limitations in the use of inequality relationships. In general, they give unambiguous information only for unitary structure factor amplitudes of about one-half or greater. If the values of the U_{hkl}'s are too small, then the ambiguity remains. As the molecular size increases, the probability of any general U_{hkl} being greater than one-half decreases, and so this reduces the applicability of inequalities to larger structures. Also, a moment's thought will show that since an inequality such as relation (14.25) can only demonstrate that a given quantity must be larger than another, we can never prove that $U_{2h\,2k\,2l}$ is associated with a negative sign. Since we know that many structure factors are associated with negative signs (if all were positive, there would be a large peak at the origin of the electron density map), we see that inequalities can in general never identify the signs of all the structure factors, and so a total solution to the phase problem is impossible.

Different inequalities may be generated for all the 230 space groups, and further examples are given in Woolfson (1997). These represent a formidable range of possible relationships to try, but certain difficulties do arise, not least the fact that the general use of inequalities is restricted to centrosymmetric crystals. The resolution of a positive/negative ambiguity is a far simpler problem than the identification of a phase angle anywhere between 0 and 2π, and no general inequalities for the non-centrosymmetric case have been found to be useful.

Equalities. One of the difficulties with the Harker–Kasper inequalities is the nature of the inequality itself—being a somewhat vague relationship, one would not expect an inequality to be of widespread application. In 1952, this problem was to a certain extent alleviated by the introduction of certain equalities which give more specific information than do the inequalities. The equality relationships to be discussed refer only to centrosymmetric crystals or centric zones of non-centrosymmetric crystals, and this represents a major limitation on their use. Since only the signs of the structure factors are to be determined, we may represent the sign associated with a structure factor F_{hkl} by using the notation $s(hkl)$.

By considering a structure of identical, resolved (i.e. non-overlapping) atoms, Sayre (1952) derived the following relationships between the signs of three intensities indexed as hkl; $h'k'l'$; and $h + h', k + k', l + l'$:

$$s(hkl) = s(h'k'l') \, . \, s(h + h', k + k', l + l')$$

and since $s(hkl) = 1/s(hkl)$,

$$s(h + h', k + k', l + l') = s(hkl) \, . \, s(h'k'l') \tag{14.26}$$

Thus knowledge of the signs $s(hkl)$ and $s(h'k'l')$ permits the sign $s(h + h', k + k', l + l')$ to be determined unambiguously. This *triple-product relationship* (TPR) between reflections forming such a *vector triplet* is true only for the hypothetical case of a structure composed of identical atoms, and has been shown to hold if the associated structure factors are large.

Probabilities. When the magnitudes of the unitary structure factors U_{hkl} are not large enough for equation (14.26) to be strictly true, it has been shown that the relation is still probably correct, although in some cases the sign equality may break down. If we adopt the symbol $\approx$ to mean 'probably equals', then, for a general centrosymmetric crystal, equation (14.26) may be rewritten in the form

$$s(hkl).s(h'k'l').s(h + h', k + k', l + l') \approx 1 \tag{14.27}$$

Various analyses have been carried out to determine the statistical interpretation of the relation (14.27) so that the probability of its being correct may be quantified.

The use of probabilities as opposed to equalities is therefore rather more general, although subject to errors. But if a set of signs can be identified, we may be able to compute an interpretable Fourier synthesis and remove the errors in the resulting electron density map during subsequent refinement.

One of the problems, however, associated with the use of probabilities is that they can give the sign of a certain structure factor only if those associated with two others are already known. Often, a few signs may be known unambiguously or, in certain circumstances, three signs may be arbitrarily allotted, which fixes the origin of the structure (Woolfson, 1997). In general, the origin-fixing reflections have to be chosen by considering their parity group. In this situation, parity refers to whether the h, k and l are even (e) or odd (o). Considering a structure with a centre of symmetry at $(0, 0, 0)$, if the origin is moved to other centres of symmetry, the signs of the structure factors can change. When the origin is at $(0, 0, 0)$, the structure factor is given by

$$[F_{hkl}]_{0\ 0\ 0} = \sum_j f_j e^{2\pi i \left(hx_j + ky_j + lz_j \right)}$$

whereas when the origin is shifted half a unit cell along **a** the structure factor becomes

$$[F_{hkl}]_{0.5\ 0\ 0} = \sum_j f_j e^{2\pi i \left(h(x_j - 1/2) + ky_j + lz_j \right)}$$

$$[F_{hkl}]_{0.500} = (-1)^h [F_{hkl}]_{0\ 0\ 0}$$

Hence if h is odd, this origin shift changes the sign of the reflection but if h is even, the sign is unaffected. Hence the signs of reflections in different hkl parity groups (*oee*, *oeo* etc.) are affected in different ways by different origin shifts, except for reflections in the *eee* parity group, which do not change sign whichever origin is used. If three strong reflections belonging to different parity groups are chosen such that the product of their parities on h, k and l does not give the *eee* parity group, then by assigning these reflections arbitrary signs (e.g. all $+1$) the origin is uniquely defined to be one of eight possible positions in the unit cell (Woolfson, 1997). This set of reflections may be used as a starting point to extend phases to other reflections that are related to the starting set by inequality or probability relationships.

One of the features of the use of probabilities (and also direct methods in general) is that it is not possible to assign phases to all structure factors simply by moving from one reciprocal-lattice point to the next. The techniques work only for sets of three structure factors $U_{hkl}, U_{h'k'l'}$ and $U_{h+h'\ k+k'\ l+l'}$ and are more likely to be valid if the magnitudes of the structure factors are large. In practice, the signs of a number of structure factors may be determined, but sooner or later the phase-determining technique becomes stuck, and we must start again for other groups of structure factors.

If it is impossible to initiate a sequence of sign determination owing to lack of knowledge of two signs with which to start the procedure, then, in order to gain as much information as possible, we may use *sign symbols* in the following manner. Suppose a certain structure factor has a large magnitude $|U_{hkl}|$ but its sign is unknown. If we allot to this $|U_{hkl}|$ a sign symbol a (which represents either $+1$ or -1), then certain other structure factors may be shown to have the same sign a or perhaps the opposite sign, $-a$. This process may be repeated as often as possible using all reflections related through equation (14.27), and eventually most of the structure factors may be given either an unambiguous sign or one of several sign symbols a, b, c, ... In *symbolic addition*, it is a matter of using all of the relationships to reduce the number of unknown sign symbols for the whole dataset (Ladd and Palmer, 2003).

The sign symbols represent one of two possibilities and, by introducing their use, we have reduced the total number of unknowns to be determined but the phase problem is still by no means solved. If there are n sign symbols, since each can take two values, there are still 2^n different possible ways in which the set of structure factors may be associated with signs. For a structure expressed in terms of three sign symbols, there are only eight combinations, and it is quite easy for a computer to calculate a Fourier synthesis for each of these eight possibilities, and so electron density maps may be drawn. Inspection of these eight maps may allow one to be selected as being consistent with other information known about the molecule. But if there are ten sign symbols, there are now 1024 different combinations, and it would take quite a long time to calculate this number of Fourier syntheses and select the correct map. As the molecular size increases, the number of independent intensities also increases, and this inevitably implies that successively more sign symbols have to be used. This is one of the reasons why direct methods become of less use as the size of the molecule under study increases.

So far, the entire discussion has been concerned with centrosymmetric structures. The problem is vastly more difficult for non-centrosymmetric molecules, but one particular probability relation, known as the *tangent formula* (Karle and Hauptman, 1956), is applicable to these cases, and has been widely used.

If the phase angle associated with the structure factor F_{hkl} is represented for brevity as ϕ_h, then

$$\tan \phi_h \approx \frac{\langle |E_{h'}||E_{h-h'}| \sin (\phi_{h'} + \phi_{h-h'}) \rangle_{h'}}{\langle |E_{h'}||E_{h-h'}| \cos (\phi_{h'} + \phi_{h-h'}) \rangle_{h'}} \tag{14.28}$$

in which the $\langle \ldots \rangle_{h'}$ notation represents that an average is taken over all values of $h'k'l'$. This equation can be used to estimate phases for reflections from a known starting set, such as might be derived from the origin-fixing set and the reflections in centric zones. This is referred to as *tangent expansion*. The phases for all reflections can then be recalculated from the previous estimates using the above formula in an iterative process known as *tangent refinement*. One of the most important developments in direct methods has been the inception of *multi-solution methods*, which have led to the demise of symbolic addition as a means of solving structures (Germain *et al.*, 1970; Sheldrick, 2008). The multi-solution approach involves giving strong reflections random phases, and the tangent formula is then used to estimates phases for the remainder of the dataset, which are then subjected to cycles of tangent refinement. The same strong reflections are then given a different set of random phases and the tangent refinement repeated. This process is repeated with as many initial random phase sets as the user chooses, which can be anything from 50 to 1000, or many more if needed. Each possible solution is given a figure of merit (see below), which allows the likelihood of it being correct to be assessed and all solutions to be ranked automatically. Generally, the solutions fall into two clusters: those with a good figure of merit, where the phase sets have converged to correct values, each giving a consistent electron density map for the structure, and those where the phases are wrong and give an incorrect map. This bimodality is an important (but not definitive) sign that the structure has been solved, and can be detected for automatic termination of the process (Xu and Weeks, 2008).

Further recent improvements in direct methods involve the use of *dual-space recycling* (Weeks *et al.*, 2001; Ealick, 1997), in which trial atoms are placed in the unit cell either at random positions or at positions which are consistent with peaks in the Patterson function. The trial structures are used to generate phases, which are refined in reciprocal space, essentially by the methods described in the previous paragraph. The resulting phases are then used to calculate a map in real space, from which peaks are picked either at random or by taking the strongest or, effectively, by taking those that are consistent with the Patterson function. The selected peaks are used to calculate phases again, which are then refined as before, and a new electron density map is calculated, from which peaks are picked again. These cycles are repeated several times over to give a putative solution for the structure. Then a completely new random structure is generated and refined in the same way, thus generating another putative solution. These methods will generate as many solutions as the user specifies and can even be left to run indefinitely. The program SHELXD (Schneider and Sheldrick, 2002) uses a superposition function to generate starting atoms that are consistent with the Patterson function, and during the recycling, a proportion of the peaks picked at random from the Fourier map are omitted from the phase calculation to help eliminate bias—correct sites should be regenerated anyhow following the phase refinement. As with conventional multi-solution methods, the putative structures are ranked by their figure of merit. SHELXD uses the correlation coefficient between E_o and E_c as the figure of merit for each putative solution, whereas SnB (short for 'shake and bake') uses a probabilistic minimal function (Weeks *et al.*, 2001) and a crystallographic R-factor (similar to equation 14.1) based on E-values. Both methods for dual-space recycling require an estimate of the number of expected sites. Other direct methods programs, for example ACORN (Yao *et al.*, 2006), are available, and SnB is now available in the BnP suite of macromolecular solving tools (Weeks and Furey, 2007).

In dual-space recycling, the use of trial positions determined from the Patterson function is well suited for heavy-atom substructure solution, whereas the use of random positions is likely to be of greater value in solving whole macromolecular structures *ab initio* using atomic-resolution data, where the Patterson function would be too convoluted to solve by superposition methods. Proteins that diffract to near-atomic resolution and possess endogenous 'heavy atoms' such as iron are highly amenable to the dual-space approach, since the heavy-atom positions are readily determinable from the native Patterson function and can serve as seeds for the process (Frazão *et al.*, 1995).

In general, with direct methods only the strongest E-values (ideally $|E| > 1.5$) are used initially, owing to the inherently greater reliability of the phases determined for these reflections. Since only a small proportion of the data are used (typically 15%), this confers great speed on the calculations even for the more computationally intensive dual-space recycling. Note that the use of E-values can cause a problem in high-resolution shells, where the signal is weak, since the noise will be inflated, and it is therefore important to truncate the data at a suitable resolution to omit measurements below the noise level.

14.7   Refinement of heavy-atom sites

The coordinates of the heavy-atom sites for any particular derivative, which can be determined either by Patterson or by direct methods, are approximate owing to various sources of error. It is possible to derive improved coordinates, the occupancy of the heavy atom at each site and even the temperature factor for each site by a number of refinement methods. In *reciprocal-space refinement*, we adjust the parameters of the atoms in the structure or, in this case, heavy-atom substructure, so that the calculated diffraction pattern agrees as well as possible with the observed one. It is also possible, and indeed very common with heavy-atom derivatives, to perform *real-space refinement*, in which the heavy-atom coordinates, occupancies and temperature factors are adjusted so that the Patterson function calculated from these parameters agrees as well as possible with the observed difference Patterson function obtained from the isomorphous or anomalous differences. In all refinement protocols using low-to-medium-resolution data, the estimated temperature factors of the heavy atoms will be highly correlated with the occupancies and it is common to refine only the heavy-atom coordinates and occupancies for this reason, especially if only low-resolution data are available ($d_{min} > 4.5$ Å).

The importance of heavy-atom refinement lies in the fact that to estimate the phases of the protein, we need to have an accurate estimate of the heavy-atom structure factor F_H. This structure factor is calculated from the heavy-atom coordinates (x_j, y_j, z_j) and temperature factors B_j using the formula described in Chapter 11, but modified to include the occupancies q_j of each heavy-atom site:

$$F_{hkl} = \sum_j q_j f_j e^{-B_j \sin^2 \theta / \lambda^2} e^{2\pi i (hx_j + ky_j + lz_j)} \tag{14.29}$$

The occupancy represents the fraction of molecules within the crystal possessing a heavy atom at that site.

The coordinates of a heavy-atom site can be determined reasonably accurately either by direct methods or by analysis of the Patterson function, and these determine the phase of its contribution to F_H. In contrast, the occupancy and temperature factor of a heavy-atom site determine the amplitude of its contribution to F_H. Hence these parameters are very important, since if they contain errors, this will give rise to errors in the calculated phases of the protein. Arguably, the occupancy is the most important parameter to be determined at this stage, since the heavy-atom coordinates derived from the Patterson function or by direct methods will most likely give a sufficiently accurate estimate of the heavy-atom phase for our present purposes, assuming that the substructure has been solved correctly. In contrast, the uncertainty in the occupancy could be much greater. For example, if the occupancy of a particular heavy-atom site is less than 50%, merely assuming that its occupancy is 100% will overestimate its contribution to F_H by more than a factor of 2. Errors in the occupancies lead to significant bias in the final electron density map, in the form of peaks or holes in the electron density for the protein at the heavy-atom positions (Bloomer, 1983).

In real-space refinement of heavy-atom positions, which is performed by a number of programs, for example VECREF (Tickle, 1991) and SOLVE (Terwilliger and Eisenberg, 1983), the Patterson function is calculated from the initial heavy-atom parameters, i.e. the coordinates of the heavy-atom sites assuming that their occupancy is 100%, and assigning the sites a trial temperature factor (e.g. 20). The calculated Patterson function is then compared with the isomorphous-difference Patterson function, and the heavy-atom parameters are adjusted to maximise the fit of the calculated Patterson function to the observed difference Patterson function. In practice, this is done by adjusting the parameters to minimise the sum of the squares of the differences between the two Patterson functions. This is an example of least-squares refinement, which is something that we will return to many times, most notably in Chapter 15, where the details will be covered extensively. The refinement algorithms take into account the fact that, as discussed in Section 14.4, the peaks in an isomorphous-difference Patterson function will be approximately half their expected values owing to the approximation made in calculating this function. Care has to be taken in handling the Patterson origin peak, which dominates the function, since a heavy atom at any position, not necessarily a correct one, will improve the correlation between the observed and calculated Patterson functions if the origin peak is considered. In principle, the origin may simply be excluded or the data can be treated to effectively remove it using methods that we described in Chapter 13, Section 13.3. The program VECREF takes the additional step of including only Patterson densities that are within a fixed radius of the peaks of the calculated Patterson function to exclude noise peaks in the observed Patterson function. One advantage of real-space refinement is that it has a large radius of convergence, i.e. it can correct for larger errors in the trial heavy-atom parameters than can be rectified by other methods; for example, for the positional parameters, the radius of convergence is estimated to be $d_{min}/\sqrt{2}$, where d_{min} is the maximum resolution of the dataset. This compares with a typical radius of convergence of $d_{min}/4$ for reciprocal-space methods.

It used to be commonplace to refine heavy-atom parameters in reciprocal space using estimated $|F_H|$ values (obtained for example by use of equation (14.20)) as experimental observations and using standard refinement procedures that will be covered in the next chapter. Whilst this method was very popular, it has been largely superseded by real-space or maximum-likelihood phase refinement methods, which have greater statistical rigour and stability.

One of the most commonly used methods of heavy-atom refinement is *phase refinement*, which works in reciprocal space and, as the name implies, involves calculating phases for the protein. The details of how the protein phases are calculated from the heavy-atom parameters will be covered in a later section, but for now we will assume that the phase for each structure factor F_P can be estimated from the heavy-atom derivative data and their associated heavy-atom coordinates. In practice, the heavy-atom derivative that we wish to refine is omitted from the phase calculation to avoid bias, and therefore phase refinement can only be done properly when more than one derivative is available. For each protein structure factor, we can use the protein phase ϕ_P and the amplitude $|F_P|$, along with the heavy-atom parameters for the derivative that is being refined ($|F_H(\text{calc})|$ and $\phi_H(\text{calc})$, obtained from equation

(14.29)), to calculate the amplitude of the protein–heavy-atom complex $|F_{PH}(\text{calc})|$. The value of $|F_{PH}(\text{calc})|$ is compared with the observed value $|F_{PH}(\text{obs})|$, and the heavy-atom parameters are adjusted iteratively to minimise the sum of the squares of the differences between $|F_{PH}(\text{obs})|$ and $|F_{PH}(\text{calc})|$ for the whole dataset whose heavy-atom parameters are being refined. Incidentally, the difference between $|F_{PH}(\text{obs})|$ and $|F_{PH}(\text{calc})|$ is referred to as the *lack-of-closure error*, which will be discussed in more detail later. Therefore phase refinement of a derivative is concerned with minimising the sum of the squares of the lack-of-closure errors for that derivative, the variables being its heavy-atom coordinates, occupancies and temperature factors. Once refinement of one derivative is complete, another heavy-atom derivative is refined in exactly the same way and the process is repeated until all available derivatives have been refined. The whole process is repeated for all derivatives until the heavy-atom parameters converge. One problem with phase refinement relates to the tendency of different heavy-atom derivatives to have common sites, which is an unavoidable feature of heavy-metal chemistry and can lead to significant bias in the refined heavy-atom parameters.

Phase refinement has been improved by the application of maximum-likelihood theory, which considers the probability of a particular set of measurements (in this case $|F_P|$ and $|F_{PH}|$) being made given the trial heavy-atom parameters being refined. The optimum set of parameters is that which maximises this probability, which is calculated as a log-likelihood. This method, which has been implemented in the program MLPHARE (CCP4, 1994), has an advantage over the earlier phase refinement protocol, which is that the protein phase is not treated as a constant; in fact, all possible values of the phase are considered and weighted according to the probability of being correct. Further significant developments of the method are available in the program SHARP (De la Fortelle and Bricogne, 1997).

14.8 Cross-phasing

One of the complications of determining the heavy-atom sites is that we need to ensure that all sites of a derivative are determined with respect to the same origin and hand. With Patterson methods, we saw above in Section 14.5 that this problem is overcome by choosing one site (usually the strongest) from the symmetry function and then using the superposition function to determine the positions of other sites which have cross-vectors to the first site. Hence all of the sites are then determined with respect to the hand and origin of the site that we initially picked from the symmetry function. Had we selected a different site from the symmetry function, all of the sites subsequently determined from the superposition function could have different coordinates from the previous set owing to being on a different hand and origin. However, this would not matter, because it represents a perfectly valid alternative solution of the Patterson function.

If we were able to use just one derivative to calculate an electron density map for the protein, then the structure of the protein would be determined with respect to the same origin and hand as that of the heavy-atom substructure used to calculate the map. Normally, however, we require several heavy-atom derivatives to calculate an interpretable electron density map for the protein. If each of these derivatives is solved, for example by Patterson methods alone, then it is very likely that they will have different hands and origins, and if we were to combine the phase information from all of the derivatives at this stage to calculate an electron density map for the protein, the resulting map would be completely uninterpretable. Hence it is of the utmost importance to ensure that all derivatives have their heavy-atom substructures determined with respect to the same hand and origin. Which hand and origin is entirely arbitrary, as long as all heavy-atom sites in all derivatives are measured with respect to the same reference frame.

There are various ways of doing this, but by far the most popular is to calculate a cross-phased Fourier map. As mentioned above, we can calculate an electron density map for the protein using any one of the derivatives, and if we happen to choose a very good derivative (e.g. *PH*1), we might be able to see interpretable features of electron density such as right-handed α-helices and β-sheets. In order to calculate this map, we need to calculate phases for the protein (ϕ_{hkl}), and we do this by methods that will be covered in the next section. However, normally the phases obtained from a single derivative are not of good enough quality to yield an interpretable electron density map for the protein. Nevertheless, we can use these phases to calculate a difference Fourier map for another derivative (e.g. *PH*2), which should contain peaks for the heavy-atom sites in the second derivative with the same hand and origin as for the derivative used to calculate the phases. The formula we use is

$$\rho(x, y, z) = \frac{1}{V} \sum_h \sum_k \sum_l \left(|F_{PH2}(hkl)| - |F_P(hkl)| \right) e^{i\phi_{hkl}} e^{-2\pi i (hx + ky + lz)} \qquad (14.30)$$

We need to repeat this process for each derivative to ensure that all heavy sites are determined relative to the same hand and origin as for the first derivative (*PH*1). As we will see in the next chapter, where difference Fourier maps are described in more detail, the quality of the difference map depends on how accurate the phases are. Hence it helps to use the best derivative as the one from which we calculate protein phases, so that we stand the best chance of finding all of the sites in the difference Fourier maps for the other derivatives. In practice, we also need to check that the sites found in each difference Fourier map agree with peaks in the Patterson function for that derivative. If the derivative that we use to calculate the protein phases has strong anomalous differences and if we use the Δ_{ano} data in the phase calculations, we should also calculate a second set of protein phases with the heavy-atom sites of that derivative inverted. Whichever phase set gives the strongest peaks in the cross-phased Fourier map for another derivative defines the absolute hand of the heavy-atom substructure for the first derivative. All derivatives subsequently solved by use of the cross-phased Fourier map will then be on the correct absolute hand, too. If we do not follow this procedure, then we can arrive at an electron density map for the protein which is a mirror image of what it should be. This can be recognised from the presence of left-handed α-helices rather than the expected right-handed ones, and can easily be corrected by inverting all heavy-atom sites used in the phasing and recalculating the protein electron density map, which should then have the correct hand. The absolute hand of a heavy-atom derivative with a strong anomalous signal can also be determined by using a phase refinement procedure and refining the occupancies using the anomalous data. If the resulting 'anomalous occupancies' refine to negative values, the heavy-atom substructure should be inverted and re-refined. Positive anomalous occupancies indicate that the absolute hand of the heavy substructure is correct. Note that with MAD or SAD phasing, the enantiomorph cannot be determined from the refined occupancies or any other phasing statistics, and must be assigned by calculating the electron density map for the protein with the substructure on both hands. The best map indicates the correct hand.

Given the complexities of these steps and the proneness of humans to error, it is sensible to use one of the automated phasing packages such as SOLVE (Terwilliger and Berendzen, 1999a), which perform the above calculations without the need for manual intervention; for example, when anomalous data are present, SOLVE automatically tests the inverse of the heavy-atom configuration to determine the correct hand of the anomalous-scattering substructure.

14.9 The isomorphous replacement method

In the preceding sections, we have looked at methods for determining the binding sites of heavy atoms along with methods for ensuring that all of the sites in all derivatives have a consistent hand and origin, and we have also looked at techniques for refining the heavy-atom parameters. In these sections we mentioned several times that, once the heavy-atom parameters have been determined and preferably refined, we can calculate phases for the protein molecule and thence reach our goal of calculating an electron density map. This section will now deal with the process of estimating the phases for the protein from the heavy-atom derivative data.

One way to visualise the problem of determining the phase ϕ_P of a protein structure factor F_P is shown in Fig. 14.4. It is important to note that F_H is known in terms of amplitude and phase at this stage, since these parameters can be calculated from the heavy-atom substructure. Following the treatment given in Blundell and Johnson, (1976), use of the cosine rule gives

$$|F_{PH}|^2 = |F_P|^2 + |F_H|^2 - 2|F_P||F_H|\cos\left(180° - (\phi_P - \phi_H)\right)$$

$$= |F_P|^2 + |F_H|^2 + 2|F_P||F_H|\cos(\phi_P - \phi_H)$$

$$\therefore \cos(\phi_P - \phi_H) = \frac{|F_{PH}|^2 - |F_P|^2 - |F_H|^2}{2|F_P||F_H|}$$

Hence

$$\phi_P - \phi_H = \arccos\left(\frac{|F_{PH}|^2 - |F_P|^2 - |F_H|^2}{2|F_P||F_H|}\right)$$

$$\therefore \phi_P = \phi_H + \arccos\left(\frac{|F_{PH}|^2 - |F_P|^2 - |F_H|^2}{2|F_P||F_H|}\right)$$

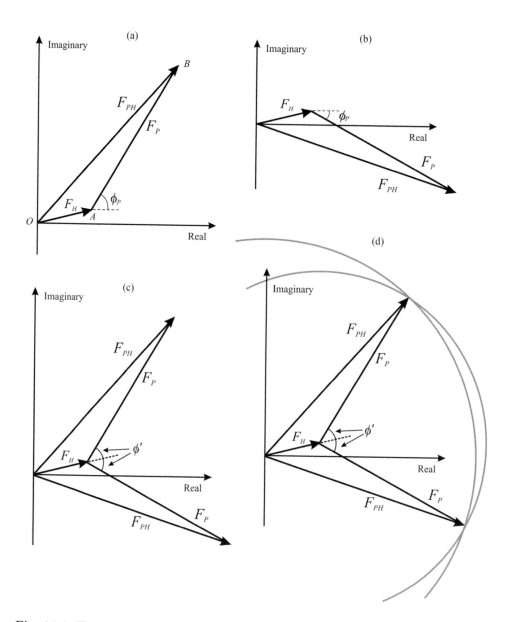

Fig. 14.4 *The alternative values of the protein phase arising in single isomorphous replacement.* (a) shows how the phasing triangle is constructed from knowledge of the amplitude and phase of F_H and only the amplitudes of F_{PH} and F_P. Note that angle OAB is $180° - (\phi_P - \phi_H)$. (b) shows an alternative and equally plausible construction. Both possible phasing triangles are shown in (c), which demonstrates that the two possible values for ϕ_P are symmetrically disposed about F_H. (d) shows a geometric method of solving the phase problem, in which a circle of radius $|F_P|$ is drawn with its centre at the head of F_H. Another circle, of radius $|F_{PH}|$, is drawn with its centre at the origin and is found to intersect the other circle at two points, each of which represents a possible solution to the unknown protein phase.

Since ϕ_H and $|F_H|$ have already been determined and the other terms in this equation have been measured ($|F_{PH}|$ and $|F_P|$), it appears that we can now calculate the protein phase ϕ_P. However, there is a slight problem arising from fundamental trigonometry, which is that if we define an angle ϕ' as

$$\phi' = \arccos\left(\frac{|F_{PH}|^2 - |F_P|^2 - |F_H|^2}{2|F_P||F_H|}\right) \tag{14.31}$$

then, because $\cos(\theta) = \cos(-\theta)$, there is an ambiguity in the sign of the angle ϕ' and the equation for the phase of the protein becomes

$$\phi_P = \phi_H \pm \phi' \tag{14.32}$$

Hence, with a single heavy-atom derivative, we have two possible values for ϕ_P, which are shown in Fig. 14.4(a) and (b), and these are symmetrically disposed about F_H, as shown in Fig. 14.4(c). Hence, with *single isomorphous replacement* (SIR), we therefore reduce the phase problem to an 'either/or' ambiguity.

An alternative way to visualise the solution of the phase problem by SIR and the resulting ambiguity in the protein phase is by geometric construction using *phasing circles*. This is shown in Fig. 14.4(d), in which we have drawn a circle of radius $|F_{PH}|$ centred at the origin and another circle of radius $|F_P|$ centred at the head of F_H. There are, of course, two positions at which these circles intersect, corresponding to the two possible values for the protein phase, which at this stage cannot be distinguished. The key to solving this ambiguity is in the preparation of another heavy-atom derivative in which the heavy atom occupies different sites in the protein molecule. If we repeat the SIR phasing exercise with this second derivative, it will give us two more independent and equally likely estimates of the protein phase, one of which should equal one of the estimates obtained with the first derivative, to within a reasonable margin of error. Thus the availability of a second derivative allows us to fully resolve the phase problem, and this is the key concept behind the method of *multiple isomorphous replacement* (MIR). We can use geometric construction to demonstrate how this works, as shown in Fig. 14.5. Since this process has to be repeated for every reflection in the dataset, which is likely to consist of several thousand diffraction spots, in practice the calculations have to be done by computer.

When there are more than two derivatives, we could in principle treat them as we have just described to obtain an estimate of the protein phase from each pair of derivatives and arrive at some sort of consensus phase for each reflection. However, in practice, current MIR phasing methods use data from all of the derivatives simultaneously to derive statistical estimates of the protein phases. The first attempt to analyse the errors in experimental phasing and derive equations that would allow all derivatives to be treated simultaneously was due to Blow and Crick (1959). These authors reasoned that, to a first approximation, the errors in the phasing triangle, such as that shown in Fig. 14.6(a), could be considered to reside in $|F_{PH}|$. Since $|F_P|$ is considered to be accurately measured, the position of point P in Fig. 14.6(a) depends only on the values of F_H and F_{PH}. The effect of errors in F_H and F_{PH} means that we will not necessarily be able to fully close the triangle, and point P will reside within

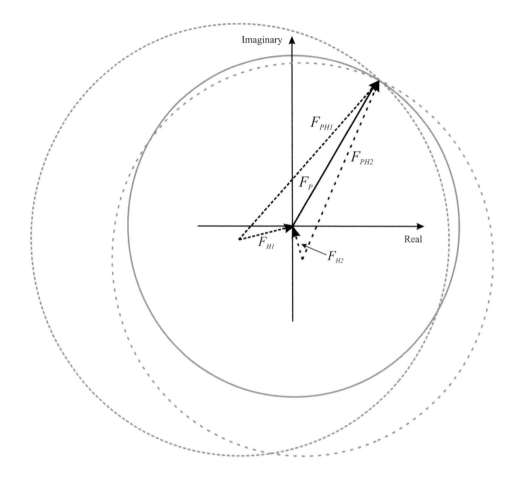

Fig. 14.5 *The phasing circles with two heavy-atom derivatives.* When two derivatives are available, F_{H1} and F_{H2} corresponding to the two derivatives can be drawn with their heads at the base of F_P. A circle is then drawn with a radius corresponding to $|F_{PH1}|$ centred at the base of F_{H1}, and another circle of radius $|F_{PH2}|$ centred at the base of F_{H2}. Where these two circles intersect with the circle of radius $|F_P|$ centred on the origin (drawn as a solid line) defines the phase of F_P uniquely.

an elliptical probability distribution with the major axis of the ellipse parallel to F_{PH} as shown in Fig. 14.6(b). Moving P along the minor axis of the ellipse does not alter the phase of F_P relative to F_H very much if F_H is small compared with the other two sides of the triangle. Hence the probability distribution may be represented by a Gaussian curve along the major axis of the ellipse as shown in Fig. 14.6(c), and the whole error may be regarded as residing in F_{PH}.

Let us assume that we have a trial value for the protein phase ϕ_P which we have obtained by some means or even simply guessed. We can then use this trial phase in combination with our knowledge of $|F_H|$, $|F_P|$ and ϕ_H to calculate a theoretical

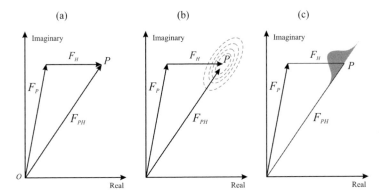

Fig. 14.6 *The lack-of-closure error.* (a) shows a typical phasing triangle, in which it is assumed that $|F_P|$ is error-free, and if F_H is small in comparison, the errors in it will have a small effect on the calculated protein phase. Consequently, the error distribution is elliptical with the long axis parallel to F_{PH}, as shown in (b). Blow and Crick proposed that the lack-of-closure error could be considered to reside in F_{PH} with a normal distribution, as shown in (c).

value for F_{PH}, referred to as F_{PH}(calc), and then compare this with the observed value F_{PH}(obs). The difference between $|F_{PH}(\text{obs})|$ and $|F_{PH}(\text{calc})|$ is referred to as the 'lack-of-closure error', or ε (shown in Fig. 14.7), and it should be low if the trial value of the protein phase is close to one of the ϕ_P values at which the phasing circles intersect. Conversely, the lack-of-closure error will be large if our trial estimate of the protein phase does not allow the phasing triangle to close. The formula for the lack-of-closure error for any F_{PH} of the jth heavy-atom derivative, ε_j, is

$$\varepsilon_j(\phi_P) = |F_{PHj}(\text{obs})| - |F_{PHj}(\text{calc}, \phi_P)| \tag{14.33}$$

and the probability that the trial protein phase angle ϕ_P for that reflection is correct, as assessed using the data for the jth derivative, is given by

$$P_j(\phi_P) = e^{[-\varepsilon_j(\phi_P)^2/2E_j^2]} \tag{14.34}$$

where E_j is a measure of the total error arising from errors in the data measurement and processing, non-isomorphism, and incorrect heavy-atom parameters. For practical purposes, E_j is taken as the probability-weighted rms lack-of-closure error in F_{PHj}, calculated for all possible values of ϕ_P. The nature of this formula is such that when the amplitude of the lack-of-closure error is low, the probability that the protein phase is correct has a high value and, conversely, when $\varepsilon_j(\phi_P)^2$ is large, the probability $P_j(\phi_P)$ has a low value, which is consistent with an unlikely phase. We can use this formula to calculate how $P_j(\phi_P)$ varies as a function of ϕ_P, giving us a *phase probability distribution*, which for a single derivative will be a bimodal curve such as that shown in Fig. 14.8(a). The peaks in this curve correspond to protein phase angles where the lack-of-closure error is lowest, and therefore these represent the two most probable values for the phase.

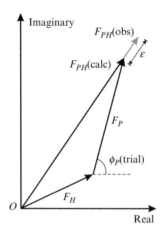

Fig. 14.7 *Estimating the lack-of-closure error as a function of the protein phase.* From knowledge of $|F_H|$ and ϕ_H (both of which are calculated from the heavy-atom substructure), we can use $|F_P|$ and a trial value of the protein phase $\phi_P(\text{trial})$ to calculate $|F_{PH}|$, which is shown as $|F_{PH}(\text{calc})|$. The difference between the observed and calculated values of $|F_{PH}|$ is referred to as the lack-of-closure error, ε, which will be small if the trial phase is near to one of the values that closes the phasing triangle, one of which is the correct phase. Conversely, ε will be large if the trial phase is a poor estimate.

Hence the Blow and Crick approach allows us to calculate a phase probability distribution for a heavy-atom derivative. We said above that when data from just one derivative are used, there are two equally likely possibilities for the protein phase, which are symmetrically disposed about the heavy-atom phase ϕ_H, and it is this effect that is reflected in the bimodal phase probability distribution. If we have another derivative in which the heavy atoms have bound at different sites, then the heavy-atom phase for that derivative will be different and the phase probability distribution, whilst still bimodal, will appear to be rather different from that of the first derivative, as shown in Fig. 14.8(b). However, one of the peaks in the phase probability distribution for the second derivative should overlap reasonably well with one of the peaks in the phase probability distribution for the first derivative, and this is the basis for resolving the phase ambiguity. We can combine the two phase probability distributions, i.e. $P_1(\phi_P)$ and $P_2(\phi_P)$, by multiplying them together, i.e. at each value of the protein phase (ϕ_P) at which we have calculated $P_1(\phi_P)$ and $P_2(\phi_P)$, we multiply the two probabilities together according to the formula

$$P(\phi_P) = P_1(\phi_P) \times P_2(\phi_P)$$

and the resulting joint probability distribution should ideally have a clear peak for the most probable phase, such as that shown in Fig. 14.8(c). This method is not restricted to just two heavy-atom derivatives, since we can combine phase information from any number of derivatives by multiplying the phase probability distributions together as

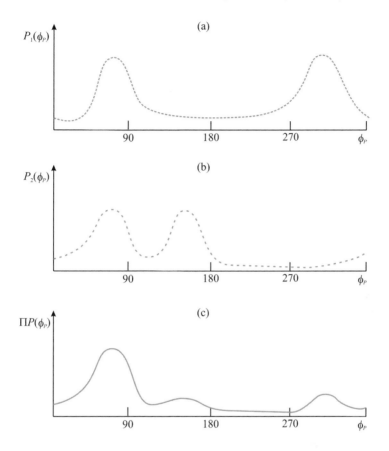

Fig. 14.8 *Phase probability distributions.* (a) shows the phase probability distribution for a typical heavy-atom derivative, with two peaks corresponding to the possible phase estimates. (b) shows the distribution obtained with a second derivative, which has one peak in common with the first derivative. Multiplying the probability distributions in (a) and (b) gives the joint phase probability distribution (c), with a large peak for the most probable phase.

represented in the following formula:

$$P(\phi_P) = \prod_j P_j(\phi_P) \tag{14.35}$$

Typically, we calculate the phase probabilities at $10°$ intervals on ϕ_P from 0 to $360°$ for each reflection of each derivative so that we can estimate the phase of each protein structure factor by analysing the joint probability distributions. We can then use the resulting phases to calculate an electron density map for the protein by Fourier synthesis. Blow and Crick showed that if we use the phase with the highest overall probability, referred to as the 'most probable phase', we give too much weight to uncertain phases in the Fourier synthesis. They showed that we can minimise the rms error in the electron density map by using the *centroid phase* instead, which is often referred to as the 'best phase', or ϕ_{best}, and by weighting each protein structure factor by a figure of merit that reflects the sharpness of the phase probability distribution. Whilst the centroid phase is used in MIR, it is not very helpful in SIR, since the protein phase will be set to ϕ_H or $\phi_H + 180°$, and the resulting map contains images of the protein superimposed on inverse copies of the molecule in a convoluted manner and is invariably uninterpretable (Grosse-Kunstleve and Adams, 2003). Likewise, with SAD phasing, use of the centroid phase leads to a map containing the protein electron density similarly convoluted with a negative inverse image. Hence, with SIR or SAD phasing, we exploit additional information in an effort to break the phase ambiguity, and these methods will be covered in the next section.

In MIR, a reflection for which all of the derivatives have phase probability distributions in which one of the two peaks overlap well will have a joint probability distribution with a sharp peak for the most probable phase. In contrast, a reflection for which all of the derivatives give conflicting phase information will have a combined distribution that is flat, or at least lacking a single dominant peak. Blow and Crick (1959) proposed the use of a *figure of merit* to weight the structure factors used in the Fourier synthesis according to the precision with which their phases have been determined. The figure of merit m can have a value ranging from a minimum of 0.0 for the least precise phases to a maximum of 1.0 for the most precisely resolved phase angles. Figure 14.9 shows a phase probability distribution drawn on an Argand diagram where the joint probability values are shown as contour lines. If the phase probability distribution has two diametrically opposite peaks of equal probability, i.e. the phase estimates are conflicting, the figure of merit will be zero. In contrast, a single sharp peak would have a figure of merit close to 1.0. By taking moments to find the centroid of the distribution, the following relationships are apparent:

$$m \cos \phi_{best} = \frac{\sum_i P(\phi_i) \cos \phi_i}{\sum_i P(\phi_i)} \tag{14.36}$$

$$m \sin \phi_{best} = \frac{\sum_i P(\phi_i) \sin \phi_i}{\sum_i P(\phi_i)} \tag{14.37}$$

$$\tan \phi_{best} = \frac{\sum_i P(\phi_i) \sin \phi_i}{\sum_i P(\phi_i) \cos \phi_i} \tag{14.38}$$

$$m = \frac{\sum_i P(\phi_i) \cos(\phi_{best} - \phi_i)}{\sum_i P(\phi_i)} \tag{14.39}$$

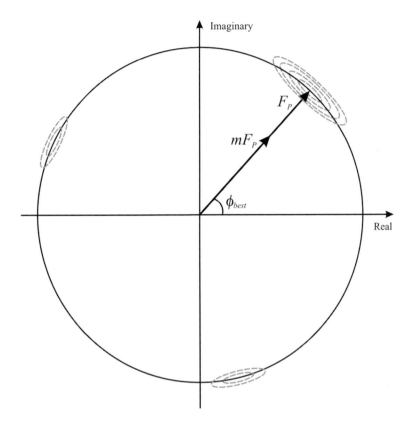

Fig. 14.9 *A typical phase probability distribution represented by an Argand diagram.* The phase probability distribution is contoured in dashed lines and the centroid of the distribution is indicated by mF_P, which has a phase denoted ϕ_{best}.

From these equations, we can determine ϕ_{best} and m for each reflection, and these are used in a Fourier synthesis of the protein electron density map, known as the 'best Fourier':

$$\rho(x, y, z) = \frac{1}{V} \sum_h \sum_k \sum_l m_{hkl} |F_P(hkl)| e^{i\phi_{best}(hkl)} e^{-2\pi i(hx+ky+lz)} \qquad (14.40)$$

Hendrickson and Lattman (1970) have devised a trigonometric expansion for the exponential term of equation (14.34) which allows the probability distribution for a particular derivative, $P_j(\phi_P)$, to be represented as a series of constants for each reflection, referred to as Hendrickson–Lattman coefficients. Data for new derivatives can be incorporated into the probability distribution for each reflection by straightforward addition of these coefficients, which is computationally faster than the Blow and Crick approach, although both yield essentially the same result. The significant advantage of the Hendrickson–Lattman coefficients is that they encapsulate the phase probability distribution in four constants that are easily stored for use in phasing attempts at a later stage, and are of much use with SIR data owing to the bimodality of the phase probability distribution, which cannot be recreated from ϕ_{best} and m alone.

To assess the quality of experimental phases, there are various parameters that can be analysed. The most commonly quoted parameter is the mean figure of merit, which will be high when the probability distributions for all reflections are sharp. However, it can also be artificially high when the phases are poor if the rms lack-of-closure error or E_j in the equation for $P_j(\phi_P)$ is underestimated. Other commonly used criteria include the phasing power (Ph.P) and the Cullis R-factor (Cullis *et al.* 1961), which are calculated for each derivative and are given by the formulae below:

$$\text{Ph.P} = \frac{\langle |F_H(\text{calc})|\rangle}{\langle \varepsilon(\phi)\rangle} \tag{14.41}$$

$$R_{Cullis} = \frac{\sum |(|F_{PH}| - |F_P|) - F_H(\text{calc})|}{\sum (|F_{PH}| - |F_P|)} \tag{14.42}$$

where $\langle \ldots \rangle$ indicates the rms value. The denominator in the equation for the phasing power is calculated using the probability-weighted lack-of-closure error (Terwilliger and Eisenberg, 1987), i.e., for a very sharp, unimodal probability distribution, this will tend towards the lack-of-closure error for the best phase, which should be low, thus giving a high value for the phasing power. Values of 0.5–1.0 are usable, whereas numbers in excess of 1.5 are very good and values greater than 2.0 (as are often obtained with multi-wavelength anomalous data) indicate that excellent phases have been obtained. The Cullis R-factor should be significantly less than 90% for a good derivative, and values around or below 70% indicate that the derivative is likely to give high-quality phase information. Of course, the ultimate criterion for assessing the success of the phasing exercise is the interpretability of the final electron density map. Note that there are alternative definitions of the Cullis R-factor, for example that used by MLPHARE (CCP4, 1994), in which the numerator in equation (14.42) is replaced by a summation of the lack of closure at the best phase.

We mentioned above that errors in the derivative-to-native scale factor and the heavy-atom occupancies can lead to bias in the final electron density map for the protein (Bloomer, 1983). This can be verified by inspection of the phasing triangle shown in Fig. 14.4(a), where it can be seen that if the F_H side of the triangle was made longer to mimic the effect of overestimating the occupancy, then because the lengths of the other sides are fixed, the angle $(\phi_P - \phi_H)$ would increase, i.e. the phases estimated for the protein would become less like those of the heavy atom. Since the electron density is always strongly dependent on the phases, a map calculated with this estimate of the protein phase would have less electron density than it should at the site of the heavy atom whose occupancy we have overestimated. Conversely, underestimating the occupancy would lead to excess electron density at the site of the heavy atom. Similar effects occur when the derivative-to-native scale factor contains errors; for instance, if F_{PH} is overestimated, we can mimic the effects of this by stretching that side of the triangle (Fig. 14.4(a)) and keeping the lengths of the other sides constant, which will decrease $(\phi_P - \phi_H)$. Therefore when the scale factor is overestimated, the phases for the protein become more like those of the heavy atom, and the resulting map will be positively biased. In contrast, the map will be negatively biased if the derivative-to-native scale factor is underestimated. To check for these effects, one can check that the distribution of $|\phi_P - \phi_H|$ is roughly flat and that the mean value of $|\phi_P - \phi_H|$ is approximately $90°$.

14.10 Exploiting anomalous scattering effects in phasing

In the preceding section, we have seen how isomorphous differences between the structure factors for a native protein and a heavy-atom derivative can yield phase information about the protein. In this section, we will look at methods for determining the protein phases that take advantage of anomalous scattering. This effect arises from absorption of the incident X-ray beam when its energy matches that of an electronic transition in one of the elements present in the crystal. Our ability to exploit these phenomena relies on tunable synchrotron radiation, since, prior to its availability, X-ray sources were restricted to fixed wavelengths. In principle, one heavy-atom derivative with a strong anomalous signal allows the phase ambiguity that arises with single isomorphous replacement to be resolved, and this is referred to as SIRAS (single isomorphous replacement with anomalous scattering) phasing . The effect of anomalous scattering can be demonstrated by inspection of Fig. 14.10. The phasing triangles for just the isomorphous difference alone are shown in Fig. 14.10(a), and those for the Friedel mates $F_{PH}(+)$ and $F_{PH}(-)$ are shown in Fig. 14.10(b) and (c). We see that whilst each phasing diagram yields two possible phases for the protein, only one protein phase is consistent in all three diagrams, shown by F_P, drawn as a thick line.

In SIRAS phasing, we can often determine the hand of the heavy-atom substructure by inverting the heavy-atom coordinates and recalculating the protein phases. The net effect of inverting the heavy-atom coordinates on the phasing diagram is that we swap $+F_H''$ and $-F_H''$. The phasing triangles are then formed again and, if the hand of the heavy-atom substructure is now wrong, the phase estimates for the protein will not be as consistent as they are in Fig. 14.10. This can be understood by imagining that the phasing triangle for $F_{PH}(+)$ shown in Fig. 14.10(b) is drawn with the base of F_P positioned at the tip of $-F_H''$ instead of being at the tip of $+F_H''$. To close the phasing triangles, the possible values for the protein phase will be less consistent; the phases will have a lower figure of merit and will yield a poorer electron density map. Having the heavy-atom substructure on the wrong hand should also be revealed during phase refinement, since the occupancies of the heavy atoms, determined from the anomalous data, will refine to negative values. If inverting the heavy-atom substructure improves the figures of merit and the map for the protein, then we know that the hand of our initial heavy-atom substructure was wrong and that the inverted one is correct, and this should now be reflected in positive anomalous occupancies for the heavy-atom sites.

The ability of anomalous scattering to discriminate between the possibilities for the phase can also be demonstrated trigonometrically (Kartha, 1975; Helliwell, 1992) by referring to Fig. 14.3 again, where the following equation arises from application of the cosine rule to the 'isomorphous' phasing triangle:

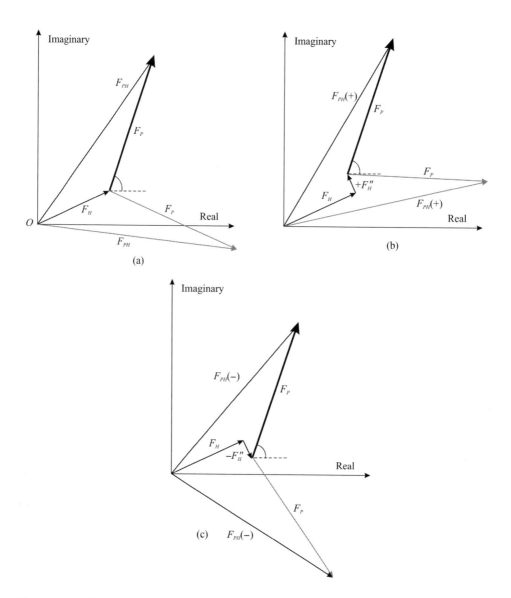

Fig. 14.10 *Phasing by single isomorphous replacement with anomalous scattering (SIRAS).*
(a) shows the two possible phasing triangles with a single heavy-atom derivative, which cannot
be distinguished. The effect of the anomalous scattering due to the heavy atom is shown in
(b) and (c), where the phasing geometry is shown for the Friedel mates $F_{PH}(+)$ and $F_{PH}(-)$,
respectively. Only one of the phase estimates for F_P, indicated as a thick line, is consistent
between (a), (b) and (c), thus breaking the ambiguity that arises in the absence of anomalous
scattering.

$$|F_P|^2 = |F_{PH}|^2 + |F_H|^2 - 2|F_{PH}||F_H|\cos(\phi_{PH} - \phi_H)$$

$$\therefore \phi_{PH} = \phi_H \pm \arccos\left(\frac{|F_{PH}|^2 + |F_H|^2 - |F_P|^2}{2|F_{PH}||F_H|}\right) \tag{14.43}$$

With reference to equation (14.17), given in Section 14.4, we can see that consideration of the 'anomalous' phasing triangle leads to the following relationship:

$$\phi_{PH} = \frac{\pi}{2} + \phi_H \pm \arccos\left(\frac{|F_{PH}(+)|^2 - |F_{PH}(-)|^2}{4|F_{PH}||F_H''|}\right) \tag{14.44}$$

Use of equations (14.43) and (14.44) to derive independent estimates of ϕ_{PH} from the isomorphous and anomalous data will give a unique solution for the phase of F_{PH}. The protein phase can then be calculated from knowledge of ϕ_{PH} using the following formula that follows again from the trigonometry of Fig. 14.3:

$$\phi_P = \arctan\left(\frac{|F_{PH}|\sin\phi_{PH} - |F_H|\sin\phi_H}{|F_{PH}|\cos\phi_{PH} - |F_H|\cos\phi_H}\right) \tag{14.45}$$

In practice, phasing with anomalous data proceeds as in isomorphous replacement by using trigonometric formulae to calculate the lack-of-closure error for both $F_{PH}(+)$ and $F_{PH}(-)$ at a series of trial values for the protein phase. The probability that each trial phase is correct is given by

$$P_j(\phi_P) = e^{\left(-[\varepsilon_j^+(\phi_P)^2 + \varepsilon_j^-(\phi_P)^2]/2E_j^2\right)}$$

where $\varepsilon_j^+(\phi_P)$ and $\varepsilon_j^-(\phi_P)$ are the lack-of-closure errors for the Friedel mates $F_{PH}(+)$ and $F_{PH}(-)$, respectively (Blow and Rossmann, 1961), although alternative forms are possible to take advantage of the lack of non-isomorphism with anomalous differences (Matthews, 1966).

In MAD phasing, we make measurements of $F_{PH}(+)$ and $F_{PH}(-)$ at several different wavelengths (usually three), which are chosen to optimise the real and imaginary components of the anomalous scattering signal, f' and f'', respectively, in equation (14.15). Since anomalous scattering is an absorption effect which reduces the intensity of the diffraction spots, f' values have negative signs. Note also that some elements which are commonly used as anomalous-scattering 'phasing vehicles', most notably selenium, should not really be regarded as 'heavy atoms', but we will retain the F_H and F_{PH} nomenclature in most of the following sections for consistency of the formulae. However, elements between Cs and Pt in the periodic table (atomic numbers 55 to 78) are commonly used in MAD analyses, with a preference for those at the high-atomic-weight end of this range since their electron transitions involving their K and L electron shells are within the accessible wavelength range of synchrotron sources (see Table 14.1).

MAD phasing relies on having a very precisely tunable source of X-rays, which is provided at many synchrotron beam lines, and optimum wavelengths for data collection have to be determined for each sample by recording an X-ray fluorescence spectrum from the crystal prior to data collection. A typical X-ray emission spectrum for a crystal containing an anomalous scatterer is shown in Fig. 14.11 along with

Table 14.1 Absorption edges for selected elements commonly used in macromolecular MAD studies, after González (2003).

Element	Atomic number	Absorption edge(s) (Å)
Fe	26	1.74
Cu	29	1.38
Zn	30	1.28
Se	34	0.98
Br	35	0.92
Pt	78	1.07, 0.93, 0.89
Au	79	1.04, 0.90, 0.86
Hg	80	1.01, 0.87, 0.84
Pb	82	0.95, 0.82, 0.78

an indication of the wavelengths at which data would normally be collected in a MAD experiment. Ideally, the emission spectrum will possess a sharp peak, or 'white line' (such as that shown in Fig. 14.11), the wavelength of which is characteristic of the anomalously scattering element. The term 'white line' was coined by Dirk Coster (1924), who first observed these phenomena in photographic measurements of X-ray absorption spectra.

The wavelength corresponding to the peak of the white line (λ_{peak}) maximises the f'' component of the anomalous scattering, and this is therefore the wavelength at which the anomalous difference will be maximal. This is normally the wavelength at which data would be collected first, for the simple practical consideration that if the diffraction pattern were to fade sufficiently to prevent the remaining datasets from being collected, then at least we would have a complete SAD dataset that should provide sufficient phase information for structure analysis. Following collection of this dataset, one would usually then collect data at a wavelength corresponding to the steepest part of the absorption edge (λ_{edge}), since this maximises the amplitude of the f' component of the anomalous scattering. Most beam lines provide a software tool for analysis of the fluorescence spectrum, for example CHOOCH (Evans and Pettifer, 2001), which determines wavelengths corresponding to the extrema of f' and f'' at which we should collect data.

The relationship between f' and f'' as a function of wavelength is shown in Fig. 14.12, where it can be seen that the maximum negative deflection of f' corresponds to the point of inflection in f''. The quantities f' and f'' are related mathematically by a relationship known as a Kramers–Kronig transform, which has exactly the property that we have just mentioned, namely that the extremity of one function corresponds to the point of inflection of the other. The physical basis of anomalous scattering is described in Blundell and Johnson (1976). Interestingly, the classical model does not account for the key feature of anomalous scattering that provides us with the

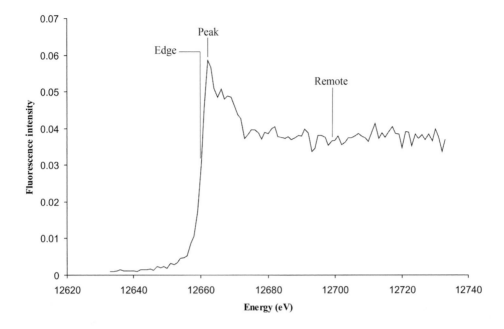

Fig. 14.11 *An X-ray fluorescence scan of a selenomethionine-containing protein crystal.* The energy of the incident beam is shown in electronvolts on the x axis, and the fluorescence intensity is shown in arbitrary units on the vertical axis. The white-line feature, or sharp peak, at about 12.66 keV is clearly visible.

strongest phasing information, namely the white line which is associated with certain electronic transitions. Not all elements show white-line phenomena; for example, lead, gold and mercury lack such a feature, but platinum, ytterbium and iridium do exhibit strong white lines. Fortuitously selenium, which can readily be incorporated into recombinant proteins during expression, does have a strong white-line feature at a very convenient wavelength for protein data collection (0.98 Å). An additional dataset is usually collected at a wavelength that is remote from the edge (λ_{remote}), usually at a lower wavelength (higher energy) so that there are still appreciable anomalous differences although they will not be as strong as at the white line (λ_{peak}). Note that the energy and wavelength of electromagnetic radiation are inversely related as shown by the following equation, which follows from the equations given in Chapter 4, Section 4.11:

$$E = \frac{hc}{\lambda} = \frac{12.3985}{\lambda} \tag{14.46}$$

where the energy E has units of kiloelectronvolts (keV) and the wavelength is in Å.

In MAD data collection, bearing in mind that anomalous differences are much smaller than isomorphous differences, high multiplicity and Friedel completeness of the measured dataset are of the essence. Many synchrotron beam lines have been optimised for maximum intensity to allow data collection to the highest possible resolution,

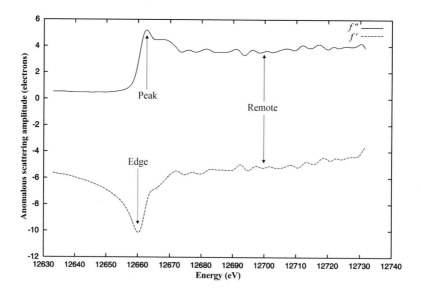

Fig. 14.12 *The anomalous scattering factors f' and f'' derived from an X-ray fluorescence spectrum.* Wavelengths that maximise the amplitude of f' (lower line) and f'' (upper line) are chosen from the graph prior to data collection to maximise the phasing information obtained.

and frequently it is found that radiation damage at such installations limits crystal lifetime. In MAD, we need only to collect low-to-medium-resolution data, albeit of high accuracy, at several wavelengths and so it is important to maximise crystal lifetime. These requirements are often well met by the lower-flux bending-magnet beam lines at many synchrotron sources, where radiation damage is markedly reduced. When data have been collected at all wavelengths, usually three, then provided the crystal is still diffracting strongly, there is much to be gained from re-collecting the data at all wavelengths at 180° away on the crystal rotation axis, since all diffraction spots measured in this second wedge of data will be Friedel mates of those collected in the first wedge. This is known as an *inverse-beam experiment,* and collecting data in this manner can significantly improve the measurements of the Δ_{ano}'s and enhance the subsequent structure analysis. There is increasing interest in measuring anomalous signals from endogenous elements such as sulphur and calcium, which have absorption edges at long wavelengths ($\lambda > 3$ Å). However, the low intensity of synchrotron beam lines at such wavelengths and other problems such as absorption and high background mean that such experiments are technically difficult, and the practical upper limit for wavelength is currently around 2 Å (González, 2003). During MAD or SAD data processing, it is important to inspect the signal-to-noise ratio of the anomalous data, which should be strongest in the low-to-medium-resolution range. This is an important indication of the data quality, particularly with heavy-atom-soaked crystals, since merely observing a good X-ray fluorescence peak at the expected wavelength for the heavy atom does not always mean that it has bound to the protein in an ordered manner, and it may simply be aggregated in the solvent surrounding the crystal.

In MAD phasing, we need to calculate the differences between the values of the structure factor amplitude of each reflection measured at the different wavelengths. These differences are referred to as *dispersive differences*, and they differ fundamentally from anomalous differences, which relate to the differences between the amplitudes of Friedel mates (see equation (14.14)). The dispersive difference of a reflection measured at two different wavelengths is directly related to the difference in f' for the anomalously scattering element at these two wavelengths. Since the edge wavelength λ_{edge} corresponds to the most negative value of f' (Fig. 14.12), then it follows that by comparing the structure factor amplitudes obtained at this wavelength with those measured at the remote wavelength we can expect to obtain dispersive differences that are larger than would be obtained by comparing any other pair of datasets. It should be appreciated that dispersive differences tend to be smaller than anomalous differences, but nevertheless it is possible to obtain useful phasing information from them.

The effects of the real and imaginary components of the anomalous scattering at different wavelengths are shown in Fig. 14.12. Since f' is negative, it will subtract from F_H and, as shown in Fig. 14.13, this effect is largest for the edge wavelength (λ_{edge}). Hence, by measuring data at three wavelengths and by determining the positions of the anomalous scatterers (and thus F_H), we can construct appropriate phasing circles, which will intersect at a unique value of the protein phase. It is also possible to derive an algebraic solution for the protein phase from MAD measurements (Hendrickson, 1991), and this method is covered extensively in other textbooks (e.g. Helliwell, 1992; Drenth, 2007). However, in practice it is more common for the data collected at different wavelengths to be treated as 'pseudo-isomorphous replacement' derivative data, and phases can be calculated with conventional phasing algorithms (Phillips and Hodgson, 1980) as covered in Section 14.9, usually by treating the remote dataset (λ_{remote}) as the 'native'.

The positions of the anomalous scatterers are usually determined by analysis of an anomalous-difference Patterson function calculated using the data collected at λ_{peak}, where the f'' component is largest, or by direct methods using the same anomalous differences. In principle, the dispersive differences could also be used either to calculate a dispersive-difference Patterson function or in direct methods, but since they tend to be smaller than the Δ_{ano}'s, the analysis is generally less successful. However, the complementary nature of isomorphous and anomalous differences, as discussed in Section 14.4 and encapsulated in equation (14.20), emphasises that combining anomalous and dispersive differences might be a better means of determining the positions of the anomalous scatterers. This is particularly important for selenomethionine MAD, since proteins can contain dozens of selenium atoms within the asymmetric unit, and the determination of very large substructures requires that we make best use of all available data. One method of preprocessing the anomalous data from all wavelengths to optimise the anomalous differences has been described by Hai-Fu *et al.* (1993) and is available in the program REVISE (CCP4, 1994). The resulting Patterson maps tend to give very clear peaks for the anomalous-scattering substructure that can be readily solved either by vector search or by direct methods. An alternative approach involves

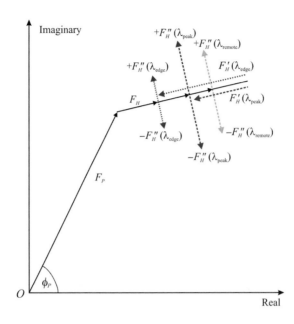

Fig. 14.13 *An Argand diagram showing the effects of anomalous scattering at different wavelengths close to an absorption edge.* Values of F_H'' at three different wavelengths, corresponding to the peak of the fluorescence spectrum, the edge and a remote wavelength, are shown. The real component of the anomalous scattering, corresponding to f', is also indicated as F_H' for the peak and edge wavelengths.

normalising the measurements and subjecting them to various statistical tests to select only the most reliable anomalous differences (Blessing and Smith, 1999). This method, which can also be applied to isomorphous-difference data, works well with dual-space recycling since phasing by direct methods is dominated by a subset of the strongest E-values.

Following determination of the sites of the anomalous scatterers, it is important that their coordinates and occupancies are refined using the MAD data. This refinement proceeds in essentially the same way as in the heavy-atom phase refinement described above for use with isomorphous differences. The remote dataset is usually treated as the 'native' and the peak and edge datasets as derivatives with common sites. In this situation, the 'derivative minus native' isomorphous differences are what we referred to earlier as dispersive differences, and these reflect the difference in f' for the two wavelengths being compared. The dispersive differences and the anomalous differences are used to refine the coordinates, occupancies and temperature factors of the anomalous scatterers in the asymmetric unit. Note that separate occupancies are determined from the dispersive and anomalous data, the dispersive differences providing us with a *dispersive occupancy*, which is a measure of the difference in f' between the two datasets being compared, and the anomalous data providing an *anomalous occupancy*, which is a measure of the f'' component of the 'derivative' dataset. An initial run of phase refinement is done using the centric reflections only (which possess no anomalous differences), to refine the coordinates and dispersive occupancies of each site with the λ_{peak} and λ_{edge} datasets, and then a second run is done using all reflections to refine both the dispersive and the anomalous occupancy factors. It is sensible to inspect the site occupancies and omit those that refine to values well below the average.

The highly automated phasing program SOLVE (Terwilliger and Berendzen, 1999a) analyses MAD data by scaling the data at all wavelengths to generate a reference or 'native' dataset, and the structure factors for the anomalous scatterers (F_A) are then calculated from the Friedel pairs. The components of each F_A parallel and orthogonal to the structure factor of the non-anomalously scattering atoms are estimated from probability distributions, and Bayesian statistics are used to estimate the 'best' value of F_A based partly on the expected number and type of anomalous scatterers present. The amplitude of F_A is used in a Patterson function to locate the anomalous scatterers, which are subjected to Patterson-based refinement. Finally, the refined anomalous-scattering substructure is used to calculate protein phases in a manner equivalent to that involved in phasing by single isomorphous replacement with anomalous scattering (SIRAS) (Terwilliger, 1994a,b). Other elegant phasing methods are available, for example SHARP (De la Fortelle and Bricogne, 1997), which essentially treat MAD phasing as a 'pseudo-SIR' or 'pseudo-MIR' problem using likelihood functions.

When phasing is being done by use of anomalous-difference measurements at a single wavelength, preferably optimised to maximise f'' (SAD), then the data available to us are somewhat different from those in MAD or MIR in that we do not have a dataset that can be treated as a reference, and this has a bearing on the trigonometry of the phasing diagram. Our measurements consist only of $|F_{PH}(+)|$ and $|F_{PH}(-)|$ and if the anomalous signal is small, as it usually is, then we can obtain an estimate of $|F_{PH}|$ from the familiar formula

$$|F_{PH}| \approx \frac{|F_{PH}(+)| + |F_{PH}(-)|}{2} \tag{14.18}$$

In SAD, we need to determine the phase of F_{PH} (rather than F_P, which is unknown) to calculate the electron density map for all atoms present in the protein derivative. The sites of the anomalously scattering atoms are located by Patterson vector search or direct methods from the Δ_{ano} data and refined. For each reflection, we can then calculate F_H and F_H'', whose real and imaginary components can be drawn on a phasing diagram such as that shown in Fig. 14.14(a). In this figure, two phasing circles with radii equal to $|F_{PH}(+)|$ and $|F_{PH}(-)|$ are drawn, and two possible positions for the heavy-atom structure factors are shown. It should be remembered that F_P is unknown in terms of magnitude and phase at this stage, but two possibilities for F_P (of different amplitude) are shown in the diagram, pointing towards the base of each F_H. Two possible F_{PH} can then be drawn extending from the origin towards the tip of each F_H. The possibilities for F_{PH} are symmetrically disposed about the direction of $+F_H''$, and hence the phase probability distribution for F_{PH} will be a symmetric bimodal one. In SIRAS phasing, the knowledge of $|F_P|$ allows us to avoid the ambiguity

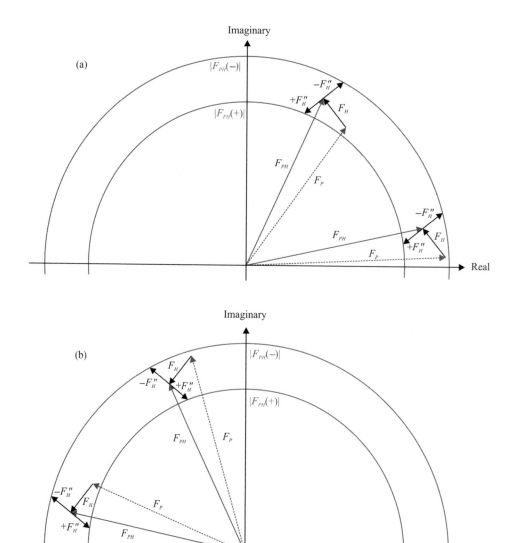

Fig. 14.14 *SAD phasing diagrams.* (a) An Argand diagram showing phasing circles with radii equal to the amplitudes of the Friedel mates $|F_{PH}(+)|$ and $|F_{PH}(-)|$. Two possible positions of the heavy-atom structure factors which are compatible with the inner and outer circles are shown. These generate two equally plausible phasing triangles, in which the possibilities for F_{PH} are shown as grey lines and the two possibilities for F_P (of different amplitude) are shown as broken grey lines. The two possibilities for F_{PH} are symmetrically disposed about the direction of F''_H. (b) shows the effect of inverting the configuration of the heavy-atom substructure, which generates two more alternative phasing triangles and thus two equally plausible phase estimates for F_{PH}. Hence it is necessary to calculate phases for the protein with the heavy-atom substructure on both possible hands and to inspect the resulting maps, since only the correct hand will give an interpretable electron density. This figure was adapted from Wang *et al.* (2007).

which arises in the SAD case, where the inherent problem is that we do not have a native dataset. However, we can crack the ambiguity in the phase that arises with SAD by various considerations, including the expectation that since the anomalous scatterers are part of the structure, ϕ_{PH} should be correlated with ϕ_H. Hence, of the two possible phases for F_{PH}, the phase that orients F_{PH} most parallel to F_H has a slightly higher probability of being correct, and this can be incorporated into the phase probability distribution (Dodson, 2003). The theory of phasing with partial structures is something that will be covered in Chapter 15. It is also possible to use direct methods to resolve the phase ambiguity that arises with SAD or SIR (Hao *et al.*, 2000), but the most powerful method of solving this problem is to use density modification methods to improve the resulting electron density map, and these will be covered in Section 14.11.

One important thing to remember with SAD (and MAD) phasing is there is a handedness problem similar to that described for SIRAS phasing at the start of this section. In a SAD phasing exercise, if we invert the heavy-atom substructure, F_H is given a phase $-\phi_H$ and the corresponding heavy-atom structure factors can be redrawn as shown in Fig. 14.4(b). This generates two alternative phasing triangles for F_{PH} and, consequently, two more possibilities for F_P, which are as plausible as those generated for the original hand of the heavy-atom substructure. The way to solve the problem is the purely practical one of calculating the phases for both possible hands of the heavy-atom substructure and inspecting the electron density maps calculated from these phases. If the hand of the heavy-atom site is wrong, then both estimates of ϕ_{PH} will be wrong and the resulting 'best phase' will give an uninterpretable map. Instead, if the hand is correct, one of the estimates of ϕ_{PH} will be correct, and this one should be 'selected' by the density modification or other methods used to resolve the SAD phase ambiguity and give an interpretable map. Hence, in SAD phasing, having the anomalous-scattering substructure on the correct hand is essential for obtaining an interpretable electron density map for the protein. Similar arguments apply in the case of MAD since, by inspection of Fig. 14.13, it is easy to see that the phasing triangles for each wavelength can be drawn in exactly the same way as for the SAD case, although in principle the additional datasets allow the correct choice of ϕ_{PH} to be made for a given hand of the heavy-atom substructure (Wang *et al.*, 2007). Again, inverting the substructure will generate another possibility for ϕ_{PH}, although only one of these possibilities will correspond to the correct protein phase and give an interpretable map. The program SOLVE (Terwilliger and Berendzen, 1999a) performs these inversion operations automatically and assesses the quality of the protein map by analysing the degree of variation in electron density throughout the asymmetric unit (Terwilliger and Berendzen, 1999b). A map calculated with random phases will have noise features distributed uniformly in all sections. In contrast, reasonable estimates of the phases will yield a promising map, with distinct volumes of highly variable electron density corresponding to the protein, as well as regions of smooth, low density corresponding to the solvent, and such a map would score highly, indicating the correct hand of the anomalous-scattering substructure.

One further interesting complication arises with space groups that form enantiomorphic pairs, for example $P4_1$ and $P4_3$, $P3_1$ and $P3_2$, $P6_1$ and $P6_5$, and $P4_12_12$

and $P4_32_12$, in which the arrangement of the motifs in one space group is the mirror image or inverse of the arrangement in the other. This can be verified from inspection of Fig. 3.17 in Chapter 3, where it can be seen that the points generated by the 3_1 and 3_2 screw axes are, in fact, related by a mirror plane perpendicular to the threefold axis, and the same applies to a number of other screw axis symmetries. These chiral space groups form pairs which cannot be discriminated by their systematic absences, although the ambiguity can be resolved by the use of anomalous data. The important facet of these chiral space groups, as far as phasing is concerned, is that when we invert the heavy-atom substructure to recalculate the protein phases, we also have to invert the space group. Whichever hand of the anomalous-scattering substructure gives an interpretable map defines the correct space group. One way to understand these effects is to think of a protein forming a lattice in a non-chiral space group (e.g. $P2_1$). If we could invert the whole lattice along with its constituent molecules, we would generate a lattice of inverted molecules that was an inverse copy of the original structure. In an Argand diagram, inverting a structure is equivalent to reflecting the structure factor through the real axis. The resultant structure factor for the inverse structure will have an amplitude that is exactly the same as that for the original structure, and hence the Patterson function (which is calculated from the square of the amplitude) will be the same for the original structure and for its inverse. If the protein is in a non-chiral space group such as $P2_1$, when the original lattice of molecules is inverted, the space group of this new arrangement is the same. However, if the protein molecules are originally arranged in a chiral space group such as $P4_1$, then if we invert this whole arrangement, we also by definition invert the space group (which would become $P4_3$ in this example), as well as inverting the molecules themselves. Hence the Patterson function of a heavy-atom substructure in a chiral space group will be the same as that formed by the inverse substructure in the inverse space group. If we were to invert a molecule in a chiral space group and keep the space group the same, then the intermolecular vectors and hence the Patterson function would be different. In contrast, with a non-chiral space group such as $P2_1$, if we invert a molecule or a heavy-atom substructure, neither the space group nor the Patterson function will change.

14.11 Density modification

The initial map from SIR, MIR, SAD or MAD will contain noise due to errors in the phases. Even with very good experimental data, these errors can be large; for example, in SIR and SAD there are two equally likely estimates of each protein phase and, if the estimates are far apart, the centroid phase will contain a significant error. 'Density modification' covers a range of methods which aim to recognise the noise in an electron density map and remove it. After the map has been 'corrected', phases for the protein are recalculated and recombined with the experimental phases, which helps to resolve the phase ambiguity that arises, particularly with SIR and SAD. The recombined phases are used to calculate a new map, which is modified again and used to calculate new phases in an iterative process that is repeated until no further improvement in the map occurs. This procedure can yield a significantly improved electron density for the protein, thus aiding the task of interpreting the map and fitting the amino acid sequence.

Noise in an electron density map can be recognised by various features, including the following:

(a) *Negative electron density.* Since the electron density from a Fourier synthesis calculated with $|F_P|$ and ϕ_P should show the electron clouds surrounding the atomic positions, it must always be positive, and any negative features will arise from errors in the amplitudes or phases used in the calculation. Therefore, the electron density in the regions where ρ is negative can simply be set to zero.

(b) *Strong electron density in solvent regions.* The solvent regions of the crystal should have a low electron density owing to the disordered nature of water. Whilst some of the water molecules close to the protein will be ordered owing to hydrogen bonding with surface amino acids, these are usually only well defined in high-resolution electron density maps. At this stage of the analysis, we are likely to have experimental phases that extend only to medium resolution ($\sim$3 Å), and even if the phases were perfect, it is unlikely that many of the ordered water molecules on the surface of the protein would be visible at this resolution. Hence, providing that we can identify the protein–solvent boundary, we can flatten any strong features of electron density in the solvent regions, for example by resetting them to the rms density for the solvent.

(c) *Differences between the electron density for identical molecules within the asymmetric unit.* With a medium-resolution map for a structure with more than one molecule in the asymmetric unit, any differences between the molecules in this group are likely to be very small and if differences are apparent in the map, it is likely that they are due to noise. Hence, if we take each point of the map for one molecule and average the electron densities at the equivalent points in the other molecules within the asymmetric unit, we can generate an 'averaged map', and in doing so we remove some of the noise. This method relies on knowing the NCS operators relating the molecules within the asymmetric unit and also on determining a separate boundary for each molecule, which is referred to as a 'mask'. The distinction between the mask and the solvent boundary is shown in Fig. 14.15. These methods also require interpolation of the electron density, since the NCS-related positions of any grid point in one monomer will most likely fall between the grid points in another.

Often the experimental phases extend only to a resolution that is much lower than the diffraction limit of the crystal. Density modification allows us to extend the phases to a resolution beyond those which are determined experimentally. This is because the procedure starts with a low-to-medium-resolution map derived from the experimental phases and improves this so that meaningful phases can be calculated to a slightly higher resolution. The new phases are used to calculate a higher-resolution map, which is again modified in the same way. Thus the phases are progressively extended and refined to the resolution limit of the native data.

In MAD and SAD phasing, density modification is used routinely to help break the phase and/or hand ambiguities that arise, and with a promising initial map, the

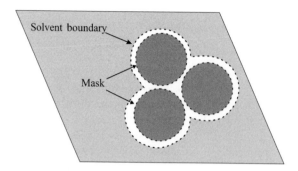

Fig. 14.15 *Solvent boundary and mask.* In density modification, we need to define the solvent boundary of the molecule, and when NCS is present (as shown in this example of a trimeric protein) a mask can also be determined to allow the electron density for each subunit to be averaged.

results can be remarkable. However, it should be appreciated that density modification is not a means of turning bad phases into good ones, but rather that it can improve promising phases and is often capable of turning good initial phase estimates into very good ones.

The routine use of density modification in phasing began with the development of an automated procedure for determining the solvent boundary of the protein which was devised by B. C. Wang (1985). Wang proposed smoothing out noise in the initial electron density map (calculated with $m|F_P|$ and ϕ_{best}) by replacing the electron density ρ_j at each grid point with a weighted average of density values within a surrounding sphere of radius R, typically 8–10 Å. The modified density ρ'_j is given by the formula

$$\rho'_j = k \sum_i w_i \rho_i \tag{14.47}$$

where w_i is the weight for each grid point and k is a constant. The weight is set to zero for points with negative electron density to ensure that these regions are removed from the calculations. For all other points with non-zero density, the weight is given by

$$w_i = 1 - \frac{r_{ij}}{R} \tag{14.48}$$

where r_{ij} is the distance of point i from the central point j. The terms in this equation are shown in Fig. 14.16. The sphere is effectively moved from one grid point to the next in the asymmetric unit of the map, and each new density value is calculated from the raw electron density values within the sphere of the original map. The electron density values are weighted by their distance from the central point such that grid points close to the centre are the most strongly weighted and those furthest out contribute very little. The new, modified density values are written into a separate map, referred to as the *smoothed map*, which is used to determine the solvent boundary. To do this, an electron density cut-off value is chosen. All points in the map with electron

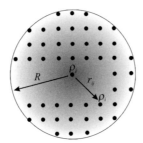

Fig. 14.16 *The sliding sphere used in smoothing an electron density map.* Typically, a sphere of radius 8–10 Å is centred on each grid point in turn, and the electron densities at surrounding points within the sphere are averaged. The density at each grid point is weighted such that those points furthest from the centre make the least contribution, as indicated by the shading.

density values lower than the cut-off are treated as solvent. Conversely, all points in the map with electron density values higher than the cut-off are treated as part of the protein. The electron density cut-off value is chosen such that the number of grid points in the unit cell which are treated as belonging to the solvent agrees with the estimated solvent content of the crystal, which can be estimated by methods covered in Chapter 11. The boundary definition is thus wholly automatic, although the results should be inspected. If the map contains small patches defined as protein spread uniformly throughout the unit cell, then this indicates that the phases used to calculate the map are poor and probably cannot be improved by density modification. In contrast, if the regions defined as protein form large discrete volumes with a well-defined solvent boundary, the phases are most likely to be good and can be improved further by density modification. The next step is known as solvent flattening and involves the density values at grid points outside the envelope of the protein being set to the average value for the solvent region. Hence the electron density in the solvent channels is set to a uniform low value, which we would expect for such regions of the crystal. It is important that all low-resolution reflections are used in density modification, since these are sensitive to the gross shape of the protein and solvent regions of the crystal.

At this stage, it is worth recapping that we have taken the experimental map and smoothed out some of the noise in it using a sliding spherical 'window' of grid points, we have removed any negative electron density, we have defined a cut-off electron density for the protein/solvent boundary which is consistent with the solvent content of the crystal and we have flattened the electron density in the solvent regions to a low value. It should be apparent that we are using additional physical information such as the crystal's solvent content and the positivity of electron density to modify the map in a knowledge-based manner and thereby reduce the random or systematic noise in it. We can now Fourier transform the modified electron density to calculate the structure factor amplitudes and phases $|F_{imp}|$ and ϕ_{imp} for the improved map. If the experimental phases were good enough to give a protein-like electron density map, the ϕ_{imp}'s should be more accurate estimates of the protein phases than the original experimental phases ϕ_{best} were. We could simply abandon the experimental phases at this point, but this would waste useful information and introduce bias into the phases. Instead, the ϕ_{imp}'s are recombined with the experimental phase information by multiplication of the phase probability distributions, much as we described for MIR phasing in Section 14.9. The phase probability distribution for the modified map is calculated by the same method that is used when a partial model is available for the protein (Read, 1986), and this will be covered in the next chapter. Density modification thus helps to resolve the severe phase ambiguity that arises with SIR and SAD, much like a second derivative would in MIR. This is particularly true if the experimental phase probability distributions are recorded as Hendrickson–Lattman coefficients, which can accurately reproduce any bimodal character. The multiplication of the phase probability distributions can be represented by the equation below:

$$P_{comb}(\phi) = P_{exp}(\phi) \times P_{imp}(\phi) \tag{14.49}$$

where the experimental phase probability distribution is given by $P_{exp}(\phi)$ and that of the modified map is given by $P_{imp}(\phi)$. Again, the centroid phase of the combined probability distribution $P_{comb}(\phi)$ can be taken and used to calculate a new map, which should be appreciably better than the initial one.

The whole process is then repeated by spherical averaging, eliminating negative density, redetermining the envelope, solvent flattening, Fourier inversion of the map and phase recombination. The cycles can be continued until no further improvement in the electron density map occurs and the phases converge. Initially, data to low resolution are used, but higher-resolution data can be added in small steps, thus extending the phases to progressively higher resolution in a process known as *phase extension*. In favourable cases, if experimental phases are available only to medium or low resolution and native data are available to high resolution (e.g. 2 Å or better), the resulting electron density map can be of extraordinarily high quality, with no manual input from the crystallographer, who is spared the effort of preparing more yet derivatives and collecting more data. There may be a slight deterioration in the quality of the map for surface amino acids that happen to lie outside the solvent boundary, and hence their electron densities are flattened, but the overall improvement in connectivity and interpretability of the map far outweighs any potential disadvantages. Whilst density modification is principally of benefit in experimental phasing rather than in molecular replacement, there have been a number of cases where it has been pivotal in solving the structures of assemblies or fusion proteins where at least one component is of known structure and can be solved by molecular replacement. Following density modification, there may be sufficient electron density for the unknown component to reveal its structure.

Leslie (1987) devised a reciprocal-space method for performing the Wang calculations which is faster than doing them in real space. The reciprocal-space method is based on the fact that the smoothed map $\rho_{av}(i, j, k)$ obtained by averaging the local electron density around each grid point (i, j, k) in the original map is a convolution of the original map with the weighting function $w(r)$, as shown in the following equation:

$$\rho_{av}(i, j, k) = \rho_{trunc}(i, j, k) * w(r) \tag{14.50}$$

where $\rho_{trunc}(i, j, k)$ is the original map after truncation of negative density to zero. In Chapter 7, we showed that the Fourier transform of a convolution is the product of the Fourier transforms of the two functions, and in this case the Fourier transform of the smoothed electron density becomes

$$T\left[\rho_{av}(i, j, k)\right] = T\left[\rho_{trunc}(i, j, k)\right] \cdot T[w(r)] \tag{14.51}$$

The Fourier transform of the truncated map, $T[\rho_{trunc}(i, j, k)]$, can be calculated by fast Fourier transform (FFT), and the Fourier transform of the weighting function, $T[w(r)]$, can also be computed readily.

To summarise, in the reciprocal-space approach to Wang density modification, the map is first calculated and its negative density truncated to zero. The structure factors are then recalculated from this map by FFT and multiplied by the Fourier transform of the weighting function. Finally, the smoothed map can be calculated from the modified F's (again by FFT) and the solvent boundary determined in the manner described above. The rest proceeds essentially as in Wang's original method, but with the benefit of reduced computation time.

Non-crystallographic symmetry averaging. For crystals with more than one molecule per asymmetric unit, the electron density for each subunit can be averaged to improve the map. This assumes that differences between molecules within the asymmetric unit are negligible, which is generally true for medium-resolution maps. The general principles are demonstrated in Fig. 14.17, from which it should be apparent that density averaging depends on two things:

(a) having a mask which defines the monomer, and
(b) knowing the NCS operators which relate all monomers within the asymmetric unit.

The mask represents the shape or boundary of the monomer and is derived computationally either from the electron density map, from skeletonising the electron density (in which dummy atoms are fitted to regions of contiguous density) or from a protein coordinate file, if available, for example from a molecular replacement solution (Kleywegt and Jones, 1994; Abrahams and Leslie, 1996). The mask represents the shape of the molecule using a 3D grid, much like an electron density map, with each grid point having a variable set to indicate whether it belongs to the protein or to the solvent; for example, all points within the protein may have a value of 1 and all points within the solvent a value of 0. Steps have to be taken to ensure that the mask of one monomer does not overlap with that of another, and any regions of overlap can be removed. In general, masks often have to be smoothed and pruned to remove spurious pits, extrusions, cavities inside the mask and isolated 'blobs' of protein in the solvent region (Kleywegt and Jones, 1994). The NCS operators can be defined by analysis of the coordinates of heavy-atom sites, and this is particularly powerful with selenomethionine MAD, where there are likely to be several anomalous-scattering sites per monomer (e.g. Lu, 1999; Terwilliger, 2002b). They can also be derived from a molecular replacement solution. The NCS operators can be refined between cycles of density modification.

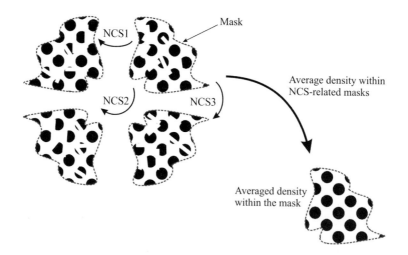

Fig. 14.17 *The removal of random noise in a map by NCS averaging.* A diagrammatic representation of a tetrameric protein with significant random noise in the initial map (top left). The NCS operators are used to average the electron density at equivalent points for each monomer within the mask, thereby smoothing out random errors and giving a much cleaner map (bottom right).

One important variation on this theme is the subject of cross-crystal averaging, in which the electron density derived from experimental phases in two or more non-isomorphous crystal forms of the same protein can be averaged. This is available in most contemporary density modification programs.

Histogram matching. This is a technique of electron density modification that has been borrowed from image processing (Zhang and Main, 1990). The distribution of electron density values present in the map is plotted as a histogram and compared with that of a standard map of similar resolution, as shown in Fig. 14.18. The method relies on the finding that the electron density histograms of highly refined protein structures of the same resolution are very similar and are independent of the fold of the protein. The histogram of the experimental map can be divided into bins containing equal numbers of density points. The aim is to modify the electron density distribution of this map such that it matches the standard distribution. To do this, parameters are determined which scale and shift (i.e. add to or subtract from) the electron density values of the points in each bin of the experimental map. Further refinements of the method such as ensuring that the shifts applied to physically adjacent points in the map are correlated are possible, but it is probable that histogram matching is a somewhat less powerful tool than solvent flattening or averaging.

A fast and robust density modification protocol has been developed by Cowtan (1994) in the program DM (CCP4, 1994), which performs all of the above procedures automatically; an example of a map 'before and after' the use of DM is shown in Fig. 14.19. An alternative approach, which allows the use of a smaller radius for the averaging sphere ($R \geq$ maximum resolution) and hence a more detailed solvent

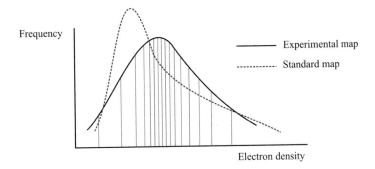

Fig. 14.18 *Histogram matching.* The electron density histogram of a map calculated from the experimental phases (thick solid line) is compared with that for a highly refined structure at a similar resolution (thick dashed line). The electron density values are divided up into bins (thin solid lines), and corrections which scale and shift them to match the standard histogram are determined.

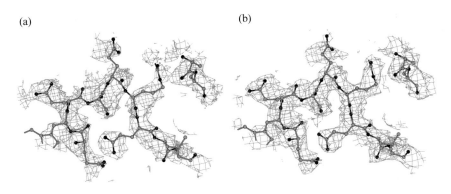

Fig. 14.19 *An example of density modification.* The initial electron density calculated at 3.0 Å with experimental phases is shown in (a), and the map resulting from density modification is shown in (b). The improved connectivity of the density for the main chain of the polypeptide (shown as balls and sticks) is clearly seen.

boundary, has been devised by Abrahams and Leslie (1996). In this method, the solvent boundary is determined from calculating another map showing the standard deviation of the electron density within the sliding sphere—regions within the protein will have a high standard deviation, unlike the solvent, which will be comparatively flat. This method also allows 'flipping' of the solvent density to negative values (rather than flattening it to a uniform low value), which was found to reduce phase error and improve map interpretability. The effectiveness of the density modification can be judged by use of a small 'reporter sphere' in a flat region of the solvent, which is excluded from the density modification procedure. If the electron density in this sphere becomes flatter as a result of density-modifying the remainder of the map, this provides an unbiased indication that the phases have improved. The beneficial effect of solvent flipping may stem from its effects at the phase recombination stage, since the probability distributions for the phases calculated from the density-modified map are estimated from the agreement between the structure factors calculated from the modified map (F_c) and the observed structure factors (F_o). Since the F_c's will be biased towards the F_o values (particularly if the map is not modified very much), the agreement between the F_c's and F_o's can be artificially good and the width of the probability distribution will be underestimated. Solvent flipping may reduce the extent to which the F_c's are biased towards the F_o's, thus broadening the probability distribution of the density-modified phases and allowing larger phase changes to occur in the phase recombination step. In DM, the progress of the density modification is assessed in reciprocal space using randomly selected 'reporter' reflections which are omitted from the initial map calculation. If the agreement between the F_o's and F_c's for these reflections improves as a result of density modification, as monitored by a 'free R-factor', this provides unbiased evidence that the phases are improving.

Further recent developments include the application of more rigorous statistical methods to all stages (Terwilliger, 2000, 2002a,b; Terwilliger *et al.*, 2008) such that the probability of each point in the map belonging to either protein or solvent is estimated, allowing a 'probabilistic solvent boundary' to be determined. A histogram is fitted to the densities in the protein and solvent regions, yielding target densities which, together with the NCS constraints, allow the overall probability of the map being correct to be estimated. This probability is assessed as a function of each protein phase angle, and this information can be recombined with the experimental phase probability distribution. The cycles are iterated until the phases converge. If a molecular replacement solution is used to provide initial phases, these phases are used only to calculate the initial map and not subsequently, ensuring that the problem of model bias, which could be experienced with earlier programs, is less likely to arise.

Summary

The derivation of the correct values of the phases for the structure factors F_{hkl} from the relative intensities $|F_{hkl}|^2$ constitutes the *phase problem*. In macromolecular crystallography, the most successful ways of solving the phase problem involve incorporation of a heavy atom into the crystal by soaking it with a suitable reagent (such as K_2PtCl_4) or adding the reagent immediately prior to crystallisation, or, in the case of selenomethionine, the 'heavy atom' is incorporated during *in vivo* synthesis of the protein. When a protein is soaked or co-crystallised with a heavy-atom reagent, it is essential that the protein conformation and crystal packing are not grossly affected by the presence of the heavy atom, and derivatives that meet this criterion are said to be isomorphous. Solving a protein with X-ray data collected from one derivative is referred to as *single isomorphous replacement* (SIR), and when two or more derivatives are used, as is usually the case, the process is referred to as *multiple isomorphous replacement* (MIR). Some heavy elements scatter anomalously at the wavelength used, or, indeed, the wavelength can be tuned to match the absorption edge of the heavy atom to maximise the effects of *anomalous diffraction* to provide additional phase information. These single-wavelength (SAD) and multi-wavelength (MAD) phasing methods have largely routinised the structure solution of selenomethionine-substituted proteins.

In all phasing methods, it is essential that the derivative data are accurately scaled to the native, or reference, dataset. This can be achieved by determining resolution-dependent or anisotropic scale factors or even by local scaling of the datasets using small volumes of reciprocal space. The success of scaling can be assessed from the R-factor, or fractional isomorphous difference,

$$R = \frac{\sum ||F_{PH}| - |F_P||}{\sum |F_P|} \tag{14.1}$$

A further refinement is Kraut scaling, in which the overall contribution of the heavy atom is considered.

The sites of heavy atoms can be determined by two methods:

(a) *Patterson vector search.* An isomorphous-difference Patterson map calculated with the coefficients

$$\Delta_{iso} = |F_{PH}| - |F_P| \tag{14.11}$$

will reveal the sites of the heavy atoms at approximately half their expected heights. Likewise, an anomalous-difference Patterson calculated with the coefficients

$$\Delta_{ano} = |F_{PH}(+)| - |F_{PH}(-)| \tag{14.14}$$

will reveal the positions of anomalously scattering elements. The most systematic way to analyse a Patterson map is by using the vector superposition function.

(b) *Direct methods.* Since the heavy-atom substructure generally consists of at most a few dozen well-resolved peaks, the positions of the atoms can be determined by use of the *tangent formula* derived for solving small-molecule structures, and using the normalised isomorphous or anomalous differences:

$$\tan \phi_{\mathbf{h}} \approx \frac{\langle |E_{\mathbf{h}'}||E_{\mathbf{h-h'}}| \sin (\phi_{\mathbf{h}'} + \phi_{\mathbf{h-h'}}) \rangle_{\mathbf{h}'}}{\langle |E_{\mathbf{h}'}||E_{\mathbf{h-h'}}| \cos (\phi_{\mathbf{h}'} + \phi_{\mathbf{h-h'}}) \rangle_{\mathbf{h}'}} \tag{14.28}$$

In this approach, strong reflections are assigned random phases and the above formula is used to estimate and refine the phases for the remainder of the dataset. The correct phase sets will reveal the positions of the heavy atoms in a Fourier map. A powerful development of this method, known as dual-space recycling, involves repeated cycles of tangent refinement in reciprocal space and selection of the strongest peaks in the resulting map, from which a new phase set is calculated. These methods are also capable of solving entire protein structures when the crystals diffract to atomic resolution (~ 1 Å).

Following determination of the heavy-atom or anomalous-scattering substructure, the coordinates and occupancies of the sites must be refined, which can be done in reciprocal or real space. It is also essential to ensure that all heavy-atom derivatives have their heavy-atom substructures determined with respect to the same hand and origin, which can be done by calculating a cross-phased Fourier map.

The protein phases are determined by calculating the phases of the heavy-atom structure factors from the refined substructure. F_H can be drawn from the origin of the Argand diagram, and a circle of radius $|F_P|$ that is centred on the tip of F_H can then be drawn. Finally, a circle of radius $|F_{PH}|$ is drawn centred on the origin, and this will intersect the F_P circle at two positions corresponding to two possible values for the protein phase. To automate these calculations, the protein phase is varied systematically from 0 to 360° in small steps, and at each trial phase the difference between $|F_{PH}(\text{obs})|$ and $|F_{PH}(\text{calc})|$, or the *lack-of-closure error*, is determined. The probability of each trial phase being correct can be calculated from its associated lack-of-closure error, giving a phase probability distribution which will be bimodal for each derivative. When several derivatives are available, their phase probability distributions can be combined by multiplication, giving a joint distribution which should be unimodal, thereby indicating the most probable phase. Blow and Crick (1959) showed that the *centroid phase* (ϕ_{best}) minimises the rms error in the final electron density map. Anomalous data also help to resolve the ambiguity of the phase probability distribution, and the success of phasing is assessed by the *mean figure of merit*, the *phasing power* (Ph.P) and the *Cullis R-factor*:

$$m = \frac{\sum_i P(\phi_i) \cos(\phi_{best} - \phi_i)}{\sum_i P(\phi_i)} \tag{14.39}$$

$$\text{Ph.P} = \frac{\langle |F_H(\text{calc})| \rangle}{\langle \varepsilon(\phi) \rangle} \tag{14.41}$$

$$R_{Cullis} = \frac{\sum |(|F_{PH}| - |F_P|) - F_H(\text{calc})|}{\sum (|F_{PH}| - |F_P|)} \tag{14.42}$$

SAD or MAD phasing involves collection of data at one or more wavelengths which optimise the real and imaginary components of the anomalous scattering. In MAD, the phases for the protein can be calculated by treating the measurements at each wavelength as *pseudo-MIR* derivative data, although the absence of a true native dataset means that there is ambiguity in the hand of the heavy-atom substructure, which has to be resolved by calculating a map with the heavy-atom sites on both possible hands, only one of which will give an interpretable map.

Experimental phases can be improved significantly by applying knowledge-based constraints to the electron density map and calculating new phases from the modified map. These techniques are referred to as *density modification* and involve smoothing the map and flattening the solvent density, truncating unrealistic density values, averaging NCS-related images of the molecule, and rescaling the electron density histogram to match the expected distribution of a highly refined structure of comparable resolution. The cycles of density modification are repeated until the phases converge.

References

Abdel-Meguid, S. S. (1995) Structure determination using isomorphous replacement. In *Crystallographic Methods and Protocols*, ed. Jones, C., Mulloy, B. and Sanderson, M. R., Methods in Molecular Biology, Volume 56, 153–71.

Abrahams, J. P. and Leslie, A. G. W. (1996) Methods used in the structure determination of bovine mitochondrial F$_1$ ATPase. *Acta Crystallogr. D* **52**, 30–42.

Blessing, R. H. (1997) LOCSCL: a program to statistically optimize local scaling of single-isomorphous replacement and single-wavelength-anomalous-scattering data. *J. Appl. Crystallogr.* **30**, 176–7.

Blessing, R. H. and Smith, G. D. (1999) Difference structure-factor normalisation for heavy-atom or anomalous scattering substructure determinations. *Acta Crystallogr.* **32**, 664–70.

Bloomer, A. (1983) Peaks and holes at heavy atom sites. Daresbury Laboratory, UK. *Info. Q. Protein Crystallogr.* **11**, 41–6,

Blow, D. M. and Crick, F. H. C. (1959) The treatment of errors in the isomorphous replacement method. *Acta Crystallogr.* **12**, 794–802.

Blow, D. M. and Rossmann, M. G. (1961) The single isomorphous replacement method. *Acta Crystallogr.* **14**, 1195–1202.

Blundell, T. L. and Johnson, L. N. (1976) *Protein Crystallography.* Academic Press, New York.

Boggon, T. J. and Shapiro, L. (2000) Screening for phasing atoms in protein crystallography. *Structure* **8**, 143–9.

CCP4 (Collaborative Computational Project Number 4) (1994) The CCP4 suite: programs for protein crystallography. *Acta Crystallogr. D* **50**, 760–3.

Coster, D. (1924) Uber die absorptionsspektren im Rontgengebiet. *Z. Phys.* **25**, 83–98.

Cowtan, K. (1994) DM: an automated procedure for phase improvement by density modification. *Joint CCP4 ESF-EACBM Newslett. Protein Crystallogr.* **31**, 34–8.

Cullis, A. F., Muirhead, H., Perutz. M. F., Rossmann, M. G. and North, A. C. T. (1961) The structure of haemoglobin. VIII. A three-dimensional Fourier synthesis at 5.5 Å resolution: determination of the phase angles. *Proc. Roy. Soc. A* **265**, 15–38.

Dauter, Z. (2002) New approaches to high-throughput phasing. *Curr. Opin. Struct. Biol.* **12**, 674–8.

De la Fortelle, E. and Bricogne, G. (1997) Maximum-likelihood heavy-atom parameter refinement for multiple isomorphous replacement and multiwavelength anomalous diffraction methods. *Methods Enzymol.* **276**, 472–94.

Dodson, E. (2003) Is it jolly SAD? *Acta Crystallogr. D* **59**, 1958–65.

Dodson, E. and Vijayan, M. (1971) The determination and refinement of heavy-atom parameters in protein heavy-atom derivatives. Some model calculations using acentric reflections. *Acta Crystallogr. B* **27**, 2402–11.

Drenth, J. (2007) *Principles of Protein X-ray Crystallography*, 3rd edn. Springer, New York.

Ealick, S. E. (1997) Now we're cooking: new successes for shake-and-bake. *Structure* **5**, 469–72.

Evans, G. and Pettifer, R. F. (2001) CHOOCH: a program for deriving anomalous-scattering factors from X-ray fluorescence spectra. *J. Appl. Crystallogr.* **34**, 82–6.

Frazão, C., Soares, C. M., Carrondo, M. A., Pohl, E., Dauter, Z., Wilson, K. S., Hervés, M., Navarro, J. A., De la Rosa, M. A. and Sheldrick, G. M. (1995) *Ab initio* determination of the crystal structure of cytochrome c_6 and comparison with plastocyanin. *Structure* **3**, 1159–69.

Germain, G., Main, P. and Woolfson, M. M. (1970) On the application of phase relationships to complex structures. II. Getting a good start. *Acta Crystallogr. B* **26**, 274–85.

González, A. (2003) Optimising data collection for structure determination. *Acta Crystallogr. D* **59**, 1935–42.

Grosse-Kunstleve, R. W. and Adams, P. D. (2003) On the symmetries of substructures. *Acta Crystallogr. D* **59**, 1974–7.

Hai-Fu, F., Woolfson, M. M. and Jia-Xing, Y. (1993) New techniques of applying multi-wavelength anomalous scattering data. *Proc. Roy. Soc. Lond. A* **442**, 13–32.

Hao, Q., Gu, Y. X., Zheng, C. D. and Fan, H. F. (2000) OASIS: a computer program for breaking phase ambiguity in one-wavelength anomalous scattering or single isomorphous substitution (replacement) data. *J. Appl. Crystallogr.* **33**, 980–1.

Helliwell, J. R. (1992) *Macromolecular Crystallography with Synchrotron Radiation.* Cambridge University Press, Cambridge.

Hendrickson, W. A. (1991) Determination of macromolecular structures from anomalous diffraction of synchrotron radiation. *Science,* **254**, 51–8.

Hendrickson, W. A. and Lattman, E. E. (1970) Representation of the phase probability distributions for simplified combination of independent phase information. *Acta Crystallogr. B* **26**, 136–43.

Howell, P. L. and Smith, G. D. (1992) Identification of heavy-atom derivatives by normal probability methods. *J. Appl. Crystallogr.* **25**, 81–6.

Karle, J. and Hauptman, H. (1956) A theory of phase determination for the four types of non-centrosymmetric space groups. *Acta Crystallogr.* **9**, 635–51.

Kartha, G. (1975) Applications of anomalous scattering studies in protein structure analysis. In *Anomalous Scattering*, ed. Ramaseshan, S. and Abrahams, S. C. Munksgaard, Copenhagen, pp. 363–91.

Kleywegt, G. J. and Jones, T. A. (1994) Halloween ... masks and bones. In *From First Map to Final Model*, ed. Bailey, S., Hubbard, R. and Waller, D. Daresbury Laboratory, Warrington, UK, pp. 59–66.

Kraut, J., Sieker, L., High, D. F. and Freer, S. T. (1962) Chymotrypsinogen: a three dimensional Fourier synthesis at 5 Å resolution. *Proc. Natl. Acad. Sci. USA* **48**, 1417–24.

Ladd, M. F. C. and Palmer, R. A. (2003) *Structure Determination by X-ray Crystallography*, 4th edn. Springer, New York.

Leslie, A. G. W. (1987) A reciprocal-space method for calculating a molecular envelope using the algorithm of B.C. Wang. *Acta Crystallogr A* **43**, 134–6.

Leslie, A. (1991) Heavy atom derivative screening. In *Isomorphous replacement and anomalous scattering*, ed. Wolf, W., Evans, P. R. and Leslie, A. G. W. Daresbury Laboratory, Warrington, UK, pp. 9–22.

Lu, G. (1999) FINDNCS: a program to detect non-crystallographic symmetries in protein crystals from heavy atoms sites. *J. Appl. Crystallogr.* **32**, 365–8.

Matthews, B. W. (1966) The extension of the isomorphous replacement method to include anomalous scattering measurements. *Acta Crystallogr.* **20**, 82–6.

Matthews, B. W. and Czerwinski, E. W. (1975) Local scaling: a method to reduce systematic errors in isomorphous replacement and anomalous scattering measurements. *Acta Crystallogr. A* **31**, 480–7.

Petsko, G. A. (1985) Preparation of isomorphous heavy-atom derivatives. *Methods Enzymol.* **114**, 147–57.

Phillips, J. C. and Hodgson, K. O. (1980) The use of anomalous scattering effects to phase diffraction patterns from macromolecules. *Acta Crystallogr. A* **36**, 856–64.

Read, R. J. (1986) Improved Fourier coefficients for maps using phases from partial structures with errors. *Acta Crystallogr. A* **42**, 140–9.

Sauer, O., Schmidt, A. and Kratky, C. (1997) Freeze-trapping isomorphous xenon derivatives of protein crystals. *J. Appl. Crystallogr.* **30**, 476–86.

Sayre, D. (1952) The squaring method: a new method for phase determination. *Acta Crystallogr.* **5**, 60–5.

Schneider, T. R. and Sheldrick, G. M. (2002) Substructure solution with SHELXD. *Acta Crystallogr. D* **58**, 1772–9.

Sheldrick, G. M. (2008) A short history of SHELX. *Acta Crystallogr.* **64**, 112–22.

Sheldrick, G. M., Dauter, Z., Wilson, K. S., Hope, H. and Sieker, L. C. (1993) The application of direct methods and Patterson interpretation to high-resolution native protein data. *Acta Crystallogr. D* **49**, 18–23.

Singh, A. K. and Ramaseshan, S. (1966) The determination of heavy atom positions in protein derivatives. *Acta Crystallogr.* **21**, 279.

Terwilliger, T. C. (1994a) MAD phasing: Bayesian estimates of F_A. *Acta Crystallogr. D* **50**, 11–6.

Terwilliger, T. C. (1994b) MAD phasing: treatment of dispersive differences as isomorphous replacement information. *Acta Crystallogr. D* **50**, 17–23.

Terwilliger, T. C. (2000) Maximum likelihood density modification. *Acta Crystallogr. D* **56**, 965–72.

Terwilliger, T. C. (2002a) Statistical density modification with non-crystallographic symmetry. *Acta Crystallogr. D* **58**, 2082–6.

Terwilliger, T. C. (2002b) Rapid automatic NCS identification using heavy-atom substructures. *Acta Crystallogr. D* **58**, 2213–5.

Terwilliger, T. C. and Berendzen, J. (1999a) Automated MAD and MIR structure solution. *Acta Crystallogr. D* **55**, 849–61.

Terwilliger, T. C. and Berendzen, J. (1999b) Discrimination of solvent from protein regions in native Fouriers as a means of evaluating heavy-atom solutions in the MIR and MAD methods. *Acta Crystallogr. D* **55**, 501–505.

Terwilliger, T. C. and Eisenberg, D. (1983) Unbiased three-dimensional refinement of heavy-atom parameters by correlation of origin-removed Patterson functions. *Acta Crystallogr. A* **39**, 813–7.

Terwilliger, T. C. and Eisenberg, D. (1987) Isomorphous replacement: effects of errors on the phase probability distribution. *Acta Crystallogr. A* **43**, 6–13.

Terwilliger, T. C., Sung-Hou, K. and Eisenberg, D. (1987) Generalised method of determining heavy-atom positions using the difference Patterson function. *Acta Crystallogr. A* **43**, 1–5.

Terwilliger, T. C., Grosse-Kunstleve, R. W., Afonine, P. V., Moriarty, N. W., Zwart, P. H., Hung, L.-W., Read, R. J. and Adams, P. D. (2008) Iterative model building, structure refinement and density modification with the PHENIX AutoBuild wizard. *Acta Crystallogr. D* **64**, 61–9.

Tickle, I. J. (1983) Local scaling of heavy-atom derivative data and the solution of heavy-atom difference Pattersons by vector-search. Daresbury Laboratory, UK. *Info. Q. Protein Crystallogr.* **11**, 5–10.

Tickle, I. J. (1991) Refinement of single isomorphous replacement heavy-atom parameters in Patterson versus reciprocal space. In *Isomorphous Replacement and Anomalous Scattering*, ed. Wolf, W., Evans, P. R. and Leslie, A. G. W. Daresbury Laboratory, Warrington, UK, pp. 87–95.

Tickle, I. J. (1992) LOCAL: local scaling program for heavy atom derivatives. Daresbury laboratory, UK. *Joint CCP4 ESF-EACM Newslett. Protein Crystallogr.* **26**, 22–6.

Wang, B. C. (1985) Resolution of phase ambiguity in macromolecular crystallography. *Methods Enzymol.* **115**, 90–112.

Wang, J., Wlodawer, A. and Dauter, Z. (2007) What happens when the signs of the anomalous differences or the handedness of the substructure are inverted. *Acta Crystallogr. D* **63**, 751–8.

Weeks, C. M. and Furey, W. (2007) Application of direct methods to macromolecular structure solution. In *Macromolecular Crystallography: Conventional and High-Throughput Methods—A Practical Approach*, ed. Sanderson, M. and Skelly, J. Oxford University Press, Oxford, pp. 129–41.

Weeks, C. M., Sheldrick, G. M., Miller, R., Usón, I. and Hauptman, H. A. (2001) *Ab initio* phasing by dual-space direct methods. In *Advances in Structure Analysis*, ed. Kuzel, R. and Hasek, J. Czech & Slovak Crystallographic Association, Prague, pp. 37–64.

Woolfson, M. M. (1997) *An Introduction to X-ray Crystallography*. Cambridge University Press, Cambridge.

Xu, H. and Weeks, C. M. (2008) Rapid automated substructure solution by Shake-and-Bake. *Acta Crystallogr. D* **64**, 172–7.

Yao, J. X., Dodson, E. J., Wilson, K. S. and Woolfson, M. M. (2006) ACORN: a review. *Acta Crystallogr. D* **62**, 901–908.

Zhang, K. Y. J. and Main, P. (1990) Histogram matching as a new density modification technique for phase refinement and extension of protein molecules. *Acta Crystallogr. A* **46**, 41–6.

15

Refinement

15.1   The necessity for refinement

Having calculated the electron density function $\rho(x, y, z)$ by a Fourier synthesis of the structure factors F_{hkl} according to equation (9.15),

$$\rho(x, y, z) = \frac{1}{V} \sum_h \sum_k \sum_l F_{hkl} e^{-2\pi i(hx+ky+lz)} \tag{9.15}$$

we may then display an electron density map. This will be a contour map of the electron density for the protein, which will have peaks centred on the atomic positions. From the electron density map, we may be able to infer all the detailed structural information concerning the secondary, tertiary and quaternary structure of the protein within the crystal. We therefore desire that the electron density map be as accurate as possible so that we may have confidence in assigning a given molecule a particular structure. Any errors in either the magnitudes or the phases of the structure factors F_{hkl} will give rise to discrepancies between the electron density as calculated using equation (9.15) and that which corresponds to the real structure, and so we wish to minimise these errors to obtain the best possible model from the experimental data. For this reason, refinement is an integral and most vital part of structure solving, and one which any practising crystallographer will spend a lot of time on.

There are many stages in the overall scheme of a crystal structure analysis which can give rise to errors in the structure factors F_{hkl}. Firstly, there are direct experimental errors in the primary data, measured usually by area detector techniques. These errors are in general random, and may be minimised by careful experimental technique. Associated with the errors in intensity data recording are those which arise from very weak diffraction maxima. If a particular reciprocal-lattice point has a structure factor of very small magnitude, the corresponding diffraction spot will be very weak, and may not be recorded by the intensity detector.

The next source of error comes in the correction of our primary data to derive the relative intensities $|F_{hkl}|^2$. As we saw in Chapter 11, our experimental observable is the integrated reflecting power, from which we derive the relative intensity $|F_{hkl}|^2$ by applying the various correction factors as described in Chapter 11.

These corrections correct explicitly for the state of polarisation of the incident beam and the geometry of the intersection of a reciprocal-lattice point with the Ewald sphere, whilst empirical corrections are applied for absorption and radiation damage. The first two effects are characterised by the well-defined polarisation and Lorentz factors, but the empirical scaling corrections for absorption, radiation damage and other factors can be very difficult to assess, and therefore they are likely to be sources of error in the relative intensities $|F_{hkl}|^2$.

The solution of the phase problem introduces a third possibility of inaccuracy. All the chemical modification methods assume that the crystals of the native molecule and the heavy-atom derivative are perfectly isomorphous, and that the structure of the native molecule is unperturbed on adding or replacing heavy atoms. These assumptions are never true, and so the phases initially derived for each of the structure factors for a non-centrosymmetric crystal inevitably contain significant errors.

In determining the phases experimentally, there will be errors if there are not enough good derivatives, and in molecular replacement, the phasing model is likely to have significant differences from the target structure. The errors introduced into the first calculated electron density function, as discussed above, all concern difficulties inherent in the estimation of the values of the magnitudes and phases of the structure factors F_{hkl}. A different and rather more subtle error will now be investigated.

In Chapter 6, during our discussion of the nature of diffraction, we found out that when an incident wave interacts with an obstacle, the diffracted waves spread out over all space. The diffraction pattern therefore extends over all space, and the information contained within it may be determined fully only if the diffraction pattern is sampled over all space. Since we monitor only a finite volume of reciprocal space, then we will not extract the total information contained within the diffraction pattern, and on performing the inverse Fourier transform to derive the structure of the obstacle, we will not be able to reproduce the obstacle exactly.

The application of these ideas to the specific case of the diffraction of X-rays by crystals concerns the fact that the Fourier transform of a real lattice is the reciprocal lattice, which is infinite in extent, and is characterised by triplets hkl in which h, k and l individually take on allowed values from minus infinity to plus infinity. Each reciprocal-lattice point is associated with a structure factor F_{hkl}, and in order to derive the electron density function exactly, we must perform the Fourier synthesis of equation (9.15),

$$\rho(x, y, z) = \frac{1}{V} \sum_h \sum_k \sum_l F_{hkl} e^{-2\pi i(hx+ky+lz)} \tag{9.15}$$

for all values of h, k and l over the total infinite range. If we assume that we have the structure factors F_{hkl} correct in both magnitude and phase, then we will generate the precise electron density function $\rho(x, y, z)$ only by taking the Fourier synthesis of an infinite number of terms.

If we use a finite series, then the electron density function we calculate will be different from the ideally correct electron density function as a result of the further terms in the Fourier synthesis of equation (9.15) which we have neglected. The error introduced by using a finite summation is referred to as a *series termination error*, in which we make reference to the fact that we terminate an infinite series prematurely. Series termination errors are inevitable, since it is impossible to collect and process an infinite amount of data. This is obvious, since the X-ray detector will only intercept diffracted waves over a restricted volume of space. With a well-designed instrument, however, it is possible to place the photon counter at almost any position with respect to the crystal, and it appears as if we are able to scan virtually all of the space around a crystal. Yet despite this, a series termination error must exist for the following reason.

In Section 8.14, we discussed the relationship of the Ewald sphere to the interpretation of diffraction patterns, and we introduced the concept of the limiting sphere, which has a radius twice that of the Ewald sphere. As a result of the limitation set on diffraction by the function $\sin\theta$ in Bragg's law, we proved that for a given incident wavelength λ, only those reciprocal-lattice points characterised by reciprocal-lattice vectors whose magnitudes S_{hkl} satisfy the inequality

$$S_{hkl} \leq \frac{2}{\lambda} \tag{8.32}$$

are able to be detected in an X-ray diffraction experiment. It is therefore the limiting sphere which defines those reciprocal-lattice points which can and cannot be sampled in any diffraction experiment. As can be seen from the inequality (8.32), for any finite wavelength, the size of the limiting sphere is finite, implying that we could never sample all reciprocal-lattice points even if we were able to scan all around the crystal. In practice, the shortest commonly available X-ray wavelength is of the order of 0.7 Å, but whatever incident radiation is used, there is always a limit on the amount of the reciprocal lattice which can be monitored, and so series termination errors always exist.

The effect of series termination errors is to cause any peak in the electron density function as calculated by the Fourier synthesis of equation (9.15) to be surrounded by what are known as *diffraction ripples*. If an infinite Fourier synthesis is used, then every electron density peak has a perfect shape, as in Fig. 15.1(a). When a finite series is used, the main peak is surrounded by minor peaks of varying sizes, as in Fig. 15.1(b). The magnitude of the satellite peaks depends on the number of terms taken in the Fourier summation. If the electron density peaks are well separated, then the diffraction ripples cause no problems in interpretation, but for closely neighbouring atoms, in adverse circumstances, it is possible that the diffraction ripples of two atoms may combine so as to cause an intermediate peak. The intermediate peak is spurious,

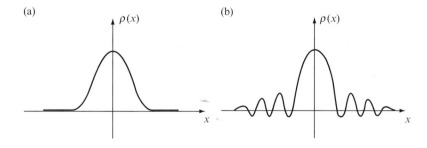

Fig. 15.1 *Series termination errors.* An infinite Fourier synthesis will give a perfect electron density peak as in (a), but a finite summation gives peaks surrounded by diffraction ripples (b).

and does not represent a real peak in the electron density function of the true structure. It is possible that such a peak may be misinterpreted as indicating the presence of a missing atom, and so we must be aware of the dangers implicit in series termination errors. In practice, with typical protein electron density maps, diffraction ripples are only significant for heavy atoms. However, series termination errors contribute significantly to the broadening of the electron density for light atoms in the structure.

15.2   Obtaining the trial structure

The errors introduced into the calculation of the electron density function concern firstly the inaccuracies in the magnitudes and phases of the structure factors F_{hkl}, and secondly those arising as a result of series termination. Initially, the accuracy of the calculated electron density function is limited considerably more by errors in the structure factor phases than by any other errors, and so our first task in refinement is to derive a more accurate set of phases for the structure factors.

In much of the subsequent discussion, we will be using many subscripts, and in order to avoid an ugly multiplicity of these, we will now drop the F_{hkl} notation in favour of simply F. We shall assume that we have completed an X-ray diffraction experiment, and possess a set of structure factors characterised by magnitudes $|F_o|$ as obtained by the observation of the diffraction intensities, and phases ϕ_o derived from one of the methods of solving the phase problem as discussed in the previous chapters. The significance of the subscript o is that this signifies parameters derived directly from experimental observation. We may now compute a Fourier synthesis of the electron density $\rho_o(x, y, z)$ according to equation (9.15), using the observed values $|F_o|$ and ϕ_o, that is essentially of the form

$$\rho(x, y, z) = \frac{1}{V} \sum_h \sum_k \sum_l |F_o| \, e^{-i[2\pi(hx+ky+lz)-\phi_o]} \tag{15.1}$$

Equation (15.1) contains the phase angle ϕ_o in the complex exponential. For the various reasons discussed above, the electron density function is unlikely to correspond exactly to the electron density function of the true structure, but we can draw an electron density map which will reveal certain features of the crystal structure. In the ideal case of an atomic-resolution map, there will be peaks which we may identify as atomic positions, but in the majority of real-life protein cases we will have a map at medium resolution (typically 3.0–2.5 Å resolution), into which we need to fit a molecular model of the amino acid sequence of the protein. This can be done either manually, using a computer graphics program, or automatically, using a specialised map interpretation package. Whichever method we choose, we may derive a set of atomic positions which constitute a *trial structure* or *phasing model*, and it is this initial model of the electron density function which we wish to refine.

As mentioned, the initial model can be constructed either by manual interpretation of the electron density map by use of a suitable graphics program such as Coot (Emsley and Cowtan, 2004), as shown in Fig. 15.2, or by automatic routines that aim to fit segments of polypeptide to the electron density, and these methods will be described later. Molecular graphics is probably the best approach when molecular replacement

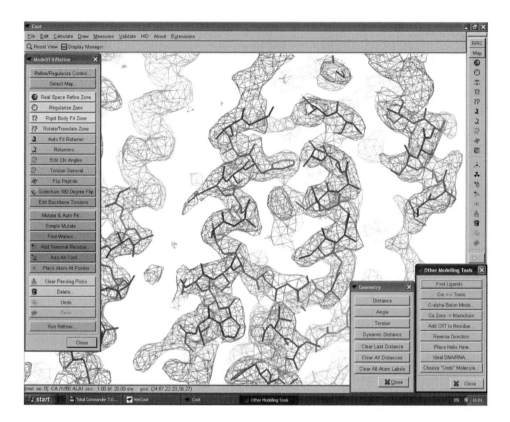

Fig. 15.2 *Electron density modelling.* A screenshot of the molecular graphics modelling program Coot (Emsley and Cowtan, 2004), showing an experimental electron density map at 3.0 Å resolution (grey lines) which has been partially interpreted by the automatic chain-tracing program MAID (Levitt, 2001). The preliminary model (dark lines) is an excellent starting point for further elaboration, and Coot has the necessary functionality to allow the crystallographer to perform this task.

has been used for the initial structure analysis, since the correctly oriented and positioned search model can be visualised along with its electron density and rebuilt. In contrast, when experimental phasing methods are used to obtain the first map, there will probably be no model to rebuild, and the fold of the molecule will have to be determined from the electron density. Graphics packages are especially important if the data extend only to low or medium resolution, since the automatic map interpretation routines do not work very well at these resolutions. The widespread use of computer graphics in map interpretation began with the development of FRODO (Jones, 1978) and required vastly expensive equipment. In contrast, today, better performance can be achieved using very modest and inexpensive computers. Computer graphics programs allow the electron density to be displayed at variable contour levels, and amino acids can be inserted into the structural model at desired locations. These programs can display a precalculated map, and others (e.g. Coot) can calculate the electron density 'on the fly' from the reflection amplitudes and phases. Tracing the initial polypeptide backbone is assisted by skeletonisation of the electron density (e.g. Greer, 1974), in which dummy atoms are placed in regions of strong electron density, and those 'bone atoms' which reside in contiguous regions are treated as putative main-chain atoms of the protein.

In the early stages of model building, it is common to build segments of the backbone initially as polyalanine, and much the same applies for the automatic model-building programs. When the side chains have been identified, the polyalanine backbone can be 'mutated' to the correct sequence. The 'mutation' of these alanine residues to the correct amino acids can readily be accomplished using a database of commonly occurring conformations to guide fitting of the side chains to the electron density. Certain amino acids can be identified more easily than others; for example, the electron-rich side chain of cysteine, particularly when forming a disulphide bridge, and the side chains of large aromatic residues are often readily identified, as are the methionines in selenomethionine MAD maps—the coordinates of the selenium sites are, by definition, already known as a prerequisite for calculating the phases (see Chapter 14). It may be anathema to purists, but secondary-structure prediction can also be very helpful in map interpretation. Symmetry-related molecules are also displayed, which is important in building the surface regions of proteins and the solvent structure to prevent overlap with neighbouring molecules.

The amino acids can be rebuilt by moving individual atoms or fragments, or by adjusting the torsion angles interactively. The latter has the advantage of preserving the local stereochemistry. If individual atoms are moved, the geometry of the amino acids will require regularisation to ensure that they conform to known bond lengths and angles, etc., and this can be performed in short zones by most molecular graphics programs, using methods that we will describe later on. It is also possible to refine the fit of the model to the electron density map, usually in small zones. Incidentally, this *real-space refinement*, which is available in most graphics programs, was originally developed as a method for refinement of the entire structure (Diamond, 1971; Fletterick and Wyckoff, 1975), but the dependence of the map quality on the phase accuracy has led to a preference for reciprocal-space refinement methods which do not rely on the experimental phases, as will be covered later.

The hardest parts of any protein to build are the flexible loop regions, where the electron density can be poor and there are no repetitive patterns such as can be found in α-helices and β-sheets. Difficult loop regions can be built by selecting appropriate fragments from a database of known structures. Most molecular graphics programs have powerful tools for validating the stereochemistry of the model and for superimposing related molecules for structural comparisons.

The wholly automated interpretation of electron density maps is now feasible owing to the development of various powerful methods that can provide the crystallographer with an excellent trial model for manual optimisation with a graphics program. The ARP/wARP method (Perrakis *et al.*, 1997) relies on fitting a number of dummy atom models to the electron density, which are then refined as free atoms. The refinement involves methods that we will describe later, interspersed with pruning of the model to remove atoms that have weak density or are too close together, etc., and the addition of new atoms. Phases are calculated from the ensemble of dummy atom models and are recombined for calculation of a new map. The averaging of the phases from the different dummy atom models reduces the noise in the resultant map, into which the polypeptide chain is traced automatically by pattern matching (Lamzin *et al.*, 2001). This method requires reasonably high-resolution data ($d < 2.7$ Å). In contrast, the program MAID (Levitt, 2001) works with both high- and medium-resolution maps by essentially emulating the steps that a crystallographer would take in fitting the chain using a graphics program and, for this reason, it is helpful to describe it in some detail.

MAID starts by skeletonising the map and fitting pieces of polypeptide to regions of helix or sheet, and the polypeptide fragments are extended as far as possible. MAID then tries to build the loop regions, firstly by looking at bone traces that connect the fitted regions and selecting the shortest possible connections with the strongest electron density. MAID then tests if the amino acid sequence can be aligned satisfactorily with the side chain electron density for two of the connected regions. If sequence assignment is possible, the side chains are added and fitted to the density by torsion angle molecular dynamics (which will be covered later). The program also uses dynamics at the alignment stage to test the fit of all 20 amino acids to the electron density for each side chain, essentially mimicking a crystallographer guessing what amino acid would be present from the shape of the side chain density. Prior assignment of the sequence helps during building of the loop region, since the conformational properties of certain amino acids are taken into account. Each new residue that is inserted into the loop region is tested in a range of allowed conformations that are refined by dynamics, and those which fit the density best are selected. The loop conformation which fits the density and overlaps the next fitted section best is selected and 'fused' by dynamics with the connecting segment. MAID continues to extend the structure as far as possible in this manner, finally making use of crystallographic symmetry to identify and extend all of the fitted regions and to 'group' the ones which are close together into separate subunits. An example of a MAID chain tracing of a medium-resolution map is shown in Fig. 15.2. A number of other automatic map-fitting routines are available, including TEXTAL (Ioerger and Sacchettini, 2002), RESOLVE (Terwilliger, 2003a,b) and QUANTA (Oldfield, 2003), and tests by Badger (2003) have confirmed their effectiveness.

15.3 Assessing the trial structure

The trial structure is usually a rather crude representation of the true structure. At the resolutions typical of most protein analyses, the electron density for neighbouring atoms tends to be overlapped, and for this reason the positioning of many atoms will be appreciably in error. Note that the model will only contain the atoms C, N, O, S and any other heavier elements that are naturally present in the molecule, since hydrogen is associated with a very weak electron density function and it is impossible to resolve it at all in X-ray analyses of proteins except when atomic-resolution data are available.

However, having identified a trial set of atomic coordinates (x_j, y_j, z_j) from the initial structure, we may calculate values for the structure factors using equation (9.21). Since these structure factors are calculated on the basis of a trial model, we represent them by F_c, where the subscript c signifies 'calculated':

$$F_c = \sum_j f_j e^{-B_j \sin^2 \theta / \lambda^2} e^{2\pi i \left(h x_j + k y_j + l z_j \right)} \tag{9.21}$$

The values of the atomic scattering factors used in this calculation are represented as f_j and the temperature factor as B_j. Each atomic scattering factor tabulated in, for example, Vol. III of the *International Tables for X-ray Crystallography* is that applicable to the hypothetical case of zero vibration. The B_j term is the temperature factor assessed on the basis of some model of atomic vibration and disorder, which are assumed to be spherically symmetric, and for the resolutions at which most proteins are analysed, this is a reasonable assumption. Use of a single parameter B_j as the temperature factor assumes that the effect of thermal motion or disorder of any atom is spherically symmetric. Since covalently bonded atoms are in general restrained by chemical bonds, it is unlikely that this spherical symmetry is realised, and a more accurate representation is to construct a model in which an atom has different freedoms of vibration in three orthogonal (mutually perpendicular) directions. In the spherically symmetric case, the atom would sweep out a sphere characterised by a single displacement parameter B_j. In the more realistic case, the atom is likely to sweep out a general ellipsoid (see Fig. 15.3), characterised by three parameters defining the shape of the ellipsoid and three specifying its orientation, giving a total of six parameters in all. A better estimation of the thermal vibrations and disorder is therefore to associate each atom with six *anisotropic displacement parameters*, but this is possible only for proteins that diffract to atomic resolution and so, for the vast majority of cases, an isotropic temperature factor will be assigned to each atom; typically, trial values of the order of 20–30 Å^2 would be suitable for a structure factor calculation.

We are now in possession of two sets of structure factor data. On the one hand are the F_o's derived directly from experiment, and on the other are the F_c's calculated from the trial structure. Since the trial structure is unlikely to be a very accurate representation of the real structure, we do not expect the F_c's to match the F_o's exactly. The reasons for this include the fact that we have usually not identified all atomic sites in the first calculated electron density function. The F_c's are therefore calculated on the basis of an incomplete structure, whereas the F_o's are derived from a physical experiment in which we know that all the atoms present contribute to the diffraction pattern. A second and equally important reason for disagreement between the F_c's and the F_o's is the likelihood that the atomic model contains significant errors.

In general, then, the observed structure factors F_o and the calculated structure factors F_c will have different values, but we would expect corresponding values to be reasonably similar. Also, we would expect that the more closely the calculated and observed structure factors agree, then the more closely does the trial structure represent the true structure. The discrepancy between the observed and calculated

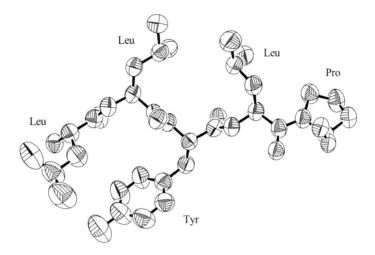

Fig. 15.3 *Thermal ellipsoids.* A small part of a protein that was refined at 0.9 Å resolution using anisotropic B-factors, which are represented as ellipsoids. The leucine and tyrosine side chains on the lower left appear to show correlated displacements. Note that proline residues, such as the one on the far right, often exhibit anisotropy at this resolution owing to puckering of the ring. This figure was prepared using the Oak Ridge thermal-ellipsoid plot program ORTEP-3 for Windows (Farrugia, 1997).

structure factor amplitudes is therefore an indication of the degree of confidence which may be ascribed to any given trial structure. In order to put this concept on a more quantitative basis, we now define the *discrepancy index*, or *residual index*, as

$$R = \frac{\sum |(|F_o| - |F_c|)|}{\sum |F_c|} \qquad (15.2)$$

in which the summation goes over all diffraction maxima. It is difficult to state the precise meaning of R, since we cannot make exact statements such as 'a value of R of such and such implies so many per cent error in the electron density function'. Rather, R represents a general indication of the correctness of the structure in a more descriptive way. In the ideal case in which the observed and calculated structure factor amplitudes agree exactly, then the value of the residual is zero. On the other hand, a totally incorrect structure in which the atoms are randomly arranged was shown by Wilson (1942) to give a value of R of 0.828 for a centrosymmetric crystal or 0.586 for a non-centrosymmetric one. Any value of R less than about 0.5 therefore indicates a certain correspondence between the trial structure and the real structure, and during the course of refinement, the value of the residual should decrease, implying that we are obtaining a better fit between the trial structure and the real one. An initial model of a protein structure that is largely correct is likely to have an R-value in the range of 0.4–0.5.

Another useful indication of the correctness of the trial structure is the correlation coefficient between F_o and F_c, which is given in equation (15.3) below, and has a maximum of 1 for a perfect structure and a value of 0 for a structure where there is no relationship between F_o and F_c. Typically, a trial model of a protein structure will have a correlation in the region of 0.3 or greater. The correlation coefficient is given by

$$CC = \frac{\sum (|F_o| - \langle |F_o| \rangle)(|F_c| - \langle |F_c| \rangle)}{\sqrt{\sum (|F_o| - \langle |F_o| \rangle)^2 \sum (|F_c| - \langle |F_c| \rangle)^2}} \tag{15.3}$$

A structure refined at high resolution should have a correlation coefficient well in excess of 90% and an R-factor around 20% or less. Both the R-factor and the correlation coefficient are commonly quoted as percentages rather than fractional values.

15.4 Least-squares refinement

Suppose that we have carried out an experiment from which we obtain data which we plot on a graph, expecting to obtain a straight line as in Fig. 15.4. Note that we assume that there are no errors in x on the graph.

Since there is inevitably some experimental random error in all our measurements, the points never lie on a perfect straight line, and we are faced with the problem of deciding which is the 'best' line which we may plot through our experimental points. What we require is some criterion by which we may judge that one particular line is a 'better' representation of our data than any other. This criterion is necessarily statistical, and the commonly accepted procedure for obtaining this 'best' line was first formulated by the French mathematician Adrien-Marie Legendre in 1806, who

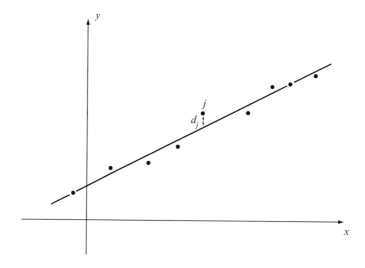

Fig. 15.4 *Least squares.* Our problem is to find the 'best' line through the experimental points. The jth point is separated from the line by the vertical distance d_j, and Legendre's criterion of the 'best' line is that which minimises the quantity $\sum_j d_j^2$.

used it originally as a method to determine the orbital elements of comets from multiple observations.

If we draw any acceptable line which reasonably passes through the experimental points, then in general, the jth point will not be on the line, but will be separated from it by some vertical distance d_j. We expect there to be roughly as many points on one side of the line as there are on the other, and so the values of d_j will be both positive and negative. The quantity d_j^2 is, however, always positive, and so if we have a total of n points, then the quantity $(1/n)\sum_j d_j^2$ represents a valid measure of the average deviation of the points from the line. Legendre's criterion of the 'best' line is to draw that line which minimises the quantity $\sum_j d_j^2$, thereby minimising the overall deviation of the points from the line. The technique by which the required 'best' line may be derived from the experimental data is called *least-squares analysis*, and this approach was first applied to crystallographic refinement by E. W. Hughes in 1941, for refinement of the small molecule melamine (Hughes, 1941). It is interesting to note that in the days when the only calculating machines were mechanical, one cycle of refinement of just 18 parameters using 105 reflections required 24 hours of work (Hughes, 1978).

One of the problems with applying any refinement technique, be it least squares or otherwise, to protein molecules concerns the fact that the electron density maps of proteins are at much lower resolutions than those of small molecules—typically somewhere between 3 and 2 Å. Hence it is doubtful if a shift of 0.1 Å of an atomic site would be very meaningful for such a map, although it may be very significant for a small molecule, where a movement of this magnitude would exceed the positional errors in the structure severalfold. For small organic molecules, commonly quoted R-values are often as low as 0.05 or less, but for proteins, where the data do not define the structure with anything like the same precision, values of 0.2 to 0.3 are regarded as acceptable for refined structures. Lower R-values and greater precision are achievable with proteins that diffract to atomic resolution.

15.5   Theory of the least-squares method

Let us assume that a variable q which we can measure is a function of three variables x, y and z which are known and three constants a, b and c which are unknown:

$$q = ax + by + cz \tag{15.4}$$

If we have measurements of q at various values of x, y and z, then we need to determine the values of a, b and c which best fit the measurements. If only three values of q have been measured at different x, y and z, then we could solve the three simultaneous equations for the three unknowns a, b and c uniquely. However, when there are more observations than unknowns, we can take advantage of the redundancy in the data to obtain more accurate values of the unknowns than could be obtained by arbitrarily picking three sets of measurements and using just these to solve for the three unknowns.

For any observation q_j, the error E_j is given by

$$E_j = ax_j + by_j + cz_j - q_j$$

Legendre's principle stipulates that the sum of the squares of the errors for all measurements of q_j (given by the equation below) must be a minimum with respect to the unknowns (a, b and c). The summation is given by

$$\sum_j E_j^2 = \sum_j \left(ax_j + by_j + cz_j - q_j\right)^2 \tag{15.5}$$

and at a minimum, the partial derivatives of equation (15.5) with respect to a, b and c must be zero. Hence

$$\frac{\partial \sum_j E_j^2}{\partial a} = \frac{\partial \sum_j E_j^2}{\partial b} = \frac{\partial \sum_j E_j^2}{\partial c} = 0$$

Differentiating equation (15.5) with respect to a gives

$$\frac{\partial \sum_j E_j^2}{\partial a} = 2 \sum_j x_j \left(ax_j + by_j + cz_j - q_j\right)$$

and this will equal zero at a minimum. Hence

$$a \sum_j x_j^2 + b \sum_j x_j y_j + c \sum_j x_j z_j - \sum_j q_j x_j = 0 \tag{15.6}$$

Note that for equation (15.5) to be at a minimum, the second derivative $\partial^2 \sum E_j^2 / \partial a^2$ must be positive. Since

$$\frac{\partial^2 \sum_j E_j^2}{\partial a^2} = 2 \sum_j x_j^2$$

we can indeed be sure that this corresponds to a minimum, because x^2 must always be positive when x is real.

By differentiating equation (15.5) with respect to b and c, we arrive at two more equations, of the form

$$a \sum_j x_j y_j + b \sum_j y_j^2 + c \sum_j y_j z_j - \sum_j q_j y_j = 0 \tag{15.7}$$

$$a \sum_j x_j z_j + b \sum_j y_j z_j + c \sum_j z_j^2 - \sum_j q_j z_j = 0 \tag{15.8}$$

Hence we now have three simultaneous equations (15.6), (15.7) and (15.8), which are known as the *normal equations* and can be solved for the three unknowns a, b and c by standard methods. The solutions to these simultaneous equations correspond to values of the unknown constants a, b and c which minimise the sum of the squares of the differences between the observed and calculated values of q. The elegance of the least-squares method is that if we have linear equations for the unknowns, such as equation (15.4), we can differentiate with respect to each of the unknown parameters and arrive at as many simultaneous equations as there are unknowns.

The method of least squares is amenable to a very elegant representation in matrix form. The original set of equations can be represented by

$$
\begin{bmatrix}
x_1 & y_1 & z_1 \\
x_2 & y_2 & z_2 \\
x_3 & y_3 & z_3 \\
x_4 & y_4 & z_4 \\
\vdots & \vdots & \vdots
\end{bmatrix}
\begin{bmatrix}
a \\
b \\
c
\end{bmatrix}
=
\begin{bmatrix}
q_1 \\
q_2 \\
q_3 \\
q_4 \\
\vdots
\end{bmatrix}
\tag{15.9}
$$

or, in simplified matrix form, as

$$\mathbf{X}\mathbf{a} = \mathbf{q}$$

If we pre-multiply both sides of the equation by the transpose of the matrix $\mathbf{X}$, i.e. $\mathbf{X}^T$, as follows,

$$\mathbf{X}^{\mathrm{T}}\mathbf{X}\mathbf{a} = \mathbf{X}^{\mathrm{T}}\mathbf{q} \tag{15.10}$$

the left-hand side of this equation can be represented in full matrix form as

$$
\begin{bmatrix}
x_1 & x_2 & x_3 & x_4 & \cdots \\
y_1 & y_2 & y_3 & y_4 & \cdots \\
z_1 & z_2 & z_3 & z_4 & \cdots
\end{bmatrix}
\begin{bmatrix}
(ax_1 + by_1 + cz_1) \\
(ax_2 + by_2 + cz_2) \\
(ax_3 + by_3 + cz_3) \\
(ax_4 + by_4 + cz_4) \\
\vdots
\end{bmatrix}
=
\begin{bmatrix}
a\sum_j x_j^2 + b\sum_j x_j y_j + c\sum_j x_j z_j \\
a\sum_j x_j y_j + b\sum_j y_j^2 + c\sum_j y_j z_j \\
a\sum_j x_j z_j + b\sum_j y_j z_j + c\sum_j z_j^2
\end{bmatrix}
$$

and the right-hand side of equation (15.10) can be represented as

$$
\mathbf{X}^T\mathbf{q} =
\begin{bmatrix}
x_1 & x_2 & x_3 & x_4 & \cdots \\
y_1 & y_2 & y_3 & y_4 & \cdots \\
z_1 & z_2 & z_3 & z_4 & \cdots
\end{bmatrix}
\begin{bmatrix}
q_1 \\
q_2 \\
q_3 \\
q_4 \\
\vdots
\end{bmatrix}
=
\begin{bmatrix}
\sum_j q_j x_j \\
\sum_j q_j y_j \\
\sum_j q_j z_j
\end{bmatrix}
$$

Hence

$$
\begin{bmatrix}
a\sum_j x_j^2 + b\sum_j x_j y_j + c\sum_j x_j z_j \\
a\sum_j x_j y_j + b\sum_j y_j^2 + c\sum_j y_j z_j \\
a\sum_j x_j z_j + b\sum_j y_j z_j + c\sum_j z_j^2
\end{bmatrix}
=
\begin{bmatrix}
\sum_j q_j x_j \\
\sum_j q_j y_j \\
\sum_j q_j z_j
\end{bmatrix}
$$

which is identical to the least-squares equations (15.6), (15.7) and (15.8) deduced above by differentiating with respect to each of the unknowns. Hence rearrangement of equation (15.10) as follows allows us to give a matrix expression for the least-squares estimates of the unknown constants:

$$\mathbf{a} = (\mathbf{X}^{\mathrm{T}}\mathbf{X})^{-1}\mathbf{X}^{\mathrm{T}}\mathbf{q} \tag{15.11}$$

In the case of the data from an X-ray diffraction experiment, our observed quantities are a set of structure factor magnitudes $|F_o|$, and we have a proposed model from which we calculate the structure factor magnitudes $|F_c|$ by use of the formula below:

$$F_c = \sum_j f_j e^{-B_j \sin^2 \theta / \lambda^2} e^{2\pi i \left(hx_j + ky_j + lz_j \right)} \tag{9.21}$$

What we wish to do is to obtain a model which gives the best possible fit between the observed and calculated structure factor amplitudes. Specifically, we wish to find those atomic coordinates (x_j, y_j, z_j) and temperature factors which, when used to calculate the structure factors, give maximal agreement with the observed structure factors, and so we attempt to find a model that minimises the quantity $\sum(|F_o| - |F_c|)^2$, which is known as the minimisation function, or M. Our purpose is to refine three positional coordinates and up to six displacement parameters for each atom to give the best fit with the experimental data. Since our structure is, at this stage, predominantly correct, any refinements to be made are small, and we may define a series of increments to the trial model position and temperature factor parameters as Δx_j, Δy_j, Δz_j and ΔB_j, which we wish to determine in order to correct the model. We may use Taylor's theorem to show that

$$|F_o| \approx |F_c| + \sum_j \left(\Delta x_j \frac{\partial |F_c|}{\partial x_j} + \Delta y_j \frac{\partial |F_c|}{\partial y_j} + \Delta z_j \frac{\partial |F_c|}{\partial z_j} + \Delta B_j \frac{\partial |F_c|}{\partial B_j} \right)$$

and therefore

$$|F_o| - |F_c| \approx \sum_j \left(\Delta x_j \frac{\partial |F_c|}{\partial x_j} + \Delta y_j \frac{\partial |F_c|}{\partial y_j} + \Delta z_j \frac{\partial |F_c|}{\partial z_j} + \Delta B_j \frac{\partial |F_c|}{\partial B_j} \right)$$

Calling $|F_o| - |F_c|$ simply ΔF gives

$$\Delta F \approx \sum_j \left(\Delta x_j \frac{\partial |F_c|}{\partial x_j} + \Delta y_j \frac{\partial |F_c|}{\partial y_j} + \Delta z_j \frac{\partial |F_c|}{\partial z_j} + \Delta B_j \frac{\partial |F_c|}{\partial B_j} \right) \tag{15.12}$$

This is analogous to equation (15.4), in which the measured parameter q was a linear function of three known quantities (x, y, z) and three unknown parameters (a, b, c):

$$q = ax + by + cz \tag{15.4}$$

In the example of X-ray refinement given in equation (15.12), ΔF, which is analogous to q in equation (15.4), is known for each reflection. Likewise, the derivatives $(\partial |F_c|/\partial x_j, \partial |F_c|/\partial y_j$ etc.), which are analogous to x, y and z in equation (15.4), are also known. The increments $(\Delta x_1, \Delta y_1$ etc.) are analogous to a, b and c in equation (15.4), since these constitute the unknown parameters that we wish to determine. It is not strictly true that the derivatives can be treated as known quantities, since they have to be calculated from the model, and we will look at how this is done later.

We showed that equation (15.4) could be represented in matrix form as

$$\mathbf{q = Xa}$$

and, by comparison with equation (15.12), we see that in the case of crystallographic refinement, the matrix $\mathbf{q}$ becomes

$$
\begin{bmatrix}
\Delta F_1 \\
\Delta F_2 \\
\Delta F_3 \\
\vdots
\end{bmatrix}
$$

while the matrix $\mathbf{X}$, which is referred to as the Jacobian (after Carl Jacobi, 1804–1851), is of the form

$$
\begin{bmatrix}
\frac{\partial |F_1|}{\partial x_1} & \frac{\partial |F_1|}{\partial y_1} & \frac{\partial |F_1|}{\partial z_1} & \frac{\partial |F_1|}{\partial B_1} & \frac{\partial |F_1|}{\partial x_2} & \frac{\partial |F_1|}{\partial y_2} & \cdots \\
\frac{\partial |F_2|}{\partial x_1} & \frac{\partial |F_2|}{\partial y_1} & \frac{\partial |F_2|}{\partial z_1} & \frac{\partial |F_2|}{\partial B_1} & \frac{\partial |F_2|}{\partial x_2} & \frac{\partial |F_2|}{\partial y_2} & \cdots \\
\frac{\partial |F_3|}{\partial x_1} & \frac{\partial |F_3|}{\partial y_1} & \frac{\partial |F_3|}{\partial z_1} & \frac{\partial |F_3|}{\partial B_1} & \frac{\partial |F_3|}{\partial x_2} & \frac{\partial |F_3|}{\partial y_2} & \cdots \\
\vdots & \vdots & \vdots & \vdots & \vdots & \vdots & \ddots
\end{bmatrix}
$$

and the matrix $\mathbf{a}$ is

$$
\begin{bmatrix}
\Delta x_1 \\
\Delta y_1 \\
\Delta z_1 \\
\Delta B_1 \\
\Delta x_2 \\
\Delta y_2 \\
\vdots
\end{bmatrix}
$$

In our treatment of the simple linear equation (15.4), we showed that we can derive the final least-squares estimates of the unknowns by solution of the equation

$$
\mathbf{X}^{\mathrm{T}}\mathbf{X}\mathbf{a} = \mathbf{X}^{\mathrm{T}}\mathbf{q} \tag{15.10}
$$

The product of the matrices $\mathbf{X}^{\mathrm{T}}\mathbf{X}$ is another matrix, which is referred to as the normal matrix and is of the form

$$
\begin{bmatrix}
\sum_j x_j^2 & \sum_j x_j y_j & \sum_j x_j z_j \\
\sum_j x_j y_j & \sum_j y_j^2 & \sum_j y_j z_j \\
\sum_j x_j z_j & \sum_j y_j z_j & \sum_j z_j^2
\end{bmatrix}
$$

The first thing to note about this is that it is a square $n \times n$ matrix, where n is the number of unknowns. Secondly, the diagonal of the normal matrix (from top left to bottom right) consists of squared terms, which must always be positive, in contrast to the cross-terms such as $x_j y_j$, which can be negative. In any of the summations, for example $\sum x_j y_j$, the presence of negative terms will of course reduce the total value of the sum. The effect of this is that in many applications, the diagonal of the normal matrix is made up of terms that are considerably larger than the off-diagonal terms, and the normal matrix is therefore said to have a dominant diagonal. A third feature of the normal matrix is that it is a symmetric square matrix and is therefore equal to its transpose.

In crystallographic least squares, the normal matrix will be given by the product

$$\mathbf{X}^{\mathrm{T}}\mathbf{X} = \begin{bmatrix} \frac{\partial|F_1|}{\partial x_1} & \frac{\partial|F_2|}{\partial x_1} & \frac{\partial|F_3|}{\partial x_1} & \cdots \\ \frac{\partial|F_1|}{\partial y_1} & \frac{\partial|F_2|}{\partial y_1} & \frac{\partial|F_3|}{\partial y_1} & \cdots \\ \frac{\partial|F_1|}{\partial z_1} & \frac{\partial|F_2|}{\partial z_1} & \frac{\partial|F_3|}{\partial z_1} & \cdots \\ \vdots & \vdots & \vdots & \ddots \end{bmatrix} \begin{bmatrix} \frac{\partial|F_1|}{\partial x_1} & \frac{\partial|F_1|}{\partial y_1} & \frac{\partial|F_1|}{\partial z_1} & \cdots \\ \frac{\partial|F_2|}{\partial x_1} & \frac{\partial|F_2|}{\partial y_1} & \frac{\partial|F_2|}{\partial z_1} & \cdots \\ \frac{\partial|F_3|}{\partial x_1} & \frac{\partial|F_3|}{\partial y_1} & \frac{\partial|F_3|}{\partial z_1} & \cdots \\ \vdots & \vdots & \vdots & \ddots \end{bmatrix}$$

and will have the form

$$\begin{bmatrix} \sum \left(\frac{\partial|F|}{\partial x_1}\right)^2 & \sum \frac{\partial|F|}{\partial x_1}\frac{\partial|F|}{\partial y_1} & \sum \frac{\partial|F|}{\partial x_1}\frac{\partial|F|}{\partial z_1} & \sum \frac{\partial|F|}{\partial x_1}\frac{\partial|F|}{\partial B_1} & \sum \frac{\partial|F|}{\partial x_1}\frac{\partial|F|}{\partial x_2} & \cdots \\ \sum \frac{\partial|F|}{\partial y_1}\frac{\partial|F|}{\partial x_1} & \sum \left(\frac{\partial|F|}{\partial y_1}\right)^2 & \sum \frac{\partial|F|}{\partial y_1}\frac{\partial|F|}{\partial z_1} & \sum \frac{\partial|F|}{\partial y_1}\frac{\partial|F|}{\partial B_1} & \sum \frac{\partial|F|}{\partial y_1}\frac{\partial|F|}{\partial x_2} & \cdots \\ \sum \frac{\partial|F|}{\partial z_1}\frac{\partial|F|}{\partial x_1} & \sum \frac{\partial|F|}{\partial z_1}\frac{\partial|F|}{\partial y_1} & \sum \left(\frac{\partial|F|}{\partial z_1}\right)^2 & \sum \frac{\partial|F|}{\partial z_1}\frac{\partial|F|}{\partial B_1} & \sum \frac{\partial|F|}{\partial z_1}\frac{\partial|F|}{\partial x_2} & \cdots \\ \sum \frac{\partial|F|}{\partial B_1}\frac{\partial|F|}{\partial x_1} & \sum \frac{\partial|F|}{\partial B_1}\frac{\partial|F|}{\partial y_1} & \sum \frac{\partial|F|}{\partial B_1}\frac{\partial|F|}{\partial z_1} & \sum \left(\frac{\partial|F|}{\partial B_1}\right)^2 & \sum \frac{\partial|F|}{\partial B_1}\frac{\partial|F|}{\partial x_2} & \cdots \\ \sum \frac{\partial|F|}{\partial x_2}\frac{\partial|F|}{\partial x_1} & \sum \frac{\partial|F|}{\partial x_2}\frac{\partial|F|}{\partial y_1} & \sum \frac{\partial|F|}{\partial x_2}\frac{\partial|F|}{\partial z_1} & \sum \frac{\partial|F|}{\partial x_2}\frac{\partial|F|}{\partial B_1} & \sum \left(\frac{\partial|F|}{\partial x_2}\right)^2 & \cdots \\ \vdots & \vdots & \vdots & \vdots & \vdots & \ddots \end{bmatrix} \quad (15.13)$$

where each summation is over all of the reflections in the dataset. We now have assembled all of the matrices that allow the unknown parameters to be determined by use of equation (15.11).

Interestingly, each summation in the normal matrix is an approximation of the second derivative of the minimisation function M with respect to the same variables, as shown below, and for this reason the normal matrix is sometimes referred to as the 'second-derivative matrix'. It is also called the Hessian, after Ludwig Hesse (1811–1874), a student of Jacobi, and is equal to

$$\begin{bmatrix} \frac{\partial^2 M}{\partial x_1^2} & \frac{\partial^2 M}{\partial x_1 \partial y_1} & \frac{\partial^2 M}{\partial x_1 \partial z_1} & \cdots \\ \frac{\partial^2 M}{\partial y_1 \partial x_1} & \frac{\partial^2 M}{\partial y_1^2} & \frac{\partial^2 M}{\partial y_1 \partial z_1} & \cdots \\ \frac{\partial^2 M}{\partial z_1 \partial x_1} & \frac{\partial^2 M}{\partial z_1 \partial y_1} & \frac{\partial^2 M}{\partial z_1^2} & \cdots \\ \vdots & \vdots & \vdots & \ddots \end{bmatrix} \quad (15.14)$$

Terms of the form $\partial^2 M/\partial x_1 \partial y_1$ are called mixed derivatives and indicate that the function is first differentiated with respect to one variable and the resulting function is then differentiated with respect to the other variable.

As we mentioned earlier, the least-squares solution for the unknowns can be represented as

$$\mathbf{a} = (\mathbf{X}^{\mathrm{T}}\mathbf{X})^{-1}\mathbf{X}^{\mathrm{T}}\mathbf{q} \quad (15.11)$$

showing that, in principle, we need to invert the normal matrix. With a typical protein of, say, 2500 atoms, there will be four parameters that we need to determine for each atom (Δx, Δy, Δz and ΔB), giving a total of 10 000 unknowns, meaning that the normal matrix will contain of the order of 10 000 × 10 000 terms. Calculating the terms of the normal matrix is very computationally intensive, and in order to solve for the unknowns we need to invert it. Inverting a matrix containing 10^8 or, often, more terms is again a very computationally significant task. However, these problems can be simplified significantly owing to the properties of the normal matrix that we outlined above, the most important one being that the normal matrix has a dominant diagonal. Thus one way of solving for the unknowns is to assume that all off-diagonal terms of the normal matrix are zero.

Referring to equation (15.13), we can see that with the diagonal approximation, the normal equations (15.10) become

$$
\begin{bmatrix}
\sum\left(\frac{\partial|F|}{\partial x_1}\right)^2 & 0 & 0 & 0 & 0 & \cdots \\
0 & \sum\left(\frac{\partial|F|}{\partial y_1}\right)^2 & 0 & 0 & 0 & \cdots \\
0 & 0 & \sum\left(\frac{\partial|F|}{\partial z_1}\right)^2 & 0 & 0 & \cdots \\
0 & 0 & 0 & \sum\left(\frac{\partial|F|}{\partial B_1}\right)^2 & 0 & \cdots \\
0 & 0 & 0 & 0 & \sum\left(\frac{\partial|F|}{\partial x_2}\right)^2 & \cdots \\
\vdots & \vdots & \vdots & \vdots & \vdots & \ddots
\end{bmatrix}
\begin{bmatrix}
\Delta x_1 \\ \Delta y_1 \\ \Delta z_1 \\ \Delta B_1 \\ \Delta x_2 \\ \Delta y_2 \\ \vdots
\end{bmatrix}
$$

$$
=
\begin{bmatrix}
\frac{\partial|F_1|}{\partial x_1} & \frac{\partial|F_2|}{\partial x_1} & \frac{\partial|F_3|}{\partial x_1} & \cdots \\
\frac{\partial|F_1|}{\partial y_1} & \frac{\partial|F_2|}{\partial y_1} & \frac{\partial|F_3|}{\partial y_1} & \cdots \\
\frac{\partial|F_1|}{\partial z_1} & \frac{\partial|F_2|}{\partial z_1} & \frac{\partial|F_3|}{\partial z_1} & \cdots \\
\vdots & \vdots & \vdots & \ddots
\end{bmatrix}
\begin{bmatrix}
\Delta F_1 \\ \Delta F_2 \\ \Delta F_3 \\ \vdots
\end{bmatrix}
$$

from which it is possible to see that for the first unknown parameter,

$$
\Delta x_1 \sum\left(\frac{\partial|F|}{\partial x_1}\right)^2 = \sum \Delta F \frac{\partial|F|}{\partial x_1}
$$

and hence the following general formula for each unknown (Δx_i) can be deduced:

$$
\Delta x_i = \frac{\sum \Delta F \frac{\partial|F|}{\partial x_i}}{\sum\left(\frac{\partial|F|}{\partial x_i}\right)^2}
\tag{15.15}
$$

Increments in all of the parameters describing the structure (Δx_i, Δy_i, Δz_i, ΔB_i) can be estimated in this way. These increments can be applied to the structure to give us a new model that should be more consistent with the experimental measurements. A new set of F_c values can be calculated from the revised model and, accordingly, the new set of ΔF or $|F_o| - |F_c|$ values can be used in another cycle of least-squares refinement to derive a new set of shifts to apply to the model. This iterative process is repeated until convergence is achieved, as assessed by the refinement R-factor and correlation coefficient given in Section 15.3. In this process, we compensate for the fact that in each cycle the calculated increments Δx_i are only approximations of the true shifts by repeating the process many times such that the atoms iteratively approach their correct positions and their temperature factors converge.

For completeness, an important refinement of equation (15.15) is that we need to weight the observed structure factors according to their precision, and hence the equations for the increments are actually of the form

$$\Delta x_i = \frac{\sum w \Delta F \frac{\partial |F|}{\partial x_i}}{\sum w \left(\frac{\partial |F|}{\partial x_i} \right)^2} \tag{15.16}$$

In classical least squares, the weight w is inversely related to the variance of the observation (in this case this is the structure factor amplitude), such that observations for which the errors are large are down-weighted in the refinement:

$$w = \frac{1}{\sigma^2 (|F|)} \tag{15.17}$$

The problem with this in X-ray analysis is that the low-resolution terms have vastly greater signal-to-noise ratio than the high-resolution reflections, and it is the latter set that are the most useful in the refinement of a detailed molecular structure. Consequently, most refinement programs use more even weighting schemes to ensure that the most important data are not down-weighted.

We will now look at how the derivatives used in the above formulae are calculated based on formulae given by Agarwal (1978) and Booth (1949). Since the derivatives are of the form $\partial |F| / \partial x_j$, we need to calculate the gradient of the *amplitude* of the structure factor with respect to the parameter we are refining. By the chain rule of differentiation, we can state that

$$\frac{\partial |F_{hkl}|^2}{\partial x_j} = 2 |F_{hkl}| \frac{\partial |F_{hkl}|}{\partial x_j} \tag{15.18}$$

$$\therefore \frac{\partial |F_{hkl}|}{\partial x_j} = \frac{1}{2 |F_{hkl}|} \frac{\partial |F_{hkl}|^2}{\partial x_j}$$

and, by ignoring anomalous scattering, we may state that

$$F_{\overline{hkl}} = F^*_{hkl}$$

and

$$|F_{hkl}|^2 = F_{hkl} F^*_{hkl} = F_{hkl} F_{\overline{hkl}}$$

$$\therefore \frac{\partial |F_{hkl}|}{\partial x_j} = \frac{1}{2 |F_{hkl}|} \frac{\partial \left(F_{hkl} F_{\overline{hkl}} \right)}{\partial x_j}$$

By the product rule of differentiation,

$$\frac{\partial |F_{hkl}|}{\partial x_j} = \frac{1}{2|F_{hkl}|}\left[F_{hkl}\frac{\partial F_{\overline{hkl}}}{\partial x_j} + F_{\overline{hkl}}\frac{\partial F_{hkl}}{\partial x_j}\right] \tag{15.19}$$

Since

$$F_{hkl} = \sum_j f_j e^{-B_j \sin^2 \theta/\lambda^2} e^{2\pi i\left(hx_j+ky_j+lz_j\right)}$$

we find that

$$\frac{\partial F_{hkl}}{\partial x_j} = (2\pi i h)\, f_j e^{-B_j \sin^2 \theta/\lambda^2} e^{2\pi i\left(hx_j+ky_j+lz_j\right)}$$

and

$$\frac{\partial F_{\overline{hkl}}}{\partial x_j} = (-2\pi i h)\, f_j e^{-B_j \sin^2 \theta/\lambda^2} e^{-2\pi i\left(hx_j+ky_j+lz_j\right)}$$

Substituting the above two derivatives and the following equations,

$$F_{hkl} = |F_{hkl}|\, e^{i\phi_{hkl}}$$

$$F_{\overline{hkl}} = |F_{hkl}|\, e^{-i\phi_{hkl}}$$

into equation (15.19) gives

$$\frac{\partial |F_{hkl}|}{\partial x_j} = \frac{1}{2|F_{hkl}|}\Big[|F_{hkl}|\, e^{i\phi_{hkl}}(-2\pi i h)\, f_j e^{-B_j \sin^2 \theta/\lambda^2} e^{-2\pi i\left(hx_j+ky_j+lz_j\right)}$$

$$+ |F_{hkl}|\, e^{-i\phi_{hkl}}(2\pi i h)\, f_j e^{-B_j \sin^2 \theta/\lambda^2} e^{2\pi i\left(hx_j+ky_j+lz_j\right)}\Big]$$

$$= \frac{1}{2}(2\pi i h)\, f_j e^{-B_j \sin^2 \theta/\lambda^2}\left[e^{-i\phi_{hkl}}e^{2\pi i\left(hx_j+ky_j+lz_j\right)} - e^{i\phi_{hkl}}e^{-2\pi i\left(hx_j+ky_j+lz_j\right)}\right]$$

$$= \frac{1}{2}(2\pi i h)\, f_j e^{-B_j \sin^2 \theta/\lambda^2}\, 2i \sin\left[2\pi\left(hx_j+ky_j+lz_j\right) - \phi_{hkl}\right]$$

$$\therefore \frac{\partial |F_{hkl}|}{\partial x_j} = (-2\pi h)\, f_j e^{-B_j \sin^2 \theta/\lambda^2}\, \sin\left[2\pi\left(hx_j+ky_j+lz_j\right) - \phi_{hkl}\right] \tag{15.20}$$

Hence the derivative can be calculated from the trial parameters of the atom (x_j, y_j, z_j, B_j) and the phase of the reflection (ϕ_{hkl}) which is calculated from the current model.

We will now look at how the derivatives with respect to the temperature factor can be calculated. Since

$$F_{hkl} = \sum_j f_j e^{-B_j \sin^2 \theta/\lambda^2} e^{2\pi i\left(hx_j+ky_j+lz_j\right)}$$

then

$$\frac{\partial F_{hkl}}{\partial B_j} = \left(-\frac{\sin^2\theta}{\lambda^2}\right) f_j e^{-B_j \sin^2\theta/\lambda^2} e^{2\pi i (hx_j + ky_j + lz_j)}$$

$$\frac{\partial F_{\overline{hkl}}}{\partial B_j} = \left(-\frac{\sin^2\theta}{\lambda^2}\right) f_j e^{-B_j \sin^2\theta/\lambda^2} e^{-2\pi i (hx_j + ky_j + lz_j)}$$

Hence, from equation (15.19), we can determine that

$$\frac{\partial |F_{hkl}|}{\partial B_j} = \frac{1}{2|F_{hkl}|} \left[|F_{hkl}| e^{i\phi_{hkl}} \left(-\frac{\sin^2\theta}{\lambda^2}\right) f_j e^{-B_j \sin^2\theta/\lambda^2} e^{-2\pi i (hx_j + ky_j + lz_j)} \right.$$

$$\left. + |F_{hkl}| e^{-i\phi_{hkl}} \left(-\frac{\sin^2\theta}{\lambda^2}\right) f_j e^{-B_j \sin^2\theta/\lambda^2} e^{2\pi i (hx_j + ky_j + lz_j)} \right]$$

$$= \frac{1}{2} \left(-\frac{\sin^2\theta}{\lambda^2}\right) f_j e^{-B_j \sin^2\theta/\lambda^2} \left[e^{i\phi_{hkl}} e^{-2\pi i (hx_j + ky_j + lz_j)} + e^{-i\phi_{hkl}} e^{2\pi i (hx_j + ky_j + lz_j)} \right]$$

$$\therefore \frac{\partial |F_{hkl}|}{\partial B_j} = \left(-\frac{\sin^2\theta}{\lambda^2}\right) f_j e^{-B_j \sin^2\theta/\lambda^2} \cos\left[2\pi (hx_j + ky_j + lz_j) - \phi_{hkl}\right] \qquad (15.21)$$

All parameters in this equation are known or can be calculated from the trial model.

An alternative way to calculate these derivatives is to refer to Fig. 9.7 in Chapter 9, which demonstrates how the structure factor is the vector sum of the contributions of each atom to the scattering. The projection of the contribution of any atom j onto the net structure factor F_{hkl} is given by

$$\text{proj}_j = f_j e^{-B_j \sin^2\theta/\lambda^2} \cos\left[2\pi (hx_j + ky_j + lz_j) - \phi_{hkl}\right]$$

The amplitude of the structure factor is thus given by the sum of the individual projections,

$$|F_{hkl}| = \sum_j f_j e^{-B_j \sin^2\theta/\lambda^2} \cos\left[2\pi (hx_j + ky_j + lz_j) - \phi_{hkl}\right]$$

and application of the chain rule for differentiation gives the same results as above, for example

$$\frac{\partial |F_{hkl}|}{\partial x_j} = (-2\pi h) f_j e^{-B_j \sin^2\theta/\lambda^2} \sin\left[2\pi (hx_j + ky_j + lz_j) - \phi_{hkl}\right] \qquad (15.20)$$

Note that intensity-based refinement, in which the sum of $(I_o - I_c)^2$ is minimised across the dataset, has become more popular and, in principle, the associated derivatives can be obtained from equation (15.18). This method is more appropriate for weak data, since calculating $|F_o|$ and the associated standard deviation $\sigma(|F_o|)$ when the intensity has a low value involves some approximations (see Chapter 11).

One feature of refinement is that the shifts calculated for the coordinate parameters tend to be considerably smaller than the estimated coordinate errors, whereas the shifts in the temperature factor parameters tend to be larger in relation to the estimated error. This means that in our trial model of the structure, the coordinates have to be more accurate than the temperature factors, since the latter have a much greater radius of convergence. Unlike the coordinates, trial temperature factors can be very significantly in error and they will still converge to the correct values during refinement.

It is possible to do refinement with experimental phase information included as observations, and this can be effective at low-to-medium resolution, where the data-to-parameter ratio would not be favourable if the structure factor amplitudes alone were used. However, in general, this is a less commonly used approach than refinement with the reflection amplitudes or intensities owing to concerns about the accuracy of the initial experimental phases.

15.6   The use of stereochemical restraints

In the most detailed cases with atomic-resolution data, we fit a total of three positional and six displacement parameters for each atom in the structure, and from this we derive a very accurate electron density map. On calculating the residual R from the final model, in favourable cases, a value considerably less than 0.1 is obtained, indicating a high correlation between the proposed model and the true structure.

However, the majority of proteins diffract to around 2.0 Å resolution and if we were to refine the structure by the methods that we have just described, the molecular model would have a chemically unreasonable geometry. This arises from the fact that refinement using the X-ray data alone pushes atoms into positions that maximise the agreement between $|F_o|$ and $|F_c|$ without regard to the bond lengths, angles, torsion angles and planarity of certain groups. With atomic-resolution data, to say 1.0 Å, the electron density map will be sufficiently precise to yield a structure with excellent geometry, since the errors in the atomic positions will be of the order of 0.01 Å. However, with a protein diffracting to only 2.0 Å resolution, the same errors will be more than an order of magnitude greater, typically around 0.2 Å, and will yield a model with a chemically unacceptable geometry. Very accurate geometry for amino acids has been determined by comparison of atomic-resolution structures of amino acids and peptides (Engh and Huber 1991), and this information has been used to construct a library of standard bond lengths, bond angles etc., which can be used as *restraints* in refinement to ensure that the atoms move to chemically reasonable positions. Since we are trying to derive a model that fits the electron density map well, the X-ray and geometric information must be weighted sensibly in the refinement so that in the initial cycles the X-ray data dominates the calculation of the shifts applied to the atoms, otherwise the restraints may dominate and the refinement simply becomes a form of geometry optimisation. With atomic-resolution data, the restraints can even be omitted altogether towards the end of the refinement and individual bond lengths can be refined freely, often revealing much about the protonation states of polar groups (Coates *et al.*, 2001; Ruiz *et al.*, 2005; Fisher *et al.*, 2008). However, for the majority of studies, restraints are of paramount importance, and a means of incorporating stereochemical information into refinement was first suggested by Jurg Waser (1963).

In restrained refinement, instead of minimising $\sum(|F_o| - |F_c|)^2$, we attempt to minimise a function of the following general form:

$$\sum w_F(|F_o| - |F_c|)^2 + \sum w_E(|E_t| - |E_c|)^2 \tag{15.22}$$

in which E_c is the calculated value of a geometric parameter such as a bond length or angle in the model, and E_t is the target value. The weighting of the X-ray and geometry terms is given by w_F and w_E. We showed above that the $|F_o| - |F_c|$ (or ΔF) terms can be approximated by Taylor's theorem:

$$\Delta F = |F_o| - |F_c| \approx \sum_j \left(\Delta x_j \frac{\partial |F_c|}{\partial x_j} + \Delta y_j \frac{\partial |F_c|}{\partial y_j} + \Delta z_j \frac{\partial |F_c|}{\partial z_j} + \Delta B_j \frac{\partial |F_c|}{\partial B_j} \right) \tag{15.12}$$

Similarly, the geometric residual can be approximated in the same way. Thus

$$\Delta E = E_t - E_c \approx \sum_j \left(\Delta x_j \frac{\partial E_c}{\partial x_j} + \Delta y_j \frac{\partial E_c}{\partial y_j} + \Delta z_j \frac{\partial E_c}{\partial z_j} + \Delta B_j \frac{\partial E_c}{\partial B_j} \right) \tag{15.23}$$

Note that this equation includes a temperature factor term because it is reasonable to expect that atoms which are bonded or close together will have correlated temperature factors.

The effect of including restraints in the minimisation function is equivalent to adding extra simultaneous equations. We saw that in the simple case of the equation

$$q = ax + by + cz \tag{15.4}$$

we could formulate the least-squares solution for the unknowns in the following form:

$$\begin{bmatrix} \sum_j x_j^2 & \sum_j x_j y_j & \sum_j x_j z_j \\ \sum_j x_j y_j & \sum_j y_j^2 & \sum_j y_j z_j \\ \sum_j x_j z_j & \sum_j y_j z_j & \sum_j z_j^2 \end{bmatrix} \begin{bmatrix} a \\ b \\ c \end{bmatrix} = \begin{bmatrix} \sum_j q_j x_j \\ \sum_j q_j y_j \\ \sum_j q_j z_j \end{bmatrix}$$

If we now make some extra observations that give us additional simultaneous equations as follows,

$$q' = ax' + by' + cz'$$

we may simply add the extra observations (x', y', z', q') to the matrix equation and, dropping the subscripts for clarity, this gives

$$\begin{bmatrix} (\sum x^2 + \sum x'^2) & (\sum xy + \sum x'y') & (\sum xz + \sum x'z') \\ (\sum xy + \sum x'y') & (\sum y^2 + \sum y'^2) & (\sum yz + \sum y'z') \\ (\sum xz + \sum x'z') & (\sum yz + \sum y'z') & (\sum z^2 + \sum z'^2) \end{bmatrix} \begin{bmatrix} a \\ b \\ c \end{bmatrix} = \begin{bmatrix} \sum qx + \sum q'x' \\ \sum qy + \sum q'y' \\ \sum qz + \sum q'z' \end{bmatrix}$$

which should permit an even more accurate determination of the unknowns a, b and c. Likewise, in the case of X-ray refinement, we can add the extra simultaneous equations due to the geometric restraints and, ignoring the weights for clarity, the equivalent matrix equation becomes

$$
\begin{bmatrix}
\left(\sum\left(\frac{\partial|F|}{\partial x_1}\right)^2 + \sum\left(\frac{\partial E}{\partial x_1}\right)^2\right) & \left(\sum\frac{\partial|F|}{\partial x_1}\frac{\partial|F|}{\partial y_1} + \sum\frac{\partial E}{\partial x_1}\frac{\partial E}{\partial y_1}\right) & \cdots \\
\left(\sum\frac{\partial|F|}{\partial y_1}\frac{\partial|F|}{\partial x_1} + \sum\frac{\partial E}{\partial y_1}\frac{\partial E}{\partial x_1}\right) & \left(\sum\left(\frac{\partial|F|}{\partial y_1}\right)^2 + \sum\left(\frac{\partial E}{\partial y_1}\right)^2\right) & \cdots \\
\left(\sum\frac{\partial|F|}{\partial z_1}\frac{\partial|F|}{\partial x_1} + \sum\frac{\partial E}{\partial z_1}\frac{\partial E}{\partial x_1}\right) & \left(\sum\frac{\partial|F|}{\partial z_1}\frac{\partial|F|}{\partial y_1} + \sum\frac{\partial E}{\partial z_1}\frac{\partial E}{\partial y_1}\right) & \cdots \\
\vdots & \vdots & \ddots
\end{bmatrix}
\begin{bmatrix}
\Delta x_1 \\ \Delta y_1 \\ \Delta z_1 \\ \Delta B_1 \\ \Delta x_2 \\ \vdots
\end{bmatrix}
$$

$$
=
\begin{bmatrix}
\sum\Delta F\frac{\partial|F|}{\partial x_1} + \sum\Delta E\frac{\partial E}{\partial x_1} \\
\sum\Delta F\frac{\partial|F|}{\partial y_1} + \sum\Delta E\frac{\partial E}{\partial y_1} \\
\sum\Delta F\frac{\partial|F|}{\partial z_1} + \sum\Delta E\frac{\partial E}{\partial z_1} \\
\vdots
\end{bmatrix}
\tag{15.24}
$$

In principle, the normal equations can then be solved for the unknown parameter shifts by inverting the normal matrix, as described above. The shifts calculated from the structure factor terms are only significantly correlated for pairs of atoms whose electron densities overlap. For this reason, the X-ray terms of the diagonal elements of the normal matrix itself are large, since each atom is effectively overlapped with itself. In contrast, the off-diagonal terms for atoms whose electron density is not directly connected will be low. The geometric terms in the normal matrix relate atoms that happen to be spatially close together but may not necessarily have overlapping electron density, and hence the off-diagonal geometric terms can be large. These effects are very important because they allow simplification of the normal matrix, which is crucial in speeding up the calculations. In situations where we refine only the atomic positions, the X-ray terms are usually taken as 3×3 blocks along the diagonal, i.e. the correlations of the shifts along x, y and z for each atom on its own are considered. We use 4×4 blocks if we refine the atomic coordinates with an isotropic temperature factor, and 9×9 blocks when coordinates and anisotropic B-values are refined. The only off-diagonal terms to be computed are the geometric ones, and this vastly reduces the computation time involved in refinement. This approach is known as the *sparse matrix method* and is used by most macromolecular refinement programs (Haneef et al., 1985; Hendrickson and Konnert, 1980; Murshudov et al., 1997; Sheldrick and Schneider, 1997; Tronrud et al., 1987).

One important aspect of the normal matrix in restrained refinement is that the structure factor terms have to be carefully weighted in relation to the geometric terms. To be strictly correct, each of the terms in the summations in the above equation (15.24) should contain the weighting factors given in equation (15.22). The structure factor data can be weighted according to the precision of the measurements, given by equation (15.17), and the weighting of the geometric data will control the stereochemical quality of the model. The relative weighting of these terms can be controlled by the user. In general, it is sensible to start a refinement round with high relative weights for the X-ray data to allow larger shifts that are dominated by the

X-ray terms. The relative weighting of the X-ray terms in relation to the geometric terms may need to be decreased as the refinement converges to give a final model with acceptable rms deviations in the bond lengths, bond angles etc.

Inspection of equation (15.15) shows that the shift in any parameter is inversely related to the corresponding diagonal element in the normal matrix. In a situation where the diagonal element for a certain parameter shift approaches zero, such as might happen if an atom has very poor electron density, the estimated shift becomes very large, especially if only the X-ray terms are used. However, assuming that the affected atom is not something such as an isolated water molecule, the geometric restraints would normally prevent such excessive shifts being applied. Nevertheless, it is common for other methods of dampening the calculated shifts to be used, such as adding a constant term to the diagonal of the normal matrix (Haneef *et al.*, 1985). The net effect of this is that when the parameter shifts are estimated, for example by means of equation (15.16), the constant (or Marquardt factor, M) is effectively added to the denominator, as shown in equation (15.25) below, thus dampening the calculated shifts, which can be important for avoiding numerical instabilities:

$$\Delta x_i = \frac{\sum w \, \Delta F \frac{\partial |F|}{\partial x_i}}{\left[M + \sum w \left(\frac{\partial |F|}{\partial x_i} \right)^2 \right]} \tag{15.25}$$

Because of the truncation of the Taylor series used in linearising the equation for ΔF, equation (15.12), and the diagonal or block diagonal approximation of the normal matrix, the shifts calculated by the solution of the normal equations do not usually correspond exactly to the minimum of $\sum (\Delta F)^2 + \sum (\Delta E)^2$. There are a number of ways of improving the estimated shifts obtained in each cycle of least squares. One such way is to calculate the value of the minimisation function at a series of steps in the unknown parameters that are applied in the direction of the calculated shifts. The step size is typically 0.2–0.4 of the calculated shift. The overall optimum factor by which to scale the calculated shifts in all parameters is then determined by fitting a parabola to the values of the minimisation function. In order for this process to be efficient, only a small proportion of the data are used to calculate the minimisation function at the trial values of the shift factor (Haneef *et al.*, 1985).

One method for solving simultaneous equations that works very well when the equations have a dominant diagonal is the *Gauss–Seidel method*, and this has been used in crystallographic refinement (Haneef *et al.*, 1985). To use this method, we initially assume that all of the unknowns in the simultaneous equations are zero. We can then obtain a better estimate of the first unknown by assuming that the off-diagonal terms are zero and using the resulting value of the first parameter to estimate the second unknown. These two values can be used to estimate the third unknown, etc., in a sequential process that is repeated until the unknowns converge.

A more efficient means of solving the normal equations is the *conjugate gradient method*, in which each calculated shift vector is estimated in successive steps where the direction and magnitude of each step are informed by those of the previous step. Of course, in the first such cycle there is no prior information to combine with the calculated parameter shifts, and so these are simply used as they are. Conjugate gradient methods have been extensively developed for general optimisation and are widely used.

Least squares is in fact an application of Newton's method of optimisation, which is a means of finding minima (and maxima) of functions. Consider the Taylor expansion of a general function $f(x)$, which is

$$f\left(x + \Delta x\right) = f\left(x\right) + \Delta x\, f'\left(x\right) + \frac{1}{2}\Delta x^2 f''\left(x\right)$$

where $f'(x)$ and $f''(x)$ are the first and second derivative, respectively. At a maximum or minimum, the derivative of the above function with respect to Δx must be zero:

$$\therefore f'\left(x\right) + \Delta x\, f''\left(x\right) = 0$$

Hence

$$\Delta x = -\frac{f'\left(x\right)}{f''\left(x\right)} \tag{15.26}$$

Thus, assuming that we have a trial solution for x that is close to a minimum, a shift in this parameter can be estimated by the above equation and can be applied to the trial value. With a quadratic function the calculated shift should be exact, but with other functions, several iterations will be required. Hence the shifted solution is used to calculate new first and second derivatives to determine another shift, which is applied and the process repeated until convergence is achieved. The Newton method is an important numerical tool for finding minima of functions and has a number of variants.

In order to speed up the calculations of the shifts given by equation (15.26), it is possible to omit calculating the $f''(x)$ term, which is simply replaced by a constant, and this is known as the *method of steepest descents*. In a one-dimensional example, such as that shown in Fig. 15.5, if we have a trial solution for x, shown as x_0, then we can calculate the gradient of the function at that starting point, $f'(x_0)$. Since the gradient is positive, we clearly need to reduce x to approach the minimum, whereas a negative gradient would indicate that x should be increased. Thus a small shift in x is calculated to move down the gradient of the function. The new value of x is given by Newton's formula as

$$x_{j+1} = x_j - \varepsilon f'\left(x_j\right) \tag{15.27}$$

where ε is a small positive constant that can be chosen. At the new position, we recalculate the gradient and apply another shift in the appropriate direction. The successive shifts are applied until we reach the minimum, where the gradient will be zero. The main drawback of this method compared with Newton's method is that, because of the additional approximation, a larger number of steps need to be taken to reach the minimum. However, in crystallographic refinement, calculating the Hessian or normal matrix and inverting it are very computationally expensive tasks, and simply replacing the inverse of this matrix with the constant term ε in equation (15.27) leads to a significant reduction in computer time—it is usually much quicker to do many

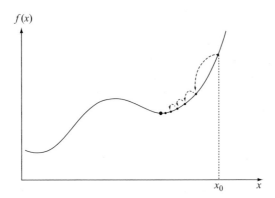

Fig. 15.5 *The method of steepest descents.* The trial solution, or starting point, for x is shown as x_0 on the right-hand side. To find a minimum of $f(x)$, successive shifts in x are calculated from the gradient $f'(x)$ and are given by $-\varepsilon f'(x)$, where ε is a constant. After a number of iterations, the closest minimum (indicated by the largest dot) is found. As the minimum is approached, the gradient approaches zero and the calculated shifts become infinitesimally small.

approximate cycles than a few very accurate ones. In steepest-descent refinement, the Hessian is treated as a diagonal matrix with all terms equal to a constant that can be estimated from the data. This approach has a good radius of convergence and is utilised in a number of refinement methods (Booth, 1949; Tronrud, 2004).

In the early stages of a refinement run, the errors in the model will be at their largest. For this reason, we need to have as large a radius of convergence as possible. One way of achieving this is to initially limit the reflection data to medium resolution (e.g. $d_{min} > 2.8$ Å) and gradually increase this in steps of, say, 0.1 Å until we reach the resolution limit of the dataset, where several more cycles are done to allow convergence with the highest-resolution data. The advantage of starting at lower resolution is that we reduce the chance of the parameter shifts becoming stuck in local minima. This can be seen by inspection of Fig. 15.5, where we showed that gradient descent minimisation of the function follows a downhill path, which in this case leads into a trough on the side of the main gradient. Since minimisation methods will always work by shifting the parameters down the gradient, if the true minimum were to be further down the main gradient, as illustrated in Fig. 15.5, we would never reach it. One way to avoid this problem is to calculate the gradient with lower-resolution reflections such that it becomes smoother and hence will have fewer false minima.

In certain situations, it can be helpful to treat the whole molecule or parts of it as rigid bodies in which the bond lengths, angles and other geometric parameters are absolutely fixed (Sussmann *et al.*, 1977). This has a number of advantages, including an increase in the radius of convergence and a significant reduction in computer time. Rigid-body refinement is particularly useful for fine-tuning molecular replacement solutions using data to medium resolution (see Chapter 13), and is available as an option in most refinement programs.

In macromolecular refinement, calculating the structure factors and their derivatives is the most time-consuming component of the process, and this has led to the development of elegant fast Fourier transform (FFT) methods (Ten Eyck 1977), which have been applied extensively to refinement methodologies, for example TNT (Tronrud *et al.*, 1987) and REFMAC (Murshudov *et al.*, 1997). The development of the FFT method was pivotal to progress in the field during the 1980s and 1990s. However, developments in computer technology now mean that conventional structure factor calculation is almost completely trivial, even with very modest computers, and refinement programs that utilise this methodology, such as SHELX-97 (Sheldrick and Schneider, 1997), are very much back in the frame and have the advantage of greater accuracy in the calculations.

Very high-resolution data complicate modelling and refinement owing to the alternative conformations of disordered side chains becoming apparent in the electron density map. Such structures can be refined readily with SHELX-97, which can also handle data from twinned crystals. Macromolecular refinement is done predominantly by conjugate gradient least squares, but calculation of the full normal matrix is possible and is recommended for the final cycles if atomic-resolution data are available. SHELX-97 allows full-matrix inversion, which gives the experimental standard deviations for the atomic coordinates, as well as for bond lengths and other geometric parameters. Occupancies and anisotropic B-factors can be refined, and 'riding' hydrogen atoms can be added automatically to the standard amino acids. The contribution of these hydrogen atoms to the scattering is calculated but their positions are not explicitly refined, i.e. the data-to-parameter ratio is not adversely affected, and their inclusion can improve the refinement statistics. Refinement of the model using intensities is done by default in SHELX, since this was found to be superior to refinement versus $|F|$, owing in part to the difficulty of estimating $|F|$ and $\sigma(|F|)$ for weak data. Rigid-group refinement is possible with low-to-medium-resolution data. Chiral volume restraints prevent inversion of $C\alpha$ atoms, as well as the $C\beta$ atoms of threonine and isoleucine, which are the only amino acids to have chiral centres in their side chains.

15.7 The benefits of non-crystallographic symmetry

A well-known problem of macromolecular refinement is the paucity of observations compared with the number of parameters. One way to improve the data-to-parameter ratio is to exploit non-crystallographic symmetry (NCS), and this is very powerful for refinement at low-to-medium resolution. For example, if we have a tetrameric protein with four monomers in the asymmetric unit, we can build a model of one subunit and refine just that subunit using the NCS operators to generate the other subunits. The calculated shifts are then the 'average' of the shifts estimated for all subunits in the asymmetric unit (Haneef *et al.*, 1985). The molecule can be refined as a rigid body or with stereochemical restraints, and the NCS operators must also be refined in this process. Individual refinement programs differ in their implementation of NCS; for example, SHELX (Sheldrick and Schneider, 1997) effectively restrains the torsion angles of NCS-related subunits to be equal, and the coordinates of all subunits must be provided, whereas the program CNS (Brünger *et al.*, 1998) allows either strict NCS constraints or NCS restraints. The latter option involves calculating atomic parameter shifts that minimise the discrepancies between the NCS-related molecules and the 'average' structure. At low-to-medium resolution, use of NCS can give significant enhancements in map quality but at high resolution, the detailed differences between the NCS-related molecules, especially in the side chains, can be problematic.

15.8 Modelling rigid-group displacement

The atoms which are chemically connected within rigid side chain groups usually move in a highly concerted manner, for example in the case of the atoms within aromatic rings. Instead of modelling this displacement by assigning each atom an isotropic or anisotropic B-factor, we can refine parameters that describe the motion of the rigid group as a whole, and these are referred to as TLS tensors, where TLS is short for 'translation, libration and screw'. Tensors are matrices that provide a generalised mathematical description of vectors and coordinate transformations, the details of which can be found in Schomaker and Trueblood (1968). The translational tensor describes the translational displacement, the libration tensor describes the rotational displacement of the group and the screw tensor describes motion involving a combination of translation and rotation. This mode of refinement, which is available in REFMAC (Murshudov *et al.*, 1997) and RESTRAIN (Haneef *et al.*, 1985), can be used either to refine the TLS tensors of small groups, for example individual side chains, when high-to-atomic-resolution data are available, or to refine the TLS tensors of the whole molecule or its constituent domains or secondary structure elements, if concerted large-scale displacement is suspected. In these situations, TLS refinement can lead to a significant improvement in refinement statistics, since it models the anisotropy in the reflection data and allows the individual isotropic B-factors to be refined simultaneously (Winn *et al.*, 2001). Since each TLS group requires 20 parameters, refinement of a limited number of TLS groups does not significantly worsen the data-to-parameter ratio.

15.9 Simulated annealing

This is a type of refinement which derives its name from the annealing of metals, which can be strengthened by heating and cooling. In macromolecular refinement, simulated annealing represents one of the most significant breakthroughs. It involves the application of molecular dynamics to simulate atomic movements at unphysiologically high temperatures, usually several thousand kelvin. During the simulation, the experimental X-ray measurements are incorporated into the calculations as restraints, much like the way that the geometric terms are used as restraints in conventional protein refinement. The molecule can then be slowly cooled to allow convergence to a state that is consistent with the X-ray and energy terms.

Molecular dynamics involves assigning random velocities to each atom and calculating the forces exerted on them at successive time intervals by solving the classical equations of motion. This is done using the Verlet algorithm (Verlet, 1967). The temperature of the system is related to its total kinetic energy, which can be controlled in the simulation by computationally adding or removing energy to or from the system. The empirical potential energy of the system is given by the following equation, in which all of the terms k are referred to as potential functions, force constants or force fields:

$$E_{\text{chem}} = \sum_{\text{bonds}} k_b \left(r_t - r_c\right)^2 + \sum_{\text{angles}} k_\theta \left(\theta_t - \theta_c\right)^2 + \sum_{\text{dihedrals}} k_\phi \cos\left(n\phi_c - d\right)$$

$$+ \sum_{\text{improper}} k_\omega \left(\omega_t - \omega_c\right)^2 + \sum_{\text{atom pairs}} \left(ar_c^{-12} + br_c^{-6} + cr_c^{-1}\right) \qquad (15.28)$$

Geometric distortions of the structure are modelled in classical physical terms; for example, stretching a chemical bond is treated just like stretching a mechanical spring. In this example, we know from Hooke's law of elasticity that the force required to stretch or compress a spring by a certain amount is directly proportional to the deformation caused, and the energy stored is therefore proportional to the square of the deformation. The corresponding constants of proportionality in molecular force fields are determined for small molecules from a combination of experiment (infrared or Raman spectroscopy) and theoretical calculations. The summations in equation (15.28) are over all bond lengths r, bond angles θ, torsion angles ϕ, improper torsion angles ω, which define the planarity and chirality of certain groups, and van der Waals contacts. In most of these terms, the actual parameters calculated from the model (e.g. r_c) are compared with ideal or target values (e.g. r_t), which are indicated by the parameter having the appropriate subscript c or t. The third term describes how the potential energy varies as a simple trigonometric function of the torsion angles; the fifth term describes the form of the van der Waals forces, which depend on $1/r^{12}$ and $1/r^6$, where r is the interatomic distance, and the electrostatic forces, which depend on $1/r$, along with the respective force constants a, b and c. The precise form and details of all of the terms are beyond the scope of this book, and the reader should follow up additional reading for further explanation. It should also be appreciated that for crystallographic purposes we are not particularly interested in performing an exact or meaningful molecular dynamics simulation; we are simply interested in using it as a method to perturb the model to a greater extent than can be achieved with least squares so as to free any parts of it which are trapped in false minima.

In conventional restrained refinement, the geometric terms should ideally be expressed in the form of the energy equation (15.28) but are sometimes simplified for computational convenience; for example, bond angles and chiral tetrahedral groups can be restrained by use of distance restraints. If an atom is bonded to another, and the second atom is bonded to a third, we can impose a distance restraint between the first and third atoms to maintain the correct bond angle subtended at the second atom. Likewise, these 'third-atom' distances can be used to retain the correct configuration of tetrahedral groups, and 'fourth-atom' distances, in principle, provide a means of restraining torsion angles. In this sort of refinement, we merely use artificial force constants or weights for the geometric terms that give the model acceptable geometry at the end of refinement.

The first successful attempt to make use of more sophisticated energy calculations in macromolecular X-ray refinement was reported by Jack and Levitt (1978), whose approach was to use the potential energies calculated by the above equation as the geometric restraint in least-squares refinement. Brünger *et al.*, (1987) first reported performing molecular dynamics at high temperature (typically 3000 K) with slow cooling, using both the potential-energy and X-ray minimisation functions to solve the equations of motion. Thus the 'pseudo-total energy' in simulated annealing can be represented as

$$E = E_{chem} + w_{X\text{-}ray}\, E_{X\text{-}ray} \qquad\qquad (15.29)$$

where $E_{X\text{-}ray}$ is the familiar crystallographic residual $\sum(|F_o| - |F_c|)^2$ or M. The relative weighting of the potential-energy and crystallographic components of the 'pseudo-energy' has to be balanced carefully to ensure that neither dominates the function excessively. This is done by doing a short simulation with the E_{chem} energies only and comparing the resulting gradient with that for the $E_{X\text{-}ray}$ terms alone. During the main simulation, the high-temperature phase is typically followed by a slow cooling phase and, finally, conventional crystallographic refinement using the X-ray data and energy restraints. Compared with the least-squares methods, where the parameter shifts always move down the gradient, the use of a high temperature in simulated annealing allows uphill shifts to occur. This gives simulated annealing a much larger radius of convergence than the methods which minimise the crystallographic residual. During the simulation, the model gradually becomes trapped by energy barriers corresponding to those of the true structure rather than local energy minima of the initial model. Thus simulated annealing is very often pivotal to the correct interpretation of poor regions of initial electron density maps in both molecular replacement and experimental phasing exercises.

It has been found that performing torsion angle dynamics, in which the bond lengths and bond angles are held fixed but rotations around the bonds are permitted, increases the radius of convergence further, particularly with low-to-medium-resolution data (Brünger and Rice, 1997). The data-to-parameter ratio is also much better with torsion angle refinement, since there are typically fivefold fewer torsion angle parameters to refine than positional (x, y, z) coordinates in Cartesian refinement (Tronrud, 2004). Typically a slow cool from 5000 K with a cooling rate of 25 K per cycle would be suitable for torsion angle dynamics.

All methods of refinement require manual intervention in rebuilding the molecule to optimise the fit to the map. However, with simulated annealing, shifts in the positions of atoms of several Å and 180° torsion angle flips can readily be achieved, meaning that the amount of manual work associated with inspecting and interpreting the electron density map is substantially reduced.

15.10   Cross-validation

Earlier in this chapter, we mentioned that the progress of refinement is usually assessed by calculating an R-factor by use of the following equation:

$$R = \frac{\sum |(|F_o| - |F_c|)|}{\sum |F_c|} \qquad\qquad (15.2)$$

During refinement, we minimise a function of the form $\sum(|F_o| - |F_c|)^2$ (or M) by using the $|F_o|$ and $|F_c|$ data to calculate shifts that are applied to the model. Hence, if all of the reflections contribute to the calculated shifts, the similarity of these two functions (R and M) suggests that the R-factor must, by definition, decrease during a refinement, provided that any stereochemical or other restraints are applied correctly. Hence this bias means that using the refinement R-factor as a criterion for judging the success of the process is, to some extent, a self-fulfilling prophecy, and this has led to the development of other techniques for judging the success, or otherwise, of macromolecular refinements.

The most commonly used approach to avoid this problem of bias is to omit a small percentage of the reflection data, typically 5 or 10%, from the refinement calculations so that they have no influence on the calculated parameter shifts (Brünger, 1997). These reflections, which form the *test set*, are randomly selected throughout reciprocal space, and the remainder of the dataset forms the *working set* of reflections that are used in the refinement. Monitoring the R-factor of the test set, known as the *free R-factor*, provides an unbiased assessment of whether the refinement is genuinely improving the accuracy of the model. In successful protein refinements, typically at 2 Å resolution, the refinement R-factor will be around 20% or slightly below and the free R-factor will usually be a few per cent higher, perhaps as much as 10% higher (Tickle *et al.*, 1998, 2000). If the difference between the R-factor and the free R-factor greatly exceeds 10%, it is likely that parts of the structure are still missing or not yet modelled correctly. If atomic-resolution data are available, the final refinement R-factor can be in the vicinity of 10% or less and, again, the R-free, as it is also called, should be only a few per cent higher. The free R-factor is highly correlated with the phase error, and a high R-free is therefore indicative of serious errors in the model such as mistracing of the polypeptide chain or a gross error in positioning the search model in molecular replacement.

In situations where the structure possesses non-crystallographic symmetry, there is an additional complication that the reflections in the test set will be related to those in the working set by the NCS operators, and thus the R-free is not truly unbiased. Since reflections which are related by symmetry in the diffraction pattern, for example F_{hkl}, $F_{\overline{hkl}}$ etc, are at the same resolution, it is possible to avoid the problem of bias by selecting the R-free set in thin spherical shells of reciprocal space. This should ensure that reflections in the test set are not related to those in the working set by NCS.

In any rebuilding and refinement exercise, there is a tendency to overinterpret the electron density map, for example by fitting water molecules into every available feature of electron density in the solvent region. Because this introduces more parameters into the model, which 'soak up' experimental errors, it usually gives a better refinement R-factor, and is known as overfitting. This effect can be seen with reference to Fig. 15.4, where we saw a straight line fitted to some experimental data. The straight line clearly gives a good fit to the data and is defined by just two parameters—its gradient and intercept. Any discrepancy between the observed y values and the straight line is most probably due to errors of measurement. If, instead, we were to fit a polynomial curve to

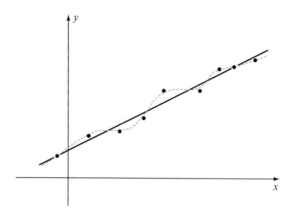

Fig. 15.6 *A polynomial fit to straight-line data.* Fitting a polynomial to the linear data shown in Fig. 15.4 yields the grey dashed line. Whilst the value of $\sum_j d_j^2$ is much lower for the polynomial fit, the improvement is artificial since we have increased the number of parameters in the model to fit the noise in the data.

the measurements (as shown in Fig. 15.6), we would obtain a much better fit to all of the experimental measurements. However, in doing so we have substantially increased the number of parameters in the model so that the curve fits the noise in the data. In macromolecular refinement, we have a vast number of parameters to describe the model and relatively few data. Overfitting does not lead to any improvements in the quality or interpretability of the electron density map and is not a good practice. It is better to judge the progress of refinement by monitoring the R-free, rather than the refinement R-factor. The R-free is also a good criterion for assessing whether some of the more sophisticated refinement options, such as TLS refinement, anisotropic B-factors and riding hydrogen atoms, make a useful contribution. Many refinement programs use Babinet's principle to model the effect of disordered solvent on the low-resolution data (Moews and Kretsinger, 1975), and this can have a beneficial effect on the R-free.

15.11   Use of Fourier maps in refinement

In structure analysis, we usually interpret the initial Fourier map to derive a model of the structure that can be refined by least squares, simulated annealing or some other means. From the refined model we may derive a new set of phases ϕ_c, and can perform a second Fourier synthesis using the observed structure factor magnitudes $|F_o|$ along with the new set of *calculated* phases ϕ_c:

$$\rho(x,y,z) = \frac{1}{V} \sum_h \sum_k \sum_l |F_o| \, e^{-i[2\pi(hx+ky+lz)-\phi_c]} \qquad (15.30)$$

This Fourier synthesis may be contrasted with the previous one, in which we used the observed magnitudes and observed phases:

$$\rho(x, y, z) = \frac{1}{V} \sum_h \sum_k \sum_l |F_o| \, e^{-i[2\pi(hx+ky+lz)-\phi_o]} \qquad (15.1)$$

Since the calculated phases ϕ_c should be more accurate than the observed phases ϕ_o, the new electron density function should be an improved representation of the molecular structure.

Much attention has been directed at determining the best form of Fourier synthesis to use in macromolecular refinement. Whilst the synthesis given in equation (15.30) uses the most accurate information that we have available to us, i.e. observed structure factor amplitudes and calculated phases, the resulting map will inevitably be biased towards the model that was used to calculate the phases. It has been shown that a Fourier synthesis of the following form, known as a $2F_o - F_c$ map, minimises this bias, i.e. incorrectly placed atoms will have less electron density and missing atoms will have more density than in the F_o Fourier map:

$$\rho(x, y, z) = \frac{1}{V} \sum_h \sum_k \sum_l (2|F_o| - |F_c|) \, e^{-i[2\pi(hx+ky+lz)-\phi_c]} \qquad (15.31)$$

For correctly placed atoms, the $2F_o - F_c$ map will yield approximately the same density as the straightforward F_o Fourier map given by equation (15.30). However, for incorrectly placed atoms, the effect of subtracting F_c in the $2F_o - F_c$ map will reduce the electron density for these atoms, which is very helpful during model building since atoms with weak density can simply be deleted or moved to their correct positions. The effect of multiplying the F_o term by 2 helps to bolster the electron density for atoms that are missing from the model. For reasons that we will examine in the next section, the missing atoms will appear at half their expected heights in the F_o electron density map calculated by use of equation (15.30) (Main, 1979). Consequently, in the $2F_o - F_c$ map, the density for the missing atoms will tend to appear more strongly than in an F_o Fourier map, and hence, in macromolecular crystallography, calculation of a $2F_o - F_c$ Fourier synthesis is standard practice.

Following the first refinement round, the $2F_o - F_c$ Fourier synthesis gives an electron density map from which we may identify new atomic positions. In general, this electron density map should be consistent with the original phasing model except in places where the model is incorrect. Most probably, density for extra amino acids will also be evident and spurious features of density in the previous map should have disappeared. Hence we need to rebuild the model using a suitable graphics program, and the new set of atomic positions that we derive must be refined again. This cyclic process of interpreting successive electron density maps interspersed with refinement in reciprocal space is continued until all features in the electron density map have been interpreted. By the time we reach this point, the R-factor and the R-free should both have converged to acceptable values.

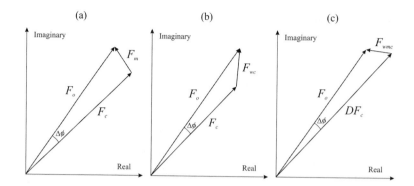

Fig. 15.7 *The effects of missing and incorrectly placed atoms on the calculated structure factor.* (a) shows the case of an incomplete structure without errors. The structure factor of the model is given by F_c, and that of the atoms missing from the model is shown as F_m. (b) shows the case of a complete structure with errors. To correct the structure factor calculated from this model for the wrongly positioned atoms, the vector F_{wc} is applied. (c) shows the case of a partial model with errors, in which the calculated structure factor can be corrected by addition of the vector F_{wmc}.

15.12 The difference Fourier synthesis

In the previous section we saw that as an adjunct to the refinement of a protein structure, it is essential to calculate $2F_o - F_c$ maps to aid the rebuilding of the model and to find missing atoms. We can examine the properties of the $2F_o - F_c$ Fourier synthesis in more detail by inspection of Fig. 15.7(a), where we consider simplistically that F_o is given by the vector sum of F_c for the current model and the structure factor of the missing atoms (F_m) (Main, 1979). Hence

$$|F_m|^2 = |F_o|^2 + |F_c|^2 - 2|F_o||F_c|\cos\Delta\phi$$

At this stage, we do not know the value of the phase error $\Delta\phi$ due to the missing atoms, but a way to estimate it has been devised by Sim (1959, Sim 1960), and this method has been adapted for macromolecular data (reviewed in Read, 1986). These methods give us an estimate of $\cos\Delta\phi$ for each reflection, which is referred to as the expectation value, or $\langle\cos\Delta\phi\rangle$, and on substituting this into the above equation, we can see that

$$\left\langle|F_m|^2\right\rangle = |F_o|^2 + |F_c|^2 - 2|F_o||F_c|\langle\cos\Delta\phi\rangle$$

We saw in the previous chapter that to minimise the rms error in an electron density map, we need to weight each structure factor by the figure of merit, which is essentially the cosine of the phase error (Blow and Crick, 1959). In the present case, we therefore can calculate a map with $\langle \cos \Delta\phi \rangle |F_o|$ as the Fourier coefficient. Rearranging the above equation, we get

$$|F_o| \langle \cos \Delta\phi \rangle = \frac{|F_o|^2}{2|F_c|} + \frac{|F_c|}{2} - \frac{\langle |F_m|^2 \rangle}{2|F_c|}$$

Hence

$$|F_o| \langle \cos \alpha \rangle e^{i\phi_c} = \frac{|F_o|^2 e^{i\phi_c}}{2|F_c|} + \frac{|F_c| e^{i\phi_c}}{2} - \frac{\langle |F_m|^2 \rangle e^{i\phi_c}}{2|F_c|}$$

Since

$$|F_o|^2 = (F_c + F_m)(F_c^* + F_m^*)$$

$$|F_o| \langle \cos \Delta\phi \rangle e^{i\phi_c} = \frac{(F_c + F_m)(F_c^* + F_m^*) e^{i\phi_c}}{2|F_c|} + \frac{|F_c| e^{i\phi_c}}{2} - \frac{\langle |F_m|^2 \rangle e^{i\phi_c}}{2|F_c|}$$

$$|F_o| \langle \cos \Delta\phi \rangle e^{i\phi_c} = \frac{\left(|F_c|^2 + F_c F_m^* + F_m F_c^* + |F_m|^2\right)}{2|F_c| e^{-i\phi_c}} + \frac{|F_c| e^{i\phi_c}}{2} - \frac{\langle |F_m|^2 \rangle}{2|F_c| e^{-i\phi_c}}$$

$$|F_o| \langle \cos \Delta\phi \rangle e^{i\phi_c} = \frac{\left(|F_c|^2 + F_c F_m^* + F_m F_c^* + |F_m|^2\right)}{2F_c^*} + \frac{|F_c| e^{i\phi_c}}{2} - \frac{\langle |F_m|^2 \rangle}{2F_c^*}$$

$$\therefore |F_o| \langle \cos \Delta\phi \rangle e^{i\phi_c} = F_c + \frac{F_m}{2} + \frac{F_m^* e^{2i\phi_c}}{2} + \frac{|F_m|^2 - \langle |F_m|^2 \rangle}{2F_c^*} \tag{15.32}$$

The first term on the right-hand side of equation (15.32) tells us that a Sim-weighted F_o Fourier synthesis will yield peaks of full height for atoms present in the model, and the second term confirms our statement in the previous section that the missing atoms will appear at half height. The third and fourth terms contribute only background noise. From the above equation, we can also see that the expression for the $2F_o - F_c$ synthesis can be given as follows:

$$(2|F_o| \langle \cos \Delta\phi \rangle - |F_c|) e^{i\phi_c} = F_c + F_m + F_m^* e^{2i\phi_c} + \frac{|F_m|^2 - \langle |F_m|^2 \rangle}{F_c^*} \tag{15.33}$$

From the terms on the right-hand side of equation (15.33), we can see that the $2F_o - F_c$ synthesis yields peaks at the expected height for the correctly placed atoms of the model and it also contains peaks at full height for the missing atoms, in contrast to the F_o synthesis, in which the missing atom peaks appeared at only half height. This accounts for the widespread adoption of the weighted $2F_o - F_c$ synthesis as the preferred map calculation.

Another form of Fourier synthesis which is very commonly used, especially to locate missing atoms, is known as the difference Fourier synthesis. This involves calculating a map with $F_o - F_c$ as the Fourier coefficients as below:

$$\rho(x, y, z) = \frac{1}{V} \sum_h \sum_k \sum_l (|F_o| - |F_c|) e^{-i[2\pi(hx + ky + lz) - \phi_c]} \tag{15.34}$$

The majority of the protein molecule will not appear in the difference map, since it should be accurately accounted for by the model. However, positive difference density will be present for atoms that are missing from the model. This property makes the difference Fourier map very important in identifying ligands bound to the protein, for example a substrate, cofactor or inhibitor, and water molecules that are not yet included in the model. If we return to equation (15.32), we can see that the $F_o - F_c$ synthesis is of the following form:

$$(|F_o| \langle \cos \Delta\phi \rangle - |F_c|) e^{i\phi_c} = \frac{F_m}{2} + \frac{F_m^* e^{2i\phi_c}}{2} + \frac{|F_m|^2 - \langle |F_m|^2 \rangle}{2F_c^*} \tag{15.35}$$

The first term on the right-hand side of equation (15.35) confirms that this map should predominantly contain positive peaks for atoms that are missing from the model. We can also see from this equation that the difference density peaks for missing atoms will be approximately half their expected heights. Thus, to give the density for missing atoms at the correct height, a difference map can be calculated with weighted $2(F_o - F_c)$ terms for acentric reflections and $F_o - F_c$ terms for centrics, where F_o and F_c are collinear (although the factor of 2 is often ignored).

One advantage of the $2F_o - F_c$ map over the F_o map is that the peaks for wrongly positioned atoms are lower, allowing them to be removed from the model. This can be confirmed by inspection of Fig. 15.7(b), where we consider that F_c contains the contributions of wrongly positioned atoms. The structure factor F_o is therefore the vector sum of F_c and a correction term F_{wc}. Therefore

$$|F_{wc}|^2 = |F_o|^2 + |F_c|^2 - 2|F_o| |F_c| \cos \Delta\phi$$

If we call the expectation value of $\cos \Delta\phi$ for this reflection $\langle \cos \Delta\phi \rangle$, we can see that

$$\langle |F_{wc}|^2 \rangle = |F_o|^2 + |F_c|^2 - 2|F_o| |F_c| \langle \cos \Delta\phi \rangle$$

$$\therefore |F_o| \langle \cos \Delta\phi \rangle = \frac{|F_o|^2}{2|F_c|} + \frac{|F_c|}{2} - \frac{\langle |F_{wc}|^2 \rangle}{2|F_c|}$$

and

$$|F_o| \langle \cos \Delta\phi \rangle e^{i\phi_c} = \left(\frac{|F_o|^2}{2|F_c|} + \frac{|F_c|}{2} - \frac{\langle |F_{wc}|^2 \rangle}{2|F_c|} \right) e^{i\phi_c}$$

$$\therefore |F_o| \langle \cos \Delta\phi \rangle e^{i\phi_c} = \frac{|F_o|^2}{2F_c^*} + \frac{F_c}{2} - \frac{\langle |F_{wc}|^2 \rangle}{2F_c^*} \tag{15.36}$$

In contrast, the $2F_o - F_c$ synthesis is equivalent to

$$(2|F_o| \langle \cos \Delta\phi \rangle - |F_c|) e^{i\phi_c} = \frac{|F_o|^2}{F_c^*} - \frac{\langle |F_{wc}|^2 \rangle}{F_c^*} \qquad (15.37)$$

In equation (15.36), the $F_c/2$ term adds to the electron density for the wrongly positioned atoms. The absence of this term in the $2F_o - F_c$ synthesis, as given by equation (15.37), means that the wrongly positioned atoms will have less electron density associated with them.

The determination of optimal weights for the Fourier synthesis has been the subject of much work (reviewed in Read, 1986). The Sim weight, which we have outlined above, corrects for missing atoms provided the known part of the model is correct. A widely used method of weighting devised by Read (1986) aims to correct the electron density map for errors in the model as well as missing atoms. In essence, the F_o term in a $2F_o - F_c$ map (or an $F_o - F_c$ map) is weighted by the Sim weight m, and the F_c terms are weighted by a term referred to as D, which was defined by Luzzati (1952) as a function of the rms fractional coordinate error, $\Delta \mathbf{r}$:

$$D = \langle \cos (2\pi \mathbf{S} \cdot \Delta \mathbf{r}) \rangle \qquad (15.38)$$

D has a minimum of 0, corresponding to a structure with infinite errors, and a maximum of 1, for a complete structure with no errors. The phasing triangle corresponding to an intermediate situation, in which the model has both errors and missing atoms, is shown in Fig. 15.7(c). The structure factor F_o is considered to be the vector sum of DF_c and a vector F_{wmc} corresponding to the lack of closure due to the wrongly placed and missing atoms. Hence

$$|F_{wmc}|^2 = |F_o|^2 + D^2 |F_c|^2 - 2D |F_o| |F_c| \cos \Delta\phi$$

and replacing $\cos \Delta\phi$ by the expectation value $\langle \cos \Delta\phi \rangle$ gives

$$\langle |F_{wmc}|^2 \rangle = |F_o|^2 + D^2 |F_c|^2 - 2D |F_o| |F_c| \langle \cos \Delta\phi \rangle$$

$$\therefore |F_o| \langle \cos \phi \rangle = \frac{|F_o|^2}{2D |F_c|} + \frac{D |F_c|}{2} - \frac{\langle |F_{wmc}|^2 \rangle}{2D |F_c|}$$

and

$$|F_o| \langle \cos \phi \rangle e^{i\phi_c} = \left(\frac{|F_o|^2}{2D |F_c|} + \frac{D |F_c|}{2} - \frac{\langle |F_{wmc}|^2 \rangle}{2D |F_c|} \right) e^{i\phi_c}$$

$$= \frac{|F_o|^2}{2DF_c^*} + \frac{DF_c}{2} - \frac{\langle |F_{wmc}|^2 \rangle}{2DF_c^*}$$

Since

$$|F_o|^2 = F_o F_o^* = F_o (DF_c^* + F_{wmc}^*)$$

$$= F_o F_{wmc}^* + F_o DF_c^*$$

then

$$|F_o| \langle \cos \Delta\phi \rangle e^{i\phi_c} = \frac{(F_o F_{wmc}^* + F_o DF_c^*)}{2DF_c^*} + \frac{DF_c}{2} - \frac{\langle |F_{wmc}|^2 \rangle}{2DF_c^*}$$

$$= \frac{F_o F_{wmc}^*}{2DF_c^*} + \frac{F_o}{2} + \frac{DF_c}{2} - \frac{\langle |F_{wmc}|^2 \rangle}{2DF_c^*}$$

$$= \frac{F_o}{2} + \frac{DF_c}{2} + \frac{\left(F_o F_{wmc}^* - \langle |F_{wmc}|^2 \rangle\right)}{2DF_c^*}$$

$$= \frac{F_o}{2} + \frac{DF_c}{2} + \frac{\left([DF_c + F_{wmc}] F_{wmc}^* - \langle |F_{wmc}|^2 \rangle\right)}{2DF_c^*}$$

$$= \frac{F_o}{2} + \frac{DF_c}{2} + \frac{F_{wmc}^*}{2} e^{2i\phi_c} + \frac{\left(|F_{wmc}|^2 - \langle |F_{wmc}|^2 \rangle\right)}{2DF_c^*}$$

Since the last two terms of this equation contribute only to noise, we can ignore them and rearrange the equation as follows:

$$F_o = 2|F_o| \langle \cos \Delta\phi \rangle e^{i\phi_c} - DF_c$$

Factorising out the exponential term and remembering that $\langle \cos \Delta\phi \rangle$ is the Sim weight m gives us

$$F_o = (2m|F_o| - D|F_c|) e^{i\phi_c} \qquad (15.39)$$

Thus maps can be calculated with the terms $2m|F_o| - D|F_c|$ and $2(m|F_o| - D|F_c|)$ for a difference map, where m and D are the Sim and Luzzati weights, respectively (note that for centric reflections, only $m|F_o|$ should be used for the $2F_o - F_c$ map and $m|F_o| - D|F_c|$ for the difference map). In practice, the weights can be calculated by the method of Read (1986) using the program SIGMAA. This derives its name from a weighting parameter σ_A that was used by Srinivasan and Ramachandran (1965) and is equivalent to D but relates instead to E-values rather than structure factor amplitudes. The values of σ_A are calculated from the normalised $|F_o|$ and $|F_c|$ values as a function of resolution, and a linear plot of $\ln(\sigma_A)$ versus $\sin^2 \theta/\lambda^2$ enables the mean coordinate error $\langle |\Delta r| \rangle$ to be determined from the gradient. In theory, D is related to $\langle |\Delta r| \rangle$ by the following equation:

$$D = e^{-\pi^3 (\langle |\Delta r| \rangle)^2 (\sin \theta/\lambda)^2}$$

assuming that the errors are normally distributed. For completeness, σ_A is related to D by the following equation:

$$\sigma_A = D \sqrt{\frac{\sum_{\text{partial}} f_j^2}{\sum_{\text{all atoms}} f_j^2}}$$

where the summation of squared scattering factors in the denominator is over all atoms in the structure and that in the numerator is for the partial structure. With macromolecular data, these terms are estimated from the F_o and F_c data rather than from the theoretical scattering factors.

As mentioned earlier, difference Fourier maps are of much value in analysing the structures of ligands bound to proteins, and for this purpose one usually calculates a difference Fourier map of the following form:

$$\rho(x, y, z) = \frac{1}{V} \sum_h \sum_k \sum_l \left(|F_{PI}| - |F_P| \right) e^{-i[2\pi(hx+ky+lz)-\phi_c]} \qquad (15.40)$$

If the ligand complex crystallises in the same form as the native protein, the F_P values can be calculated from the known structure of the protein component of the complex. If the ligand complex (PI) is not isomorphous, the protein component has to be oriented and positioned first using molecular replacement (see Chapter 13) and refinement before the structure factors F_P are calculated. Usually, the water molecules (and any ligands) should be removed from the protein model for calculating structure factors to avoid subtracting their electron density from that of the bound ligand. The difference map (such as that shown in Fig. 15.8) should possess positive peaks for the ligand, a model of which can then be built and refined using the $|F_{PI}|$ data. Appropriate stereochemical restraints to aid refinement and model building of the ligand can be obtained from published chemical crystallographic studies or from a number of Web servers dedicated to this task, for example PRODRG (Schuettelkopf and van Aalten, 2004).

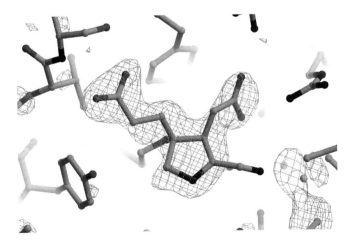

Fig. 15.8 *A difference Fourier map for a bound ligand.* An example of a difference electron density map (pale lines), showing a cyclic ligand molecule bound to one of the catalytic residues at an enzyme active site. The difference density has been calculated at 1.6 Å resolution and is contoured at 3 standard deviations above the mean density; at lower resolutions, a less stringent contour level of 1.5–2.0 rms would suffice.

Since the difference Fourier map represents the difference $\rho_o(x, y, z) - \rho_c(x, y, z)$, it emphasises the discrepancies between the proposed and true structures. If the proposed structure is in fact identical to the true structure, then the difference Fourier map will be featureless other than having a possible background fluctuation due to random errors. But if, as is always the case, the proposed structure differs from the true structure, then the difference Fourier map will have certain features which give us useful information. As an example, we may consider the case in which we have identified the position of an atom at some point, when that atom in reality is centred at a very closely neighbouring point. The atomic electron density functions for the true and proposed nuclear positions are shown in Fig. 15.9(a). The two electron density functions overlap, but their peaks are displaced. On subtracting the calculated electron density function (which corresponds to the proposed position) from the observed electron density function (which corresponds to the true position), we obtain a curve which is negative close to the proposed location, but is positive towards the true location. In a difference Fourier map, this appears as a sharp peak which has a very steep gradient on one side leading into a depression, as in Fig. 15.9(b). In order to obtain a more correct structure, we must shift the proposed atomic position towards the true position, implying that we must move the atomic site in the direction corresponding to a motion 'up the hill' in the difference Fourier map. Identification of regions in a difference Fourier map with an appearance such as that in Fig. 15.9(b) therefore allows us to refine atomic positions with considerable accuracy.

Another use of the difference Fourier map concerns the estimation of the occupancies and temperature factors of atoms. On calculating the structure factors from any

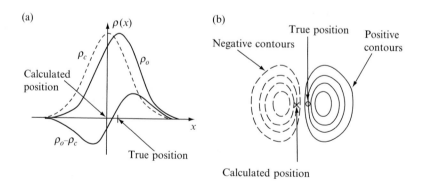

Fig. 15.9 *Misplaced atoms.* The electron density function $\rho_o(x, y, z)$ is centred on a point very close to that from which we derived the function $\rho_c(x, y, z)$. On subtracting the calculated from the observed electron density function to obtain the difference Fourier map, we derive a curve with a minimum towards the calculated position, and a maximum close to the true atomic position as in (a). In a difference Fourier map, this has the appearance of (b). To refine the model, we must move the proposed atomic position up the slope of the difference Fourier map.

proposed model, we include with each atomic scattering factor f_j an occupancy factor q_j and a correction for thermal and disorder effects B_j of the following form:

$$F_{hkl} = \sum_j q_j f_j e^{-B_j \sin^2 \theta / \lambda^2} e^{2\pi i (hx_j + ky_j + lz_j)} \qquad (14.29)$$

For atoms of a protein, the occupancy is usually fixed at unity, but it can be varied for ligand groups, and the difference Fourier map can indicate errors in the occupancies of such groups. If we have overestimated the occupancy of a ligand, the resulting difference map will possess negative peaks for the ligand atoms, much like the effects of incorrectly placed atoms. Underestimating the occupancy would give a difference map with positive peaks for the ligand. An equivalent effect often arises if there is uncertainty about the chemical nature of the ligand, for example if what should be an electron-dense atom in the model has been incorrectly assigned as one with a much lower atomic number. Similar effects occur if we have overestimated the parameter B_j; in that case, the electron density peak corresponding to that particular atom in the calculated structure is rather lower and broader than the true case. This is depicted in Fig. 15.10(a), in which are shown the electron density functions of the real and proposed atoms. Calculating the difference Fourier map is tantamount to performing the subtraction $\rho_o(x, y, z) - \rho_c(x, y, z)$, and we obtain a curve which is positive in the centre and negative around the outside. In a difference Fourier map, this appears as a peak surrounded by a trough, as in Fig. 15.10(b). Conversely, underestimation of the parameter B_j gives a trough surrounded by a crest in the difference Fourier map. These effects become particularly apparent with very high-resolution protein data.

The electron density function of an atom vibrating with spherical symmetry will be circular in plan, whereas an anisotropic vibrator will have a more elliptical electron density function as in Fig. 15.11(a). On subtraction to form $\rho_o(x, y, z) - \rho_c(x, y, z)$, we obtain a quatrefoil shape with two positive and two negative lobes as in Fig. 15.11(b). The existence of such contours on a difference map can, consequently, give us

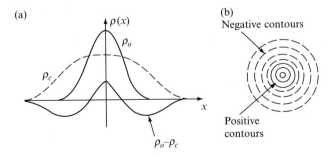

Fig. 15.10 *Overestimation of temperature factors.* If the temperature factor B_j of an atom is overestimated, then the calculated electron density function is lower and broader than the true electron density function, as in (a). On deriving the difference Fourier map, we obtain a peak surrounded by a trough as in (b).

(a)   (b)

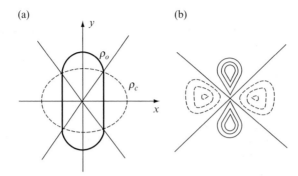

Fig. 15.11 *Anisotropic displacement parameters.* The curve ρ_o in (a) represents a plan of the electron density function of an atom which has more freedom of vibrational motion along the y axis than along the x axis; ρ_c represents the electron density calculated at the current stage of refinement. On subtraction to form the difference Fourier synthesis, we obtain a quatrefoil shape in the difference Fourier map as in (b).

information on the anisotropy of the displacement parameters. If a protein diffracts to atomic resolution, then a lot of the noise in the difference map arising from these effects can be removed by refinement of anisotropic B-factors.

The overall result of a difference Fourier synthesis is potentially fourfold. Firstly, we may make refinements to atomic sites. Secondly, missing atoms may be located and incorporated into the model by virtue of the positive difference density, and thirdly, wrongly positioned atoms may be removed by virtue of the negative peaks associated with them. Fourthly, the temperature factors and occupancies may be checked, and with very high-resolution data the possibility of anisotropic oscillations may be investigated. As a result of the knowledge derived from the difference Fourier synthesis, we may alter our proposed model, giving us a structure which is more likely to represent the true structure. We may now commence another round of reciprocal-space refinement based on the new model. The aim of alternating real-space rebuilding using weighted $2F_o - F_c$ and $F_o - F_c$ maps with refinement in reciprocal space is that we will eventually arrive at a model for which the difference map is featureless, or at least possesses no peaks bigger than three standard deviations. Whilst this is never fully attainable, it is an important goal of the refinement process.

15.13   The maximum-likelihood method in refinement

In least-squares refinement, we adjust the parameters of the model to minimise the weighted sum of $(|F_o| - |F_c|)^2$ for all reflections in the dataset. In the least-squares method, it is assumed that the errors are random and uncorrelated and have equal, finite variance. This will not be strictly true, for example if a significant part of the model is missing, although it is a reasonable approximation for a good model. If the conditional probability of the observations ($|F_o|$ or I_o) is a known function of the parameters of the model, we can apply another form of function minimisation, known as maximum likelihood. In statistics, the term 'conditional probability' refers to the probability of some event given some other event or, in our case, the probability of having measured a particular set of intensity values given a particular model of the crystal structure. In maximum-likelihood refinement, the parameters of the model are adjusted to maximise the above probability. This overall conditional probability is calculated as the product of the probabilities for each reflection, which can be estimated from the corresponding $|F_o|$, F_c and D (or σ_A) of each diffraction spot by expressions given by Murshudov *et al.* (1997). The probability of the model can also be assessed from its consistency with the ideal stereochemistry such as the bond lengths and angles. In statistics, this is referred to as prior knowledge of the system, and with proteins the prior stereochemical knowledge allows us to calculate a prior probability distribution from the current model parameters. It is then possible to combine the conditional and prior probability distributions to calculate the overall probability of the model, which is strictly known as the posterior probability. Combining the two probability distributions (prior and conditional) is done by using Bayes' Theorem, which is a form of statistical inference—essentially, the two distributions are multiplied. Our aim is to adjust the model to maximise the posterior probability, and to do this, for mathematical convenience, we need to minimise the negative logarithm of the posterior probability. Shifts in the parameters of the model which minimise this function are determined essentially by the methods that we outlined above for least-squares refinement, and the corresponding first and second derivatives for the Jacobian and Hessian matrices are given by Murshudov *et al.* (1997). Interestingly, for a good model with a good dataset, the likelihood function is equivalent to the least-squares minimisation function—the least-squares method will give the same solution as the maximum-likelihood method in the case of a normal distribution. However, maximum likelihood has a larger radius of convergence and performs better when the model parameters are at the outer reaches of the radius of convergence. Maximum-likelihood refinement can be performed with a number of widely used programs, including REFMAC (Murshudov *et al.*, 1997), CNS (Brünger *et al.*, 1998) and BUSTER/TNT (Tronrud *et al.*, 1987).

15.14   Validation and deposition

Following refinement of the structure, it is important to verify that it does not possess improbable stereochemistry. It is conventional to monitor bond lengths, angles, the planarity of certain groups, van der Waals contacts and torsion angles. The latter can be visualised most easily by the Ramachandran plot, which displays the main-chain ϕ and ψ torsion angles and allows immediate identification of residues that reside outside the favourable regions for α-helix and β-sheet. A number of programs are available to apply these tests, including PROCHECK (Laskowski *et al.*, 1993) and MOLPROBITY (Lovell *et al.*, 2003). Note that glycine lacks a side chain and therefore has greater torsional freedom than other amino acids. In contrast, proline has its ϕ angle restricted to approximately $-60°$ by virtue of its side chain, which is covalently linked with the main-chain nitrogen. Side chains tend to adopt staggered conformations, and this should be checked visually during model building; any remaining outliers will be highlighted at the validation stage when the model's side chain torsion angles (χ_1, χ_2 etc.) are analysed.

The geometry of flexible loop regions is often poorer than that of buried residues, which are better defined by the electron density. Regions of the molecule which have poor electron density often have poorer geometry than the rest of the structure owing to the difficulty of interpreting the map in these regions. Analysis of the average temperature factor versus residue number is also extremely helpful for identifying regions of the molecule that need more attention. The exposed loops are often disordered, and it is to be expected that residues at the surface of the protein will have higher than average B-factors. However, should any buried regions have high B-factors, this almost certainly indicates errors in the chain tracing. It may be helpful and is sometimes very revealing to search the database of known structures for other proteins with similar folds, for example by use of DALI (Holm and Sander, 1995).

It is also important the check that the amino acids make favourable interactions in the structure. For example, the hydrophobic residues should be buried in the core away from the water molecules that surround the protein, and the majority of surface residues will have polar side chains for favourable hydrogen bonding with the solvent. Charged and polar residues do occur in the core, but they invariably pair up with other such residues to form electrostatic or hydrogen-bonding interactions. The side chains of some amino acids are often difficult to fit unambiguously to the electron density even if it is of high quality, for example the amide groups of asparagine and glutamine and the imidazole group of histidine. In favourable cases, these can be oriented correctly by inspection of the geometry of the hydrogen-bonding environment; for example, a main-chain carbonyl oxygen can only form a hydrogen bond with the nitrogen of an amide group and not with the oxygen. Post-translational modifications (such as phosphorylations and glycosylations) can also become apparent, as can unexpected reactions during purification or crystallisation, such as the reaction of cysteine residues with mercaptoethanol in the mother liquor.

Finally, the structure can be deposited with the Protein Data Bank (PDB) via the Internet (www.wwpdb.org), and a number of validation tests will be run on it prior to it being accepted for public release, which is a requirement for publication by most journals. Once released, the coordinates will be available in PDB format, which carries information about the experimental methods used to determine and refine the structure, as well as the crystallographic unit cell and symmetry, which are important for displaying the crystal packing. The coordinates of each atom are given in units of Å with respect to orthogonal axes, rather than fractional crystallographic coordinates, but the two coordinate systems are readily interconverted. The temperature factor of each atom is also given, usually as a B_{iso}, but sometimes as the mean square displacement, or $\overline{u_{iso}^2}$, often referred to simply as U_{iso}. The factor U_{iso} has the advantage over B_{iso} of being a more intuitive measure of the displacement, but the two can be readily interconverted by the following relationship:

$$B_{iso} = 8\pi^2 \overline{u_{iso}^2}$$

from which it can be seen that the B_{iso} values will exceed the equivalent U_{iso} values by a factor of $8\pi^2$, which is approximately 79. Both B_{iso} and U_{iso} have units of Å^2.

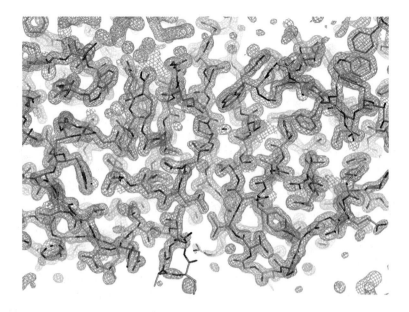

Fig. 15.12 *An example of a refined electron density map at 1.75 Å resolution.* The $2F_o - F_c$ map for this protein is shown in thin grey lines. In addition to the amino acids, which are displayed as shaded sticks, well-defined peaks of electron density for numerous water molecules are visible at the surface of the protein (courtesy of Dr R. Hussey).

It is conventional to deposit the structure factors at the same time as the coordinate data, so that others have the possibility of calculating the electron density map, and this, of course, represents the ultimate criterion by which any structure should be judged. An example of a high-resolution protein electron density map obtained after several rounds of refinement and rebuilding is shown in Fig. 15.12 and is typical of what can be achieved with strongly diffracting crystals.

Summary

The inaccuracies inherent in the observed structure factor magnitudes and phases, coupled with series termination errors, imply that the first electron density map derived is often not a very accurate representation of the true structure. Accordingly, the structural model derived from this will contain errors. The process by which we obtain an improved model of the structure and an electron density map in which we have more confidence is known as *refinement*.

The *trial structure*, or *phasing model*, with which we start refinement can be obtained by use of a molecular graphics program and/or one of the automatic chain-tracing routines. Whilst the fit of the model to the electron density can be optimised by *real-space refinement*, further refinement is usually done in reciprocal space using the diffraction amplitudes or intensities, since the relative errors in these are much smaller than those associated with the experimental phases.

In reciprocal-space refinement by the least-squares method, the model is adjusted to maximise the agreement between the observed and calculated structure factor amplitudes by minimising the function $\sum(|F_o| - |F_c|)^2$. The mathematical model of the structure consists of three positional coordinates for each atom and a temperature factor that describes the extent of any thermal or other disorder. With atomic-resolution data, six displacement parameters can be refined for each atom to give the best fit with the experimental data. The progress of refinement is assessed by calculation of an R-factor and (optionally) a correlation coefficient (CC):

$$R = \frac{\sum |(|F_o| - |F_c|)|}{\sum |F_c|} \tag{15.2}$$

$$CC = \frac{\sum (|F_o| - \langle|F_o|\rangle)(|F_c| - \langle|F_c|\rangle)}{\sqrt{\sum (|F_o| - \langle|F_o|\rangle)^2 \sum (|F_c| - \langle|F_c|\rangle)^2}} \tag{15.3}$$

An initial model of a protein structure is likely to have an R-value in the range of 0.4–0.5, and this will decrease to around 0.2 over several rounds of reciprocal-space refinement and manual rebuilding of the model. An unbiased assessment of the convergence of the refinement is provided by omitting a small proportion of the reflection data from the refinement calculations (typically 5%) and calculating the R-factor for this group, known as the R-free, separately.

For computational expedience, the complexity of macromolecular refinement is reduced by simplifying the least-squares normal matrix such that only the large X-ray terms along or close to the diagonal are calculated. This is known as the *sparse matrix method*, and further simplifications such as the *method of steepest descents* are possible. The paucity of the diffraction data with macromolecules entails that stereochemical restraints must be applied to the model of the structure during refinement to ensure that it retains acceptable molecular geometry, and these restraints are incorporated as off-diagonal terms in the normal matrix.

The exploitation of non-crystallographic symmetry is a powerful tool in macromolecular refinement and can lead to significant enhancements in the quality of the resulting electron density. Where diffraction data are not available to sufficient resolution to refine anisotropic temperature factors, it may prove useful to refine TLS tensors, which describe the rigid-group displacement in the structure, and this approach can lead to significant improvements in refinement statistics.

The use of high-temperature molecular dynamics calculations which include the X-ray diffraction data has provided an especially powerful means of refining structures where the model possesses errors that are greater than can be corrected by least squares. This is known as *simulated annealing*, and further developments involving the application of maximum-likelihood statistics have increased the power of both this and conventional refinement methods.

Since each refinement cycle involves calculating an approximation of shifts in the atomic positions and temperature factors, it is an iterative process, requiring several cycles to reach convergence. An electron density map is then calculated and must be inspected using a graphics program, such that the model can be rebuilt. The improved model is then subjected to further cycles of refinement, followed by manual rebuilding. To aid interpretation of the electron density maps, weights (m and D) are usually applied to the F_o and F_c terms that enhance the electron density for missing atoms and reduce the density for wrongly positioned ones:

$$\rho(x, y, z) = \frac{1}{V} \sum_h \sum_k \sum_l \left(2m \, |F_o| - D \, |F_c|\right) e^{-i[2\pi(hx+ky+lz)-\phi_c]}$$

A difference Fourier map is of much value in locating missing atoms and analysing the structures of ligands bound to proteins:

$$\rho(x, y, z) = \frac{1}{V} \sum_h \sum_k \sum_l \left(m \, |F_o| - D \, |F_c|\right) e^{-i[2\pi(hx+ky+lz)-\phi_c]}$$

A number of validation tools are available to assess the stereochemical quality of the model during and after refinement. The final result of refinement should be an electron density map which is very highly correlated with the true structure.

References

Agarwal, R. C. (1978) A new least-squares refinement technique based on the fast Fourier transform algorithm. *Acta Crystallogr. A* **34**, 791–809.

Badger, J. (2003) An evaluation of automated model-building procedures for protein crystallography. *Acta Crystallogr. D* **59**, 823–7.

Blow, D. M. and Crick, F. H. C. (1959) The treatment of errors in the isomorphous replacement method. *Acta Crystallogr.* **12**, 794–802.

Booth, A. D. (1949) The refinement of atomic parameters by the technique known in X-ray crystallography as 'the method of steepest descents'. *Proc. Roy. Soc. A* **197**, 336–55.

Brünger, A. T. (1997) Free *R* value: cross-validation in crystallography. *Methods Enzymol.* **277**, 366–96.

Brünger, A. T. and Rice, L. M. (1997) Crystallographic refinement by simulated annealing: methods and applications. *Methods Enzymol.* **277**, 243–69.

Brünger, A. T., Kuriyan, J. and Karplus, M. (1987) Crystallographic *R* factor refinement by molecular dynamics. *Science* **235**, 458–60.

Brünger, A. T., Adams, P. D., Clore, G. M., DeLano, W. L., Gros, P., Grosse-Kunstleve, R. W., Jiang, J.-S., Kuszewski, J., Nilges, N., Pannu, N. S., Read, R. J., Rice, L. M., Simonson, T., and Warren, G. L. (1998) Crystallography and NMR system (CNS): A new software system for macromolecular structure determination. *Acta Crystallogr. D* **54**, 905–21.

Coates, L., Erskine, P. T., Crump, M. P., Wood, S. P. and Cooper, J. B. (2001) Five atomic resolution structures of endothiapepsin inhibitor complexes: implications for the aspartic proteinase mechanism. *J. Molec. Biol.* **318**, 1405–15.

Diamond, R. (1971) A real-space refinement procedure for proteins. *Acta Crystallogr. A* **27**, 436–52.

Emsley, P. and Cowtan, K. (2004) Coot: model-building tools for molecular graphics. *Acta Crystallogr. D* **60**, 2126–32.

Engh, R. A. and Huber, R. (1991) Accurate bond and angle parameters for X-ray protein structure refinement. *Acta Crystallogr. A* **47**, 392–400.

Farrugia, L. J. (1997) Ortep-3 for Windows. *J. Appl. Crystallogr.* **30**, 565.

Fletterick, R. J. and Wyckoff, H. W. (1975) Preliminary refinement of protein coordinates in real space. *Acta Crystallogr. A* **31**, 698–700.

Fisher, S. J., Helliwell, J. R., Khurshid, S., Govada, L., Redwood, C., Squire, J. M. and Chayen, N. E. (2008) An investigation into the protonation states of the C1 domain of cardiac myosin-binding protein C. *Acta Crystallogr. D* **64**, 658–64.

Greer, J. (1974) Three-dimensional pattern recognition: an approach to automated interpretation of electron density maps of proteins. *J. Molec. Biol.* **82**, 279–301.

Haneef, I., Moss, D. S., Stanford M. J. and Borkakoti, N. (1985) Restrained structure-factor least-squares refinement of protein structures using a vector processing computer. *Acta Crystallogr. A* **41**, 426–33.

Hendrickson, W. A. and Konnert, J. H. (1980) Incorporation of stereochemical information into crystallographic refinement. In *Computing in Crystallography*, ed. Diamond, R. Indian Institute of Science, Bangalore, India, Chapter 13, pp. 1–23.

Holm, L. and Sander, C. (1995) Dali: a network tool for protein structure comparison. *Trends Biochem. Sci.* **20**, 478–80.

Hughes, E. W. (1941) The crystal structure of melamine. *J. Am. Chem. Soc.* **63**, 1737–52.

Hughes, E. W. (1978) The crystal structure of melamine. *Current Contents* **40**, 13–13.

Ioerger, T. R. and Sacchettini, J. C. (2002) Automatic modeling of protein backbones in electron density maps via prediction of $C\alpha$ coordinates. *Acta Crystallogr. D* **58**, 2043–54.

Jack, A. and Levitt, M. (1978) Refinement of large structures by simultaneous minimization of energy and R factor. *Acta Crystallogr. A* **34**, 931–5.

Jones, T. A. (1978) A graphics model building and refinement system for macromolecules. *J. Appl. Crystallogr.* **11**, 268–72.

Lamzin, V., Morris, R. and Perrakis, A. (2001) Automatic building and refinement of protein crystal structures. In *Advances in Structure Analysis*, ed. Kuzel, R. and Hasek, J. Czech & Slovak Crystallographic Association, Prague, pp. 73–80.

Laskowski, R. A., MacArthur, M. W., Moss, D. S. and Thornton, J. M. (1993) PROCHECK: A program to check the stereochemical quality of protein structures. *J. Appl. Crystallogr.* **26**, 283–91.

Levitt, D. G. (2001) A new software routine that automates the fitting of protein X-ray crystallographic electron-density maps. *Acta Crystallogr. D* **57**, 1013–9.

Lovell, S. C., Davis, I. W., Arendall, B., de Bakker, P. I. W., Word, J. M., Prisant, M. G., Richardson, J. S. and Richardson, D. C. (2003) Structure validation by $C\alpha$ geometry: ϕ, ψ, and $C\beta$ deviation. *Proteins Struct. Funct. Genet.* **50**, 437–50.

Luzzati, V. (1952) Traitement statistique des erreurs dans la determination des structures cristallines. *Acta Crystallogr.* **5**, 802–10.

Main, P. (1979) A theoretical comparison of the β, γ' and $2F_o - F_c$ syntheses. *Acta Crystallogr. A* **35**, 779–85.

Moews, P. C. and Kretsinger, R. H. (1975) Refinement of the structure of carp muscle calcium-binding parvalbumin by model building and difference Fourier analysis. *J. Molec. Biol.* **91**, 201–25.

Murshudov, G. N., Vagin, A. A. and Dodson, E. J. (1997) Refinement of macromolecular structures by the maximum-likelihood method. *Acta Crystallogr. D* **53**, 240–55.

Oldfield, T. J. (2003) Automated tracing of electron-density maps of proteins. *Acta Crystallogr. D* **59**, 483–91.

Perrakis, A., Sixma, T. K., Wilson, K. S. and Lamzin, V. S. (1997) wARP: improvement and extension of crystallographic phases by weighted averaging of multiple-refined dummy atom models. *Acta Crystallogr. D* **53**, 448–55.

Read, R. J. (1986) Improved Fourier coefficients for maps using phases from partial structures with errors. *Acta Crystallogr. A* **42**, 140–9.

Ruiz, F., Hazemann, I., Mitschler, A., Joachimiak, A., Schneider, T. R., Karplus, M. and Podjarny, A. D. (2005) The crystallographic structure of aldose reductase-IDD552 complex shows direct proton donation from tyrosine 48. *Acta Crystallogr. D* **60**, 1347–54.

Schuettelkopf, A. W. and van Aalten, D. M. F. (2004) PRODRG – a tool for high-throughput crystallography of protein–ligand complexes. *Acta Crystallogr. D* **60**, 1355–63.

Schomaker, V. and Trueblood, K. N. (1968) On the rigid-body motion of molecules in crystals. *Acta Crystallogr. B* **24**, 63–76.

Sheldrick, G. M. and Schneider, T. R. (1997) SHELXL: high-resolution refinement. *Methods Enzymol.* **277**, 319–43.

Sim, G. A. (1959) The distribution of phase angles for structures containing heavy atoms. II. A modification of the normal heavy-atom method for non-centrosymmetrical structures. *Acta Crystallogr.* **12**, 813–5.

Sim, G. A. (1960) A note on the heavy-atom method. *Acta Crystallogr.* **13**, 511–2.

Srinivasan, R. and Ramachandran, G. N. (1965) Probability distribution connected with structure amplitudes of two related crystals. V. The effect of errors in the atomic coordinates on the distribution of observed and calculated structure factors. *Acta Crystallogr.* **19**, 1008–14.

Sussmann, J. L., Holbrook, S. R., Church, G. M. and Kim, S.-H. (1977) A structure-factor least-squares refinement procedure for macromolecular structures using constrained and restrained parameters. *Acta Crystallogr. A* **33**, 800–804.

Ten Eyck, L. F. (1977) Efficient structure-factor calculation for large molecules by the fast Fourier transform. *Acta Crystallogr. A* **33**, 486–92.

Terwilliger, T. C. (2003a) Automated main-chain model building by template matching and iterative fragment extension. *Acta Crystallogr. D* **59**, 38–44.

Terwilliger, T. C. (2003b) Automated side-chain model building and sequence assignment by template-matching. *Acta Crystallogr. D* **59**, 45–9.

Tickle, I. J., Laskowski, R. A. and Moss, D. S. (1998) R_{free} and the R_{free} ratio. I. Derivation of expected values of cross-validation residuals used in macromolecular least squares refinement. *Acta Crystallogr. D* **54**, 547–57.

Tickle, I. J., Laskowski, R. A. and Moss, D. S. (2000) R_{free} and the R_{free} ratio. II. Calculation of the expected values and variances of cross-validation statistics in macromolecular least-squares refinement. *Acta Crystallogr. D* **56**, 442–50.

Tronrud, D. E. (2004) Introduction to macromolecular refinement. *Acta Crystallogr. D* **60**, 2156–68.

Tronrud, D. E., Ten Eyck, L. F. and Matthews, B. W. (1987) An efficient general-purpose least squares refinement program for macromolecular structures. *Acta Crystallogr. A* **43**, 489–501.

Verlet, L. (1967) Computer 'experiments' on classical fluids. I. Thermodynamical properties of Lennard-Jones molecules. *Phys. Rev.* **159**, 98–103.

Waser, J. (1963) Least-squares refinement with subsidiary conditions. *Acta Crystallogr. A* **16**, 1091–4.

Wilson, A. J. C. (1942) Determination of absolute from relative X-ray intensity data. *Nature (Lond.)* **150**, 151–2.

Winn, M. D., Isupov, M. N. and Murshudov, G. N. (2001) Use of TLS parameters to model anisotropic displacements in macromolecular refinement. *Acta Crystallogr. D* **57**, 122–33.

16
Complementary diffraction methods

In this chapter, we will look predominantly at diffraction methods that involve neutrons, rather than X-rays, owing to the unique properties of neutrons that complement an X-ray structure analysis. However, we will touch on the X-ray Laue method in greater detail than has been given in previous sections, for example Chapter 11, Section 11.7, where we introduced the principles briefly. The latter method has great potential for studies of enzyme catalytic intermediates owing to the speed with which data can be collected. Interestingly, its counterpart in the neutron field (neutron Laue) represents one of the most important methods by which data can be collected efficiently.

16.1   Finding hydrogen atoms in X-ray structures

Hydrogen atoms and protons are responsible for many of the biological functions of proteins, such as the specific recognition of other molecules and catalysis of biochemical reactions. Hence there is much interest in locating hydrogen atoms in protein structures, not least because they constitute around 50% of the atoms in a typical protein. However, the fact that hydrogen has only one electron, or none in the case of a proton, means that the hydrogen atoms are usually 'invisible' to X-rays. Consequently, in electron density maps, the hydrogen atom positions have to be inferred from the molecular geometry of the C, N, O and other atoms that are 'visible' to X-rays in the structure. For non-exchangeable hydrogens, their positions are determined reasonably accurately by the geometry of 'visible' atoms with which they are bonded. In contrast, the positions of exchangeable hydrogens may only be inferred from the surrounding polar groups, and often there is much ambiguity as to which groups are protonated and which are not.

Experimental information on the location of hydrogen atoms in proteins can be obtained from X-ray cryocrystallographic studies if atomic-resolution data can be collected ($d_{min} < 1.2$ Å). Hydrogen atom positions can be determined for particularly well-ordered regions of the main chain and side chains (Fig 16.1), but disorder and high temperature factors can still render many of the more labile hydrogen atoms invisible. In addition, hydrogen atoms involved in low-barrier hydrogen bonds (short, strong interactions often found at enzyme active sites) will be distributed between two positions around 0.5 Å apart (Cleland *et al.*, 1998) and may therefore have high apparent temperature factors. There is evidence for these effects in neutron studies of carboxylic acid dimers (Wilson, 2001). Studies have shown that a number of functional groups in proteins, including carboxylates, are particularly vulnerable to radiolytic damage at the radiation doses that might be required to collect atomic-resolution X-ray data from protein crystals (Burmeister, 2000; Ravelli and McSweeney, 2000). Aspartate and glutamate residues, including those involved in buried salt-bridge interactions, are found to become disordered rapidly and appear to suffer from decarboxylation of the side chain even at 100 K. These effects could make the process of locating hydrogen atoms on active-site carboxylate groups with X-ray data difficult.

However, atomic-resolution X-ray crystallography provides an alternative means of defining the protonation states of carboxylate groups (Fig. 16.1). Neutral carboxyl groups have a significant difference between the C–OH and C=O bond lengths (typically 1.20 Å for the C=O bond and 1.30 Å for the C–OH bond), whereas ionised carboxylates have identical C–O bond lengths (typically 1.25 Å) owing to resonance. Thus unrestrained refinement using atomic-resolution data is a very powerful tool for determination of the protonation state of each residue and avoids the difficulties of locating weak electron density due to the hydrogens themselves (Helliwell *et al.*, 2002).

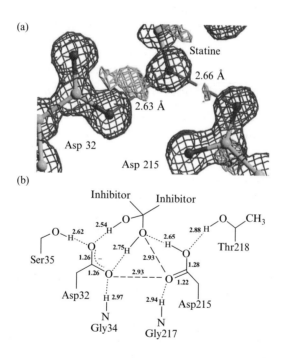

Fig. 16.1 *Atomic-resolution analysis of an enzyme active site.* (a) shows the electron density of the catalytic residues with a bound transition state analogue at 0.9 Å resolution. The dark lines show the $2F_o - F_c$ density (contoured at 1 rms) and the pale contours show the $F_o - F_c$ density (contoured at 2 rms), providing evidence for key hydrogen atoms at the catalytic centre. (b) shows the results of unrestrained refinement of a different complex at a similar resolution to define the protonation states of the catalytic carboxyl groups based on their C–O bond lengths (shown in Å). Figures courtesy of Professor L. Coates (Oak Ridge National Laboratory) and reproduced from Coates *et al.* (2008) with permission.

However, it must be remembered that the apparent bond lengths will be influenced by the temperature factors of the atoms involved, and this type of analysis is therefore most reliable for atoms in well-ordered parts of proteins.

16.2 Neutron protein crystallography

Crystallography using neutrons instead of X-rays provides a rather more powerful tool for locating the elusive hydrogen atoms or protons in a structure (Myles, 2006). Unlike X-rays, which are scattered by electrons, neutrons are scattered by the nuclei of atoms. The net scattering is the sum of two components, namely *potential scattering*, which is essentially the elastic scattering component, and *resonance scattering*, which involves absorption of energy from the incident beam and its re-emission by quantum transitions (Bacon, 1962). Potential scattering is proportional to the nuclear radius, or to the cube root of the mass of the nucleus (the atomic mass number). In contrast to potential scattering, which is always positive, resonance scattering can become negative and large enough to outweigh the potential term. This can occur when nuclear resonance energies are close to the energy of the incident beam, causing a negative scattering amplitude. The latter effect is due to an additional phase change of $180°$, or π radians, upon scattering and, importantly, this is a feature of the scattering by hydrogen atoms, giving rise to negative density for hydrogens in a neutron Fourier map. This is a well-known feature of neutron density maps which allows hydrogens to be distinguished from C, N and O atoms—the latter give positive density of approximately twice the value expected for hydrogen. In contrast to X-ray diffraction, where the scattering factors of elements increase progressively with atomic number, neutron scattering factors are relatively uniform across the periodic table within a factor of 2 or 3, and they vary somewhat unpredictably owing to resonance scattering effects (see Fig. 16.2). Note that neutron scattering factors are more commonly referred to as scattering lengths.

Atomic nuclei are small (radii 10^{-4} Å) compared with the wavelength of neutrons (e.g. 1 Å) and therefore behave as point scatterers with no dependence of the scattering amplitude on θ. This is in stark contrast to the situation with X-rays, where the scattering amplitude decreases appreciably with increasing θ, as emphasized in Sections 9.7 and 11.13. Consequently, the high-θ data does not fade off as much with neutrons as it does with X-rays (see Fig. 16.3). This is of much benefit, given than neutron fluxes are usually several orders of magnitude below those achievable at synchrotron X-ray sources. Flux considerations generally entail that the crystal volume has to be of the order of several cubic millimetres with a normal protein sample.

Neutrons are produced either in a *nuclear reactor* or at a *spallation source*, and a number of such facilities around the world have beam lines for protein crystallographic work. The majority of neutron sources are of the fission reactor type. A spallation source involves bombarding a metal target with protons from a particle accelerator, or synchrotron. The neutrons produced by the target of the spallation source or the core of the reactor then enter various experimental beam lines where they are monochromated to varying degrees, collimated and directed at a sample. Spallation sources tend to achieve higher instantaneous neutron fluxes, and further improvements

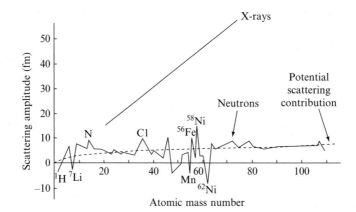

Fig. 16.2 *Neutron scattering lengths.* The variation of neutron scattering length b with atomic mass number is shown. The dashed line indicates the contribution of potential scattering, and the deviations from this line which occur with certain elements are due to resonance scattering. The equivalent relationship for X-rays at a typical θ angle is also indicated. Adapted from Bacon (1962).

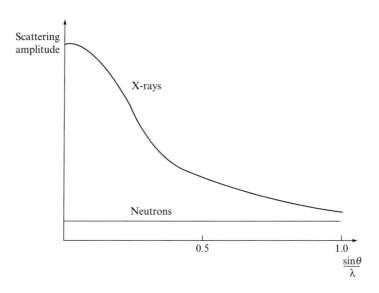

Fig. 16.3 *A comparison of the X-ray and neutron scattering amplitudes as a function of θ.* Whilst the X-ray scattering factor decreases appreciably with increasing $\sin\theta/\lambda$, the neutron scattering amplitude is independent of the scattering angle.

will be achievable using the liquid mercury targets that are being installed at the newest sources (Teixeira *et al.*, 2008). The protons hitting the target arrive in pulses at a relatively low frequency, for example 20–60 Hz, and radiation (including neutrons) of a range of wavelengths is produced at each pulse. After a proton pulse, the time taken for an individual neutron to reach the detector of an experimental station can be recorded and this allows the wavelength of the neutron to be determined from its measured velocity. This approach, which is known as *time-of-flight analysis*, relies on the de Broglie relationship, equation (16.1) below. This states the important reciprocal relationship between the velocity and wavelength of neutrons

$$\lambda = \frac{h}{mv} \tag{16.1}$$

where h is Planck's constant, m is the mass of a neutron and v is its velocity.

As neutrons are not charged, they are difficult to detect directly, so their detection requires the sensing of an event generated by a neutron hitting a detector. Common forms of neutron detectors use either a charged multi-wire system to detect the ionization of ^{3}He or the light emitted from a lithium fluoride-based scintillator to detect neutron events. These electronic detectors have excellent time resolution, which is essential for time-of-flight analysis, since the detector records not only the position but also the arrival time of each neutron in the nanosecond or microsecond range. Diffracted neutrons can be recorded using image plate technology (as covered in Chapter 11) with special image plates that have been doped with gadolinium (Gd_2O_3) specifically to render them sensitive to neutrons—when a neutron strikes a gadolinium ion, γ-rays are emitted, and it is these that are recorded by the image plate (Amemiya and Miyahara, 1988). However, the sensitivity of neutron image plates to γ-rays prohibits their use close to a reactor and, as they have no time resolution, they are not optimal for use at a spallation source.

If neutrons are involved in multiple collisions with atoms in a certain material, the energy of the neutrons will equilibrate with that of the surrounding medium. If the material has a temperature T, the wavelength of 'thermal neutrons' emitted by the material is inversely proportional to the square root of the absolute temperature T (Bacon, 1962). Thermal neutrons in equilibrium with a medium at ambient temperature have a wavelength that is conveniently around 1.5 Å, and such neutrons will be travelling at velocities of the order of several kilometres per second. At higher temperatures, 'hot neutrons' of shorter wavelength are produced, and at lower temperatures, 'cold neutrons' of longer wavelength. In practice, neutrons of the desired wavelength range are obtained by passage of the high-energy neutrons from a reactor or spallation source through a 'moderator' (usually water, heavy water (D_2O) or hydrogen), which slows them down appropriately; the moderator is said to *thermalise* the neutrons passing through it. In essence, thermal neutrons are of low energy, typically 0.025 eV, compared with X-rays, which have energies of the order of 10 keV, and neutrons have the important advantage of being non-ionising, thereby virtually eliminating radiation damage to the sample under study. This usually obviates the need to freeze the crystal, which is an advantage because larger crystals, such as those needed for a neutron analysis, tend to become disordered to a greater extent than smaller ones upon freezing. Nevertheless, the majority of neutron protein crystallography instruments permit the sample to be frozen to liquid nitrogen or liquid helium temperatures to enable freeze-trapping studies of short-lived intermediates.

Whilst crystal monochromators may be used with neutrons, generally there is a huge loss in the intensity of the beam. However, the modest speeds with which thermal neutrons travel (several km/s) compared with the near-relativistic speeds of X-rays allow us to select a range $\Delta\lambda$ of neutron wavelengths using mechanical devices known as choppers. These consist of rotating discs containing an aperture which allows neutron transmission. The discs are arranged as shown in Fig. 16.4 and driven at high speed. Only those neutrons which are of the correct speed to pass both discs of the chopper are selected. At spallation sources, choppers need to be tightly synchronized (or phase-locked) with respect to the proton pulse rate, firstly to block the harmful high-energy neutrons and γ-rays that are produced each time the proton pulse hits the target, and secondly to block the low-energy neutrons that are last to leave the target, which could interfere with neutrons produced by the next pulse (Langan *et al.*, 2004).

The neutron atomic scattering factor of an element has dimensions of length and, accordingly, is referred to as the scattering length, b. The neutron scattering length of any element is of the order of the nuclear radius. The probability of a neutron being scattered when passing through a system of N nuclei is $N\sigma$, where σ is the scattering cross-section of a nucleus. The formula for σ was given by Bacon (1962) and is shown in the equation below:

$$\sigma = \frac{\text{scattered neutron flux}}{\text{incident neutron flux}} = 4\pi b^2 \qquad (16.2)$$

At the start of Chapter 9, we described the two types of scattering, namely coherent scattering, in which there is a definite phase relationship between the incident and

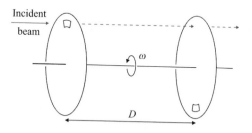

Fig. 16.4 *A neutron beam chopper.* Neutrons pass through the slot in the left-hand disc and travel towards the right-hand disc. Only those neutrons that are travelling at a certain velocity v pass through the slot in the right-hand disc. For these neutrons, $v = D/t$, where t is the time taken for the second disc to rotate so that its slot is in line with the beam. Only those neutrons which cross the gap between the discs in the same time t will pass through the second slot. If α is the angular separation of the slits ($180°$ in the example shown) and ω the angular velocity of the discs, then $\omega = \alpha/t$, and substituting this into the equation for velocity gives $v = \omega D/\alpha$. We may then derive the wavelength of the selected neutrons from the de Broglie equation $\lambda = h/mv$ (equation (16.1)).

scattered radiation, and incoherent scattering, where this relationship breaks down. With neutrons, incoherent scattering arises because a neutron has a spin quantum number of 1/2 which can interact with the spin of the nucleus. Some elements or isotopes have nuclei with zero spin, and these give rise only to coherent scattering. In general, since both coherent and incoherent effects are significant with neutron scattering, the total scattering cross-section of a nucleus is the sum of the cross-section for coherent scattering (σ_{coh}) and the cross-section for incoherent scattering (σ_{incoh}):

$$\sigma = \sigma_{coh} + \sigma_{incoh} \tag{16.3}$$

For certain elements, most notably hydrogen ^{1}H, the total scattering cross-section is dominated by the incoherent component, which gives a high background in the diffraction pattern. The problem can be circumvented by replacing as many hydrogen atoms as we can with the isotope deuterium. In contrast to hydrogen, which has a spin of 1/2, deuterium has a spin of 1 and therefore it predominantly scatters neutrons coherently. Hence replacing as many hydrogens in a protein as possible with deuterium minimizes the incoherent scattering. This can be achieved by growing or soaking the crystal in a mother liquor made from D_2O rather than ordinary water—more details are given later. Incidentally, D_2O is sometimes used as a crystallization additive because proteins have significantly lower solubility in it than in H_2O.

16.3 Neutron data collection

The relatively low flux of neutron beams from reactor sources has until recently restricted neutron crystallography to only a few proteins which produce very large crystals. Whilst neutron analysis imparts no observable radiation damage to biological samples, the long timescales required to collect data by conventional monochromatic means have made most studies prohibitive. To address these concerns, one of the first developments was the neutron Laue diffractometer LADI, developed by the EMBL-Grenoble Outstation and the Institut Laue-Langevin (Cipriani et al., 1996; Myles et al., 1998). The LADI diffractometer allows data to be collected relatively quickly by combining a quasi-Laue incident beam (produced by a Ni–Ti multilayer filter) with a cylindrical, neutron-sensitive image plate detector that surrounds the sample. The relatively long wavelengths used mean that high-resolution reflections will have 2θ angles greater than $90°$. Hence, for data collection, the crystal is placed at the centre of a metal drum coated with image plate material which completely encircles the sample (Fig. 16.5). The first generation of this device had the image plate on the outer surface of the drum, although since the drum was made of thin aluminium, the absorption of the diffracted neutrons was small. At the end of each exposure, the drum is spun on its cylindrical axis so that it can be scanned by a read-head that moves along the surface of the drum in the manner of a phonograph. Data is collected as a series of 'still images' with the crystal fixed at different regularly spaced orientations to sample different parts of reciprocal space. An example image is shown in Fig. 16.6. Later generations of the machine have the image plate 'coating' on the inside of the drum, which can therefore be made of much thicker metal to reduce the background.

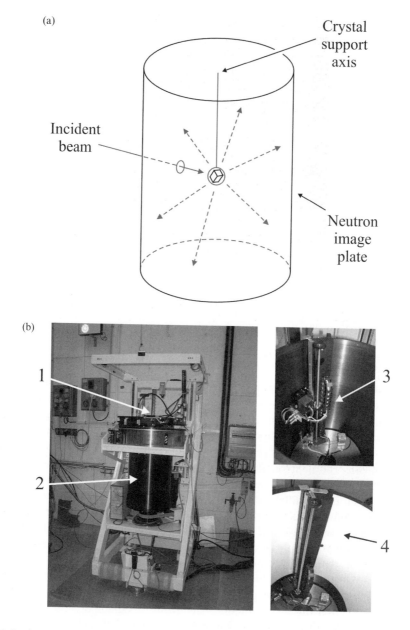

(a)

Crystal
support
axis

Incident
beam

Neutron
image
plate

(b)

1

2

3

4

Fig. 16.5 *A neutron Laue image plate diffractometer.* (a) A schematic of a neutron image plate detector. The drum-shaped detector surrounds the sample in order to collect high-θ reflections. (b) The LADI-3 diffractometer at the Institut Laue-Langevin (ILL, Grenoble, France), with views of the sample mounting cryostat (1), the detector drum (2), the scanner during construction (3) and the image plate lining the drum (4) (courtesy of Dr M. P. Blakeley, ILL).

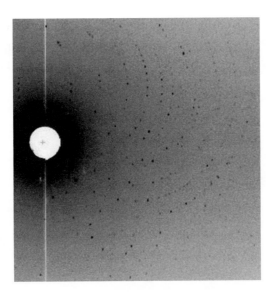

Fig. 16.6 *A neutron Laue image.* A protein diffraction image obtained from the LADI diffractometer at ILL, showing diffraction spots to approximately 2.0 Å resolution.

The scanner is therefore located inside the drum and moves out of the active area of the image plate during an exposure (see Fig. 16.5).

One significant problem with the Laue method is that a proportion of the diffraction spots are spatially overlapped owing to the large number of diffraction maxima that are recorded on a single image. In addition, there is another source of spot overlap due to the fact that harmonics of any reflection can appear in the same place on the detector. For example, if the crystal is at an orientation such that a reflection of index (h, k, l) diffracts X-rays of wavelength λ, then reflection $(2h, 2k, 2l)$ will diffract X-rays of wavelength $\lambda/2$ at exactly the same Bragg angle, and the same applies to higher-order harmonics. In a monochromatic experiment, this is not a problem, since only one wavelength is used. However, in the Laue method, it is a significant effect owing to the fact that a spectrum of X-rays is incident on the crystal. This affects the low-resolution spots most, and gives rise to the well-known 'low-resolution hole' which afflicts data collected in this manner. However, this problem is overcome in neutron Laue by use of a relatively narrow wavelength range $\Delta\lambda$, or band-pass (2.7–3.6 Å), thus ensuring that the vast majority of spots recorded are not harmonically overlapped. With time-of-flight instruments, this effect is not a concern, since the components of a harmonic overlap will be recorded independently because the spectrum of the diffracted beam will be recorded at each pixel of the detector.

The successors of the LADI instrument (VIVALDI and LADI-III) have provided a significant gain in efficiency that is partly due to the image plates being located on the inside of the drum. These instruments have made feasible studies of larger biological complexes and smaller crystals than were previously possible (Teixeira *et al.*, 2008). Ultimately, neutron Laue instruments have enabled the location of important

Fig. 16.7 *The Protein Crystallography Station at the Los Alamos Neutron Science Center.* The goniometer holding the crystal can be seen on the right, along with the cryostat. The curved detector is clearly visible, extending from the left to the rear right (reproduced with permission from Langan *et al.*, 2004).

hydrogen/deuterium atom positions in a number of systems at the resolutions typical of the majority of X-ray protein structure analyses (Blakeley *et al.*, 2004; Bon *et al.*, 1999; Habash *et al.*, 1997, 2000; Hazemann *et al.*, 2005; Niimura *et al.*, 1997).

Whilst Laue diffractometry takes advantage of many more of the available neutrons at reactor sources than do monochromatic instruments, further gains can be realized at spallation sources, which have the advantage that time-of-flight methods allow the wavelength of each recorded neutron to be determined. Not only does this allow the components of a harmonic overlap to be resolved, it also allows a bigger wavelength range to be used. Such a system is operated at the Los Alamos Neutron Science Center spallation source (Langan *et al.*, 2004), where a wavelength range of 0.6–6.0 Å is used for data collection at the Protein Crystallography Station (PCS) (see Fig. 16.7). Further instruments are planned at the Spallation Neutron Source at Oak Ridge, USA (Schultz *et al.*, 2005), at the Japan Atomic Energy Research Institute (Tanaka *et al.*, 2009) and at ISIS2 in the UK.

16.4   Neutron applications

As mentioned earlier, neutrons offer a powerful adjunct to X-ray analysis at the resolutions typical of most X-ray protein structure determinations $(d > 1.5\,\text{Å})$ by enabling hydrogen atom positions and details of solvent structure to be revealed (Schoenborn, 1994). Hydrogen and/or deuterium atoms may be located more easily in a neutron analysis than with X-rays because the scattering lengths of hydrogen (-3.7×10^{-15} m, or -3.7 fm) and deuterium ($+6.7$ fm) for neutrons are more similar to those of other biologically occurring atoms, i.e. carbon ($+6.6$ fm), nitrogen ($+9.4$ fm), oxygen ($+5.8$ fm) and sulphur ($+2.8$ fm). As emphasized earlier, hydrogens appear as characteristic negative density features in neutron maps, whilst deuteriums have density that is virtually indistinguishable from that of carbon atoms. Hydrogen and deuterium-labelled or exchanged positions can therefore be discriminated in the structure even at resolutions around 2.0 Å. Nitrogen and oxygen atoms can also be distinguished owing to their appreciably different scattering lengths, allowing the side chain carbonyl oxygen and amide nitrogen of asparagine and glutamine residues to be properly assigned. Whilst cancellation of neighbouring positive and negative scattering density can reduce the visibility of some hydrogen-rich groups in medium-resolution analyses (e.g. $H_2O(-1.7$ fm) and $-CH_3(-4.6$ fm)), the opposite is true for deuterium-labelled groups, where strong reinforcement from neighbouring positively scattering atoms can greatly enhance their visibility (e.g. $D_2O(+19.1$ fm) and $-CD_3(+26.7$ fm)). This is particularly important in D_2O solvent structure analysis, where scattering from both the deuterium and the oxygen atoms can allow the orientation and geometry of well-ordered water molecules to be determined and can render disordered or mobile water molecules more visible than in a corresponding X-ray analysis (Blakeley et al., 2004). H_2O/D_2O exchange is also needed to reduce the otherwise large hydrogen incoherent-scattering background.

To exploit the beneficial effects of deuteration, protein crystals are usually soaked in a mother liquor made from D_2O or equilibrated with it by vapour diffusion. The latter is a safer method if one wishes to avoid the risk of the crystal being damaged on transfer to a surrogate mother liquor solution. The exchange procedure enables all solvent-accessible O–H and N–H groups within the protein to undergo H–D exchange. Since neutron maps readily discriminate between H and D density in the structure, analysis of the pattern and extent of this H–D exchange can provide an elegant measure of the accessibility and mobility of each group and of the dynamics of the protein.

The uniform nature of neutron scattering lengths (for example, carbon and deuterium scatter as strongly as many of the heavy metals used in X-ray isomorphous replacement, such as lead or platinum) suggests that neutron diffraction is unlikely to succeed as a method of *de novo* protein phasing (Kossiakoff, 1983). Consequently, neutron analysis has been used as a follow-up technique for structures that have already been solved by X-ray diffraction. Most neutron analyses therefore involve refinement of a model based on the X-ray structure using the neutron data, and this refinement can be performed with the neutron data on its own or in combination with the X-ray intensities (Wlodawer and Hendrickson, 1982; Adams et al., 2009). Note that interpretation of neutron maps requires some care, owing to the fact that hydrogen atoms have negative density. For example, a negative peak in a difference Fourier map could be interpreted as a hydrogen atom when in fact it may simply be due to an error in the position of one of the neighbouring C, N, O or S atoms. Nevertheless, by careful refinement of alternative models with due consideration of the local molecular geometry, such ambiguities can usually be eliminated. Neutrons maps can seem unfamiliar to crystallographers more accustomed to X-ray electron density

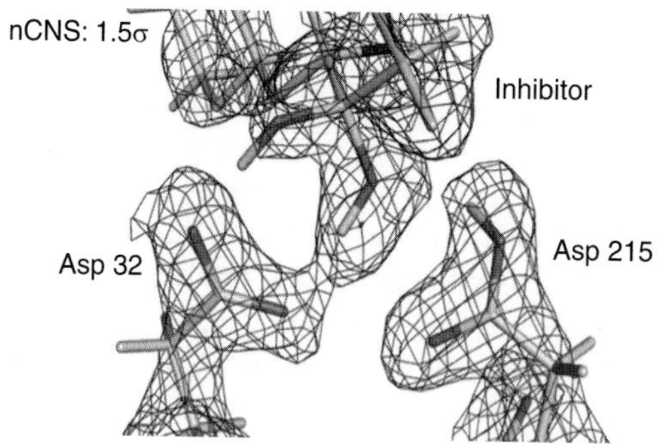

Fig. 16.8 *A neutron density map at 2.0 Å resolution.* Neutron density for the deuterium atoms (pale) at the catalytic centre of an enzyme can be seen. Figure courtesy of Professor L. Coates (Oak Ridge National Laboratory) (reproduced with permission from Coates *et al.*, 2008).

maps. For example, the density for aliphatic groups (e.g. $-CH_3$) in neutron maps tends to be weak owing to cancellation of the carbon and hydrogen scattering lengths, which are of opposite sign, and sulphur atoms, which tend to be very pronounced in electron density maps, have significantly lower neutron density than carbon or oxygen. Hence interpretation of a neutron map requires that the positions of such groups have been defined by a prior X-ray analysis, as is nearly always the case. Part of a neutron density map for a protein at 2 Å resolution is shown in Fig. 16.8.

The ability of neutrons to visualise and discriminate individual hydrogen atoms— and, more especially, the deuterium isotope at resolutions > 1.5 Å and at room temperature—is of much potential value in the study of enzymatic processes. The use of neutron diffraction in understanding enzyme mechanisms at these resolutions was pioneered by Kossiakoff and Spencer (1980), whose analysis of trypsin contributed significantly to current understanding of serine proteinase catalysis. Similar work on lysozyme (Mason *et al.*, 1984) and ribonuclease (Wlodawer *et al.*, 1983) has established the protonation states of the catalytic groups, with important implications for their mechanisms of action. Improvements in neutron instrumentation have led to further studies of the mechanism of aspartic proteinases (Adachi *et al.*, 2009; Coates *et al.*, 2001), dihydrofolate reductase (Bennett *et al.*, 2006), aldose reductase (Hazemann *et al.*, 2005), diisopropyl fluorophosphatase (Blum *et al.*, 2009) and many other enzyme families (for a review, see Blakeley *et al.*, 2008).

16.5 Advantages of perdeuteration

Expression of proteins with all hydrogens replaced by deuterium is possible, and can be undertaken by adapting *E. coli* to growing in minimal medium with a progressively higher D_2O content and the use of a deuterated carbon source such as glycerol, glucose or succinate (e.g. Petit-Haertlein *et al.*, 2009). It is also possible to make a perdeuterated medium for *E. coli* expression by use of photosynthetic algae grown in D_2O. This organism (*Scenedesmus obliquus*) does not need a carbon source, owing to its ability to fix CO_2 by photosynthesis, and the resulting algal hydrolysate is an excellent medium for expression of perdeuterated proteins in *E. coli* (Langan *et al.*, 2008). The principal advantage of perdeuteration is that smaller crystals can be used, owing to the cooperative effects of the positive scattering from bonded C and D atoms as opposed to the conflicting scattering densities of C and H atoms in a normal hydrogenated protein sample. The lack of hydrogen in the sample also lowers the background scattering. Perdeuterated crystals as small as 0.013–0.15 mm^3 in volume have been used (Petit-Haertlein *et al.*, 2009; Hazemann *et al.*, 2005) and, as mentioned earlier, minimising crystal volume aids in cryocooling of the sample.

16.6   X-ray Laue diffraction

The speed with which Laue data can be collected stems from the use of the full white beam, which allows a significant proportion of the diffraction pattern to be recorded in one image. At the most intense synchrotron sources, an exposure of 10^{-10} s produces interpretable diffraction data from protein crystals, thus creating the possibility of being able to do 'kinetic crystallography' of enzyme–substrate intermediates. In principle, enzyme-catalysed reactions can be followed as a function of time provided that all enzyme molecules in the crystal (typically 10^{14}) can be triggered synchronously. Photoactivable substrates are commonly used for this type of work, since the reaction can be triggered by 'flashing' the crystal with an intense light source, such as a laser, and subsequently collecting Laue images at known time intervals. There are many factors to consider in this type of experiment, such as the penetration depth of the light into the crystal, incomplete initiation and possible structural heterogeneity of the reaction intermediates. These experiments usually involve mutants that turn over more slowly than the wild-type enzyme, and sometimes chemically modified forms of the substrate (or enzyme) which act relatively slowly. For these reasons, intermediates can sometimes be freeze-trapped after soaking the crystal in substrate for some time or it may be possible to use a capillary as a flow cell, in which substrate is pumped over the crystal slowly, to study steady-state intermediates that accumulate. Indeed, by using cryosolvents in a flow cell, it has been possible to soak substrates or substrate analogues into an enzyme crystal below its glass transition temperature, which, as covered in Chapter 11, is usually around 200 K (Ding *et al.*, 1994; Rasmussen *et al.*, 1992; Tilton *et al.*, 1992). In this state, the enzyme does not turn over, but the reaction can be triggered by warming above the glass transition temperature, and intermediates can then subsequently be trapped by cooling the crystal. Redox reactions may be triggered by exposure of the crystal to the X-ray beam, which produces photoelectrons in the sample. These effects can be monitored in the crystal by use of a microspectrophotometer on the beam line. For reviews of these methods and case studies, the reader is referred to Stoddard (1998) and Messerschmidt (2007). Clearly, the work-up time for this type of analysis is substantial, requiring much effort in the design and testing of experiments.

The use of white Laue radiation in time-resolved experiments at modern synchrotron sources can be very damaging to the samples. For reactions which are reversible (under the conditions of the experiment) or cyclic, there is an alternative approach that reduces radiation damage and relies on the ability to synchronise the photoactivation of the reaction in the crystal with its exposure to the X-ray beam (Bourgeois *et al.*, 2007). This can be achieved by running the synchrotron in single-bunch mode (which delivers X-ray pulses typically 1 μs apart) or by using multiple bunches with an X-ray beam chopper. Thus, by phase-locking the laser pulses to the incident beam pulses, diffraction images can be recorded at a known time interval after each cycle of the reaction has been initiated by a laser pulse. This process is repeated as necessary to accumulate diffraction data of sufficient signal-to-noise ratio for structure analysis of the enzyme-bound intermediate to proceed. The use of a narrow band-pass (quasi-Laue or pink Laue) reduces radiation damage to the sample and yields data with fewer harmonic and spatial overlaps. The narrow band-pass also requires that data are collected at a greater number of orientations of the crystal, but its greater longevity in the beam allows this to be undertaken. Note that capillary-mounted (not cryocooled) crystals are preferable for this type of work, and several thousand laser pulses are required. Hence, whilst X-ray damage is reduced by the narrow band-pass, the repetitive nature of the laser activation increases the potential for laser damage to the sample, and several crystals may be required to complete the dataset. In addition, the narrow wavelength range introduces *partial Laue reflections* into the dataset, although these can be handled by modern Laue processing software. After data processing, difference Fourier maps are calculated for each time delay, allowing movements of atoms during the reaction to be visualised. The reader is referred to the excellent paper by Bourgeois *et al.* (2007), where more details of the method and many references to successful time-resolved analyses can be found.

16.7   Laue data processing

The processing of Laue data to yield background-corrected intensities generally proceeds along the lines outlined in Chapter 11, with some differences at various stages. For instance, in the automatic-indexing step, the wavelength of each spot has to be determined in addition to its h, k and l indices. Part of a typical neutron Laue image is shown in Fig. 16.6, where it can be seen that the spots form extended arcs instead of the lunes of regularly spaced spots that we are familiar with in the rotation method (see e.g. Fig. 8.18). The spots are often close together and can suffer from spatial as well as harmonic overlap, as we have discussed already. Reflections at the intersections of the arcs are referred to as 'nodals' and will have low-order indices such as (1, 1, 1) or (1, 1, 2), such that simply guessing the indices of a few nodal spots correctly can allow the remainder of the diffraction spots to be indexed. More details of the indexing algorithms are given in Ravelli *et al.* (1996) and Bourgeois *et al.* (2007).

During scaling, in addition to determining a scale factor and a temperature factor for each image, we have to consider the fact that each reflection is recorded at a different wavelength. The intensity of the primary beam will vary as a function of wavelength, as will the response of the detector and the absorption of the sample. These effects are corrected for by determination of a wavelength normalization curve (or λ-curve), which shows the variation of the scale factor as a function of wavelength. A typical example is shown in Fig. 16.9. The λ-curve can be determined during scaling

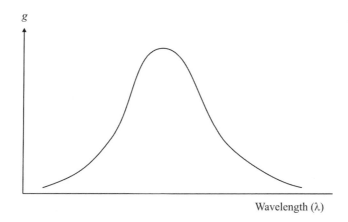

g

Wavelength (λ)

Fig. 16.9 *The λ-curve for a Laue dataset.* A plot of the scale factor versus wavelength (note that this is equivalent to the scale factor *g* as described in Chapter 11, Section 11.14). The curve will peak at a wavelength characteristic of the beam line. Wavelengths that are a long way from the peak will give rise to weak diffraction spots, and hence 'soft' wavelength limits have to be chosen during data processing to maintain overall data quality.

because there will be reflections in the dataset which have symmetry mates recorded at different wavelengths in other images. The scaling model therefore allows any wavelength dependence of the intensities to be normalised out. It is also possible to use the measurements of symmetry-related reflections recorded at different wavelengths to deconvolute the energy-overlapped intensities; for example, a reflection which is harmonically overlapped in one image may be recorded as a singlet in another image. In the past, an ingenious practical approach to deconvoluting the harmonic overlaps in X-ray Laue measurements consisted of using a number of X-ray films separated by metal foil in a device analogous to a toast rack (Helliwell, 1992). The short-wavelength X-rays penetrate deeper than those of longer wavelength, thus allowing the intensities of harmonically overlapped reflections to be estimated numerically from the intensities recorded on the successive layers of film.

Summary

Hydrogen atoms are poorly defined in X-ray analyses owing to the low scattering factor of this element, although atomic-resolution data may allow their positions to be determined experimentally in well-ordered regions of the structure. Unrestrained refinement at atomic resolution can also reveal group protonation states.

In contrast, neutrons are scattered by the nuclei of atoms and are more sensitive to hydrogens than are X-rays. In particular, the isotope deuterium scatters neutrons as strongly as do carbon atoms. Deuterium can be introduced into proteins by vapour diffusion or by simply soaking crystals in mother liquor solution made with D_2O instead of water, and this has the added benefit of reducing the background scatter caused by hydrogen. Non-exchangeable hydrogen atoms can be replaced by expressing the protein in deuterated growth media, and the resulting perdeuterated proteins allow neutron data to be collected from much smaller crystals.

Neutrons are produced at reactor or spallation sources and travel much slower than X-rays, allowing them to be partially monochromated by mechanical beam choppers. The relatively low flux of neutron sources is overcome by using a range of wavelengths, which is known as the Laue method. Neutron detectors consist of either a neutron-sensitive image plate or an electronic area detector. The latter allows time-of-flight studies, i.e. the wavelength of each neutron recorded by the detector is determined, which allows the harmonic or energy overlaps to be deconvoluted. With image plate devices, the harmonic overlap problem is overcome by using a smaller band-pass. Longer exposures than those that are used with X-rays are possible, since neutrons impart no radiation damage to the sample.

The X-ray Laue method has allowed studies of a number of enzyme catalytic mechanisms, owing to the speed with which the data can be collected following initiation of the reaction. Recent advances include the study of laser-triggered reactions, where the light pulses can be synchronised with the X-ray beam by use of high-speed X-ray beam choppers.

Processing Laue data requires a correction for the wavelength dependence of the diffraction intensities.

References

Adachi, M., Ohhara, T., Kurihara, K., Tamada, T., Honjo, E., Okazaki, N., Arai, S., Shoyama, Y., Kimura, K., Matsumura, H., Sugiyama, S., Adachi, H., Takano, K., Mori, Y., Hidaka, K., Kimura, T., Hayashi, Y., Kiso, Y. and Kuroki, R. (2009) Structure of HIV-1 protease in complex with potent inhibitor KNI-272 determined by high-resolution X-ray and neutron crystallography. *Proc. Natl. Acad. Sci. USA* **106**, 4641–6.

Adams, P. D., Mustyakimov, M., Afonine, P. V. and Langan, P. (2009) Generalized X-ray and neutron crystallographic analysis: more accurate and complete structures for biological macromolecules. *Acta Crystallogr. D* **65**, 567–73.

Amemiya, Y. and Miyahara, J. (1988) Imaging plate illuminates many fields. *Nature* **336**, 89–90.

Bacon, G. E. (1962) *Neutron Diffraction*, 2nd edn. Oxford University Press, London.

Bennett, B., Langan, P., Coates, L., Mustyakimov, M., Schoenborn, B., Howell, E. E. and Dealwis, C. (2006) Neutron diffraction studies of *Escherichia coli* dihydrofolate reductase complexed with methotrexate. *Proc. Natl. Acad. Sci. USA* **103**, 18493–8.

Blakeley, M. P., Kalb, A. J., Helliwell, J. R. and Myles, D. A. (2004) The 15-K neutron structure of saccharide-free concanavalin A. *Proc. Natl. Acad. Sci. USA* **101**, 16405–10.

Blakeley, M. P., Langan, P., Niimura, N. and Podjarny, A. (2008) Neutron crystallography: opportunities, challenges, and limitations. *Curr. Opin. Struct. Biol.* **18**, 593–600.

Blum, M.-M., Mustyakimov, M., Rüterjans, H., Kehe, K., Schoenborn, B. P., Langan, P. and Chen, J. C.-H. (2009) Rapid determination of hydrogen positions and protonation states of diisopropyl fluorophosphatase by joint neutron and X-ray diffraction refinement. *Proc. Natl. Acad. Sci. USA* **106**, 713–18.

Bon, C., Lehmann, M. S. and Wilkinson, C. (1999) Quasi-Laue neutron diffraction study of the water arrangement in crystals of triclinic lysozyme from hen egg-white. *Acta Crystallogr. D* **55**, 978–87.

Bourgeois, D., Schotte, F., Brunori, M. and Vallone, B. (2007) Time-resolved methods in biophysics. 6. Time-resolved Laue crystallography as a tool to investigate photo-activated protein dynamics. *Photochem. Photobiol. Sci.* **6**, 1047–56.

Burmeister, W. P. (2000) Structural changes in a cryo-cooled protein crystal owing to radiation damage. *Acta Crystallogr. D* **56**, 328–41.

Cipriani, F., Castagna, J. C., Wilkinson, C., Oleinek, P. and Lehmann, M. S. (1996) Cold neutron protein crystallography using a large position-sensitive detector based on image-plate technology. *J. Neutron Res.* **4**, 79–85.

Cleland, W. W., Frey, P. A. and Gerlt, J. A. (1998) The low-barrier hydrogen bond in enzymatic catalysis. *J. Biol. Chem.* **273**, 25529–32.

Coates, L., Erskine, P. T., Wood, S. P., Myles, D. A. A. and Cooper, J. B. (2001) A neutron Laue diffraction study of endothiapepsin: implications for the aspartic proteinase mechanism. *Biochemistry* **40**, 13149–57.

Coates, L., Tuan, H.-F., Tomanicek, S., Kovalevsky, A., Mustyakimov, M., Erskine, P. T. and Cooper, J. B. (2008) The catalytic mechanism of an aspartic proteinase explored with neutron and X-ray diffraction. *J. Am. Chem. Soc.* **130**, 7235–7.

Ding, X., Rasmussen, B. F., Petsko, G. A. and Ringe, D. (1994) Direct structural observation of an acyl-enzyme intermediate in the hydrolysis of an ester substrate by elastase. *Biochemistry* **33**, 9285–93.

Habash, J., Raftery, J., Weisgerber, S., Cassetta, A., Lehmann, M. S., Hoghoj, P., Wilkinson, C., Campbell, J. W. and Helliwell, J. R. (1997) Neutron Laue diffraction study of concanavalin A: the proton of Asp 28. *J. Chem. Soc. Faraday Trans.* **93**, 4313–17.

Habash, J., Raftery, J., Nuttall, R., Price, H. J., Wilkinson, C., Kalb (Gilboa), A. J. and Helliwell, J. R. (2000) Direct determination of the positions of the deuterium atoms of the bound water in concanavalin A by neutron Laue crystallography. *Acta Crystallogr. D* **56**, 541–50.

Hazemann, I., Dauvergne, M. T., Blakeley, M. P., Meilleur, F., Haertlein, M., Van Dorsselaer, A., Mitschler, A., Myles, D. A. A. and Podjarny, A. (2005) High-resolution neutron protein crystallography with radically small crystal volumes: application of perdeuteration to human aldose reductase. *Acta Crystallogr. D* **61**, 1413–17.

Helliwell, J. R. (1992) *Macromolecular Crystallography with Synchrotron Radiation.* Cambridge University Press, Cambridge.

Helliwell, J. R., Price, H. J., Deacon, A., Raftery, I. and Habash, J. (2002) Complementarity of neutron and ultrahigh resolution synchrotron X-ray protein crystallography studies: results with concanavalin A at cryo and room temperature. *Acta Physica Polonica A* **101**, 583–8.

Kossiakoff, A. A. (1983) Neutron protein crystallography: advances in methods and applications. *Ann. Rev. Biophys. Bioeng.* **12**, 159–82.

Kossiakoff, A. A. and Spencer, S. A. (1980) Neutron diffraction identifies His 57 as the catalytic base in trypsin. *Nature* **288**, 414–16.

Langan, P., Greene, G. and Schoenborn, B. P. (2004) Protein crystallography with spallation neutrons: the user facility at Los Alamos Neutron Science Center. *J. Appl. Crystallogr.* **37**, 24–31.

Langan, P., Fisher, Z., Kovalevsky, A., Mustyakimov, M., Sutcliffe Valone, A., Unkefer, C., Waltman, M. J., Coates, L., Adams, P. D., Afonine, P. V., Bennett, B., Dealwis, C. and Schoenborn, B. P. (2008) Protein structures by spallation neutron crystallography. *J. Synchrotron Radiation* **15**, 215–18.

Mason, S., Bentley, G. A. and McIntyre, G. (1984) Deuterium exchange in lysozyme at 1.4 Å resolution. In *Neutrons in Biology*, ed. Schoenborn, B. Plenum, New York, pp. 323–34.

Messerschmidt, A. (2007) *X-ray Crystallography of Biomacromolecules: A Practical Guide.* Wiley-VCH, Weinheim.

Myles, D. A. A. (2006) Neutron protein crystallography: current status and a brighter future. *Curr. Opin. Struct. Biol.* **16**, 630–7.

Myles, D. A. A., Bon, C., Langan, P., Cipriani, F., Castagna, J. C., Lehmann, M. S. and Wilkinson, C. (1998) Neutron Laue diffraction in macromolecular crystallography. *Physica B*, **241**, 1122–30.

Niimura, N., Minezaki, Y., Nonaka, T., Castagna, J. C., Cipriani, F., Hoghoj, P., Lehmann, M. S. and Wilkinson, C. (1997) Neutron Laue diffractometry with an imaging plate provides an effective data collection regime for neutron protein crystallography. *Nature Struct. Biol.* **4**, 909–14.

Petit-Haertlein, I., Blakeley, M. P., Howard, E., Hazemann, I., Mitschler, A., Haertlein, M. and Podjarny, A. (2009) Perdeuteration, purification, crystallization and preliminary neutron diffraction of an ocean pout type III antifreeze protein. *Acta Crystallogr. F* **65**, 406–409.

Rasmussen, B. F., Stock, A. M., Ringe, D. and Petsko, G. A. (1992) Crystalline ribonuclease A loses function below the dynamical transition at 220 K. *Nature* **357**, 423–4.

Ravelli, R. B. G. and McSweeney, S. M. (2000) The fingerprint that X-rays leave on structures. *Structure* **8**, 315–28.

Ravelli, R. B. G., Hezemans, A. M. F., Krabbendam, H. and Kroon, J. (1996) Towards automatic indexing of the Laue diffraction pattern. *J. Appl. Crystallogr.* **29**, 270–8.

Schoenborn, B. (1994) *Neutrons in Biology*, Basic Life Sciences. Volume 27, Plenum, New York.

Schultz, A. J., Thiyagarajan, P., Hodges, J. P., Rehm, C., Myles, D. A. A., Langan, P. and Mesecar, A. D. (2005) Design of the next generation neutron macromolecular diffractometer at the spallation neutron source. *J. Appl. Crystallogr.* **38**, 964–74.

Stoddard, B. L. (1998) New results using Laue diffraction and time-resolved crystallography. *Curr. Opin. Struct. Biol.* **8**, 612–18.

Tanaka, I., Kusaka, K., Tomoyori, K., Niimura, N., Ohhara, T., Kurihara, K., Hosoya, T. and Ozekic, T. (2009) Overview of a new biological neutron diffractometer (iBIX) in J-PARC. *Nucl. Instrum. Methods Phys. Res. A.* **600**, 161–3.

Teixeira, S. C. M., Zaccai, G., Ankner, J., Bellissent-Funel, M. C., Bewley, R., Blakeley, M. P., Callow, P., Coates, L., Dahint, R., Dalgliesh, R., Dencher, N. A., Forsyth, V. T., Fragneto, G., Frick, B., Gilles, R., Gutberlet, T., Haertlein, M., Hauß, T., Häußler, W., Heller, W. T., Herwig, K., Holderer, O., Juranyi, F., Kampmann, R., Knott, R., Krueger, S., Langan, P., Lechner, R. E., Lynn, G., Majkrzak, C., May, R. P., Meilleur, F., Mo, Y., Mortensen, K., Myles, D. A. A., Natali, F., Neylon, C., Niimura, N., Ollivier, J., Ostermann, A., Peters, J., Pieper, J., Rühm, A., Schwahn, D., Shibata, K., Soper, A. K., Strässle, T., Suzuki, J., Tanaka, I., Tehei, M., Timmins, P., Torikai, N., Unruh, T., Urban, V., Vavrin R. and Weiss, K. (2008) New sources and instrumentation for neutrons in biology. *Chemical Physics* **345**, 133–51.

Tilton, R. F., Jr, Dewan, J. C. and Petsko, G. A. (1992) Effects of temperature on protein structure and dynamics: X-ray crystallographic studies of the protein ribonuclease-A at nine different temperatures from 98 to 320K. *Biochemistry* **31**, 2469–81.

Wilson, C. C. (2001) Hydrogen atoms in acetylsalicylic acid (aspirin): the librating methyl group and probing the potential well in the hydrogen bonded dimer. *Chem. Phys. Lett.* **335**, 57–63.

Wlodawer, A. and Hendrickson, W. A. (1982) A procedure for joint refinement of macromolecular structures with X-ray and neutron diffraction data from single crystals. *Acta Crystallogr. A* **38**, 239–47.

Wlodawer, A., Miller, M. and Sjolin, L. (1983) Active site of RNase – neutron diffraction study of a complex with uridine vanadate, a transition state analogue. *Proc. Natl. Acad. Sci. USA* **80**, 3628–31.

Review III

In Chapter 10, we looked at the experimental methods for expressing and purifying proteins and described the process of screening for crystallisation conditions and subsequently optimising them. The crystals thus obtained have to be mixed with a cryoprotectant and mounted in loops before being frozen to liquid nitrogen temperatures for data collection and/or storage.

In Chapter 11, we described the generation of X-rays with both conventional sources and synchrotron facilities. The latter are extremely intense X-ray sources that operate by circulating electrons at high energy in an evacuated storage ring. The emission of X-rays is caused by the deflection of the electron beam using magnetic devices. All X-ray sources require optical components to monochromate, focus and collimate the X-ray beam.

Routine data collection is performed by slowly rotating the crystal about an axis perpendicular to the beam, and the diffraction spots are recorded by an electronic area detector. Radiation damage to the crystal is minimised by cooling it to liquid nitrogen temperatures, usually 100 K.

Analysing data collected by these instruments requires that the unit cell and the orientation matrix of the crystal are determined by autoindexing in order to predict the positions of all spots in the images. The final set of diffraction intensities (I_{hkl}) can be obtained by direct integration or profile fitting. The intensities thus obtained depend on a number of instrumental parameters, for which corrections are applied. Corrections for other physical factors, such as absorption and radiation damage, are usually made empirically during scaling of the data. The end product of data processing will be a set of unique structure factor amplitudes and their associated standard deviations, $|F_{hkl}|$ and $\sigma(|F_{hkl}|)$.

The derivation of the correct values of the phases for the structure factors F_{hkl} from the relative intensities $|F_{hkl}|^2$ constitutes the *phase problem*. In Chapter 12, we covered the use of the Patterson map as a means of locating the positions of certain heavy atoms within the unit cell, which is vital for solving the phase problem experimentally. The Patterson map also is of great importance in the molecular replacement method, in which we use the known structure of a molecule as a search model to solve an unknown structure. The associated theory and methods were covered in Chapter 13.

In Chapter 14, we covered experimental methods of solving the phase problem which involve incorporation of a heavy atom into the protein by soaking or co-crystallisation; or, in the case of selenomethionine phasing, the 'heavy atom' is incorporated during *in vivo* synthesis of the protein. The positions of the heavy atoms in the unit cell have to be determined by Patterson or direct methods and then refined, which allows the phases for the protein to be calculated. These experimental phases can be improved significantly by applying certain knowledge-based constraints to the electron density map and calculating new phases from the modified map in an iterative procedure known as density modification. Single-wavelength (SAD) and multi-wavelength (MAD) phasing methods which exploit anomalous scattering have largely routinised the structure solution of selenomethionine-substituted proteins.

The initial electron density map should allow us to build a model for at least part of the structure. This model will contain errors due to errors in the electron density and errors of interpretation. However, we can adjust the model so that it agrees as well as possible with the experimental data in another iterative process, known as refinement, as described in Chapter 15. Usually the model is adjusted to maximise the agreement between the observed and calculated structure factor amplitudes (or intensities) or to maximise its probability. The success or otherwise of refinement is assessed by a variety of criteria and, most importantly, by inspection of a new electron density map, calculated with phases from the refined model. A number of different Fourier syntheses and weighting schemes have been devised with the aim of minimising the bias of the map towards the model and enhancing the electron density of missing parts of the structure, to aid rebuilding. Several rounds of refinement and rebuilding are usually needed to complete the structure analysis.

In Chapter 16, we looked at complementary diffraction methods which yield important information on how proteins perform their key biological functions.

General bibliography

Introductory textbooks

Blow, D. *Outline of Crystallography for Biologists*. Oxford University Press, Oxford, 2002.

Rhodes, G. *Crystallography Made Crystal Clear: A Guide for Users of Macromolecular Models*, 3rd edn. Academic Press, Burlington, MA, 2006.

Reference works

Blundell, T. L. and Johnson, L. N. *Protein Crystallography*. Academic Press, New York, 1976.

Cantor, C. R. and Schimmel, P. R. *Biophysical Chemistry*, Part II: *Techniques for the Study of Biological Structure and Function*. Freeman, San Francisco, 1980.

Drenth, J. *Principles of Protein X-ray Crystallography*, 3rd edn. Springer, New York, 2007.

Giacovazzo, C., Monaco, H. L., Artioli, G., Viterbo, D., Ferraris, G., Gilli, G., Zanotti, G. and Catti, M. *Fundamentals of Crystallography*, 2nd edn, International Union of Crystallography Texts on Crystallography. Oxford University Press, Oxford, 2002.

Helliwell, J. R. *Macromolecular Crystallography with Synchrotron Radiation*. Cambridge University Press, Cambridge, 1992.

Ladd, M. F. C. and Palmer, R. A. *Structure Determination by X-ray Crystallography*, 4th edn. Plenum, New York, 2003.

Messerschmidt, A. *X-Ray Crystallography of Biomacromolecules: A Practical Guide*. Wiley, Weinheim, 2007.

Sanderson, M. R. and Skelly, J. V. *Macromolecular Crystallography: Conventional and High-Throughput Methods*. Oxford University Press, Oxford, 2007.

Stout, G. H. and Jensen, L. H. *X-ray Structure Determination: A Practical Guide*, 2nd edn. Wiley, New York, 1989.

Woolfson, M. M. *An Introduction to X-ray Crystallography*, 2nd edn. Cambridge University Press, Cambridge, 1997.

Definitive source

International Tables for Crystallography, Volumes A–G. International Union of Crystallography, Chester, UK, 2006.

Index